Climate Change and Socie

Climate Change and Society
Sociological Perspectives

EDITED BY

Riley E. Dunlap
and Robert J. Brulle

Report of the American Sociological
Association's Task Force on Sociology
and Global Climate Change

OXFORD
UNIVERSITY PRESS

OXFORD
UNIVERSITY PRESS

Oxford University Press is a department of the University of Oxford. It furthers the
University's objective of excellence in research, scholarship, and education by publishing
worldwide. Oxford is a registered trademark of Oxford University Press in the UK
and certain other countries.

Published in the United States of America by Oxford University Press
198 Madison Avenue, New York, NY 10016, United States of America

The contributors to this volume received no financial remuneration from the American
Sociological Association or any other entity for preparing their chapters or editing the volume.
Royalties from sales of the volume go to the American Sociological Association.

Library of Congress Cataloging-in-Publication Data
Climate change and society : sociological perspectives / edited by Riley E. Dunlap and Robert
J. Brulle.
     pages cm
Includes bibliographical references and index.
ISBN 978–0–19–935610–2 (hardcover : alk. paper) — ISBN 978–0–19–935611–9
(pbk. : alk. paper) 1. Climatic changes—Social aspects. 2. Environmental sociology.
I. Dunlap, Riley E. II. Brulle, Robert J.
QC903.S625 2015
304.'5–dc23   2015004447

This book is dedicated to the memories of Eugene Rosa and JoAnn Carmin, two members of the Task Force Steering Committee who passed away before the completion of this volume. JoAnn was a pioneering scholar of environmental governance and urban climate change adaptation. She made influential contributions to these areas, and was a lead author on both the Fifth Assessment Report of the IPCC and the third U.S. National Climate Assessment. Gene was a highly innovative scholar who made landmark contributions in the areas of energy, risk, and structural human ecology. He was also a member of the National Research Council's Board on Environmental Change and Society. Their contributions to understanding the social dimensions of climate change, reflected in their chapters in this volume, and to the field of environmental sociology and social science more generally, were substantial and path-breaking. They are missed.

Steering Committee

Task Force on Sociology and Global Climate Change
American Sociological Association

Riley E. Dunlap, Chair
Robert J. Brulle, Associate Chair
Robert J. Antonio
Beth Shaefer Caniglia
JoAnn Carmin
Karen Ehrhardt-Martinez
Kari Marie Norgaard
David N. Pellow
J. Timmons Roberts
Eugene A. Rosa
Kathleen Tierney

Contents

Acknowledgments

Many individuals contributed to the American Sociological Association's (ASA) Task Force on Sociology and Global Climate Change. We are indebted to the dozens of scholars who originally volunteered their services but did not find a niche, who attended planning sessions at ASA meetings, or who submitted suggestions for topics and approaches when calls went out to all ASA Sections. We apologize for not being able to list all of these individuals.

Special thanks are due Lee Herring and Margaret Vitullo who served as our liaisons with the ASA Executive Office, with Margaret's long-term service being especially helpful. Thanks also to John Logan, Sarah Soule, and Monica Prasad, our three liaisons with ASA Council.

Finally, we thank the following individuals for serving as reviewers for one or more chapters, as their efforts improved the quality of the volume: Alison Alkon, Robert Antonio, Steve Brechin, Michael Carolan, John Foran, Randy Haluza-Delay, Larry Hamilton, Craig Jenkins, Aaron McCright, Paul Mohai, Charles Perrow, Simone Pulver, Timmons Roberts, Tom Rudel, John Shandra, Rachael Shwom, William Solecki, Miranda Spencer, Andrew Szasz, Kathleen Tierney, and Richard York.

Last, but far from least, thanks are due Deborah Sweet, word-processing support specialist for Oklahoma State University's Department of Sociology. Her expert assistance was absolutely critical in preparing the final manuscript for this volume, and always delivered in a congenial fashion.

List of Contributors

Wokje Abrahamse is a Lecturer in Environmental Studies at Victoria University of Wellington, New Zealand, and an expert on interventions to encourage behavior change.

Alison Hope Alkon is Associate Professor of Sociology at the University of the Pacific and a specialist in sustainable food systems and environmental justice.

Robert J. Antonio is Professor of Sociology at the University of Kansas. He specializes in social theory and theories on the nexus of economic growth, economic inequality, and environment.

Jonn Axsen is Assistant Professor in the School of Resource and Environmental Management at Simon Fraser University. He specializes in consumer behavior, environmental technology, and climate policy.

Shannon Elizabeth Bell is an Assistant Professor of Sociology at the University of Kentucky. Her research focuses on environmental justice and the energy sector.

Steven R. Brechin is Professor of Sociology at Rutgers University, New Brunswick. His research focuses on topics that engage environmental, political, and organizational sociology.

Jeffrey Broadbent is Professor of Sociology at the University of Minnesota and leader of the international project "Comparing Climate Change Policy Networks" (COMPON).

Keith Brown is Assistant Professor of Sociology at Saint Joseph's University and a specialist in fair trade and ethical consumption.

Robert J. Brulle is a Professor of Sociology and Environmental Science at Drexel University. He specializes in analysis of the role of civil society organizations in the development of climate change policy.

Beth Schaefer Caniglia is Associate Professor of Sociology, Environmental Science, and International Studies at Oklahoma State University. She specializes in environmental policy, social movements, and environmental justice.

JoAnn Carmin was Associate Professor of Environmental Policy and Planning at the Massachusetts Institute of Technology. She was a pioneering scholar of environmental governance and urban climate adaptation and was a lead author for the Fifth Assessment Report of the IPCC.

Eric Chu is a doctoral candidate at the Massachusetts Institute of Technology, where he studies the politics of climate adaptation and governance. He has been a contributing author to the Fifth Assessment Report of the IPCC and the Third U.S. National Climate Assessment.

Brett Clark is Associate Professor of Sociology and Sustainability Studies at the University of Utah, specializing in the political economy of global environmental change and the philosophy, history, and sociology of science.

Thomas Dietz is Professor of Environmental Science and Policy and Sociology at Michigan State University. His research focuses on environmental decision making and structural human ecology. He was Vice Chair of the U.S. National Research Council's Committee on Advancing the Science of Climate Change and currently serves on the NRC Committee to Advise the U.S. Global Change Research Program.

Riley E. Dunlap is Dresser Professor and Regents Professor of Sociology at Oklahoma State University. A long-time specialist in environmental sociology, his current work focuses on the sociopolitical controversies surrounding climate change.

Karen Ehrhardt-Martinez is Senior Research Associate in the Sociology Department at Colorado State University and Founder and Director of Human Dimensions Research Associates, where she provides applied research on energy and climate issues for a range of public and private organizations.

Lawrence C. Hamilton is Professor of Sociology and Senior Fellow of the Carsey School of Public Policy at the University of New Hampshire. His interests include surveys, integrated research methods, and human–environment interactions in the Arctic.

Sharon L. Harlan is Professor of Sociology and Senior Sustainability Scientist at Arizona State University. She specializes in urban vulnerabilities to climate change.

William G. Holt is Urban Environmental Studies Program Coordinator and Assistant Professor of Sociology at Birmingham-Southern College. He focuses on climate change and resilience.

Lori M. Hunter is Professor of Sociology at the University of Colorado Boulder and an expert in connections between the environment and demographic processes such as migration.

Andrew K. Jorgenson is Professor of Sociology and Environmental Studies at Boston College. His primary areas of expertise are the political economy and human ecology of global environmental change.

Sandra T. Marquart-Pyatt is Associate Professor of Sociology and Environmental Science and Policy at Michigan State University, specializing in interrelations among environmental issues, outlooks, and concerns; climate change; sustainability; and democracy.

Aaron M. McCright is Associate Professor of Sociology at Michigan State University and a specialist in organized climate change denial and public understanding of climate change.

Joane Nagel is University Distinguished Professor of Sociology at the University of Kansas and studies gender and climate change.

Kari Marie Norgaard is Associate Professor of Sociology and Environmental Studies at the University of Oregon and a specialist in the social organization of climate denial and tribal environmental justice.

David N. Pellow is the Dehlsen Chair in Environmental Studies at the University of California, Santa Barbara, and specializes in environmental justice studies.

Charles Perrow, an organizational sociologist, is Emeritus Professor of Sociology at Yale University.

Simone Pulver is Associate Professor of Environmental Studies at the University of California at Santa Barbara, and her research focuses on private-sector responses to climate change.

J. Timmons Roberts is Professor of Environmental Studies and Sociology at Brown University, and his work focuses on equity, climate finance, and North–South politics at the UN negotiations. He is a member of the Board on Environmental Change and Society of the National Research Council.

Eugene A. Rosa was Edward R. Meyer Distinguished Professor of Natural Resources and Environmental Policy, Boeing Distinguished Professor of Environmental Sociology and Regents Professor, all at Washington State University. His research focused on structural human ecology and on risk. He served on the National Research Council's Board on Environmental Change and Society.

Thomas K. Rudel is Distinguished Professor in the Departments of Sociology and Human Ecology at Rutgers University. His research focuses on the dynamics of landscape change in the Americas.

Juliet B. Schor is Professor of Sociology at Boston College. She studies consumer society, worktime, and sustainable consumption.

Linda Shi is a doctoral candidate at the Massachusetts Institute of Technology, where she researches metropolitan and regional climate adaptation planning and governance.

Rachael L. Shwom is Assistant Professor of Human Ecology at Rutgers University. Her research focuses on how different groups of people in society make sense of and seek to address energy and environmental problems.

Dale Southerton is Professor of Sociology and Director of the Sustainable Consumption Institute at the University of Manchester. He specializes in the study of consumption and societal change.

Andrew Szasz is Professor and Chair of Environmental Studies at the University of California at Santa Cruz. He has written on the politics of environmental regulation, grassroots toxics movements, green consumption, environmental justice, and the sociology of climate change.

Kathleen Tierney is a Professor in the Department of Sociology and the Institute of Behavioral Science and Director of the Natural Hazards Center at the University of Colorado Boulder. She is a member of the National Research Council's Committee to Advise the U.S. Global Change Research Program.

Harold Wilhite is Professor of Social Anthropology at the University of Oslo's Centre for Development and the Environment and has written extensively on the theory and policies of energy consumption and savings.

Richard York is Professor of Sociology and Environmental Studies at the University of Oregon, and his research focuses on analyzing the social structural forces driving environmental change.

Climate Change and Society

1

Sociology and Global Climate Change

Introduction

Robert J. Brulle and Riley E. Dunlap

INTRODUCTION

Rapid anthropogenic climate change[1] is dramatically disrupting the bio-physical conditions that make Earth a suitable home for all natural species, including humans, and thus threatens the future of society. It is associated with increases in "natural" disasters; precipitously shifting weather patterns; threats to availability of potable water, food, and shelter; shifts in the range and prevalence of disease; species extinction; and the destabilization of ecosystems on which we depend (AAAS Climate Science Panel 2014). As with many environmental threats, the most immediate and severe effects of rapid climate change—often termed "climate disruption" (e.g., Pimm 2009)—are likely to fall upon the most socially vulnerable communities in both the United States and globally—those that are already experiencing economic, political, and cultural marginalization. These impacts include increasing conflicts over natural resources, social destabilization, population migration, and extensive adverse health consequences.

Our knowledge base about the global climate system has expanded quickly, led by the natural sciences community. Since the inception of the International Geosphere-Biosphere Program (IGBP) in 1986, and the subsequent formation of the Intergovernmental Panel on Climate Change (IPCC) in 1988, a vast array of data has been collected and research from numerous scientific disciplines integrated into a comprehensive field known as Earth Systems Science (Mooney, Duraiappah, and Larigauderie 2013). The sociologist Bruno Latour (2011:6) characterized the body of knowledge regarding climate change as "probably one of the most beautiful, sturdy and complex ever assembled." The results of these efforts, as regularly summarized by the IPCC, have served to inform and focus the global community on the sources, nature, and impacts of climate change.

In contrast, the social sciences have had at best a marginal role in this enterprise (Bjurström and Polk 2011). Consequently, there remains a significant gap in our understanding of the many facets of climate change. Despite the development of an extensive empirical literature that addresses the social dimensions of climate change, the social sciences—particularly sociology—have not been well integrated into reports produced by the IPCC and other agencies. One likely reason is that sociological research on climate change emerged only recently (beginning in the 1990s and increasing rapidly in the past decade), is spread across a wide variety of academic journals and books, and is neither easily identifiable by nor accessible to the wider intellectual community.

We seek to address this concern and to foster the integration of sociological research into the wider field of global climate change research. Over the past several years, sociology has developed a body of literature that can contribute to a deeper understanding of the human dimensions of climate change and its social, institutional, and cultural dynamics. This book provides a broad overview of crucial sociological perspectives on climate change. It is not an encyclopedic summary of all existing sociological research, but rather a collection of essays focused on areas where a considerable body of knowledge has coalesced to form definable topics. The chapters range from theoretical discussions on transforming social theory to bridge the nature/society divide to empirical analyses of the driving forces of climate change, along with a range of other climate change–related topics such as public opinion and activism, the inequitable nature of its negative impacts, and the difficulties of developing both successful mitigation and adaptation policies to address it. Our hope is that this volume will demonstrate the value of incorporating sociological analyses into the study of climate change, facilitate the integration of such work into wider research programs, and stimulate greater attention to climate change within the sociological community.

This introductory chapter traces the development of social science engagement with climate change. We then examine the limitations of the most widely used social science approaches to the topic and discuss sociology's potential contributions to furthering our understanding of climate change. Finally, we provide brief overviews of the subsequent chapters.

SOCIAL SCIENCE AND CLIMATE CHANGE

It is well recognized that social science approaches to climate change are a minor component of the major national and international reports on the phenomenon (Bjurström and Polk 2011). This omission is due in part to how scientific research on climate change developed, with a history rooted firmly in the natural sciences. In 1983, the NASA Advisory Council established an Earth System Sciences Committee chaired by geophysicist Francis Bretherton. The committee subsequently produced an important report titled *Earth System Science: A Program for Global Change* (Earth System Sciences Committee 1986). Included in the report was a conceptual model showing the interlinking relationships between components of Earth's natural systems, summarized in a diagram that has become known as the Bretherton Model.

Developed using a natural sciences perspective, the Bretherton Model shown in Figure 1.1 conceptualized the functioning of "the Earth system in time scales of decades to centuries where human forces have become prominent" (Mooney et al. 2013:3666). This part of the *Earth System* report was highly influential in the development of the research programs that now constitute the comprehensive field of climate science. Especially notable is how the role of humans was conceptualized. Among the sixteen different processes included in the model ("Ocean Dynamics," "Marine Biogeochemistry," etc.), all of "Human Activities" are collected into a single area—a literal black box. Thus, from the very beginning of the climate change research effort, the social sciences have been and continue to be marginalized. Consequently, from a natural science perspective, human activities remain in their black box, largely inaccessible.

Though incredibly complex, human activities are hardly irrelevant. Yet, in their analysis of the Third Assessment Report of the IPCC, Bjurström and Polk (2011:15) show that it was dominated by natural science perspectives,

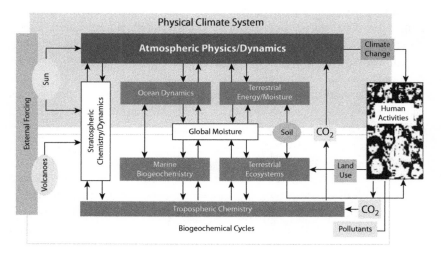

Figure 1.1 The Bretherton Model

with the result that climate change is framed "as a global environmental problem ... detached from its social contexts." Hulme (2011) agrees, and attributes the dominance of natural science to the fact that climate change research efforts were based in a community of scientists employing computer-based simulation models of Earth's climate system. These efforts began to emerge in the 1960s and were well established by the late 1980s.[2] Individuals from this community then played an important role in the formation and expansion of the overall climate research effort. By the 1990s, projections from their increasingly sophisticated climate models had gained wide acceptance; during the same period, the social sciences were largely ignoring climate change. The result, claims Hulme (2011:265), was the emergence of "climate reductionism." Climate reductionism, "in its crudest form, asserts that if social change is unpredictable, and climate change predictable, then the future can be made known by elevating climate as the primary driver of change."

The natural sciences certainly played an invaluable role, demonstrating that humans have taken on the role of a geological force capable of altering Earth's natural systems and cycles—including its climate. As a consequence of human actions, changes in natural systems that were once slow and incremental, forming a relatively constant backdrop against which social dynamics played out, are now happening at an increasingly rapid pace, with concurrent major impacts on human societies. Thus, human social and earth system dynamics have entered a new era, termed the Anthropocene, in which "natural forces and human forces became intertwined, so that the fate of one determines the fate of the other" (Zalasiewicz et al. 2010:2231). The arrival of the Anthropocene era has also brought recognition that the natural sciences are incapable of fully conceptualizing Earth system dynamics without the contributions of the social sciences. For example, a central insight (emphasized by sociologists early on [Crenshaw and Jenkins 1996]) is that the drivers of anthropogenic climate change are deeply rooted in the routines of everyday life and the social structure of modern societies (Reusswig and Lass 2010:156).

Indeed, a wide range of social science research on climate change has emerged over the past two decades (Dryzek, Norgaard, and Schlosberg 2011; Lever-Tracy 2010), giving impetus to efforts to incorporate social science into the study of this topic and environmental change more broadly. However, recent analyses (Hackmann and St. Clair 2012; Weaver et al. 2014) have argued that these efforts are inadequate, and have offered various critiques of three of the dominant intellectual currents within contemporary social science approaches to climate change: (1) Coupled Human–Natural Systems/Sustainability Science, (2) Individual-Level Analysis, and (3) Post-Political Framing of Climate Change. In the following sections, we provide an overview of these approaches, along with discussion of their limitations as noted by various critics. We then discuss how a sociological perspective can supplement these approaches to help provide a more complete understanding of the human and social dimensions of climate change.

Coupled Human–Natural Systems/Sustainability Science

The first and arguably most influential application of social science to climate change takes the form of Coupled Human–Natural Systems (CHANS). CHANS emerged out of the effort to integrate the social sciences into the global climate change research program. This effort was initiated by the formation in 1996 of the International Human Dimensions Program, a project of the International Social Science Council (ISSC) (Mooney et al. 2013). This program has continued to develop and has been integrated into the newly created Future Earth research program (O'Riordan 2013). CHANS serves as a core integrating concept for this program. In this approach, the global environmental system is seen as a complex series of interactions between human systems and natural systems. As characterized by Liu et al. (2007:1513), "in coupled human and natural systems, people and nature interact reciprocally and form complex feedback loops." These interactions constitute complex, adaptive human–environment systems whose study requires the use of interdisciplinary approaches involving both the natural and social sciences. CHANS research generally takes the form of specific place-based studies of reciprocal interactions between human activities and natural systems, particularly the impacts of the former on the latter, and the resulting feedback of ecological changes on human communities. The key questions involved in CHANS research focus on "the demographic, economic and technological drivers of such systems; vulnerability and resilience as emergent properties of such systems; their propensity for non-linear, threshold, or irreversible behaviors; and above all the ways in which their behaviors as systems emerge from adaptive actions by their constituent agents, interacting on a spatially heterogeneous tableau at multiple scales" (Clark 2010:62).

Further development of the CHANS framework led to the creation of a disciplinary approach known as Sustainability Science. This perspective aims to "understand the fundamental character of the interactions between nature and society" (Kates et al. 2001:641). In an important essay on the development of Sustainability Science, Clark (2007:1737) noted a need for new research on the "fundamental properties of the complex, adaptive human-environment systems that are the heart of sustainability science." Since then, Sustainability Science has emerged as a definable disciplinary perspective that has resulted in numerous publications, and it forms a key component of the research programs of the U.S. National Academy of Science (Wilbanks et al. 2013) and the U.S. Global Change Research Program (National Science and Technology Council 2012).

Despite considerable progress in the integration of the social and natural sciences via the interrelated CHANS and Sustainability Science frameworks in global environmental change research programs, Mooney et al. (2013:3668) note that "there is still much work to be done to include macroeconomists, sociologists, and behavioral scientists, and many other social scientists to develop further the knowledge base for global environmental change." Furthermore, two primary critiques of the CHANS/Sustainability Science approach have emerged.

The first critique highlights the continued framing of research questions from a natural science perspective, and the continual marginalization of the social sciences, in CHANS and Sustainability Science. The CHANS/Sustainability Science framework defines the core research questions to be addressed as those dealing with the integration of the social and natural sciences. While the distinct contributions of specific natural science disciplines are well recognized, social science perspectives are deemed relevant only if they contribute to multidisciplinary research efforts. This ignores the unique contributions that the different social sciences can make via their own disciplinary perspectives, and thus subordinates social science research to fit within natural science approaches. A recent analysis of the CHANS/Sustainable Science approach by the ISSC (Hackmann and St. Clair 2012) suggests that this practice has not "served society well," in that attempts to foster interdisciplinary research across the natural and social sciences have failed to yield global environmental change research that clearly contributes to improved efforts at climate change mitigation and adaptation. Consequently, efforts to address climate change either through mitigation or adaptation are unlikely to succeed without greater knowledge of human behavior and societal dynamics supplied by social science (Barnes et al. 2013). Mooney et al. (2013:3670) note that "The framing of the climate problem from a deterministic and mechanical perspective by the natural sciences community leaves very little room for the social sciences to explore human dimensions issues pertinent to their research agendas" (also see Castree et al. 2014). Within the social science community, this framing has been seen as the subordination of social research to the natural sciences, and the restriction of social research to a form of applied behavioral engineering (Shove 2010a, 2011:262).[3]

Not surprisingly, the ISSC assessment also emphasizes that the existing global change research agendas do not address the concerns, research interests, or skills of social science researchers such as sociologists. The

CHANS/Sustainability Science perspectives were developed largely outside of the sociological community, contributing to their having had limited social science input in general (Mooney et al. 2013:6). Recent analyses (ISSC 2013:493–496; Waltman 2013) show that only 3 percent of the publications dealing with global environmental change have come from sociologists. Thus, current efforts at integrating the social and natural sciences into a unified global change research program have not yet appreciably incorporated sociological (or anthropological, historical, or psychological) perspectives, despite growing awareness of the need for social science perspectives beyond economics. Because the social sciences have not been seen as essential within the global environmental change community, their contributions are not yet seen as highly relevant. Conversely, relatively few social scientists have seriously engaged with global change research. The result is a vicious cycle in which the most of the social sciences, including sociology, remain poorly integrated into the global environmental change research agenda (Hackmann and St. Clair 2012:12).

The second major critique of the CHANS/Sustainability Science perspective has been the central role of systems theory in its conceptual framework. The notion of an integrated social system, as conceptualized in a good deal of interdisciplinary work on climate change, is considered highly problematic within both sociology and the humanities. In their analysis, Palsson et al. (2013:7) note that "Although the rise of 'ecosystem thinking' and 'resilience thinking' represented an important advance for the natural sciences, this is sometimes seen as too restrictive for social sciences and the humanities. Social sciences and humanities generally view systemic boundaries as suitable constructs for a mechanistic world created by humans for analytical purposes and projected onto reality, rather than as an intrinsic property of the observed world." Consequently, systems theory acts as a barrier to other approaches that hold very different views of the nature of social systems. Specifically, the systems approach to social analysis (and the theoretical perspective of structural functionalism more generally) principally views societies as consensual and adaptive. This approach was highly debated in sociology during the 1960s and 1970s and was ultimately displaced by perspectives acknowledging the pervasive role of social conflict stemming from the inevitable competition among diverse interests within societies (Ritzer 1988:Chapter 7). The consequence of an emphasis on systems theory and neglect of other theoretical approaches is that analyses of global environmental change "rarely discuss global power relations, the functional aspects of the economic systems, or the fundamental issues of value formation" (Palsson et al. 2013:8). Thus, by uncritically adopting a systems approach, the CHANS/Sustainability Science perspective ignores sociological work that adopts a much wider variety of theoretical approaches and unnecessarily reduces the scope of knowledge that can be brought to bear in analyses of climate change.

Individual-Level Analysis

Another major area of criticism of existing applications of social science to climate change is the predominance of individual-level analyses, primarily via economics and psychology. The core critique is that the individual-level focus of these disciplinary approaches tends to neglect institutional, societal, and cultural perspectives, and thus limits the range of analyses. By focusing on individual behaviors and neglecting the broader factors that shape such behaviors, the individualist approaches offer a partial and constrained understanding of human behavior and social change. While this practice reflects a disciplinary division of labor, critics point out that the focus on individualistic analyses has had the effect of promoting political and policy responses that emphasize individual action and leave social, political, and economic institutions underexamined and unaccountable (Shove 2010a).

An analysis of the peer-reviewed social science literature on climate change finds that economics is the most widely represented social science discipline in climate change research (ISSC 2013:598–599). Fundamental to economic analyses of climate change is the "rational actor" model embedded in the discipline. As noted by Urry (2011:10), "the object of analysis is the 'individual' and the decisions and principles that each individual brings to the marketplace." Given the widespread societal influence of economics, it comes as no surprise that it has been highly influential in climate change research. In fact, Szerszynski and Urry (2010:3) note that economics "has largely monopolized the way that the social is conceived in discourses on climate change" and add that this "has led to a focus on human practices as individualistic, market-based, and calculative, and has thus helped to strengthen a tendency towards a certain set of responses to climate change, ones based on individual calculation, technology and the development of new markets."

The critique of this approach centers on the limits that the rational-actor economic assumption places on analyses. Foretelling the need for a sociological perspective, Szerszynski and Urry (2010:3) noted that

> Most of the time most people do not behave as individually rational, as separate economic consumers maximizing their individual utility from the basket of goods and services they can purchase and use, given fixed and unchanging preferences. People are creatures of social routine and habit, and of fashion and fad. These patterns of routine and fashion stem from how people are locked into and reproduce many different kinds of social institutions, both old and new. These include families, households, social classes, genders, work

groups, schools, ethnicities, age cohorts, nations, and scientific communities, NGOs and so on. But people are also locked into wider systems, including cultural worldviews and technological systems, that shape people's sense of what is permissible, desirable and possible.

Despite the obvious limitations of economics' emphasis on rational individual actors, economists have clearly been the most influential social science contributors to the IPCC, with the result that the IPCC "emphasizes economic aspects of climate change in comparison to other societal aspects" (Bjurström and Polk 2011:10).

Of course, this disciplinary influence stems not only from the predominance of microeconomic analyses of individuals' behaviors, but also from the prevalence of more macro-level economic approaches—such as using incentives and disincentives and carbon emissions trading schemes to promote carbon-reduction steps among corporations and across nation-states and by employing cost/benefit analyses to assess such options (Nordhaus 2008; Stern 2007). Besides still resting on the assumption of rational actors, policies such as emissions trading treat all forms of carbon emissions as equivalent. The result is that the "CO_2 emissions saved by building more efficient coal-fired power stations are equalized with that saved by building windmills," even though the former locks us into additional fossil fuel use and the latter represents a step toward sustainability (Kenis and Lievens 2014:541). Further, the results of cost/benefit analyses of the impacts of climate change and efforts to ameliorate them are heavily dependent on assumptions about proper discounting rates, leading to widely divergent estimates of their costs. More fundamentally, the appropriateness of employing cost/benefit analyses for dealing with the multifaceted nature of the long-term negative impacts of climate change is debatable (Jamieson 2014). Jamieson (2014:143) sums up his insightful review of economic approaches to climate change by noting that "Economics alone cannot tell us what to do in the face of climate change. Not all of the calculations can be performed, and even if they could, they would not tell us everything we need to know. At its best economics is a science and therefore cannot tell us what to do. At its worst it is an ideology, a normative outlook disguising itself as a report on the nature of things."[4] Finally, not only do economic perspectives limit our search for effective ways of lowering carbon emissions, but many of the policies proposed by economists are not implementable due to policy stalemates—especially in the United States (Jamieson 2014).

Psychology also focuses primarily on individuals, but operates with more inclusive and realistic models than the rational-choice model common in economics by incorporating "bounded rationality" (Simon 1985) and related views of the limitations of human decision making (Tversky and Kahneman

1986). In the past decade or so, there have been significant advances in psychological analyses of climate change. This trend is marked by the formation in 2009 of an American Psychological Association Task Force on the Interface Between Psychology and Global Climate Change (Swim et al. 2011). This work has provided valuable insights into how individuals think and feel about climate change; the role of individual behaviors in generating greenhouse (GHG) emissions and how they might be modified to reduce emissions; the factors that will likely influence reactions to climate change mitigation and adaptation; and the potential psychosocial impacts of climate change itself (Swim et al. 2011). Social psychologists in particular have contributed to the accumulating body of research on these topics while placing emphasis on analyzing what influences climate change attitudes and beliefs, facilitators and barriers to personal action to deal with climate change, and strategies for influencing climate change attitudes, beliefs, and behaviors—including the importance of determining how climate change is best "framed" (Fielding, Hornsey, and Swim 2014).

Although a good deal of psychological research on climate change has been sensitive to social and structural influences on individual thought and especially behaviors (e.g., Gifford 2011; Stern 2011), Swim et al. (2011:245) noted that "The most obvious contributions of psychology are at the individual level." In this sense psychological perspectives on climate change—like those of economics—can easily be (mis)used to reinforce the societal tendency to focus on individuals as both the primary cause of, and solution to, climate change. Thus, the critiques of the individualistic approaches that dominate psychology and economic analyses tend not to focus on the internal content of the research. Rather, they center on how these approaches are used in the formulation of policies and recommendations for action. Since individual behavior is seen to be the primary cause of climate change, "addressing the human dimensions of climate change is, in essence, a matter of incentivizing, persuading and encouraging individuals to do their bit and to 'kick the habit' of excessive resource consumption" (Shove 2010c:2).

This approach leads to an emphasis on addressing climate change by changing individual behavior via financial incentives or disincentives or through various communications efforts aimed at promoting lifestyle changes that reduce carbon emissions. A wide range of proposals seek to change individual behavior by, for example, providing more accurate information regarding the nature of global warming (on the assumption that increasing the public's scientific literacy will lead to attitudinal and behavioral shifts) (Stoknes 2014), using carefully crafted messages framed to motivate individuals to adopt behaviors that are more "sustainable" (Amsler 2009; O'Brien 2012), or employing well-designed social marketing campaigns to persuade people that it is in their best interest to recognize that climate change is a serious problem and to alter their behavior accordingly (Maibach, Roser-Renouf, and Leiserowitz 2008). No matter how well designed, such approaches ultimately rest on the premise that factually correct and well-packaged information, drawn from climate science, can help generate desired attitudinal and behavioral changes.

Because these strategies focus only on individual or small-group dynamics, they fail to adequately consider the larger social structure that forms the context in which individuals think and act. Accordingly, these individualistic communications approaches have a highly problematic conception of the role of information in the decision-making process. Instead of appealing to potential constituents in a manner that resonates with their normative expectations and everyday lived experiences, they offer technical, scientific discourse and related information, which has been shown to be ineffectual (Moser and Berzonsky 2014). They are also politically naïve, in that they assume that educational efforts focused on neutral "facts" and scientific evidence will carry the debate without taking into account the role of political power and the heavy influence of vested interests (Blühdorn 2000; Taylor 1992).

Emphasizing the role of individuals in generating carbon emissions—who are thus held responsible for reducing them—fits well with the Western and particularly American emphasis on individualism (Lipset 1997). The notion of autonomous individuals responsible for their personal choices is widely held among U.S. policymakers, the media, and the general public, and is of course quite compatible with the "rational actor" model that pervades economics. One also finds a strong emphasis on individual values, behaviors, and decision making when natural scientists deal with the "human dimensions" of global environmental change and climate change in particular (Ehrlich and Kennedy 2005; Palmer and Smith 2014). These trends suggest that the Bretherton Model's black box of human activities remains poorly illuminated in the natural sciences.

The stress on individual behavior and change thus leaves the institutions that structure everyday life and individual practices unexamined (Maniates 2001:33). In a highly insightful critique, Shove (2010a:1274) argues that focusing on the role of individuals "is a political and not just a theoretical position in that it obscures the extent to which governments sustain unsustainable economic institutions and ways of life, and the extent to which they have a hand in structuring options and possibilities." Accordingly, Shove argues that the focus on individualistic communication approaches and the support for such approaches by foundations and government "are part of an interlocking landscape of thought which constrains and prevents policy imagination of the kind required" (Shove 2010a:1282). As such, it serves to maintain the status quo. In accordance with this critique, many social scientists have advocated for the development of a wider and more integrative approach that can move beyond the limitations of individualism (Castree et al. 2014).

Post-Political Critique

A third aspect of climate change research, known as the post-political approach, has also come under growing criticism by social scientists. The post-political perspective refers to the nature of the dominant framing of climate change as evidenced in official reports, including those issued by the IPCC, the U.S. National Academy of Sciences, and Future Earth. This frame depoliticizes the discussion of climate change and, in so doing, reinforces the existing socio-politico-economic status quo. Hence, climate change discourse is said to take on a post-political form.

As discussed earlier, climate change is framed primarily through the natural sciences as an environmental problem, thus marginalizing analyses of the social processes that create and perpetuate it (Crist 2007:35). This marginalization of social science research results in the detachment of climate change from its social-structural context, as if it were devoid of political, cultural, or economic contestations (Bjurström and Polk 2011; Featherstone 2013:46; Hulme 2011). The dominant climate change discourse has thus become "scientized" (Habermas 1970:68) by the creation of a technocratic, value-neutral discourse that takes the form of "an impersonal, apolitical, and universal imaginary" (Jasanoff 2010:233) that removes moral and political considerations from the discussion. Swyngedouw (2010:215) characterizes the post-political frame as being "structured around the perceived inevitability of capitalism and a market economy as the basic organizational structure of the social and economic order, for which there is no alternative. The corresponding mode of governmentality is structured around dialogical forms of consensus formation, technocratic management and problem-focused governance" (also see Kenis and Lievens 2014). Thus the dominant framing of climate change ignores the "inevitably political character of our climate views and choices" (Reusswig and Lass 2010:167). Instead, climate change is presented as "a thoroughly depoliticized imaginary, one that does not revolve around choosing one trajectory rather than another, one that is not articulated with specific political programs or socio-ecological projects" (Swyngedouw 2011:263–264).

Reflecting their acceptance of a post-political orientation, major assessment reports accordingly fail to analyze critically the value systems, power relationships, and institutional processes that have resulted in climate change. One can, for example, read the entire set of IPCC assessment reports and not encounter any analyses of the widespread efforts to cast doubt on climate science. Another prominent example of this sort of thinking can be found in the National Research Council's *America's Climate Choices*, a path-breaking set of publications that present the latest scientific analyses of climate change in an accessible manner for policymakers and the public. In discussing different mitigation options, it notes that "Curbing U.S. population growth (either through policies to influence reproductive choices or immigration), or deliberately curbing U.S. economic growth, almost certainly would reduce energy demand and GHG emissions. Because of considerations of practical acceptability, however, this report does not attempt to examine strategies for manipulating either of these factors expressly for the purpose of influencing GHG emissions" (ACC 2010:52). While it may be politically expedient, failing to deal with the impacts of either population or economic growth clearly limits the range of options for ameliorating climate change to those that are compatible with current social, economic, and cultural systems (e.g., carbon emissions trading).

The post-political critique encompasses both the CHANS/Sustainability Science and Individualistic frameworks. The focus on consensual systems in the CHANS/Sustainability Science approach submerges the ubiquitous roles of power and conflict in social systems. Instead, it suggests the notion of a global "Earth System" that is amenable to science-based, top-down technocratic governance. Analogously, the emphasis on the role of individuals yields a prevailing narrative of climate change as stemming from the production of GHG emissions by a collection of anonymous individuals. In this scenario, there is no connection between human behavior and existing social relations and practices or institutional structure and dynamics (Jasanoff 2010:235; Kenis and Lievens 2014; Swyngedouw 2011). Instead, climate change is viewed as rooted in the behavior of an apolitical and universal "humanity" (Maniates 2001:43). Hence, all humans are seen to be contributing to the problem. This viewpoint is exemplified in former *New York Times* environmental reporter Andrew Revkin's declaration: "We grew up in the twentieth century when everything about the environment was either 'woe is me' or 'shame on you'. It is all coming undone and it's your fault. That is so antiquated when you look at a problem like climate change, where there is no whodunit. We all dunit" (Revkin 2013).

The post-political perspective unduly narrows the range of intellectual inquiry into climate change (Castree et al. 2014). Most notably, policy options to address climate change are limited "to the horizons of a liberal-capitalist order that is beyond dispute" (Swyngedouw 2011:264). This narrow view in turn constrains the range of alternatives under consideration to policies that are in accord with existing social, political, and economic relationships and structures. Adoption of the post-political frame is seen as having significant political advantages in that it obscures the institutional and structural roots of climate change, limiting political action in favor of consensual approaches such as individual behavioral change and market-oriented ones like emissions trading—thereby avoiding more conflictual strategies (Kenis and Lievens 2014). Social critic Naomi Klein acknowledges that considering policies that challenge the current economic system is deemed "politically heretical" (Klein 2014:19). Yet, ignoring them narrows the range of solutions that are considered and forecloses policy options (MacGregor 2014:628–629; Swyngedouw 2011). The result is that the dominant ideology of free-market fundamentalism remains unexamined and invisible (Klein 2014). The value and necessity of economic growth are taken for granted, and the current neoliberal economy is assumed to be a fixed and immutable system whose imperatives are logical necessities (Crist 2007:53–54) upon which all climate change policy options are to be formulated (Anderson and Bows 2012).

Fredrick Jameson (1998:50) captured the situation when he noted that "It seems to be easier for us today to imagine the thoroughgoing deterioration of the earth and of nature than the breakdown of late capitalism," and then added that "perhaps that is due to some weakness in our imaginations." Naomi Klein (2014:89) has recently reiterated this view, writing that "changing the earth's climate in ways that will be chaotic and disastrous is easier to accept than the prospect of changing the fundamental, growth-based, profit-seeking logic of capitalism." In this context, suggesting as Klein does that dealing with climate change will require fundamental alterations to the current, dominant economic system is indeed likely to be treated as "heretical." Such is the power of the near-hegemonic acceptance of the post-political stance in today's world.

The post-political critique highlights how the dominant scientific framings of climate change—including those of both social and natural scientists—are partial and incomplete, thus leading to inadequate policy proposals. While scientific analyses of environmental problems have clearly documented the negative ecological impacts of our current social order, they are not sufficient for providing an understanding of the social origins and potential resolution of climate change and global environmental change more generally. Indeed, natural science analyses obscure the social and cultural origins of climate change and thereby limit the range of possible ameliorative actions under consideration. Rather than providing relevant information to guide a major societal transition toward sustainability (O'Riordan 2013), the dominant framing of climate change has developed in accord with the governing principles of late modern society (Blühdorn 2000:30). In this sense, the dominant, post-political climate change discourse is seen as diverting attention from the societal origins of climate change and hindering recognition of the enormous challenge it poses.

SOCIOLOGICAL ANALYSES OF CLIMATE CHANGE

The foregoing demonstrates that there is growing recognition that natural science perspectives are increasingly inadequate for examining global climate change, and environmental change more broadly. This recognition has driven a shift in climate change research toward increasing receptivity for the social sciences, and a concomitant willingness by social scientists, and sociologists in particular, to incorporate climatic and other environmental factors into their research and theorizing (Antonio 2009; Dietz, Rosa, and York 2010; Murphy 2011). There is also a growing recognition of the need to better incorporate social science analyses into climate change research efforts (Castree et al. 2014; Weaver et al. 2014).

Fulfilling this need requires the expansion and integration of social science research on climate change. Hackmann, Moser, and St. Clair (2014:654) call for a shift in which "people and societies are no longer viewed as external to (or merely a simplistic black box within) the Earth system but as an integral and differentiated part of it—creating the problems and holding the key to their solution." Similarly, Reid et al. (2010:917) call for a "step change in research on fundamental questions of governance, economic systems, and the assumptions, beliefs, and values underlying human behavior." Finally, Weaver et al. (2014:656) see a need for further social science research dealing with the underlying causes of climate change, including "the behaviours and interactions of individuals, communities, markets, nations and all types of institutions."

There are a number of ongoing efforts to move past this intellectual impasse. The largest and most formalized project in this area has been developed by the ISSC. A critical report, *Transformative Cornerstones of Social Science for Global Change*, authored by Hackmann and St. Clair (2012) and published by the ISSC, developed a series of research cornerstones that define global environmental change (including climate change) research as a core social science concern. The report provides a research agenda focused specifically on "understanding processes of climate change and global environmental change as social processes embedded in specific social systems, past and present" (Hackmann and St. Clair 2012:16–20). This effort represents a significant departure from the frameworks of Sustainability Science and CHANS and those emphasizing the role of individuals. The ISSC report places the social sciences at the center of global environmental change research (see also Hackmann et al. 2014) and marks a major step forward in envisioning the role of the social sciences in global climate change research.

The view of environmental problems as inherently linked to social structure has long been a key component of environmental sociology (Schnaiberg

1980). Defined as the study of societal–environmental interactions or relationships (presaging later interdisciplinary developments such as CHANS), the field was established in the last half of the 1970s (Dunlap and Catton 1979). From the outset, environmental sociologists critiqued the larger discipline for being dominated by an anthropocentric approach, labeled the Human Exemptionalist Paradigm (HEP). This paradigm narrowed sociological inquiry to the analysis of social (as opposed to psychological, biological, and physical) facts and thus generated disciplinary neglect of the inherent interrelationships between human societies and their biophysical environments—particularly societal impacts on natural systems and the resulting effects of such ecological changes on human societies (Redclift and Benton 1994, Chapters 1–3; Urry 2011:7–8, 16). A new ecological paradigm (NEP), in which human societies are viewed as embedded in and dependent upon ecosystems, was proposed to overcome this disciplinary blindness (Catton and Dunlap 1980).[5]

However, until fairly recently, sociology continued to marginalize work seeking to integrate environmental and social concerns. Like the rest of the social sciences, sociology assumed that the natural world is a more or less passive stage on which the human project unfolds, and focused its attention on analyses of social phenomena (Dunlap 1980). The natural sciences in turn mirrored this societal–environmental distinction by focusing only on "natural" facts, or physical and biological phenomena. However, it is now clear that the well-entrenched divide between the natural sciences and social sciences reflects a nonexistent world, and maintaining this division is untenable if we are to understand contemporary problems like climate change. This realization has led a growing number of prominent scholars to recognize the need to move beyond anthropocentric social sciences (Urry 2011:8, 16; Žižek 2010:333), resulting in increased efforts to integrate social and natural science perspectives in analyses of climate change.

To examine the social dimensions of climate change, sociology brings two distinct and advantageous approaches (Driessen et al. 2013:3). First, sociology is well equipped to examine the causes and consequences of, and potential solutions to, climate change, and can therefore provide considerable insight into these phenomena. The primary driving forces of global climate change are embedded in social structure and institutions, cultural values and beliefs, and social practices. Thus, efforts to ameliorate and adapt to global warming will require understanding of these social processes at various scales from the global to the local, all fundamental domains of sociology. Sociology can thus contribute to our understanding of climate change not only through interdisciplinary engagement via CHANS/Sustainability

Science, but also through the analyses of discipline-specific questions regarding social processes and climate change.

A second role for sociology lies in providing a form of social critique. A common limitation of existing analyses of climate change is their restriction by near-hegemonic belief systems, such as the idea that market-based policies are the only feasible options for reducing carbon emissions within the current global neoliberal political-economic system. These restrictions create blind spots in societal responses to climate change and limit the range of conceivable actions (Klein 2014). A crucial component of social learning and reflexivity is the ability of a society to critically reflect upon its organizing principles in response to new conditions, and then modify them if necessary (Brulle 2015; Habermas 1970:62–80; 1998:359–387; Tabara 2013). Sociology can therefore play the vital role of moving beyond post-political thinking to examine and question the taken-for-granted belief systems that reinforce our current socioeconomic institutions and practices. By critiquing the ideas and ideologies that constitute the dominant and unchallenged understanding of climate change's causes, seriousness, and solutions, sociology can illuminate the constructed nature of these belief systems. In turn, it can explicate how such beliefs are used to sustain particular interests and limit the range of policy options. This approach takes the form of "public sociology" (Burawoy 2004, 2005).

While leading sociologists have long played the valuable role of social critics (e.g., Mills 1959), recent years have seen the emergence of public sociology as a deliberate effort to engage "publics beyond the academy in dialogue about matters of political and moral concern" (Burawoy 2004:1607). A sociological critique that engages the public can unveil the taken-for-granted nature of ideological blinders and open intellectual space for alternative viewpoints that might lead to innovative and more effective strategies for dealing with climate change. As Shove (2010b:284–285) notes, "policy-makers might do better if they had ready access to a richer repertoire of social theory, and if they were routinely encouraged to think about the conceptual foundations of what they do and the assumptions they make. By implication, failure to engage represents a lost opportunity in that potentially relevant methods and ideas are left sitting on the shelf." A public sociology of climate change can illustrate how the causes of climate change are interwoven into our current social, political, and economic order and open up new perspectives on the nature of the changes necessary for ameliorating and adapting to climate change. It could also move beyond supplementing the work of natural sciences and informing decision-makers to expanding the public and democratic dialogue on climate change to empower citizens to meaningfully participate in the formation of our collective future.[6]

AMERICAN SOCIOLOGICAL ASSOCIATION'S TASK FORCE
ON CLIMATE CHANGE AND SOCIOLOGY

In response to the perceived limitations of existing approaches to climate change, the U.S. sociology community has begun to focus increasing attention on the topic. To foster disciplinary efforts in this area, the Environment and Technology Section of the American Sociological Association (ASA) proposed the convening of a Task Force on Sociology and Global Climate Change, which was accepted by the ASA Council. Following ASA approval, and appointment of a Task Force Chair, a Steering Committee was established. Its first goal was to identify crucial aspects of climate change for which American sociology is poised to offer significant contributions, based on extant research and/or the relevance of theoretical and methodological perspectives. Following careful deliberation over the choice of topics, writing teams were then created for each one, with individuals (including both junior and senior scholars) selected based on proven track records of relevant research and scholarship—records that have often attained national and international visibility.

This book is the result of the ASA Task Force's efforts. It offers a relatively comprehensive and in-depth presentation of sociological analyses of crucial aspects of climate change. While the chosen set of topics is necessarily selective and emphasizes the research interests of American sociologists, it covers the essential dimensions of climate change—driving forces, impacts, mitigation, and adaptation—as well as a range of issues reflecting the sociopolitical dimensions of current efforts to recognize and deal with the reality of increasing global warming. Overall, the chapters reflect areas where sociology provides particularly useful perspectives and insights that have been virtually ignored or at least downplayed in analyses of climate change. By assembling them in this volume, we hope to demonstrate the relevance of such analyses for expanding our understanding of climate change—especially its inherent linkages to social structure and societal dynamics.

Chapter Descriptions

The chapters deal with four broad topic areas.[7] Chapters 2 to 4 focus on the factors that drive carbon emissions and situate these factors firmly within social structure and processes. Chapters 5 to 7 examine the impacts of climate change and how strategies to create just and equitable mitigation and adaptation efforts can be informed by sociological perspectives. Chapters 8 to 10 examine the factors that influence how society responds to climate change, including the movements that advocate for or against climate action and public opinion on climate change. Chapters 11 and 12 have a more disciplinary focus: first, an overview of debates within social theory about the significance of climate change and how to address it; second, a review of innovative methodological approaches for studying the relationship between societal and climatological phenomena. The concluding chapter takes stock of the sociological work on climate change presented in the book, emphasizing how it contributes to fulfilling the need for more social science research on climate (and global environmental) change while also pointing to the importance of further sociological engagement with climate change.

Chapters 2 Through 4

The first set of chapters deals in various ways with the driving forces of climate change. Within the natural science community, analyses of driving forces typically rely on the IPAT formula (also known as the Kaya Identity in climate change research). In this equation, Environmental Impact (I) is a function of Population (P), the level of consumption—or Affluence (A) and the nature of Technology (T) used in economic production. These three chapters extend this analysis with a consideration of the institutional, societal, and cultural driving forces that are not dealt with adequately by the natural science community.

Chapter 2 provides an overview of the structural characteristics of societies and the global economy that drive climate change. In this analysis, when authors refer to the anthropogenic "driving forces" of climate change, they mean the societal characteristics, such as demographic and economic factors, that are responsible for activities that contribute to GHG emissions (e.g., fossil fuel combustion) and changing landscapes that affect albedo (the fraction of solar radiation reflected back into space by Earth's surface). The chapter presents a structural human ecology model, STIRPAT, that extends the IPAT equation and is suitable for rigorous empirical testing, and then highlights the contrast between two theoretical models used to examine the dynamics of the driving forces of climate change. One model is ecological modernization theory, which sees a political and economic dynamic at work in the modernization process that creates both technological advances and political demand for an effective response to environmental issues. Contrasted to this approach are a group of critical political-economic theories, including the treadmill of production, which focus on the growth dynamics of capitalist economic systems and how these processes drive an ever-expanding economy and concomitantly expanding GHG emissions. Finally, the chapter considers how social practices regarding land development, energy production, and energy use also drive carbon emissions.

Chapter 3 focuses on market organizations (e.g., corporations) and their roles as major institutional sources of carbon emissions. Examining market organizations as both independent agents and entities influenced by their socioeconomic environments, the chapter first considers the actions of organizations as initiators of action based on the internal dynamics of market actors (e.g., marketing ecologically harmful products, lobbying against regulations, or choosing among different energy-production technologies). It then considers organizational behavior as the result of external economic forces and governance efforts, examining the factors that influence corporate action on climate change, including government regulations and competitive and stakeholder pressures. The chapter concludes with a discussion of the limited viability of actions being taken voluntarily by market actors to reduce carbon emissions.

Chapter 4 examines in detail the social factors that influence consumption. Rather than focusing on the "rational" decision making of individuals, this chapter lays out an approach that emphasizes the importance of social factors in driving consumption. It highlights the role of status and identity maintenance in both the levels and nature of consumption—especially high-carbon versus "green" consumption patterns. It also discusses the role of habitual consumption routines and shows that energy efficiency programs that bring these considerations into account have an increased possibility of success. Thus, a sociological perspective reveals that eliciting behavioral changes requires more than new attitudes and higher prices for carbon-intensive goods. Attention to factors such as social context, status dynamics, temporal factors, and normative practices can yield more powerful and durable reductions in consumption.

Chapters 5 Through 7

The next set of chapters focuses on the impacts of climate change, how they are connected to social processes, and how strategies to bring about effective and equitable mitigation and adaptation efforts can be informed by sociological perspectives.

Chapter 5 is devoted to the issue of climate justice. Building on the well-established literature on social inequality, this chapter summarizes research on how such inequalities in the production and consumption of energy between and within nations result in vast disparities between those that produce the greatest amounts of carbon emissions and those that feel the negative impacts of those emissions. While the rich nations of the world have been historically responsible for producing most GHG emissions, the adverse impacts of these emissions are borne by lesser-developed countries. Focusing on human vulnerabilities to the impacts of climate change, the chapter goes on to show the nature of the unequal burdens borne by communities in "sacrifice zones" of energy extraction. Finally, it examines policies to mitigate or adapt to climate change, raising serious questions about their distributional and procedural justice. The chapter concludes with specific research proposals on climate change that could use sociological knowledge and insights on inequality and social justice to offer a clearer vision of how to address our future climate.

The topic of Chapter 6 is adaptation to climate change, with an emphasis on how sociological knowledge sheds light on the social dynamics that shape individual, community, organizational, and governmental decisions and actions regarding adaptation. It begins with an examination of the societal and environmental impacts of climate change and describes regulatory, economic, political, and community-based approaches to adaptation. It next discusses global adaptation challenges such as the unequal burdens borne by poorer nations that have contributed little to GHG emissions, political disagreements over adaptation financing, and the need to integrate climate change adaptation into international development projects. It points to sociological knowledge on several topics that can improve our understanding of adaptation processes. These topics include societal responses to natural hazards, decision making under uncertainty, institutional persistence and change, population dynamics and migration, and urban governance—including civil society mobilization and participation. Beyond these topics, the chapter suggests that insights can also be gained from research in other areas, such as social and environmental movements, political sociology, and science and technology studies.

Mitigation of climate change is the focus of Chapter 7. Adopting a sociological perspective, this chapter is organized to reflect the nested nature of social structures, beginning with a discussion of sociological insights at the micro level (individuals and households). It first shows that individuals and households can play an important role in climate change mitigation efforts by taking action to reduce their own emissions, shifting their consumption practices, and acting as agents of broader political and social change via participation in social movements and political processes. Next, it emphasizes that mitigation at the meso level (organizations, companies, local government) is influenced by a number of factors, including the political and economic context in which the organization is operating, the business cycle, and the nature of the relevant network that forms the operating environment of the organization. Finally, it analyzes mitigation at the macro level, focusing on international efforts through an examination of the interactions of social movements, corporations, and nation-states. How international policies are shaped, their implications for various participants, and their effectiveness in changing outcomes are examined. The chapter ends with a brief discussion of the interplay between levels.

Chapters 8 Through 10

The next set of chapters turns to sociopolitical actors and processes crucial to societal recognition of climate change and efforts to deal with it.

Chapter 8 focuses on the efforts of social movement organizations and civil society more broadly in advocating for action on climate change. Social movements have historically been and continue to be major forces for the types of broad-scale reforms required to address problems of this magnitude. The chapter provides an overview of the existing social movement efforts regarding climate change. It starts with a theoretical overview of the role of civil society, followed by a summary of the existing literature on the historical development and worldviews of the U.S. national climate change movement. The chapter concludes with a discussion of social movement efforts to address climate change at both the international and cross-national levels, the role of religion in climate change movements, and what type of future research will be needed to advance our understanding in this area.

The public's views on climate change are generally seen as a key factor when explaining societal responses to climate change—or lack thereof. Chapter 9 examines sociological (and related) research on public opinion about climate change in order to better understand the larger social, economic, cultural, political, and environmental factors that influence the formation and modification of public views on climate change. It finds that public opinion on climate change is dynamic and differentiated. In general, awareness of climate change around the world is growing, but cross-national studies demonstrate that levels of awareness, knowledge, and concern vary greatly across and within developing and industrialized nations. Key determinants of public concern include media coverage, economic conditions, political beliefs, and the public actions of political elites. Also, there is a growing literature regarding the extent to which experiencing extreme weather events may influence individual opinion on climate change. The chapter concludes with a call for research on how the public's climate change beliefs may influence household behavior, participation in collective action, corporate behavior, and policy outcomes.

Chapter 10 examines the efforts of the climate change denial "countermovement" to delay action on climate change. It outlines the evolution of organized climate change denial, highlighting its key actors, strategies, tactics, and impacts. It documents the key roles that fossil fuel (and other) corporations, conservative foundations, and conservative think tanks have played in launching and supporting organized denial, and describes how their efforts have been stimulated and enhanced by the growth of free-market fundamentalism. It then demonstrates how these key actors have been assisted by a small number of contrarian scientists, conservative media, conservative politicians, and, more recently, self-styled "citizen-scientist" bloggers. The key strategy employed in the quarter-century–long effort to deny the reality and significance of climate change is to "manufacture uncertainty," continually raising questions about the validity and legitimacy of climate science in order to negate calls for reducing carbon emissions. The success of the denial movement is apparent in the widespread skepticism toward climate change among sectors of the American public and among conservative politicians.

Chapters 11 and 12

The next two chapters have both an internal disciplinary focus and the goal of demonstrating the value of sociology to other disciplines. They examine the implications of climate change for sociological theorizing and empirical research, as well as the utility of theorizing about climate change and applying innovative methodological approaches to study its relationship to societal phenomena—for both sociologists and other analysts.

Chapter 11 provides an overview of the major divisions in theorizing about climate change. One of the crucial epistemological divides is the "realist–constructivist" debate over climate science and climate policy. This debate centers on the dual nature of climate change—it is both a biophysical reality and a socially defined and constructed phenomenon. Here, realists stress the need for inquiry into climate change's biophysical facets and the social forces that drive global warming, while constructivists focus on the actors who define climate change as problematic and how their definition influences research, policymaking, and societal reaction. This debate has sparked disagreements over climate science certainty, globalization's impact on climate and climate policies, and state-driven versus market-oriented response strategies. The second and perhaps more intense debate centers on the dynamics of capitalism and the possibility of reforming it to create a sustainable economy and society—a debate that obviously has major implications for assessing the likelihood of effective societal responses to climate change. The chapter concludes with a discussion of the role social theory can play in structuring both empirical inquiry and normative discussions regarding the changes needed to create and maintain a sustainable future.

Chapter 12 provides an overview of key methodological approaches and techniques employed by sociologists doing empirical research on climate change. The chapter begins with a brief historical overview of environmental sociology's call to study societal–environmental relationships and the methodological challenges of empirically examining the linkages between environmental and social phenomena. Then the authors take note of the growing availability of data on environmental "variables" and a range of sophisticated methodological techniques for analyzing their relationship to social phenomena. The chapter describes analytical approaches currently employed by sociologists conducting empirical studies of climate change, with an emphasis on analyzing spatial and temporal variation as core components of societal–environmental interactions. The conclusion identifies existing methodological challenges and future opportunities and needs, including the need for more integrated social/natural science research based on statistical modeling and qualitative and multimethod approaches.

The concluding Chapter 13 briefly takes stock of the earlier chapters and focuses on the development of a future research agenda for sociological inquiry into the topic of climate change. It revisits the recent major initiatives seeking to integrate the social sciences into the study of climate change, particularly the efforts of the ISSC. It also indicates how work presented in the prior chapters, which is typically not confined by a post-political perspective, represents at least a partial response by sociologists. It emphasizes the need for far more research and efforts to demonstrate the value of sociological perspectives not only to the broader climate change (and global environmental change) community and relevant policymakers, but also to the rest of the sociological community.

CONCLUSION

Human-caused climate change is clearly one of the most important issues of the twenty-first century, and understanding climate change—its sources, impacts, and potential amelioration—is an inherently sociological concern. It is well established that the primary drivers of global climate change are social-structural and sociocultural phenomena. Since sociology possesses considerable knowledge of social and cultural systems, it has a great deal to offer in helping understand the societal origins of climate change, as well as how social, economic, political, and cultural factors are likely to affect efforts to both mitigate and adapt to climate change. Sociology can also make important contributions to clarifying the adverse social impacts of increasing climate change, such as forced migration, increased social conflict, and growing levels of injustice. It is hoped that this volume provides a good sense of what sociology can contribute and why sociological and other social science perspectives need to be incorporated more fully into future research on climate change.

NOTES

1. Hereafter we will typically refer simply to "climate change" or "global warming," but it should be understood that we are referring to anthropogenic or human-caused climate change/global warming.

2. While an "epistemic community" of climate modelers emerged fairly quickly, social scientists have provided important analyses of competing perspectives and approaches among climate modelers (see Yearley 2009 and references therein) and divergence between the modelers and practitioners of other approaches to weather and climate phenomena (Lahsen 2013).

3. Although sociologists in particular have long complained about the dominance of natural science in global/climate change research (Redclift and Benton 1994), early sociological work was limited primarily to analyses of the social construction of global change and global warming. Such analyses were often based on a rather skeptical view of natural science "evidence" and thus were not seen as all that relevant to the larger global environmental change research agendas (Dunlap and Catton 1994). This skeptical orientation stemmed in part from lingering fears of being accused of resurrecting "environmental determinism"—a long-discredited perspective in the social sciences (Judkins, Smith, and Keys 2008; Stehr and von Storch 1999)—if one saw climate change as likely to have major societal impacts.

4. Illustrating the impact of economics, climate scientists Anderson and Bows (2012:64) suggest that "the tendrils of economics have permeated into climate science" so deeply that "climate change analyses are being subverted to reconcile them with the orthodoxy of economic growth."

5. The HEP/NEP distinction has become a foundational perspective within environmental sociology (York 2008), presented in nearly all textbooks in the field, and has also been employed by a range of other social scientists, including anthropologists (Kopnina and Shoreman-Ouiment 2011), historians (Merchant 1992), and political scientists (Barry 1999). For a recent overview of the evolution and current status of environmental sociology see Dunlap (2015).

6. For a very strong endorsement of this approach, see Swedish sociologist Stellan Vinthagen's (2013:174) call for social scientists to "become an active and integrated part of the climate justice movement."

7. Zehr's (2014) recent overview of sociological research on global climate change covers the same key topics, albeit with somewhat differing emphases relative to this volume. Similarly, McNall's (2011) short book, while very selective in coverage, also touches on most of the topics here.

REFERENCES

AAAS Climate Science Panel. 2014. *What We Know: The Reality, Risks and Response to Climate Change*. Washington, DC: American Association for the Advancement of Science.

America's Climate Choices (ACC). 2010. *Limiting the Magnitude of Future Climate Change.* Washington, DC: National Research Council.

Amsler, Sarah S. 2009. "Embracing the Politics of Ambiguity: Towards a Normative Theory of 'Sustainability'." *Capitalism, Nature, Socialism* 20:111–125.

Anderson, Kevin and Alice Bows. 2012. "A New Paradigm for Climate Change." *Nature Climate Change* 2:639–640.

Antonio, Robert. 2009. "Climate Change, the Resource Crunch, and the Global Growth Imperative." *Current Perspectives in Social Theory* 26:3–73.

Barnes, Jessica, Michael Dove, Myanna Lahsen, Andrew Mathews, Pamela McElwee, Roderick McIntosh, Frances Moore, Jessica O'Reilly, Ben Orlove, Rajindra Puri, Harvey Weiss, and Karina Yager. 2013. "Contributions of Anthropology to the Study of Climate Change." *Nature Climate Change* 3:541–544.

Barry, John. 1999. *Environment and Social Theory.* London & New York: Routledge.

Bjurström, Andreas and Merritt Polk. 2011. "Physical and Economic Bias in Climate Change Research: A Scientometric Study of IPCC Third Assessment Report." *Climatic Change* 108:1–22.

Blühdorn, Ingolfur. 2000. *Post-ecologist Politics: Social Theory and the Abdication of the Ecologist Paradigm.* London: Routledge.

Brulle, Robert J. 2015. "Engaging Civil Society." In *Handbook of Global Environmental Pollution: Volume 1, Global Environmental Change*, edited by Bill Freedman. Dordrecht, Germany: Springer.

Burawoy, Michael. 2004. "Public Sociologies: Contradictions, Dilemmas, and Possibilities." *Social Forces* 82:1603–1618.

Burawoy, Michael. 2005. "For Public Sociology: 2004 Presidential Address, American Sociological Association." *American Sociological Review* 70:4–28.

Castree, Noel, William M. Adams, John Barry, Daniel Brockington, Bram Büscher, Esteve Corbera, David Demeritt, Rosaleen Duffy, Ulrike Felt, Katja Neves, Peter Newell, Luigi Pellizzoni, Kate Rigby, Paul Robbins, Libby Robin, Deborah Bird Rose, Andrew Ross, David Schlosberg, Sverker Sörlin, Paige West, Mark Whitehead, and Brian Wynne. 2014. "Changing the Intellectual Climate." *Nature Climate Change* 4:763–768.

Catton, William R., Jr. and Riley E. Dunlap. 1980. "A New Ecological Paradigm for Post-Exuberant Sociology." *American Behavioral Scientist* 24:15–47.

Clark, William. 2007. "Sustainability Science: A Room of Its Own." *Proceedings of the National Academy of Science USA* 104:1737–1738.

Clark, William. 2010. "Sustainable Development and Sustainability Science." Pp. 55–63 in *Toward a Science of Sustainability: Report from Toward a Science of Sustainability Conference*, edited by S. Levin and W. Clark. Cambridge, MA: Center for International Development at Harvard University.

Crenshaw, Edward M. and J. Craig Jenkins. 1996. "Social Structure and Global Climate Change: Sociological Propositions Concerning the Greenhouse Effect." *Sociological Focus* 29:341–358.

Crist, Eileen. 2007. "Beyond the Climate Crisis: A Critique of Climate Change Discourse." *Telos* 141:29–55.

Dietz, Thomas, Eugene A. Rosa, and Richard York. 2010. "Human Driving Forces of Global Change: Examining Current Theories." Pp. 83–132 in *Threats to Sustainability: Understanding Human Footprints on the Global Environment*, edited by E. A. Rosa, A. Diekmann, T. Dietz, and C. Jaeger. Cambridge, MA: MIT Press.

Driessen, Peter, Jelle Behagel, Dries Hegger, Heleen Mees, Lisa Almesjö, Steinar
    Andresen, Fabio Eboli, Sebastian Helgenberger, Kirsten Hollaender, Linn
    Jacobsen, Marja Järvelä, Jeppe Laessoe, Sebastian Oberthür, David Avelar, Ulrich
    Brand, Achim Brunnengräber, Harriet Bulkeley, Daniel Compagnon, Simin
    Davoudi, Heide Hackmann, Jörg Knieling, Corrine Larrue, Björn-Ola Linnér,
    Orla Martin, Karen O'Brien, Saffron O'Neill, Marleen van Rijswick, Bernd
    Siebenhuener, Carl von Ossietzky, Catrien Termeer, and Aviel Verbruggen.
    2013. "Societal Transformations in the Face of Climate Change: Research
    Priorities for the Next Decade." Paper Prepared for the Joint Programming
    Initiative: Connecting Climate Knowledge for Europe. Stockholm, Sweden: JPI
    Climate.
Dryzek, John S., Richard B. Norgaard, and David Schlosberg, eds. 2011. *The Oxford
    Handbook of Climate Change and Society*. Oxford: Oxford University Press.
Dunlap, Riley E. 1980. "Paradigmatic Change in Social Science: From Human
    Exemptionalism to an Ecological Paradigm." *American Behavioral Scientist*
    24:5–14.
Dunlap, Riley E. 2015. "Environmental Sociology." Pp. 796–803 in In *International
    Encyclopedia of the Social and Behavioral Sciences*, 2nd ed., vol. 7, edited by J.
    D. Wright. Oxford: Elsevier.
Dunlap, Riley E. and William R. Catton Jr. 1979. "Environmental Sociology." *Annual
    Review of Sociology* 5:243–273.
Dunlap, Riley E. and William R. Catton Jr. 1994. "Struggling with Human
    Exemptionalism: The Rise, Decline and Revitalization of Environmental
    Sociology." *American Sociologist* 25:5–30.
Earth System Sciences Committee. 1986. *Earth System Sciences Overview: A Program
    for Global Change*. Report of the Earth System Sciences Committee of the NASA
    Advisory Council. Washington, DC: NASA.
Ehrlich, Paul R. and Donald Kennedy. 2005. "Millennium Assessment of Human
    Behavior." *Science* 309:562–563.
Featherstone, David. 2013. "The Contested Politics of Climate Change and the Crisis
    of Neo-liberalism." *ACME: An International E-Journal for Critical Geographies*
    12(1):44–64.
Fielding, Kelly S., Matthew J. Hornsey, and Janet K. Swim. 2014. "Editorial: Developing
    a Social Psychology of Climate Change." *European Journal of Social Psychology*
    44:413–420.
Gifford, Robert. 2011. "The Dragons of Inaction: Psychological Barriers That Limit
    Climate Change Mitigation and Adaptation." *American Psychologist* 66:290–302.
Habermas, Jürgen. 1970. *Toward a Rational Society*. Boston: Beacon.
Habermas, Jürgen. 1998. *Between Facts and Norms*. Cambridge, MA: MIT Press.
Hackmann, Heidi, Susanne Moser, and Asuncion Lera St. Clair. 2014. "The Social
    Heart of Global Environmental Change." *Nature Climate Change* 4:653–655.
Hackmann, Heidi and Asuncion Lera St. Clair. 2012. *Transformative Cornerstones of
    Social Science Research for Global Change*. Paris: ISSC.
Hulme, Mike. 2011. "Reducing the Future to Climate: A Story of Climate Determinism
    and Reductionism." *OSIRIS* 26:245–266.
International Social Science Council (ISSC). 2013. *World Social Science
    Report: Changing Global Environments*. New York: UNESCO Publishing.
Jameson, Fredrick. 1998. "The Antinomies of Postmodernity." Pp. 50–72 in *The
    Cultural Turn: Selected Writings on the Postmodern, 1983–1998*. New York: Verso.

Jamieson, Dale. 2014. *Reason in a Dark Time*. New York: Oxford University Press.

Jasanoff, S. 2010. "A New Climate for Society." *Theory, Culture & Society* 27(2–3):233–253.

Judkins, Gabriel, Marissa Smith, and Eric Keys. 2008. "Determinism Within Human-Environment Research and the Rediscovery of Environmental Causation." *Geographical Journal* 174:17–29.

Kates, R., W. Clark, R. Corell, J. Hall, C. Jaeger, I. Lowe, J. McCarthy, H. Schellnhuber, B. Bolin, N. Dickson, S. Faucheux, G. Gallopin, A. Grubler, B. Huntley, J. Jager, N. Jodha, R. Kasperson, A. Mabogunje, P. Matson, H. Mooney, B. Moore, T. O'Riordan, and U. Svedlin. 2001. "Sustainability Science." *Science* 27:641–642.

Kenis, Anneleen and Matthias Lievens. 2014. "Searching for 'The Political' in Environmental Politics." *Environmental Politics* 23:531–548.

Klein, Naomi. 2014. *This Changes Everything: Capitalism Vs. the Climate*. New York: Simon and Schuster.

Kopnina, Helen and Eleanor Shoreman-Ouiment. 2011. *Environmental Anthropology Today*. London & New York: Routledge.

Lahsen, Mayanna. 2013. "Anatomy of Dissent: A Cultural Analysis of Climate Skepticism." *American Behavioral Scientist* 57:732–753.

Latour, Bruno. 2011. "Waiting for Gaia. Composing the Common World through Art and Politics." A Lecture at the French Institute for the Launching of SPEAP. London, November 2011 (http://www.bruno-latour.fr/sites/default/files/124-GAIA-LONDON-SPEAP_0.pdf).

Lever-Tracy, Constance (Ed.). 2010. *Routledge Handbook of Climate Change and Society*. Abingdon, UK: Routledge.

Lipset, Seymour M. 1997. *American Exceptionalism: A Double-Edged Sword*. New York: W. W. Norton & Company.

Liu, J., T. Dietz, S. Carpenter, M. Alberti, C. Folke, E. Moran, A. Pell, P. Deadman, T. Kratz, J. Lubchenco, E. Ostrom, Z. Ouyang, W. Provencher, C. Redman, S. Schneider, and W. Taylor. 2007. "Complexity of Coupled Human and Natural Systems." *Science* 317:1513–1516.

MacGregor, Sherilyn. 2014. "Only Resist: Feminist Ecological Citizenship and the Post-politics of Climate Change." *Hypatia* 29:617–633.

Maibach, Edward W., Connie Roser-Renouf, and Anthony Leiserowitz. 2008. "Communication and Marketing as Climate Change-Intervention Assets." *American Journal of Preventative Medicine* 35:488–500.

Maniates, Michael F. 2001. "Individualization: Plant a Tree, Buy a Bike, Save the World?" *Global Environmental Politics* 1:31–52.

McNall, Scott G. 2011. *Rapid Climate Change: Causes, Consequences, and Solutions*. New York & London: Routledge.

Merchant, Carolyn. 1992. *Radical Ecology: The Search for a Livable World*. New York: Routledge.

Mills, C. Wright. 1959. *The Sociological Imagination*. New York: Oxford University Press.

Mooney, Harold A., Anantha Duraiappah, and Anne Larigauderie. 2013. "Evolution of Natural and Social Science Interactions in Global Change Research Programs." *Proceedings of the Natural Academy of Sciences USA* 110 (Suppl 1):3665–3672.

Moser, Susan and Carol Berzonsky. 2014. "There Must Be More: Communication to Close the Cultural Divide." In *The Adaptive Challenge of Climate Change*, edited by K. O'Brien and E. Seboe. Cambridge, UK: Cambridge University Press.

Murphy, Raymond. 2011. "The Challenge of Anthropogenic Climate Change for the Social Sciences." *International Review of Social Research* 1:167–181.

National Science and Technology Council. 2012. *The National Global Change Research Plan 2012–2021: A Strategic Plan for the U.S. Global Change Research Program (USGCRP).* Washington, DC: U.S. Government.

Nordhaus, William D. 2008. *A Question of Balance: Weighing the Options on Global Warming.* New Haven, CT: Yale University Press.

O'Brien, Karen. 2012. "Global Environmental Change III: Closing the Gap Between Knowledge and Action." *Progress in Human Geography* 37:587–596.

O'Riordan, Timothy. 2013. "Future Earth and Tipping Points." *Environment* 55(5):31–40.

Palmer, Paul I. and Matthew J. Smith. 2014. "Model Human Adaptation to Climate Change." *Nature* 512:365–366.

Palsson, Gisli, Bronislaw Szerszynski, Sverker Sörlin, John Marks, Bernard Avril, Carole Crumley, Heide Hackmann, Poul Holm, John Ingram, Alan Kirman, Mercedes Pardo Buendía, and Rifka Weehuizen. 2013. "Reconceptualizing the 'Anthropos' in the Anthropocene: Integrating the Social Sciences and Humanities in Global Environmental Change Research." *Environmental Science & Policy* 28:3–13.

Pimm, Stuart L. 2009. "Climate Disruption and Biodiversity." *Current Sociology* 19:R595–R901.

Redclift, Michael and Ted Benton (Eds.). 1994. *Social Theory and the Global Environment.* London & New York: Routledge.

Reid, W., D. Chen, L. Goldfarb, H. Hackmann, Y. Lee, K. Mokhele, E. Ostrom, K. Raivio, J. Rockstrom, H. Schellnhuber, and A. Whyte. 2010. "Earth System Science for Global Sustainability: Grand Challenges." *Science* 330:916–917.

Reusswig, Fritz and Wiebke Lass. 2010. "Post-Carbon Ambivalences: The New Climate Change Discourse and the Risks of Climate Science." *Science, Technology & Innovation Studies* 6:156–181.

Revkin, Andrew. 2013. "Can We Respond to Problems like Global Warming Where There's 'No Simple Villain'." Available online *DOT Earth*, December 14 (http://dotearth.blogs.nytimes.com/2013/12/14/can-we-respond-to-problems-like-global-warming-where-theres-no-simple-villain/).

Ritzer, George. 1988. *Sociological Theory*, 2nd ed. New York: Knopf.

Schnaiberg, Allan. 1980. *The Environment: From Surplus to Scarcity.* New York: Oxford University Press.

Shove, Elizabeth. 2010a. "Beyond the ABC: Climate Change Policy and Theories of Social Change." *Environment and Planning A* 42:1273–1285.

Shove, Elizabeth. 2010b. "Social Theory and Climate Change: Questions Often, Sometimes and Not Yet Asked." *Theory, Culture & Society* 27(2–3):277–288.

Shove, Elizabeth. 2010c. "Sociology in a Changing Climate." *Sociological Research Online* 15(3) (http://www.socresonline.org.uk/15/3/12.html).

Shove, Elizabeth. 2011. "On the Difference Between Chalk and Cheese—a Response to Whitmarsh et al.'s Comments on "Beyond the ABC: Climate Change Policy and Theories of Social Change." *Environment and Planning A* 43:262–264.

Simon, Herbert A. 1985. "Human Nature in Politics: The Dialogue of Psychology with Political Science." *American Political Science Review* 79:293–304.

Stehr, Nico and Hans von Storch. 1999. "Climate Works: An Anatomy of a Disbanded Line of Research." Pp. 137–185 in *Wissenschaftlicher Rassismus. Analysen einer*

Kontinuität in den Human- und Naturwisenschaften, edited by H. Kaupen-Haas and C. Saller. Frankfurt am Main: Campus.

Stern, Nicholas. 2007. *The Economics of Climate Change: The Stern Review.* Cambridge: Cambridge University Press.

Stern, Paul C. 2011. "Contributions of Psychology to Limiting Climate Change." *American Psychologist* 66:303–314.

Stoknes, Per Espen. 2014. "Rethinking Climate Communications and the 'Psychological Climate Paradox'." *Energy Research & Social Science* 1:161–170.

Swim, Janet K., Paul C. Stern, Thomas J. Doherty, Susan Clayton, Joseph P. Reser, Elke U. Weber, Robert Gifford, and George S. Howard. 2011. "Psychology's Contributions to Understanding and Addressing Global Climate Change." *American Psychologist* 66:241–250.

Swyngedouw, Erik. 2010. "Apocalypse Forever? Post-political Populism and the Spectre of Climate Change." *Theory, Culture & Society* 27(2–3):213–232.

Swyngedouw, Erik. 2011. "Depoliticized Environments: The End of Nature, Climate Change and the Post-Political Condition." *Royal Institute of Philosophy Supplement* 69:253–274.

Szerszynski, Bronislaw and John Urry. 2010. "Changing Climates: Introduction." *Theory, Culture & Society* 27(2–3):1–8.

Tabara, J. David. 2013. "Social Learning to Cope with Global Environmental Change and Unsustainability." Pp. 253–265 in *Routledge International Handbook of Social and Environmental Change,* edited by S. Lockie, D. A. Sonnenfeld, and D. R. Fisher. London & New York: Routledge.

Taylor, Bob Pepperman. 1992. *Our Limits Transgressed: Environmental Political Thought in America.* Lawrence: University Press of Kansas.

Tversky, Amos and Daniel Kahneman. 1986. "Rational Choice and the Framing of Decisions." *Journal of Business* 59:S251–S278.

Urry, John. 2011. *Climate Change & Society.* Malden, MA: Polity.

Vinthagen, Stellan. 2013. "Ten Theses on Why We Need a 'Social Science Panel on Climate Change'." *ACME: An International E-Journal for Critical Geographies* 12(1):155–176.

Waltman, Ludo. 2013. "Annex B1: Bibliometric Analysis of Social Science Research into Climate and Global Environmental Change." Pp. 584–606 in *World Social Science Report; Changing Global Environments,* edited by International Social Science Council. New York: UNESCO Publishing.

Weaver, C., S. Mooney, D. Allen, N. Beller-Simms, T. Fish, A. E. Grambsch, W. Hohenstein, K. Jacobs, M. A. Kenney, M. A. Lane, L. Langner, E. Larson, D. L. McGinnis, R. H. Moss, L. G. Nichols, C. Nierenberg, E. A. Seyller, P. C. Stern, and R. Winthrop. 2014. "From Global Change Science to Action With Social Sciences." *Nature Climate Change* 4:656–659.

Wilbanks, Thomas, Thomas Dietz, Richard Moss, and Paul Stern. 2013. "Chapter 14: The Social Sciences and Global Environmental Change in the United States." Pp. 133–141 in *World Social Science Report; Changing Global Environments,* edited by International Social Science Council. New York: UNESCO Publishing.

Yearley, Steven. 2009. "Sociology and Climate Change after Kyoto." *Current Sociology* 57:389–405.

York, Richard. 2008. "Introduction to the Symposium on Catton and Dunlap's Foundational Work Establishing an Ecological Paradigm." *Organization & Environment* 21:446–448.

Zalasiewicz, Jan, Mark Williams, Will Steffen, and Paul Crutzen. 2010. "The New World of the Anthropocene." *Environmental Science and Technology* 44:2228–2231.

Zehr, Stephen. 2014. "The Sociology of Climate Change." *WIREs Climate Change.* doi: 10.1002/wcc.328.

Žižek, Slavoj. 2010. *Living in the End Times.* New York: Verso.

2

The Human (Anthropogenic) Driving Forces of Global Climate Change

Eugene A. Rosa, Thomas K. Rudel, Richard York,
Andrew K. Jorgenson, and Thomas Dietz

INTRODUCTION

It is well established that increases in the concentrations of greenhouse gas (GHG) emissions in the atmosphere and changes in the amount of sunlight reflecting back into space (the albedo) due to changes in land cover are the major factors leading to changes in global climate dynamics in the current era (Intergovernmental Panel on Climate Change [IPCC] 2007, 2013; U.S. National Research Council 2010). Furthermore, it is clear that human activities are primarily responsible for the increase in GHGs and alteration of land cover. However, one major challenge for addressing climate change is understanding in more detail the characteristics of societies and their social, economic, and political factors that drive GHG emissions and land use change. This is the challenge that calls for sociological research. The main questions for sociology are as follows: (1) What are the specific human activities and societal characteristics—the "driving forces" in the climate literature—principally responsible for increasing the concentration of GHGs and changing land surfaces, and (2) What can be done to alter these factors so as to curb the human impact on the climate?

A basic and all-too-obvious answer to question 1 is that burning fossil fuels and clearing forests are two of the main things that humans are doing that lead to climate change, and a similarly obvious answer to question 2 is that we need to move away from fossil fuel use and increase forest cover (IPCC 2014). However, the less obvious but more relevant answers to these questions—from the perspective of those who wish to change human activities so as to mitigate human impact on the climate—lie more deeply in the structure of societies. We refer to the characteristics of societies that have the most substantial influence on the global climate as "driving forces." We need to understand why societies produce and consume the levels of energy they do and why they manage land as they do. Without understanding how and why societies use resources and manage their environments, we cannot formulate ways to reduce fossil fuel use and deforestation, as well as to change the variety of other human activities that drive climate change. Fortunately, sociological research has helped to develop an understanding of human–environment interactions, which can point the way to how anthropogenic impacts on the climate can be curtailed.

While climate change is the focus of this volume, it is important to keep in mind that it is just one system in a dynamic of global environmental and social changes that interact with each other in complex loops of cause and effect. The social dynamics that are generating climate change are, in general, the same dynamics that are leading to loss of biodiversity, alteration of key biogeochemical cycles, and the widespread dispersal of persistent pollutants, as well as other public health threats. In turn, these environmental changes interact with each other and have begun to have substantial impacts on human well-being. In 2005 over 1,000 of the world's leading environmental scientists released a report assessing the state of the world's ecosystems, the *Millennium Ecosystem Assessment*. It concluded: "The changes that have been made to ecosystems have contributed to substantial net gains in human well-being and economic development, but these gains have been achieved at growing costs in the form of the degradation of many ecosystem services, increased risks of nonlinear changes, and the exacerbation of poverty for some groups of people. These problems, unless addressed, will substantially diminish the benefits that future generations obtain from ecosystems." (Reid et al. 2005:1). Many researchers are concerned that human actions are already placing stress on the environment beyond the limits of what is sustainable (Röckstrom et al. 2009). As the United Nations recently stated in its *Global Environmental Outlook* report: "The scale, spread and rate of change of global drivers are without precedent. Burgeoning populations and growing economies are pushing environmental systems to destabilizing limits" (United Nations Environment Programme 2012:4).

WHAT CAN SOCIOLOGY OFFER?

Some of the key driving forces of global climate change are recognized as especially important by natural scientists and sociologists alike (Rosa and Dietz 1998). In particular, the size and growth of the human population, the scale and structure of economic production and consumption, and the types of technologies societies develop and use are broadly acknowledged to be key forces influencing human impacts on the environment, including the global climate (Blanco, Gerlagh, and Suh 2014; IPCC 2007, 2013). However, despite broad agreement about these drivers, one of the important and distinctive contributions of sociology is its ability to understand the underlying forces that shape demographic, economic, and technological conditions, and in turn drive climate change.

As a discipline that had its origin in understanding how economic change, shifts in population scale and concentration, and secularization were reshaping the structures of society, sociology traditionally offered large-scale analyses and critiques of the effects of existing institutions, cultural assumptions, and structural arrangements (Durkheim 1933; Weber 1968). Thus sociology has much to contribute to an ongoing understanding of growth, market-based decision making, consumerism, technological development, and the full array of geopolitical and economic arrangements dominant in the world at present. These larger orienting arguments in turn inform empirical analyses that ground general arguments in data from particular places and times. As for climate change drivers, especially the drivers of GHG emissions, most work is based on comparing political units such as nation-states across recent decades. In addition to advancing our understanding of the relative importance of various drivers, this work yields estimates of the malleability of drivers that can be useful in policy discussions. Recent work has begun to address issues of key importance to policy, such as the degree of effectiveness of renewable energy in displacing GHG emissions (York 2012) and the extent to which human well-being can be advanced while minimizing harm to the environment (Dietz, Rosa, and York 2009; Jorgenson 2014). Thus the contributions of sociology are both theoretical (identifying structural, institutional, and cultural factors that influence human drivers of environmental stress) and methodological (providing tools that allow theories and policy proposals to be tested empirically).

There are a variety of key methodological challenges to analyzing human–environment interactions (see Chapter 13 in this volume). It is important to recognize that climate and other coupled human and natural systems occur at multiple scales. Units of analysis of the driving forces of such systems, including climate, range across all scales, from the individual human to the entire biosphere. The levels of analysis are to some degree nested. Individuals are nested into households, human communities, networks, local/regional political economies, nations, and the world system (Chase-Dunn and Jorgenson 2003). They are also part of populations, ecological communities, watersheds and air sheds, landscapes, biotic provinces, and the biosphere. However, the nesting of social units is not strict (e.g., social networks are not restricted to communities or a nation). To make matters more complex, social and ecological units rarely have the same boundaries (Dietz, Rosa, and York 2010). One of the grand challenges for understanding driving forces of environmental change is to integrate across levels, learning how the context of nation or community, network linkages, and other structural factors influence individual actions and how individual actions and interactions aggregate to produce larger emergent social structures and processes.

At present, most social and economic research on drivers has been done at one of two levels of analysis, the individual and the nation-state. A substantial body of ongoing work examines environmentally consequential action on the part of individuals, and several recent, quite accessible reviews summarize what is known about this level (Dietz, Stern, and Weber 2013; Stern 2011), including Chapter 4 in this volume. In this chapter we focus on the nation-state as the unit of analysis. The driving forces manifest themselves across a range of scales, but certainly nation-states have played important roles in generating the high rates of emissions observed during the past century, and the easy availability of data on GHG emissions and the potential drivers of emissions over the past quarter-century has facilitated analyses at this scale. The focus here on the national level of analysis continues a tradition of macro-comparative analysis that extends to the origins of sociology and draws on quantitative approaches to comparing societies that first emerged in the 1970s. Thus it forms a strong link between environmental sociology and other macro-sociological approaches, building on and extending their insights.

Anthropogenic driving forces, broadly conceptualized, are the social, economic, political, and cultural factors that generate direct and indirect pressures on the environment. This collection of factors can be visualized as a hierarchical system where the factors are nested according to their level of aggregation by unit of analysis. Individual decisions, the basic level of action, are nested in and conditioned by institutional, political, economic, and cultural contexts. Similarly, institutional, political, and economic contexts are conditioned by broad historical trends and global processes. Historical processes and their globalizing impacts are conditioned by events, cultural continuities, and path dependencies rooted in previous decisions at the individual, institutional, and political levels of aggregation.

Another useful way to organize a discussion of human influences on the environment is to differentiate the scale of human activity, the content of consumption, and the techniques (technology) used to produce goods and services (Rosa and Dietz 2012). Scale can be broken into the size of the human population and per capita affluence. Scale, composition, and technique are in turn influenced by institutions, culture, political economy, and a variety of other factors, so we can distinguish between direct drivers—scale, composition, and technique—and the indirect drivers that shape them.

In the remainder of the chapter we will review what is known about the effects of various driving forces. Our analysis focuses on GHG emissions, since they are the most important and most researched driver of climate change. But we will also briefly summarize what is known about the drivers of land use change, which influences both GHG emissions and albedo.

POPULATION, AFFLUENCE, AND TECHNOLOGY

The debate about the effect of direct drivers in producing pressure on the environment can be traced to classical writers and is often identified with economist Thomas Robert Malthus (Dietz and Rosa 1994). But it was a debate between Barry Commoner on the one hand and Paul Ehrlich and John Holdren on the other that has shaped much recent work on the issue (Blanco et al. 2014; Chertow 2001; Dietz et al. 2010; Dietz and Rosa 1994; Van Vurren et al. 2011). The IPAT accounting equation, which is also called the Kaya Identity in the climate change literature, where it is used in formulating scenarios of GHG emissions (Kaya 1990; Kaya and Yokobori 1997), emerged from that debate. IPAT stands for

$$I_{mpacts} = P_{opulation} * A_{ffluence} * T_{echnology}.$$

P and A together capture variations in the magnitude of human activities, while T captures all aspects of composition and technique and the forces that shape them. The original form of IPAT presupposes that P, A, and T each have "unit elasticity"—meaning a 1 percent change in any one of them produces a 1 percent change in Impacts (I). Traditionally, the emissions scenarios that underpin the climate projections of the IPCC use a form of the IPAT unit elasticity logic to predict future emissions. Emissions are assumed to be a simple multiplicative function of population, affluence per capita, energy use per unit of affluence, and emissions per unit of energy use.

Sociologists have reformulated the IPAT into a statistical form where parameters are not assumed but are estimated and thus can vary according to the context of model application (Dietz and Rosa 1994). STIRPAT, the name of the reformulation, has generated considerable amount of empirical research on the human drivers of GHGs and on environmental impacts more generally (Dietz and Rosa 1997; York, Rosa, and Dietz 2003). The STIRPAT formulation does not assume unit elasticity; rather, it allows the elasticity for each driver to be estimated based on empirically observed relationships. In most applications of IPAT and STIRPAT, P is the size of the human population and A is measured as gross domestic product (GDP) per person—that is, as per capita economic activity. This means that PA is the size of the economy, an approximation of the scale of human activity. If the coefficients for P and A are both positive, then the larger the scale of the economy the greater the stress on the environment, all other things held constant. And T, while labeled "technology," is really all other things, such as culture, institutional practices, and political processes. The multiplicative specification of the model recognizes that the effects of P, A, and T do not generate impacts independently of one another, but rather the effect of one factor depends on the others.

Population and consumption have consistently been shown to be principal direct driving forces of GHG emissions and other environmental pressures. Longitudinal research suggests that the effect of population size on total carbon emissions and total methane emissions is large in magnitude in both developed and less-developed countries, and the size of the estimated effect is relatively stable through time over recent decades for both groups of countries (Jorgenson and Birkholz 2010; Jorgenson and Clark 2010).

In principle, policy and other social changes can lead to changes in population and the scale of human consumption and production, in the content of consumption, or in the techniques used in production. One contribution of sociological research has been to raise issues about various aspects of scale, content, and technique—the degree to which changes in one of these factors actually lead to changes in stress on the environment. Researchers are beginning to use this approach to examine the efficacy of policies intended to reduce human pressure on the environment. For example, York (2012) finds that the development of renewable energy does not proportionally displace fossil fuel use (e.g., when other factors driving energy demand are controlled, a one-unit increase in renewables in the energy portfolio of a nation yields a reduction in fossil fuel use of considerably less than one unit). Sociologists have also noted the importance of plasticity—the ease or difficulty and the feasible speed of changing a driving force (York, Rosa, and Dietz 2002). For example, the factors that slow growth in human population size—women's education, access to contraception, improved maternal and child health—are well understood, but substantial changes in population size can take years to realize. So in the short run, population size has low plasticity, but its plasticity is high on generational time scales. The idea of plasticity, developed in macro-sociological research, has been introduced to micro-level analyses of individual and household behavioral change, where it calls attention to differences in the ease with which such behaviors can be changed (Dietz, Gardner, et al. 2009).

It is important to recognize that since World War II, most population growth has occurred in the Global South, which, due to its modest level of industrialization and consumption, has not until recently contributed a great deal to global GHG emissions. Of course, the dramatic population growth in many low-income nations has contributed to increases in emissions in those nations, but in absolute terms, the scale of emissions coming from those nations remains very modest compared to the emissions from affluent nations (Jorgenson and Clark 2012; Roberts and Parks 2006). Also, population growth in affluent nations has contributed to rising emissions there, but since the population growth rate has been low in affluent nations, other factors have had more of an influence on changes in GHG emissions.

As this example shows, the major lesson of the IPAT/STIRPAT formulation is that it is not meaningful to consider the effect of population size and growth of a given population without considering its level of affluence and consumption. A modest amount of population growth for a population that has very high levels of consumption can place tremendous pressure on the environment while substantial growth in a population that has only very modest levels of consumption may increase environmental pressure less substantially. However, it is important to note that over the last few decades, changes in patterns of consumption have blurred this binary distinction between rich and poor societies. In most middle-income countries and even in some low-income countries "middle classes" have emerged that consume at rates comparable to what is seen in the most affluent nations (Myers and Kent 2003). This has led to rapid increases in demand for consumer goods, such as animal protein and automobiles, that place substantial pressure on the environment (Jorgenson 2006a; York 2003; York and Gossard 2004). The most noteworthy cases in this regard, due to their very large and growing emissions of GHGs, are China and India, where consumption and production are growing rapidly for an extremely large number of people. Chapter 4 in this volume discusses shifting consumption patterns and the forces that drive them in more detail.

GROWTH AND MODERNIZATION

There are two dominant theoretical orientations on how social forces influence human pressure on the environment. The first of these, the treadmill of production theory (TOP), was developed by Allan Schnaiberg (Schnaiberg 1980; Schnaiberg and Gould 1994). Its basic argument is that, at least since World War II, "producers" (i.e., the owners of the means of production, such as factories) and elite managers, especially corporate leaders, have sought to expand production and consumption as rapidly as possible so as to increase profits. Since production requires natural resources, such as energy and materials, and generates pollution, rising production leads to escalating environmental problems. Furthermore, producers also invest in automation, which has the potential to displace workers unless the scale of production expands rapidly enough to generate new jobs. Due to this dynamic, organized labor has also supported rapid growth, despite the harms it generates. Similarly, governments support growth because it generates tax revenues, which allow those in government (who are often social elites connected to corporate power) to implement policies and programs to legitimize their rule. Ironically, many of these policies and programs are aimed at fixing the problems that economic growth itself has generated, such as pollution and unemployment. Thus, economic growth generates environmental and social problems, and those in power push for further growth to solve these problems in an ongoing "treadmill." An analogous process shaped changes in urban landscapes. Growth coalitions of bankers, builders, landowners, and construction workers, sometimes referred to as "growth machines," transformed landscapes and profited from these transformations (Logan and Molotch 1987). Thus both TOP and growth machine theories argue that contemporary capitalist political economies promote economic growth, prioritizing it over concerns about social inequality and environmental protection.

In addition to the changes in scale of production that the TOP examines, the capitalist system has contributed to transforming material and nutrient cycles through both growth and how it restructures production processes to maximize profit. This restructuring often disrupts natural systems, creating a "metabolic rift," undermining sustainability (Foster 1999; Foster, Clark, and York 2010). In the classic example presented by Karl Marx, rapid urbanization driven by industrialization separated people from the land, so that nutrients from agricultural production were shipped to cities in the form of food and fiber, where they ultimately ended up in landfills and sewers instead of being returned to soils as nutrients as they were in traditional societies (Foster 1999). Similar processes play out with other biogeochemical systems, where natural metabolic processes are disrupted to expand profits, such as when the carbon cycle is altered by the clearing of forests (carbon sinks) and adding the emission of long-sequestered carbon by the burning of fossil fuels (Clark and York 2005; Foster et al. 2010). The TOP and the metabolic rift theories suggest that the dynamics of capitalist systems, particularly the drive for endless growth and increased profitability, are the principal forces driving environmental degradation, and, therefore, the only lasting solution to environmental crises, such as global climate change, is to fundamentally alter political-economic systems so that growth in profits does not trump efforts for social justice and environmental protection.

Other theoretical perspectives suggest that in principle, shifts in the content of consumption or in the techniques of production could compensate to some degree for the overall increase in the scale of economic activities. Such shifts to reduce pressure on the environment even as the scale of human activity expands are at the center of the major alternative to the TOP and metabolic rift theories—ecological modernization theory (EMT). EMT and two closely related arguments, the environmental Kuznets curve (EKC) thesis and the postmaterialist theory, suggest a socioeconomic dynamic that reduces pressure on the environment. They argue that high levels of affluence lead to shifts in politics and consumer demand that ameliorate the negative effects of increased scale. EMT is sociology's contribution to this approach (Mol 1995, 2001). For EMT, rationality is a central feature of modernity, and, as part of the modernizing process, rationality infiltrates all aspects of society. As modernization proceeds, ecological rationality purportedly gains momentum, environmental problems are taken seriously, and environmental concerns come to rival economic ones as central features of decision making, thus allowing economic growth without environmental degradation. The rise of ecological rationality leads the institutions of modernity, including those connected with both the public and private sectors, to transform from within. A similar dynamic has been hypothesized to be at work in agriculture. The Borlaug hypothesis, named after the green revolution pioneer Norman Borlaug, argues that increased agricultural productivity will concentrate production on smaller land areas and allow other lands to be used to provide ecological services (Rudel et al. 2009). Such transformations might facilitate a relative decoupling between environmental harms (e.g., GHG emissions) and modernization (i.e., economic development).

In some aspects EMT makes similar predictions to those made by neoclassical economists who propose that, while a nation's early stages of economic development are associated with rising pollution and other environmental problems, continued growth eventually leads to environmental protection as the public comes to place greater value on the aesthetic properties of nature, outdoor recreation, and the health benefits of a clean environment (Grossman and Krueger 1995; Selden and Song 1994). This hypothetical relationship is represented by the EKC, an inverted U-shape where GDP per capita is the scale on the x-axis and a measure of pollution or other type of environmental impact is the scale on the y-axis. Altering the scope of this economic view, EMT adherents have suggested that there may be an EKC for urbanization rather than for economic growth more generally (Ehrhardt-Martinez 1998). After a turning point in urban development is reached, further urbanization may lead to environmental reform, since cities foster the development of the institutions of modernity (Ehrhardt-Martinez 1998). In political science, Ronald Inglehart has argued that as material needs and national security are taken for granted, postmaterialist values emerge that support environmental protection (Inglehart 1995), a theory that provides a political parallel to the EMT and the EKC.

There are many case studies showing that with economic growth some firms, sectors, and nations seem to follow a pattern of reducing stress on the environment via environmental reforms. Quantitative studies show some support for the modernization argument for pollutants having local impacts, such as many forms of air and water pollution that have human health effects. However, the core claim of the EMT/EKC is that overall stress on the environment will be reduced (York, Rosa, and Dietz 2010). But there is no consistent evidence for an EMT/EKC pattern for stresses that affect large systems or the global commons, such as aggregate resource consumption and GHG emissions; in fact, there is substantial empirical evidence indicating that the key aspects of modernization, particularly economic growth and urbanization, are generally associated with rising GHG emissions as well as having other harmful effects on the environment (Jorgenson 2006a; Jorgenson, Auerbach, and Clark 2014; York and Rosa 2003; York et al. 2010).

The claim from EMT that technological advances that come with modernization are likely to reduce human impact on the environment also has been contradicted by a substantial amount of empirical research. For example, there is clear evidence that improvements in efficiency often have failed to have substantial effects on the overall scale of environmental problems; in fact, it is commonly the case that as nations become more energy or carbon efficient at the macro-economic level (i.e., reduce their use of energy or carbon emissions per dollar of GDP), their total energy use rises (Jorgenson 2009; York 2006, 2010; York, Rosa, and Dietz 2009). This work supports longstanding arguments for the existence of the rebound effect and Jevons paradox, where improvements in energy efficiency are connected with increased levels of production and consumption, so that greater efficiency is not always associated with as much reduction in total energy use as might be expected (rebound), and in extreme cases is connected with an overall increase in energy use because production and consumption rise faster than efficiency improves (Jevons paradox) (Foster et al. 2010; York 2006). The causal connections in these patterns are likely complicated, and the association between rising efficiency and escalating consumption may not be direct, but the frequency with which cases suggesting rebound or the Jevons paradox are found in empirical research suggests that relying on improvements in efficiency to help curb GHG emissions may be misguided. Similarly, as noted above, other research has found that the development of non–fossil fuel energy sources has not displaced fossil fuel energy at a one-for-one rate (York 2012); this finding suggests that simply developing "green" technologies, without changing the context in which they are applied, may not be highly effective at reducing environmental problems.

As noted above, ecological modernization theory posits that even though economic development first harms the environment, the magnitude of the link decreases over the course of development—a form of relative decoupling. In contrast, the TOP theory argues that the ironclad relationship between environmental harms and economic development remains constant or possibly increases through time for both developed and less-developed countries. Jorgenson and Clark (2012) evaluate these competing propositions through the use of interactions between economic development and time in cross-national panel analyses of three measures of carbon dioxide emissions: total emissions, per capita emissions, and emissions per unit of production. The results vary across the three outcomes as well as between developed and less-developed countries, providing mixed and unbalanced support for both theoretical perspectives. But overall the results are more consistent with the propositions of the TOP theory (see also Jorgenson and Clark 2011; Knight and Schor 2014).

Shwom (2011) has argued that the TOP and EMT can be thought of as a spectrum rather than as opposing theories. In some societies at some points in history economic growth is promoted without consideration of its environmental consequences. In other places and contexts, ecological limits are recognized and both consumption and production are modified to ameliorate harm to the environment. Where a society sits on the spectrum between TOP and EMT depends on the distribution of political power between those who champion unfettered growth and those who seek reform. For example, the United States clearly occupies a different position than Japan or the Netherlands, given the differences between the three countries in their political economies, with the United States on a TOP trajectory and Japan and the Netherlands somewhat closer to the trajectory posited by the EMT, although still having high per capita GHG emissions by global standards. This comparison draws attention to the importance of various institutions, aspects of culture, and forms of political economic organization that can moderate the effects of the scale of human activity on the environment. The balance between environmental modernization and TOP is, therefore, subject to active political struggle (McCright and Dunlap 2010).

However, it is important to note that even societies like Japan and the Netherlands, which follow a path that in some respects appears to be along the lines of environmental modernization, have very large and unsustainable impacts on the environment, including per capita GHG emissions well above the global average and total emissions well above what is necessary to avoid dramatic climate change. Furthermore, as Ehrlich and Holdren (1971) noted decades ago when they coined the term the "Netherlands fallacy" to refer to the erroneous assumption that the environmental problems generated by a nation due to its resource consumption necessarily occur within its national borders, affluent nations like Japan and the Netherlands have been able to protect their own environments in part because they import many of the natural resources they consume and export hazardous waste and/or offshore polluting industries. The point the Netherlands fallacy emphasizes is that many nations that appear to be "ecologically modernizing" may in fact simply be shifting the environmental impacts they generate to other locales via global trade networks. This emphasizes the importance of using consumption-based measures of environmental stress, such as the ecological footprint, to complement measures based on production, in analyses of GHG emissions.

Hooks and Smith (2004), examining the military's ecological footprint, characterize the expansionary dynamics and profound environmental impacts associated with militarism as the "treadmill of destruction." Treadmill of destruction theory is largely inspired by the TOP theory but incorporates a previously ignored institution: the military. Even in the absence of armed conflict, the world's military institutions and their activities consume vast amounts of nonrenewable energy and other resources for research and development, maintenance, and operation of the overall military infrastructure. It is important to note here that the United States has the largest military budget in the world by far and is therefore responsible for a large share of military-related environmental impacts (Hooks and Smith 2004). Other researchers have situated the TOP into an international-comparative perspective. Cross-national longitudinal research suggests that increasing the scale of national militaries leads to increases in fossil fuel consumption and carbon dioxide emissions (Clark, Jorgenson, and Kentor 2010; Jorgenson, Clark, and Kentor 2010; York 2008).

CULTURE

Perhaps no term in the social sciences has as many definitions and conceptualizations as "culture." A very substantial literature examines the role of values, beliefs, norms, and other social-psychological aspects of culture that shape human decisions with consequences for the environment (Dietz, Fitzgerald, and Shwom 2005). Such decisions come in at least three roles: as a consumer, as citizen, and for some, as an activist. It is only during the past few decades, accelerated by the National Research Council study *Environmentally Significant Consumption* (Stern et al. 1997: see also Princen, Maniantes, and Conca 2002), that social scientists began to address the connections between values, beliefs, norms, and consumption practices, and work on the subject is still emerging. Consumption and climate change are reviewed in Chapter 4. Examining what influences individuals to act as citizens and as activists is a longstanding tradition in sociology; this work is reviewed in Chapter 8. Here we focus briefly on the macro-comparative aspects of culture as a driving force of consumption and decision making, while Chapter 4 discusses in detail the creation and maintenance of a consumer culture.

To sustain itself the TOP must find or create a complementary institutional form: a treadmill of consumption. There has to be market demand for what is produced; otherwise, the system of expanding economic activity would collapse. The treadmill of consumption, by ensuring that the goods produced are consumed, keeps extraction, production, and consumption relatively in balance and ever expanding. Of course, on the consumer side buyers must have the money necessary to purchase the goods, so consumption is also related to the distribution of income both within and across societies.

Political economists also raise the question of where primary responsibility for driving economic expansion lies. As noted above, the rising number of middle-class consumers is widely recognized as a driver of increased GHG emissions. However, the process of rising consumption is driven not by consumer desire per se, but by the profit-seeking interests of the capitalist class, who control corporations and the major means of production. Consumer demand is created by producers, who invest extensively in marketing operations (Foster et al. 2010; Schnaiberg 1980). The important point here is that to bring about genuine environmental reform, the focus should be on structural factors, the actions of major corporations and the capitalist class more generally, in addition to consumer choices per se. In this view culture is seen not as an exogenous and constant force but as something that changes over time, in substantial degree in response to intentional efforts to shape culture by those in power.

People have long acquired goods through transactions in markets, but markets were transformed and became a particularly salient aspect of modernity's social structures. Analysts have long examined "market societies," and there has been considerable debate about the degree to which market rationality is pervasive across cultures or is of more limited scope (Polanyi 1944). Research on the effects of engagement with markets on decision making, while provocative, is just emerging (Henrich et al. 2010). Over the last few decades, a near consensus has emerged among governments that market mechanisms and logic should play a central role in the allocation of resources, a trend usually labeled neoliberalism. A corresponding emphasis on the individual and individual actions emerges in society with a corresponding de-emphasis on collective action of all kinds. The rational-actor model, which dominates economics and much of policy analysis, has also become influential throughout the social sciences (Dietz 1994; Dietz and Stern 1995; Jaeger et al 2001).

The dominance of the neoliberal approach to macro policy, of the rational-actor model as the first principle of much policy analysis, and the continued dependence of most people on market transactions for income and sustenance and increasingly for psychological well-being and even identity have made it difficult to question the paradigm of economic growth as the key goal of contemporary societies, despite the environmental harm attendant with this growth. Thus growth has become a part of the taken-for-granted assumptions in most contemporary cultures. Shifts in the composition of consumption and in the techniques of production could in theory compensate for the growth in scale, but, as we have noted, there are many strong systematic forces, political, economic, and cultural, that prevent or subvert such transitions.

GLOBALIZATION, WORLD SYSTEMS, AND WORLD SOCIETY

Two areas of contemporary sociological research assess the ways in which forms of economic globalization or structural integration contribute to GHG emissions. These include research on (1) ecologically unequal exchange and (2) the transnational organization of production. Both traditions have roots in world-systems analysis (Chase-Dunn and Grimes 1995; Roberts and Grimes 2002).

The theory of ecologically unequal exchange considers how particular aspects of international trade (e.g., the "vertical flow" of raw material exports from the less-developed to the more-developed nations) allow for higher-income nations to substantially externalize their environmental impacts, including GHG emissions, to lower-income nations (Hornborg 1998). The vertical flow of exports refers to the flow of raw materials from poor nations to rich nations and developing nations with growing consumer markets. Cross-national longitudinal studies indicate that growth in GHG emissions in developing countries is positively associated with a higher percentage of exports to developed or high-income countries, and such ecologically unequal relationships have increased in magnitude through time (see Jorgenson 2012; Jorgenson and Clark 2011; Rice 2007; Roberts and Parks 2006). Such trade dynamics are partly a function of broadening and deepening global production networks, with subsidiaries of global corporations and domestic firms in developing nations producing goods for the consumer markets of high-income and rapidly developing nations. While greater integration in these production and trade networks may provide economic and social benefits, they also involve a variety of environmental consequences (Jorgenson 2012). These trends extend to the agricultural sector. For example, a large volume of agricultural exports was connected with high rates of tropical deforestation between 2000 and 2005 (DeFries et al. 2010; Jorgenson 2006b). Indeed, some have suggested that any analysis of coupled human and natural systems must take account of the social, ecological, and geophysical process that lead local systems to be "telecoupled" across the globe (Liu et al. 2013).

The tradition of research on the transnational organization of production emphasizes the global outsourcing of various technologies and manufacturing processes. In recent decades the real or perceived paths of world economic upward mobility for developing nations partially stem from outsourcing from firms headquartered in high-income nations. Lower-income nations often serve as attractive sites of "industrial migration" via foreign direct investment for transnational firms, with most of the goods produced intended for export to higher-income nations. For years there has been substantial debate about the effects of foreign direct investment on the environment. Some theorists argue that there will be a race to the bottom in which countries will compete to have the lowest environmental standards in order to attract foreign investment. Others have argued that many multinationals adopt uniform standards, practices, and technologies across countries; hence, facilities in low-income nations will not pollute substantially more than those in rich countries (Mol 2001). Whatever the overall patterns are for other pollutants, strands of sociological research show that foreign direct investment flowing to developing nations contributes to growth in various GHG emissions (e.g., Grimes and Kentor 2003; Jorgenson 2006a, 2007, 2009; Jorgenson, Dick, and Mahutga 2007; York 2008).

In addition to the economic dynamics of globalization, there are also important cultural aspects of globalization. The rise of international organizations, which work across societies, has been identified as an important feature of changing global dynamics. World society theory posits that international organizations play an important role in constituting and reinforcing global cultural norms (e.g., Meyer et al. 1998). International nongovernmental organizations are characterized as carriers of world culture. These organizations diffuse progressive global models that are then adopted by local actors. Meyer and his colleagues (1998) describe the existence of the "world environmental regime," composed of environmental international nongovernmental organizations (EINGOs) and other sorts of civil society groups who promote universal adoption of environmental policies, programs, and standards. Haas had argued that the international epistemic community of environmental scientists had a key role in shaping environmental reforms (Adler and Haas 1992; Haas 1993). Busch has noted the importance of national and international standards as an aspect of this kind of norm diffusion (Busch 2011). While the majority of research in this strand of world society scholarship focuses on the emergence of EINGOs as well as other civil society organizations and state agencies (e.g., Frank, Hironaka, and Schofer 2000), recent cross-national longitudinal studies indicate that EINGO presence is associated with lower rates of growth in carbon dioxide emissions (Schofer and Hironaka 2005). Other research indicates that a stronger presence of EINGOs can mitigate the impact of forms of economic globalization on growth in emissions, especially in developing nations (Jorgenson, Dick, and Shandra 2011; Shandra et al. 2004).

LAND AND INFRASTRUCTURE DEVELOPMENT POLICIES

National and international policies play an important and intertwined role with the economic and cultural dynamics we have discussed. This can be seen clearly in land and infrastructure development. Growth in the scale of the human enterprise during the twentieth century introduced important changes in landscapes that spurred increases in GHG emissions. Deforestation reduces the volume of vegetation that can absorb carbon dioxide. Land use change also changes the albedo of the earth—the ability of the landscape to reflect solar energy back into space—but this effect is small compared to that of the effect of land use change on emissions. According to the 2014 IPCC report (Blanco et al. 2014), land use changes accounted for approximately 24 percent of total annual global GHG emissions.

The changes in the drivers of land use emissions we will describe stem in large part from changes in states' postures toward metropolitan and agricultural expansion during the twentieth century (Rudel 2009). For the first forty years of the century, colonial policies played an important role in changing land use practices in Africa, Asia, and, to a lesser degree, Latin America. Government-supported settler societies confined indigenous peoples to particular tracts of land, often in mountainous or desert-like "refuge regions," and reserved the most fertile agricultural lands for the use of settlers or the state. Upon independence, during the 1950s and 1960s many of these reserved lands became parks (Guha 2000; Neumann 1998). Of more immediate import, growing populations in the newly demarcated indigenous reserves found themselves in natural resource–degrading poverty traps, a circumstance in which the degradation of natural resources creates a self-reinforcing loop in which the confined populations become poorer over time with the decline in available natural resources. Often their only viable option for short-term economic survival was to clear the land of almost all forest for agricultural use. This pattern over time produced a strong association between poverty and deforestation (Rudel 2005). Within the limits set by colonial states, poverty spurred deforestation and deforestation maintained poverty.

During the immediate postwar era both affluent nations and newly independent governments, energized in some instances by the expansion of state powers during World War II, embarked on large road construction projects. The state in affluent societies constructed highways between cities while newly independent nations built penetration roads that linked peripheral, often forest-covered, regions with urban centers. In both instances concerns over national security played an important role in the expanded programs of road building. The new roads spurred the suburban expansion of metropolitan areas in North America. In the Global South it led to agricultural expansions and associated destruction of tropical forests. Both of these changes in human settlement patterns accelerated GHG emissions after the mid-twentieth century (Rudel 2005).

The drivers of land use change shifted again in a significant way during the 1980s when the growing strength of neoliberal perspectives, coupled with debt crises, caused states in the Global South to curtail their road building and settlement programs. At the same time the increasing prevalence of neoliberal policy agendas among elites (often called the Washington consensus) spurred international trade in natural resources, many of which came from tropical forested regions (Jorgenson 2006b). In this dynamic process large commercial enterprises became the chief drivers of land use changes. The activities of oil palm plantations, cattle ranches, and logging companies accounted for the bulk of the GHG emissions from land use changes after 1990 (De Fries et al. 2010). Some of these deforested lands, especially in topographically rugged areas, have since reverted to secondary forests and, at least in the tropics, reduced net carbon emissions (Pan et al. 2011).

FUTURE RESEARCH NEEDS AND RECOMMENDATIONS

Perhaps the largest and most consequential gap in our knowledge about social systems and climate change is in how to construct a realistic global socioeconomic system that addresses human needs without degrading the environment. While there is a considerable and growing body of knowledge identifying the forces that drive GHG emissions, comparable knowledge about alternatives does not exist. Several key features of modern societies and the broader world system such as unconstrained economic growth, economic globalization, population growth, urbanization, and current forms of consumerism clearly appear to be incompatible with low GHG emissions, pointing to the need for substantial social transformation. However, while we know some of the key features of societies that need to change, we know less about what they should be changed to and how to accomplish these changes.

One important move in this direction is research examining the link between ecosystems services, human well-being, and stress on the environment. The Millennium Ecosystem Assessment emphasized the substantial dependence of billions of people around the globe on ecosystem services and the risks to those services (Reid et al. 2005). Recently, the macro-comparative tradition in environmental sociology has begun to look at the relationship between environmental degradation and well-being. The first analysis of this sort showed that quality of life had "decoupled" from energy consumption (Mazur and Rosa 1974). Recent work has examined the "payoff" in human well-being from placing stress on the environment at the macro level (Dietz, Rosa, and York 2007, 2009, 2012; Knight and Rosa 2010) and generally found that nations derive little benefit from activities that substantially stress the environment.

Building on this recent macro work, current sociological research is focusing specifically on the amount of carbon emissions emitted or energy consumed per unit of human well-being. In this new research, particular attention is being given to potential structural pathways that can lead to reductions in this relationship, which is referred to as the carbon intensity of well-being (e.g., Jorgenson 2014; Jorgenson, Alekseyko, and Giedraitis 2014). This emerging research stream on the ecological and carbon intensity of well-being offers additional opportunities for sociology to contribute to our understanding of the dynamic relationships between environmental change and the well-being of humanity (Dietz and Jorgenson 2014). This work will be of growing interest in policy discussions as major reviews have questioned the overarching focus on GDP and economic growth as the dominant measure of social progress (Frey 2008; Stiglitz, Sen, and Fitoussi 2009). Research aimed at assessing what types of political-economic order and what types of policies and programs within nations lead to socially desirable outcomes—such as reduction of inequalities and poverty and improvements in the quality of life of publics—without relying on high levels of resource consumption is highly important.

In addition to the need for work on how to improve the human condition without relying on fossil fuel use, it is important examine how changes in various aspects of society end up affecting GHG emissions. In social processes, the outcomes from one change are often counterintuitive. For example, as we have noted, research suggests that energy efficiency does not necessarily lead to dramatic reductions in energy use (York 2006) and that the development of non–fossil fuel energy sources does not necessarily displace fossil fuel use to a high degree (York 2012). This raises questions about what types of contexts (e.g., political-economic, policy, cultural) are more likely to facilitate the realization of reductions in energy use and/or fossil fuels from purportedly "green" technological developments. Thus, it is important to develop further research on how technological developments affect energy use and carbon emissions.

It is also necessary to develop a greater understanding of how processes are linked from the micro to the macro levels, and how processes around the world are interconnected. Since, for example, actions in one society may have consequences in other societies, it is not always easy to determine what the global impact will be of changes observed in a particular nation. As we have noted, for example, the Netherlands fallacy and research examining unequal ecological exchange suggest that reductions in carbon emissions in some nations may not actually indicate that processes in those nations are leading to reduction in global carbon emissions, since some nations may reduce their emissions through processes that increase emissions elsewhere. While measures such as the ecological footprint have been helpful in tracking these types of processes and examinations of trade and investment networks have made important headway, more refined measures and analyses of how trade, investment, technological exchange, and development practices are linked with fossil fuel use and carbon emissions (as well as with other types of pollution emissions and the use of other natural resources) are needed.

POLICY LESSONS

The consistent effects on climate change of the best-established drivers—population, consumption, and land use change—provide a good guide to formulating policies to reduce GHG emissions in the near term. However, efforts to alter patterns of procreation and consumer purchasing face difficulties because both sets of practices are embedded in a variety of deeply held cultural and political values. The potential for redirecting the population driver may lie not in direct measures but indirect ones. With respect to population, research across many fields over the past several decades has given the heartening finding that the best way to reduce fertility rates is to generally improve human quality of life, health care, and reproductive freedom, while advancing the status of women. These results have made obsolete concerns about the potential need for draconian "population control" methods. In particular, the provision of educational resources for women, access to contraception, and improved maternal and child health have all been shown to slow rates of population growth (Bryant 2007; Cohen 1996; Neumayer 2006).

Reducing the magnitude and composition of consumption and the magnitude and techniques of production poses a difficult challenge. One of the most deeply held values originating in the Western world, but now a global value, is a commitment to continuous growth. Economic growth is typically measured as an increase in GDP, which is the size of the economy. Nearly all nations have increasing the size of the economy as a major goal. In addition, consumption patterns that have intense effects on the environment are spreading from the Western industrial nations to the globe as a concomitant to the growing affluence of many nations and sectors within nearly every nation. There is evidence that increased efficiency in production and consumption can reduce stress on the environment, but there is no clear evidence that these shifts in scale and composition compensate for growth in population and affluence. Indeed, the evidence is just the contrary (United Nations Environment Programme 2011). Part of this is due to the Jevons paradox or rebound effect. Further, it is probably partly due to technological and cultural inertia, as evident in the "energy efficiency gap" where most households and businesses do not make investments in energy efficiency that would be economically as well as environmentally beneficial (Dietz et al 2009). And part of it is due to the active efforts of those with political interests in existing patterns of production and consumption.

The dynamics are complex. Changes in technology (technique) can lead to either increased or decreased pressure on the environment from a given pattern of consumption. Changes in preferences, which are influenced by corporate advertising and also by values, norms, beliefs, social networks and cultural images, can lead to rapid shifts in consumer behavior. But those shifts can be toward greater pressure on the environment (e.g., more consumption of animal protein, bottled water) or toward less pressure (e.g., increased household energy efficiency, increased use of public transportation). How these dynamics unfold is a major priority for future research. Certainly environmental sociologists agree with economists that we lack a full cost life cycle accounting system that takes into account environmental externalities of individual, household, and organizational choices (Davis and Caldeira 2010; U.S. National Research Council 2010:143–144; York et al. 2011). One promising route to reducing consumption is focusing more on improving more direct measures of human quality of life and well-being, rather than strictly economic measures. The research on the carbon intensity of well-being we mentioned earlier (e.g., Jorgenson 2014) points to the potential to improve both human health, as indicated by life expectancy, and subjective well-being without relying on economic growth and rises in material consumption.

State efforts to reform production processes hold out considerable promise as some have been very effective. The recent sharp declines in Brazilian deforestation provide a suggestive direction for future policy reforms to eliminate additional emissions from landscape changes. As noted above, large-scale commercial enterprises have become the chief drivers of tropical deforestation during the past two decades. At the same time these drivers have become "financialized," responsive to changes in ease of access to borrowed money. The growing importance of financial fluctuations in determining deforestation rates is well illustrated by the 1995 peak in Brazilian deforestation rates. It occurred primarily because financial reforms in Brazil drastically lowered interest rates, which in turn encouraged large landowners to borrow money to deforest more land and buy more cattle. Similarly, the Brazilian government discovered, beginning in 2005, that they could reduce illegal deforestation by cutting off all access to government-subsidized credit in municipalities with concentrations of illegal clearings. In this context "financializing" the Reducing Emissions from Deforestation and Forest Degradation (REDD+) program holds out real promise as a policy lever for reducing tropical deforestation and its associated GHG emissions (Boucher, Roquemore, and Fitzhugh 2013). And while global policy on the full suite of GHG emissions has stalled, controls on chlorofluorocarbons, which are potent greenhouse gases, remain highly effective (Parson 2003). Further, a number of nations and some cities and regions have been highly effective at promoting energy efficiency and renewable energy (Gallagher 2013), although as York (2006, 2012) has demonstrated, these efforts often do not displace emissions with high effectiveness.

One of the most important overall lessons of sociological research on the anthropogenic forces driving global climate change is that it is necessary to look beyond technical fixes and consider the social, political, and economic structures that condition human behavior and resource exploitation. Developing more efficient and less polluting technologies without also altering institutions and social structures may not be sufficient to substantially reduce GHG emissions, since political-economic systems have many dynamic feedbacks that may prevent technological fixes from having their intended effects. The rich literature on the commons suggests that it is possible to build institutional arrangements that are sustainable, but the task is challenging and the successful reforms are often complex (Dietz, Ostrom, and Stern 2003; Ostrom 2010).

The increasing integration of all societies throughout the world, especially during the recent and ongoing upsurge in forms of economic and political and even ecological globalization (Chase-Dunn, Kawano, and Brewer 2000; Dreher et al. 2008; Liu et al. 2013), underscores the importance in recognizing that society–nature relationships are embedded within a global system. Forms of economic integration allow for more powerful and wealthier nations to externalize portions of their environmental impacts to less-powerful, poorer nations. In addition, the environmental and public health consequences tied to climate change disproportionally affect poor nations and the vulnerable populations in all nations. However, it appears that world society configurations, such as EINGOs, can partly mitigate the environmental harms of global economic processes, although we have not yet seen much success from these processes with regard to climate change, compared to, for example, ozone depletion (Parson 2003). Furthermore, recent scholarship suggests that world society integration can increase the likelihood of environmental concern at the individual level in nations across the economic strata (Givens and Jorgenson 2013). This could lead to a stronger world environmental regime more committed to structural and behavioral changes around the globe, resulting in a reduction of anthropogenic GHG emissions. In addition, there are movements to promote sustainable production and consumption and even movements that question the desirability of economic growth as a fundamental metric of societal advance (Stiglitz et al. 2009). Sociological research can make important contributions to each of these developments.

REFERENCES

Adler, Emanuel and Peter Haas. 1992. "Conclusion: Epistemic Communities, World Order, and the Creation of A Reflective Research Program." *International Organization* 46:367–390.

Blanco, Gabriel, Reyer Gerlagh, and Sangwon Suh. 2014. "Chapter 5: Drivers, Trends, and Mitigation." In Intergovernmental Panel on Climate Change. Working Group III—Mitigation of Climate Change.

Boucher, Doug, Sarah Roquemore, and Estrellita Fitzhugh. 2013. "Brazil's Success in Reducing Deforestation." *Tropical Conservation Science* 6(3):426–443.

Bryant, John. 2007. "Theories of Fertility Decline and the Evidence from Development Indicators." *Population and Development Review* 33(1):101–127.

Busch, Lawrence. 2011. *Standards: Recipes for Reality*. Cambridge, MA: MIT Press.

Chase-Dunn, Christopher and Peter Grimes. 1995. "World Systems Analysis." *Annual Review of Sociology* 21:387–417.

Chase-Dunn, Christopher and Andrew K. Jorgenson. 2003. "Regions and Interaction Networks: An Institutional-Materialist Perspective." *International Journal of Comparative Sociology* 44:433–450.

Chase-Dunn, Christopher, Yukio Kawano, and Benjamin Brewer. 2000. "Trade Globalization since 1795: Waves of Integration in the World-System." *American Sociological Review* 65:77–95.

Chertow, Marion. 2001. "The IPAT Equation and Its Variants: Changing Views of Technology and Environmental Impact." *Journal of Industrial Ecology* 4:13–29.

Clark, Brett, Andrew K. Jorgenson, and Jeffrey Kentor. 2010. "Militarization and Energy Consumption: A Test of Treadmill of Destruction Theory in Comparative Perspective." *International Journal of Sociology* 2:23–43.

Clark, Brett and Richard York. 2005. "Carbon Metabolism: Global Capitalism, Climate Change, and the Biospheric Rift." *Theory and Society* 34(4):391–428.

Cohen, Joel E. 1996. *How Many People Can the Earth Support?* New York: W.W. Norton and Company.

Davis, Steven J. and Ken Caldeira. 2010. "Consumption-based Accounting of CO$_2$ Emissions." *Proceeding of the National Academy of Sciences USA.* doi: 10.1073/pnas.0906974107.

DeFries, Ruth, Thomas Rudel, Maria Uriarte, and Matthew Hansen. 2010. "Deforestation Driven by Urban Population Growth and Agricultural Trade in the Twenty-First Century." *Nature Geoscience* 3:178–181.

Dietz, Thomas. 1994. "'What Should We Do?' Human Ecology and Collective Decision Making." *Human Ecology Review* 1:301–309.

Dietz, Thomas, Amy Fitzgerald, and Rachel Shwom. 2005. "Environmental Values." *Annual Review of Environment and Resources* 30:335–372.

Dietz, Thomas, Gerald T. Gardner, Jonathan Gilligan, Paul Stern, and Michael P. Vandenbergh. 2009. "Household Actions Can Provide a Behavioral Wedge to Rapidly Reduce U.S. Carbon Emissions." *Proceedings of the National Academy of Sciences USA* 106(44):18452–18456.

Dietz, Thomas and Andrew K. Jorgenson. 2014. "Towards a New View of Sustainable Development: Human Well-Being and Environmental Stress." *Environmental Research Letters* 9:031001.

Dietz, Thomas, Elinor Ostrom, and Paul C. Stern. 2003. "The Struggle to Govern the Commons." *Science* 301(5652):1907–1912.

Dietz, Thomas and Eugene A. Rosa. 1994. "Rethinking the Environmental Impacts of Population, Affluence and Technology." *Human Ecology Review* 1:277–300.

Dietz, Thomas and Eugene A. Rosa. 1997. "Effects of Population and Affluence on CO2 Emissions." *Proceedings of the National Academy of Sciences USA* 94(1):175–179.

Dietz, Thomas, Eugene A. Rosa, and Richard York. 2007. "Driving the Human Ecological Footprint." *Frontiers in Ecology and Environment* 5(1):13–18.

Dietz, Thomas, Eugene A. Rosa, and Richard York. 2009. "Environmentally Efficient Well-Being: Rethinking Sustainability as the Relationship between Human Well-Being and Environmental Impacts." *Human Ecology Review* 16(1):114–123.

Dietz, Thomas, Eugene A. Rosa, and Richard York. 2010. "Human Driving Forces of Global Change: Examining Current Theories." Pp. 83–132 in *Threats to Sustainability: Understanding Human Footprints on the Global Environment*, edited by E. A. Rosa, A. Diekmann, T. Dietz, and C. Jaeger. Cambridge, MA: MIT Press.

Dietz, Thomas, Eugene A. Rosa, and Richard York. 2012. "Environmentally Efficient Well-Being: Is There a Kuznets Curve?" *Journal of Applied Geography* 32(1):21–28.

Dietz, Thomas and Paul C. Stern. 1995. "Toward a Theory of Choice: Socially Embedded Preference Construction." *Journal of Socio-Economics* 24:261–279.

Dietz, Thomas, Paul C. Stern, and Elke Weber. 2013. "Reducing Carbon-Based Energy Consumption through Changes in Household Behavior." *Daedalus* 142(1):78–89.

Dreher, Axel, Noel Gatson, Willem Jozef Meine Martens, and Pim Martens. 2008. *Measuring Globalization: Gauging Its Consequences.* New York: Springer.

Durkheim, Emile. 1933. *The Division of Labor in Society.* New York: Macmillan.

Ehrhardt-Martinez, Karen. 1998. "Social Determinants of Deforestation in Developing Countries: A Cross-National Study." *Social Forces* 77(2):567–586.

Ehrlich, Paul and John Holdren. 1971. "Impact of Population Growth." *Science* 171:1212–1217.

Foster, John Bellamy. 1999. "Marx's Theory of Metabolic Rift: Classical Foundations for Environmental Sociology." *American Journal of Sociology* 105:366–405.

Foster, John Bellamy, Brett Clark, and Richard York. 2010. *The Ecological Rift: Capitalism's War on the Earth.* New York: Monthly Review Press.

Frank, David John, Ann Hironaka, and Evan Schofer. 2000. "The Nation-State and the Natural Environment over the Twentieth Century." *American Sociological Review* 65:96–116.

Frey, Bruno S. 2008. *Happiness: A Revolution in Economics.* Cambridge, MA: MIT Press.

Gallagher, Kelly Sims. 2013. "Why and How Governments Support Renewable Energy." *Daedalus* 142:59–77.

Givens, Jennifer E. and Andrew K. Jorgenson. 2013. "Individual Environmental Concern in the World Polity: A Multilevel Analysis." *Social Science Research* 42:418–431.

Grimes, Peter and Jeffrey Kentor. 2003. "Exporting the Greenhouse: Foreign Capital Penetration and CO2 Emissions 1980–1996." *Journal of World-Systems Research* 9(2):261–275.

Grossman, Gene M. and Alan B. Krueger. 1995. "Economic Growth and the Environment." *Quarterly Journal of Economics* 110:353–377.

Guha, Ramachandra. 2000. *The Unquiet Woods: Ecological Change and Peasant Resistance in the Himalaya.* Berkeley: University of California Press.

Haas, Peter. 1993. "Epistemic Communities and the Dynamics of International Environmental Cooperation." Pp. 168–201 in *Regime Theory and International Relations,* edited by V. Rittberger. Oxford: Oxford University Press.

Henrich, Joseph, Jean Ensminger, Richard McElreath, Abigail Barr, Clark Barrett, Alexander Bolyanatz, Juan Camilo Cardenas, Michael Gurven, Edwins Gwako, Natalie Henrich, Carolyn Lesorogol, Frank Marlowe, David Tracer, and John Ziker. 2010. "Markets, Religion, Community Size, and the Evolution of Fairness and Punishment." *Science* 327(5972):1480–1484.

Hooks, Gregory and Chad L. Smith. 2004. "The Treadmill of Destruction: National Sacrifice Areas and Native Americans." *American Sociological Review* 69(4):558–575.

Hornborg, Alf. 1998. "Towards an Ecological Theory of Unequal Exchange: Articulating World System Theory and Ecological Economics." *Ecological Economics* 25:127–136.

Inglehart, Ronald. 1995. "Public Support for Environmental Protection: Objective Problems and Subjective Values in 43 Societies." *PS: Political Science and Politics* 15:57–71.

Intergovernmental Panel on Climate Change. 2007. *Climate Change 2007: Synthesis Report. Contribution of Working Groups I, II, and II to the Fourth Assessment Report of the Intergovernmental Panel on Climate Change.* Geneva, Switzerland.

Intergovernmental Panel on Climate Change. 2013. *Climate Change 2013: The Physical Science Basis: Summary for Policymakers.* Geneva, Switzerland.

Intergovernmental Panel on Climate Change. 2014. *Climate Change 2014: Synthesis Report. Contributions of Working Groups I, II, and III to the Fifth Assessment Report of the Intergovernmental Panel on Climate Change.* Retrieved December 2, 2014 (https://www.ipcc.ch/report/ar5/).

Jaeger, Carlo, Ortwin Renn, Eugene A. Rosa, and Thomas Webler. 2001. *Risk, Uncertainty and Rational Action*. London: Earthscan.

Jorgenson, Andrew K. 2006a. "Global Warming and the Neglected Greenhouse Gas: A Cross-National Study of the Social Causes of Methane Emissions Intensity." *Social Forces* 84:1779–1798.

Jorgenson, Andrew K. 2006b. "Unequal Ecological Exchange and Environmental Degradation: A Theoretical Proposition and Cross-National Study of Deforestation, 1990–2000." *Rural Sociology* 71:685–712.

Jorgenson, Andrew K. 2007. "Does Foreign Investment Harm the Air We Breathe and the Water We Drink? A Cross-National Study of Carbon Dioxide Emissions and Organic Water Pollution in Less-Developed Countries, 1975–2000." *Organization & Environment* 20:137–156.

Jorgenson, Andrew K. 2009. "The Transnational Organization of Production, the Scale of Degradation, and Ecoefficiency: A Study of Carbon Dioxide Emissions in Less-Developed Countries." *Human Ecology Review* 16:64–74.

Jorgenson, Andrew K. 2012. "The Sociology of Ecologically Unequal Exchange and Carbon Dioxide Emissions, 1960–2005." *Social Science Research* 41:242–252.

Jorgenson, Andrew K. 2014. "Economic Development and the Carbon Intensity of Human Well-Being." *Nature Climate Change* 4:186–189.

Jorgenson, Andrew K., Alina Alekseyko, and Vincent Giedraitis. 2014. "Energy Consumption, Human Well-Being and Economic Development in Central and Eastern European Nations: A Cautionary Tale of Sustainability." *Energy Policy* 66:419–427.

Jorgenson, Andrew, Daniel Auerbach, and Brett Clark. 2014. "The (De-) Carbonization of Urbanization, 1960–2010." *Climatic Change* 127:561–575.

Jorgenson, Andrew K. and Ryan Birkholz. 2010. "Assessing the Causes of Anthropogenic Methane Emissions in Comparative Perspective, 1990–2005." *Ecological Economics* 69:2634–2643.

Jorgenson, Andrew K. and Brett Clark. 2010. "Assessing the Temporal Stability of the Population/Environment Relationship in Comparative Perspective: A Cross-National Panel Study of Carbon Dioxide Emissions, 1960–2005." *Population and Environment* 32(1):27–41.

Jorgenson, Andrew K. and Brett Clark. 2011. "Societies Consuming Nature: A Panel Study of the Ecological Footprints of Nations, 1960–2003." *Social Science Research* 40(1):226–244.

Jorgenson, Andrew K. and Brett Clark. 2012. "Are the Economy and the Environment Decoupling? A Comparative International Study, 1960–2005." *American Journal of Sociology* 118:1–44.

Jorgenson, Andrew K., Brett Clark, and Jeffrey Kentor. 2010. "Militarization and the Environment: A Panel Study of Carbon Dioxide Emissions and the Ecological Footprints of Nations, 1970–2000." *Global Environmental Politics* 10:7–29.

Jorgenson, Andrew K., Christopher Dick, and Matthew Mahutga. 2007. "Foreign Investment Dependence and the Environment: An Ecostructural Approach." *Social Problems* 54:371–394.

Jorgenson, Andrew K., Christopher Dick, and John M. Shandra. 2011. "World Economy, World Society, and Environmental Harms in Less Developed Countries." *Sociological Inquiry* 81:53–87.

Kaya, Yoichi. 1990. *Impact of Carbon Dioxide Emission Control on GNP Growth: Interpretation of Proposed Scenarios*. Paris: IPCC Energy and Industry Subgroup, Response Strategies Working Group.

Kaya, Yoichi and Keiichi Yokobori. 1997. *Environment, Energy, and Economy: Strategies for Sustainability*. Tokyo and New York: United Nations University Press.
Knight, Kyle and Eugene A. Rosa. 2010. "The Environmental Efficiency of Well-Being: A Cross-National Analysis." *Social Science Research* 40:931–949.
Knight, Kyle and Juliet Schor. 2014. "Economic Growth and Climate Change: A Cross-National Analysis of Territorial and Consumption-Based Carbon Emissions in High-Income Countries." *Sustainability* 6:3722–3731.
Liu, Jianguo, Vanessa Hull, Mateus Batistella, Ruth DeFries, Thomas Dietz, Feng Fu, Thomas W. Hertel, Roberto Cesar Izaurralde, Eric F. Lambin, Shuxin Li, Luiz A. Martinelli, William McConnell, Emilio F. Moran, Rosamond Naylor, Zhiyun Ouyang, Karen R. Polenske, Anette Reenberg, Gilberto de Miranda Rocha, Cynthia A. Simmons, Peter H. Verburg, Peter Vitousek, Fusuo Zhang, and Chunquan Zhu. 2013. "Framing Sustainability in a Telecoupled World." *Ecology and Society* 18:26.
Logan, John R. and Harvey L. Molotch. 1987. *Urban Fortunes: The Political Economy of Place*. Berkeley: University of California Press.
Mazur, Allan and Eugene A. Rosa. 1974. "Energy and Life-Style: Massive Energy Consumption May Not Be Necessary to Maintain Current Living Standards in America." *Science* 186:607–610.
McCright, Aaron M. and Riley E. Dunlap. 2010. "Anti-Reflexivity: The American Conservative Movement's Success in Undermining Climate Science and Policy." *Theory, Culture, and Society* 27:100–133.
Meyer, John, John Boli, George Thomas, and Francisco Ramirez. 1998. "World Society and the Nation-State." *American Journal of Sociology* 103:144–181.
Mol, Arthur. 1995. *The Refinement of Production: Ecological Modernization Theory and the Chemical Industry*. Utrecht, The Netherlands: Van Arkel.
Mol, Arthur. 2001. *Globalization and Environmental Reform*. Cambridge, MA: MIT Press.
Myers, Norman and Jennifer Kent. 2003. "New Consumers: The Influence of Affluences on the Environment." *Proceedings of the National Academy of Sciences USA* 100:4963–4968.
Neumann, Roderick P. 1998. *Imposing Wilderness: Struggles Over Livelihoods and Nature Preservation in Africa*. Berkeley: University of California Press.
Neumayer, Eric. 2006. "An Empirical Test of a Neo-Malthusian Theory of Fertility Change." *Population and Environment* 27:327–336.
Ostrom, Elinor. 2010. "Polycentric Systems for Coping with Collective Action and Global Environmental Change." *Global Environmental Change* 20:550–557.
Pan, Yude, Richard A. Birdsey, Jingyun Fang, Richard Houghton, Pekka E. Kauppi, Werner A. Kurz, Oliver L. Phillips, Anatoly Shvidenko, Simon Lewis, Josep G. Canadell, Philippe Ciais, Robert B. Jackson, Stephen W. Pacala, A. David McGuire, Shilong Piao, Aapo Rautiainen, Stephen Sitch, and Daniel Hayes. 2011. "A Large and Persistent Carbon Sink in the World's Forests." *Science* 333:988–993.
Parson, Edward A. 2003. *Protecting the Ozone Layer: Science and Strategy*. New York: Oxford University Press.
Polanyi, Karl. 1944. *The Great Transformation: The Political and Economic Origins of Our Time*. New York: Rinehart.
Princen, Thomas, Michael Maniantes, and Ken Conca. 2002. *Confronting Consumption*. Cambridge, MA: MIT Press.
Reid, Walter V., Harold A. Mooney, Angela Cropper, Doris Capistrano, Stephen R. Carpenter, Hanchan Chopra, Partha Dasgupta, Thomas Dietz, Anantha Kumar

Duraiappah, Rashid Hassan, Roger Kasperson, Rik Leemans, Robert M. May, Tony (A. J.) McMichael, Prabhu Pingali, Cristián Samper, Robert Sholes, Robert T. Watson, A. H. Zakri, Zhao Shidong, Neville J. Ash, Elena Bennett, Pushpam Kumar, Marcus J. Lee, Ciara Rausepp-Hearne, Henk Simons, Jillian Thonell, and Monica B. Zuerk. 2005. *Ecosystems and Human Well-Being: Synthesis.* Washington, DC: Island Press.

Rice, James. 2007. "Ecological Unequal Exchange: International Trade and Uneven Utilization of Environmental Space in the World System." *Social Forces* 85(3):1369–1392.

Roberts, J. Timmons and Peter Grimes. 2002. "World-Systems and the Environment: Toward a New Synthesis." Pp. 167–196 in *Sociological Theory and the Environment: Classical Foundations, Contemporary Insights,* edited by R. E. Dunlap, F. H. Buttel, P. Dickens, and A. Gijswijt. Lanham, MD: Rowman & Littlefield.

Roberts, J. Timmons and Bradley C. Parks. 2006. *A Climate of Injustice: Global Inequality, North-South Politics, and Climate Policy.* Cambridge, MA: MIT Press.

Röckstrom, Johan, Will Steffen, Kevin Noone, Asa Persson, F. Stuart Chapin, Eric F. Lambin, Timothy M. Lenton, Marten Scheffer, Carl Folke, Hans Joachim Schellnhuber, Bjorn Nykvist, Cynthia A. de Wit, Terry Hughes, Sander van der Leeuw, Sverker Sorlin, Peter K. Snyder, Robert Costanza, Uno Svedin, Malin Falkenmark, Louise Karlberg, Robert W. Corell, Victoria J. Fabry, James Hansen, Brian Walker, Dianna Liverman, Katherine Richardson, Paul Crutzen, and Jonathan A. Foley. 2009. "A Safe Operating Space for Humanity." *Nature* 461:472–475.

Rosa, Eugene A. and Thomas Dietz. 1998. "Climate Change and Society: Speculation, Construction and Scientific Investigation." *International Sociology* 13:421–455.

Rosa, Eugene A. and Thomas Dietz. 2012. "Human Drivers of National Greenhouse Gas Emissions." *Nature Climate Change* 2:581–586.

Rudel, Thomas K. 2005. *Tropical Forests: Regional Paths to Destruction and Regeneration in the Late Twentieth Century.* New York: Columbia University Press.

Rudel, Thomas K. 2009. "How Do People Transform Landscapes? A Sociological Perspective on Suburban Sprawl and Tropical Deforestation." *American Journal of Sociology* 115:129–154.

Rudel, Thomas K., Laura Schneider, Maria Uriarte, B. L. Turner II, Ruth DeFries, Deborah Lawrence, Jacqueline Geoghegan, Susanna Hecht, Amy Ickowitz, Eric F. Lambin, Trevor Birkenholtz, Sandra Baptista, and Ricardo Grau. 2009. "Agricultural Intensification and Changes in Cultivated Areas, 1970–2005." *Proceedings of the National Academy of Sciences USA* 106(49):20675–20680.

Schnaiberg, Allan. 1980. *The Environment: From Surplus to Scarcity.* New York: Oxford University Press.

Schnaiberg, Allan and Kenneth A. Gould. 1994. *Environment and Society: The Enduring Conflict.* New York: St. Martin's Press.

Schofer, Evan and Ann Hironaka. 2005. "The Effects of World Society on Environmental Protection Outcomes." *Social Forces* 84:25–47.

Selden, Thomas M. and Daqing Song. 1994. "Environmental Quality and Development: Is There a Kuznets Curve for Air Pollution Emissions?" *Journal of Environmental Economics and Management* 27:147–162.

Shandra, John M., Bruce London, Owen P. Whooley, and John B. Williamson. 2004. "International Nongovernmental Organizations and Carbon Dioxide Emissions in the Developing World: A Quantitative, Cross-National Analysis." *Sociological Inquiry* 74:520–545.

Shwom, Rachael L. 2011. "A Middle Range Theorization of Energy Politics: The U.S. Struggle for Energy-Efficient Appliances." *Environmental Politics* 20:705–726.

Stern, Paul C. 2011. "Contributions of Psychology to Limiting Climate Change." *American Psychologist* 66:303–314.

Stern, Paul C., Thomas Dietz, Vernon W. Ruttan, Robert H. Socolow, and James Sweeney (Eds.). 1997. *Environmentally Significant Consumption: Research Directions.* Washington, DC: National Academy Press.

Stiglitz, Joseph, Amartya Sen, and Jean-Paul Fitoussi. 2009. *The Measurement of Economic Performance and Social Progress Revisited.* Retrieved April 7, 2015 (http://www.stiglitz-sen-fitoussi.fr/documents/overview-eng.pdf).

United Nations Environment Programme. 2011. *Decoupling Natural Resource Use and Environmental Impacts from Economic Growth, A Report of the Working Group on Decoupling to the International Resource Panel.* Nairobi.

United Nations Environment Programme. 2012. *Global Environmental Outlook.* New York.

U.S. National Research Council. 2010. *Advancing the Science of Climate Change.* Washington, DC: National Academic Press.

Van Vuuren, Detlef P., Jae Edmonds, Mikiko Kainuma, Keywan Riahi, Allison Thomson, Kathy Hibbard, George C. Hurtt, Tom Kram, Volker Krey, Jean-Francois Lamarque, Toshihiko Masui, Malte Meinshausen, Nebojsa Nakicenovic, Steven J. Smith, and Steven K. Rose. 2011. "The Representative Concentration Pathways: An Overview." *Climatic Change* 109:5–31.

Weber, Max. 1968. *Economy and Society.* New York: Bedminster Press.

York, Richard. 2003. "Cross-National Variation in the Size of Passenger Car Fleets: A Study in Environmentally Significant Consumption." *Population and Environment* 25:119–140.

York, Richard. 2006. "Ecological Paradoxes: William Stanley Jevons and the Paperless Office." *Human Ecology Review* 13:143–147.

York, Richard. 2008. "De-Carbonization in Former Soviet Republics, 1992–2000: The Ecological Consequences of De-Modernization." *Social Problems* 55:370–390.

York, Richard. 2010. "The Paradox at the Heart of Modernity: The Carbon Efficiency of the Global Economy." *International Journal of Sociology* 40:6–22.

York, Richard. 2012. "Do Alternative Energy Sources Displace Fossil Fuels?" *Nature Climate Change* 2:441–443.

York, Richard, Christina Ergas, Eugene A. Rosa, and Thomas Dietz. 2011. "It's a Material World: Trends in Material Extraction in China, India, Indonesia and Japan." *Nature and Culture* 6:103–122.

York, Richard and Marcia Hill Gossard. 2004. "Cross-National Meat and Fish Consumption: Exploring the Effects of Modernization and Ecological Context." *Ecological Economics* 48:293–302.

York, Richard and Eugene A. Rosa. 2003. "Key Challenges to Ecological Modernization Theory." *Organization and Environment* 16:273–288.

York, Richard, Eugene A. Rosa, and Thomas Dietz. 2002. "Bridging Environmental Science with Environmental Policy: Plasticity of Population, Affluence and Technology." *Social Science Quarterly* 83:18–34.

York, Richard, Eugene A. Rosa, and Thomas Dietz. 2003. "Footprints on the Earth: The Environmental Consequences of Modernity." *American Sociological Review* 68:279–300.

York, Richard, Eugene A. Rosa, and Thomas Dietz. 2009. "A Tale of Contrasting Trends: Three Measures of the Ecological Footprint in China, India, Japan, and the United States, 1961–2003." *Journal of World Systems Research* 15:134–146.

York, Richard, Eugene A. Rosa, and Thomas Dietz. 2010. "Ecological Modernization Theory: Theoretical and Empirical Challenges." In *The International Handbook of Environmental Sociology*, edited by M. R. Redclift and G. Woodgate. Cheltenham, UK: Edward Elgar Publishing Limited.

3

Organizations and Markets

Charles Perrow and Simone Pulver

INTRODUCTION

Through the direct emission of greenhouse gases (GHGs), or through the indirect encouragement of behaviors that result in GHG emissions, organizations are responsible for most of the world's carbon pollution (McKibben 2012). Of course, human choices are at the root of everything, but they are shaped and given effect by larger organizational structures, from the family and primary groups to corporations and nation states. Consider the transformation of individual desires to eat a beef hamburger into a pollutant source that is estimated to account for as much as 20 percent of global GHG emissions (Environmental Protection Agency 2008). Food corporations aggressively promote beef consumption, since steaks and hamburgers are more profitable than lentils. To meet the demand for beef, they incentivize rural farmers to turn forests into grassland, in order to raise beef that is then processed, shipped, and retailed in rich nations. This transformation generates both profits for the food corporations and GHGs in the form of carbon dioxide and methane (the latter being over twenty times more powerful a warming gas than carbon dioxide).

Our analysis of the role of private sector, for-profit organizations (i.e., market organizations) in climate change emphasizes two dynamics of organizational action. First, we review the contributions of market organizations to environmental degradation, particularly carbon pollution. However, in the climate arena, market organizations have also played a prominent role in attempts to mitigate climate change. To date the leading, and mostly ineffective, approaches for dealing with climate change—carbon markets, carbon capture and storage, and clean technology innovation—all privilege action by the private sector. Therefore, our review of market organizations and climate change offers both an analysis of the organizational drivers of GHG pollution and an evaluation of organizational action in carbon mitigation.

Second, our review considers market organizations both as independent agents and as the subjects of larger forces. Market organizations operate in an organizational field, defined as "sets of organizations that, in the aggregate, constitute a recognized area of institutional life; key suppliers, resource and product consumers, regulatory agencies, and other organizations that produce similar services or products" (DiMaggio and Powell 1983:149). Within this framework, some scholars emphasize the agency of individual organizations—that is, the ability of the most powerful organizations to shape their environment to their liking (Perrow 1986, 2011). Others emphasize the structure of the organizational field in explaining organizational forms, behavior, and outcomes (Scott 2004). Our review incorporates both perspectives. Market organizations can be the *independent* variable, manufacturing and advertising products that are ecologically harmful, lobbying for exemptions from regulations, sponsoring climate denier campaigns, or choosing techniques of energy production that are cheaper but more harmful to the environment. Such choices may emerge from the internal dynamics of the organization itself (e.g., its structure, goals, reward system, routines, culture, past experience, etc.). However, it is also the case that organizational behavior is the result of external economic forces and governance efforts, where organizational behavior is the *dependent* variable. For example, an organization's response to climate change will depend in part on applicable government regulations that limit GHG pollution or on competitive and stakeholder pressures. As a result, our review of the role that organizations and markets play in global warming belongs as much to the subfields of political and economic sociology as to the more narrow area of organizational sociology.

The rest of the chapter is divided into four sections. The first section, "A Sociological Critique of Market Approaches in Mainstream Climate Policy," explores the pitfalls of assuming there is anything like a free-market system and highlights why a sociological lens on market action is necessary for effective climate policy. The second section, "Market Organizations," focuses on organizations as independent variables and reviews sociological research on organizational dynamics, GHG emissions, and corporate lobbying. In "Market Environments," we consider organizations as dependent variables and analyze how distinct national economic and governance environments have shaped organizational carbon strategies. The concluding section, "What Is to Be Done" applies the insights of organizational, economic, and political sociology to considering pathways forward. We conclude that little can be expected of independent action by market organizations, and while there are ways to reconfigure market environments in the United States, the European Union, and emerging economies, these pathways are vulnerable to capture by large market organizations interested in maintaining the status quo.

A note before proceeding: We use the word "market" with some misgivings in this chapter. Labeling an economic system as a "market" economy disguises the fact that it is a capitalist economy, not just concerned with trade and exchange, the essence of markets, but concerned with profit maximization and economic growth, the essence of capitalism (Newell and Paterson 2010). Capitalism, in this argument, is not a system that is based upon the free exchange of goods between economic actors who are equal in power and resources. Rather, capitalist market relations are imbued with inequities and contests over power and resources. Nevertheless, convention prompts the use of the term "market."

A SOCIOLOGICAL CRITIQUE OF MARKET APPROACHES IN MAINSTREAM CLIMATE POLICY

Graphs of historical emissions clearly show that the largest upswing in global GHGs came with the Industrial Revolution, when fossil fuel energy was first harnessed on a large scale to support industrial production. This time in history also coincides with the development of capitalism (Marx 1867 [1990]), with industrial organization (Chandler Jr. 1990), and with the rise of markets as the dominant mode for structuring economic relations (Smith 1776 [1991]). Despite this clear linkage, action by capitalist, industrial organizations in markets has been only a peripheral focus in analyzing the problem of global warming. The practices and politics of polluting organizations are ignored in favor of abstract discussions of markets, economies, game theory, and the consumption choices of individuals. When market organizations are mentioned, it is mostly in the context of voluntary mitigation initiatives.

For example, the most recent Summary for Policymakers of "The Mitigation of Climate Change," the Intergovernmental Panel on Climate Change (IPCC) Working Group III contribution to the Fifth Assessment Report (AR5), acknowledges that "in many countries the private sector plays central roles in the processes that lead to emissions as well as mitigation" (IPCC 2014a:30). Unfortunately, the AR5 chapter linked to this summary statement does little to elaborate the private sector's contribution to emissions. The major polluting organizations whose profitability is directly tied to ongoing reliance on fossil fuels and their efforts to maintain the status quo go unmentioned. Instead, the report describes market organizations as contributors to carbon mitigation efforts, profiling public–private partnerships and private-sector governance initiatives (IPCC 2014b).

The IPCC's 2014 Assessment Report reproduces patterns established in earlier reports. In the 2007 IPCC Working Group I contribution to the Fourth Assessment Report, "The Physical Science Basis," human drivers of climate change are generically described as "human activities" without mention of the social structures, including organizations and markets, that channel most humans' consumptive and productive activities (IPCC 2007a:135). The 2007 Working Group III report on mitigation offers some elaboration, identifying the "market" alongside the "state" and "civil society" as one of three social actors that need to cooperate to address climate change (IPCC 2007b). However, the report's discussion of the market neglects to mention pollution by market organizations and the obstacles to action presented by the current market system. Instead, it highlights corporate greening, opportunities for eco-efficiencies, stakeholder pressure, voluntary disclosure, and corporate social responsibility and concludes that "although there has been progress, the private sector can play a much greater role in making development more sustainable" (IPCC 2007b:713).

A failure to identify market actors as polluters and optimism about the private sector's ability to cope with climate change are also echoed in U.S. climate policy discussions. For example, in the U.S. Global Change Research Program's 2014 National Climate Assessment, profit-seeking behavior by organizations in current market systems is not analyzed as a key driver of GHG emissions in the United States. Instead, market organizations are mentioned most often alongside nonprofit organizations and individuals in the context of taking "voluntary actions" (Jacoby et al. 2014).

In sharp contrast to the abstractions and optimism offered in mainstream climate assessments, the more critical perspective presented here is much less sanguine about corporate initiatives and markets' abilities to cope with climate change. We advance three primary critiques of mainstream economic thinking about organizations, markets, and climate change, grounded in organizational and economic sociology. First, we emphasize the need to focus on the core driver of GHG emissions, namely market organizations that are locked into and promote the carbon economy. Second, we argue against the tendency of economic theorizing to abstract from biophysical and social contexts. Third, we reject approaches to analyzing markets that promote a single, standardized solution.

Organizations operate in market systems that embody power relationships (Granovetter 1985) and that are embedded in institutions that guide their function (Dobbin 2001; Polanyi 1944 [1957]). In particular, present-day market systems and organizations are governed by institutions that for most of their history have ignored the environmental impacts of production. Broad-scale environmental concern is a phenomenon of the last fifty years (Dunlap and Mertig 1992), and corporate environmentalism is at most three decades old (Schmidheiny 1992). This suggests that most economic, regulatory, and normative structures governing market organizations still promote—or at minimum allow—environmental pollution. Empirical evidence of pro-pollution institutions pervade the economy. They range from the broadly accepted justifications of pollution catalogued by Freudenburg (2005), to the multiple pathways leading to the concentration of polluting emissions in poorer communities of color (Grant et al. 2010), to the ways in which information about corporate environmental practices is "shaded" and "distanced" via global commodity chains (Princen 1997). The prevalence of such pro-pollution institutions suggest skepticism rather than confidence in market solutions to climate change. Moreover, efforts to change market institutions quickly reveal the power of some market actors to prevent any changes in market rules. Organizations have successfully blocked initiatives to transition to a low-carbon economy at national and global levels (Levy and Egan 1998). While economic theory can propose pricing carbon as a solution to climate change (Krugman 2010), only a model of markets that confronts the instrumental and structural power of polluting organizations can offer insight into the social and political processes for putting a carbon tax or cap-and-trade system in place. Modeling transitions to a greener economy must take into account organizational winners and losers in order to map out a viable politics of transition (Skocpol 2012).

Sociologists also challenge the economic practice of abstracting organizations and markets from biophysical and social contexts (Catton and Dunlap 1978). Such abstractions produce assessments of climate change that operate at the level of disembodied economies and predict limited impacts and low-cost solutions (Nordhaus 2009). As the late biologist and climate scientist Steven Schneider put it, economists "accept the paradigm that society is almost independent of nature" (1997:129). However, such a paradigm can lead to perverse outcomes. For example, in *The Ecological Rift*, Foster, Clark, and York (2010) recount that Nordhaus and other environmental economists declare that while agriculture will suffer the greatest from global warming, "there is no way" that the failure of agriculture can have a large effect since it represents only 3 percent of the U.S. economy. Such an analysis ignores that a drastic reduction in agricultural productivity just might leave millions starving, despite overall gains in per capita income. Mainstream economic assessments of adaptation to climate change are equally problematic. Some economists contend that the predicted growth rates of at least 2 percent a year for the world economy should allow us to easily pay for adaptation to climate change (Nordhaus 2009), ignoring that weather disturbances, famines, flooding, and high temperatures might lead to a substantial decline in gross domestic product (GDP) and extensive human suffering.

Third, sociological perspectives challenge the tendency among mainstream economists to promote a single, standardized solution, regardless of national context, what Evans (2004) terms "institutional monocropping." Sociologists emphasize the variety of rationalities that structure markets and explain organizational behavior. Patterns in national efforts—for example, the relative (but small) success of European nations in regulating polluting organizations as compared to the United States—may be explained by the particular institutional arrangements among the state, capital, and labor that characterize national market systems (Hall and Soskice 2001). Moreover, market organizations vary within and across industries, in terms of organizational structures, operational characteristics, and corporate cultures, and these differences can have significant consequences on organizational environmental performance. The world of organizations and their environments is rich in variety, and any consideration of pathways forward should reflect this.

Our review of the sociology of organizations, markets, and climate change incorporates these three critiques. The next section analyzes how organizations contribute to carbon pollution and to maintaining a fossil fuel–based economy and society. We then evaluate the effects of different market contexts on organizational decisions about carbon in the United States, Europe, and emerging economies. Finally, we consider what can be done given the constraints of market institutions and organizational power.

MARKET ORGANIZATIONS

Organizational Determinants of GHG Pollution and Climate Capitalism

A range of organizational characteristics shape firms' environmental pollution profiles. One of the first empirical explorations of the environmental impacts of large organizations examined the release of toxic chemicals and found that the larger the organization, the larger its releases per quantity of chemical used or stored onsite (Grant, Jones, and Bergesen 2002). This put to rest the idea that it was small "alley shop" chemical plants that were more likely to have unwanted releases while the large plants, presumably subject to more stringent inspections, were cleaner. A second key variable linked to higher pollution is the presence of multiple subsidiaries, which protect the parent company from liabilities and taxes; this is the recently adopted form of governance by most large corporations (Grant and Jones 2003). Grant et al. (2010) also carefully investigated the joint effects of community features and facility characteristics in creating risky emissions in poor and minority neighborhoods. They found that branch plants correlated with highly risky emissions. Prechel and Zheng (2012) expanded the inquiry into organizational structure and pollution with a remarkable study using data from the Fortune 500 corporations from 1994 through 2001. They find that "corporations with more complex structures, greater capital dependence and those headquartered in a state with lower environmental standards have higher pollution rates" (Prechel and Zheng 2012:947). The literature on organizational determinants of environmental pollution also finds that in most industries, there are a few extreme polluters—what Freudenburg (2005) termed "disproportionality"—whose emissions exceed industry averages, even when controlling for firm size and production volumes. Disproportionality research suggests that across the economy and within industries the majority of environmental pollution can be ascribed to a few egregious polluting organizations.

Unfortunately, research focusing on organizational determinants of GHG pollution specifically is in its infancy, in part because firms have been collecting and disclosing GHG emissions data only for the past few years and in a haphazard way. For example, while 70 percent of the 500 largest corporations on the FTSE Global Equity Index series reported on their GHG emissions in 2008, only 50 percent of these corporations had their emissions data externally verified (PriceWaterhouseCoopers 2009). One comprehensive study of GHG pollution by electric utilities seeks to identify plant-level drivers of carbon emissions. The study is still under way, but the researchers have documented clear patterns of disproportionality. They find that a small number of energy plants in the top polluting countries account for the lion's share of energy-based carbon emissions (Grant, Jorgenson, and Longhofer 2013).

In the absence of comprehensive, facility-level, time-series GHG pollution data, scholars have focused on explaining a range of other climate-related organizational behaviors, such as awareness of climate change, GHG emissions disclosure, investments in operational GHG reductions, development of new "climate-friendly" products and markets, and corporate governance structures that address climate change. Such changes in organizational practices are the basis of what Newell and Paterson (2010) call "climate capitalism," in which organizations rally to the effort of decarbonizing the global economy. Three general trends can be extracted from existing empirical studies on climate capitalism (Begg, van der Woerd, and Levy 2005; Pulver 2011a; Sullivan 2008). First, organizational awareness of climate change is uniformly high across the globe, especially among large corporations. For example, a 2008 survey of India's 500 leading corporations found that 96 percent self-assessed as having a fair (38 percent) or deep (58 percent) understanding of climate change

(Emergent Ventures India and The Financial Express 2008). Global surveys, focusing on corporations in industrialized countries, show even higher rates of awareness (Enkvist and Vanthournout 2007). Second, while large companies may be aware of climate change, not all of them track and publish GHG emissions data. The Carbon Disclosure Project (CDP), an organization dedicated to promoting public disclosure of corporate GHG emissions data, documented that in 2013, 81 percent of the world's 500 largest companies listed on the FTSE Global Equity Index Series reported on their GHG emissions (CDP 2013). However, reporting rates vary dramatically from country to country. National-level CDP data from 2011 show response rates varying from 10 percent in China to 90 percent in Europe (Pulver and Benney 2013).

Third, rhetoric regarding carbon targets, emissions reductions, and climate-related business opportunities outpaces actual changes in business practices (Jones and Levy 2007). CDP data show that despite various corporate policies and emissions reduction initiatives, GHG emissions from the fifty largest emitters in their survey increased by 1.65 percent between 2009 and 2013 (CDP 2013). Most corporate climate initiatives constitute marginal efficiency improvements. Transformative corporate responses to climate change (i.e., the development of new climate-friendly products/markets) are rare, limited to between 5 and 10 percent of firms in most studies (Kolk and Pinkse 2005).

Organizational research offers insight into the characteristics of firms likely to be climate capitalists. Firm size is a predictor of carbon disclosure (Kolk and Pinske 2007) and of investments to mitigate GHG emissions (Schneider, Hoffmann, and Gurjar 2009). Cogan's (2006) survey identifies CEO support as central to proactive corporate climate governance. Others point to internal learning and decision-making processes (Berkhout, Hertin, and Gann 2006; Lowe and Harris 1998). Firms' access to climate expertise is also a key factor driving variation. In case-study research on oil companies' responses to climate change, a key organizational difference between ExxonMobil and its European rivals was that the former has a large, in-house scientific research program, whereas BP and Royal Dutch/Shell outsource most of their scientific research. As a result, BP and Royal Dutch/Shell integrated the widely accepted IPCC scientific consensus on climate change into their corporate strategy, while ExxonMobil built a climate policy response based on in-house skeptical climate science (Levy and Kolk 2002). Parallel studies also highlight the importance of corporate culture, organizational histories, and core competencies. Pulver (2007a, 2007b) demonstrates that an alignment with preexisting core competencies made some companies more likely to adopt an emissions reductions target and establish an internal emission trading system.

Organizational Agency in Climate Politics

Market organizations act as agents in climate politics through their operational practices. They also exert agency when attempting to shape the political, legal, social, cultural, and informational contexts in which they operate. The influence of market organizations on their environments is effected via structural, instrumental, and discursive power (Levy and Egan 1998). Structural power refers to the central role of market organizations in maintaining economic growth (Schnaiberg and Gould 1994). Instrumental power refers to the ability of organizational elites to affect political outcomes through financial support, lobbying activities, and social and business connections (Domhoff 2002). Finally, discursive power refers to the capacity of market organizations to frame issues and influence public opinion through advertising and other reputation-building activities. Freudenburg (2005) describes discursive power as "privileged accounts," pointing to the disproportional influence of market organizations in framing the problems of and solutions to environmental degradation. These three forms of power are exerted via direct influence in political arenas or indirectly via industry associations. The latter have been particularly prominent in the United Nations climate negotiations process (Newell 2000; Pulver 2002).

Empirical case studies examining the influence of market organizations in the climate policy arena abound. In the United States, the structural and instrumental power of energy organizations is illustrated by the example of hydraulic, horizontal fracturing for natural gas in the shale deposits in the Midwest and the East, and possibly California. Despite a range of environmental concerns—aquifers have been polluted; water from household wells can be flammable; the air around the drilling sites is poisonous; rural roads are torn up; streams and landscapes are breached; half of the millions of gallons of water needed for each well returns to the surface and must be extensively treated to remove dangerous chemicals as well as radioactive substances that had been miles deep in the ground—fracking operations have proliferated (Howarth, Ingraffen, and Engelder 2011; Tollefson 2012). Some argue that the climate benefits of generating electricity from natural gas outweigh these other concerns. However, the GHGs released from fracking have been estimated to be equivalent to that of a coal-fired plant when the whole cycle is considered, largely because of the escape of methane, a gas that is over twenty times as powerful a GHG as carbon dioxide (Sanderson 2009; Shindell et al. 2009). Nevertheless, regulation of fracking is scant because of legislation passed in 2005 (the "Halliburton loophole"), and efforts to tighten regulations have run into stiff opposition in Congress and from state governments that receive tax revenues and have surges in employment figures (Kinne et al. 2014).

In addition to their structural and instrumental power, the U.S. coal and oil industries also offer case studies of discursive power. Take for example Bell and York's (2010) analysis of the coal industry, a key actor in U.S. climate politics. Bell and York document the parallel decline in the contribution to employment of West Virginia's coal industry and the increase in corporate efforts to link the economic identity of local communities to coal mining. The influence exerted by coal companies is in the creation of an ideology that legitimates destructive industry practices in the face of ecologically based resistance from local communities. Corporations have also used their discursive power to frame debates about climate change. Several companies have made climate change a core issue in advertising campaigns (Kolk and Pinkse 2007). Others have sought nongovernmental organization (NGO) partners to lend credibility to their corporate climate change initiatives (Lowe and Harris 1998). Others still attempt to shape climate debates by funding climate science, both mainstream and skeptical. Mainstream initiatives include a British Petroleum-funded research program at Princeton University and ExxonMobil's contributions to the Global Climate and Energy Project at Stanford University (Lovell 2010). However, greater attention has been given to the links between opponent corporations and funding for the group of scientists who challenge the IPCC consensus on climate change (Leggett 1999). McCright and Dunlap (2003) trace the affiliations between conservative U.S. think tanks, U.S. climate change skeptics, and fossil fuel industry funding, pointing to the conservative countermovement as a key cause of U.S. inaction on climate change (see Chapter 10 in this volume).

Research on climate policy in Europe also offers numerous examples of structural, instrumental, and discursive power exercised by market organizations. Industry opposition is blamed for the failure of European Union carbon tax policies in the early 1990s (Ikwue and Skea 1994). Industry influence has also been central to the structure and functioning of the European Union Emissions Trading Scheme (EU ETS) (Boasson and Wettestad 2013). A key choice in designing emissions trading systems is the allocation of emissions permits, both the quantity of permits allocated and the means of allocation. In the EU ETS, corporate lobbying, time pressure, and limited data led to the over-allocation of permits in Phase 1 (2005–2007), which caused the carbon price to drop to nearly zero in early 2007 and removed any incentive to reduce emissions (Betz and Sato 2006). However, there was little governments could do. As one observer noted, "if the state went back to the companies and took away the certificates they had been allotted there would be an uproar" that "no politician could withstand" (Waldermann 2009). Organizations also successfully pushed for the free allocation of permits, based on historical emissions records, resulting in a zero net cost of pollution for polluters. At the same time, utilities pushed to legally raise customer rates to compensate for the "cost" of what were free allowances, generating windfall profits for major polluters, financed by electricity consumers (Schiermeier 2008). RWE, a major German power company and Europe's largest carbon emitter, received a windfall of about US$6.4 billion in the first three years of the system (Kanter and Mouawad 2008). The auctioning of emissions allowances (i.e., making polluters bid for the right to emit) would resolve the over-allocation and windfall profits issues. However, energy companies have actively resisted the inclusion of such provisions in Phases 2 and 3 of the EU ETS on grounds of international competitiveness concerns (Schiermeier 2008).

MARKET ENVIRONMENTS

While large market organizations may have some say in choosing their carbon strategies and some influence on policy environments, for most organizations patterns of carbon pollution and responses to climate change are determined by the institutional environments in which they operate. As Hoffman argues in his seminal study of environmental practices in the oil and chemical industries: "How the issue of environmental protection is defined within the organization is dependent on how it is defined outside the organization" (2001:7). Environmental sociology offers two competing theories of the institutional contexts that guide the environmental behavior of market organizations in capitalist economies. Ecological modernization theories focus on the potential of greening capitalist economies via radical resource productivity driven by a partnership between firms and regulators (Hawken et al. 1999; Mol 1995; Porter and Van der Linde 1995). In contrast, treadmill of production theorists (Schnaiberg and Gould 1994) argue that the industrial logic of capitalist economies creates conditions under which firms will continually increase their impacts on the environment, in the form of resource withdrawals and waste additions. Any efforts to enhance resource productivity will be overwhelmed by expanded production. Moreover, governments also rely on expanded production and thus are reluctant regulators. Global and cross-national comparative studies of carbon emissions trajectories suggest that the treadmill logic predominates (see Chapter 2). For example, worldwide GHG emissions increased 6 percent from 2009 to 2010, worse than the worst-case scenario predicted by the IPCC (Plumer 2011). The increase is attributed to economic growth, with the United States and China accounting for more than half of the worldwide increase.

Despite the broad trend of increasing GHG emissions, there is variety in the underlying market environments that drive organizational choices about GHG emissions. In the next section, we contrast organizational responses to climate change in three general types of market environ-ments: *neoliberal market economies*, in which the treadmill logic appears to dominate, with the United States as the representative case; *coordinated market economies*, such as the major EU nations, which have been most associated with ecological modernization policies; and *rapidly emerging economies*, including Russia and the economic leaders of the Global South (Bockman 2013; Hall and Soskice 2001). These economies differ in the form and stringency of government regulation of climate change (see Chapter 7), the nature of business–government relations, the energy cul-ture of the economy (e.g., the degree to which strategies of accumulation are based on cheap fossil fuels versus efficiency and energy productivity), and the degree of stakeholder mobilization on climate issues. For-profit organizations have dominated climate policy in neoliberal market econo-mies and have been forced to share power by government agencies in the coordinated market economies. In the emerging economies, states have few incentives to limit the GHG pollution of market organizations, justify-ing continued pollution in pursuit of economic growth. Our analysis of market environments also examines the economies of carbon created by the three major international carbon trading systems: the EU ETS, and the Clean Development Mechanism (CDM) and Reduced Emissions from Deforestation and Degradation (REDD+). Much organizational investment in GHG emissions reductions has occurred within the context of theses carbon market systems.

The United States: A Neoliberal Market Economy

In the United States, the major sources of carbon pollution are the coal, oil and gas, electric utility, and agricultural industries. All of them have power-ful lobbies in Congress. It is not extreme to say that within the U.S. context, market organizations are the principals and government is the agent, with respect to policies regarding global warming (Perrow 2010a). Possibilities exist to limit GHG emissions from the fossil fuel, electric utility, and agri-cultural industries, but without a price on carbon there is little incentive for these alternatives. Pricing carbon would involve either a carbon tax or a cap-and-trade program. However, in the United States such legislative efforts have been blocked. The single effort to pass cap-and-trade legislation in the United States failed (Skocpol 2012), and the more recent Environmental Protection Agency rules to limit GHG emissions from existing power plants (and refineries) take effect only later in this decade and are under legal and congressional attack. Even among liberal market economies, the United States is an outlier in the lack of GHG reduction efforts by the central gov-ernment. While Australia, another liberal market economy, has even higher per capita rates of GHG emissions than the United States, Australian regulators passed a carbon tax in 2011 (Peters et al. 2012).

The U.S. coal and oil industries offer clear examples of organizational power used to resist regulation of carbon pollution. Coal is the most carbon intensive of all fossil fuels, and to impose any limits on carbon dioxide emissions, coal should be the largest target. It accounts for about 37 percent of U.S. carbon dioxide emissions concentrated in a few hundred electric utility plants (Perrow 2010a). Next to coal, fully 44 percent of U.S. carbon dioxide emissions are from burning petroleum products in gasoline, heating oil, kerosene, diesel oil products, aviation fuels, and heavy fuels (McKibben 2012). In contrast to coal, carbon emissions from oil are distributed over millions of stationary and mobile sources. Both industries are dominated by large, highly profitable companies. Peabody Energy is the world's largest private-sector coal company. In 2008 it had profits of nearly US$1 billion and spent most of this income buying back its shares to boost its stock price (Perrow 2010a). Most oil production in the United States comes from the "Big Five" independent oil companies (BP, Chevron, ConocoPhillips, ExxonMobil, Royal Dutch/Shell). Both coal and oil companies could use their tremendous profits to increase efficiency, reduce emissions in production, and invest in renewable energy, but they have done little to even consider these options. They have not even invested much in exploration, which should be in their long-term interests (though not in the interest of the planet). Instead, in recent decades, they have consolidated, cut back on exploration, contributed to price rises, used their extravagant profits to increase shareholder value, spent almost nothing on renewables even in the face of projected declines in the availability of oil, even managed to increase their government subsidies, and actively fought to block GHG regulation in the United States (Perrow 2010a).

The success of fossil fuel lobbies in blocking GHG regulation in the United States is tied to both the structure of the U.S. legislative system and the role of ideology in U.S. climate politics. The United States has a federated government system in which power is shared between the three branches of the central government and further shared between the central government and individual states. As a result, regulatory power over business and industry is highly dispersed (Vogel 1989). In addition, the structure of the U.S. Senate is vulnerable to industry lobbying, because states with small populations have as much voting power as large states, with larger populations. The coal industry in particular has leveraged influence through the Senate. Four of the major coal-mining states are very small. Their senators, Republicans and Democrats alike, are indebted to coal companies for handsome campaign contributions—in the 2000 election, Peabody Energy contributed US$846,000 to the Republican Party (Perrow 2010a)—and many have very conservative values, reflecting the views of their constituents. There are enough senators from the major coal states (Wyoming, West Virginia, Kentucky, Pennsylvania, Montana, Ohio, Indiana, and Illinois) to block either a carbon tax or a cap-and-trade bill (Eilperin 2010).

Further complicating the politics, placing a price on carbon raises equity issues. The western and eastern states, relying more on hydroelectric power and natural gas, and urban areas, with a lower carbon footprint, would be less affected by a price on carbon. The Midwest, relying on energy from coal, and rural areas, where sparse populations travel more and use more energy to heat and cool, would suffer the most. While a tax of US$50 per ton of carbon would raise the price of gasoline by 26 percent and household natural gas prices by 25 percent, electricity from coal plants would see huge price rises, over 300 percent according to one estimate (Stavins 2008:Table 3). In research comparing U.S. representatives voting on a number of climate bills, Cragg et al. (2013) confirmed that those opposed to climate change legislation were predominantly from states and districts that were poor, more rural, and more dependent upon coal for electricity. Only strong rebates of carbon tax revenues to poor citizens would address this equity issue, and since the rich states already pay more in federal taxes than they receive in benefits, they may resist this redistribution. Cragg et al. also found that representatives with conservative ideologies (e.g., against big government and taxes) were opposed to carbon legislation regardless of the carbon footprint of their state or district. Ideology was a better predictor of legislators' voting behavior than economic interest.

In sum, the institutional environment in liberal market economies favors limited restrictions on market organizations. In the United States, the fossil fuel industries have been able to block and delay GHG regulation since initial efforts to pass a carbon tax in the 1990s, with little concern for consequences. ExxonMobil, the largest U.S. oil company, seemed not even to worry when it became public knowledge that it was one of the principal financiers of the global warming denier campaign. As one Exxon manager stated, "they cannot ignore us anyway; we are the big elephant at the table" (Levy and Kolk 2002:291).

The European Union: A Coordinated Market Economy

In coordinated market economies, the state has a stronger role in regulating and subsidizing market organizations, and both labor and consumers are represented to a greater degree in national policies than they are in liberal market economies (Hall and Soskice 2001). More market coordination has advanced the climate agenda in Europe compared to the United States. European governments and the EU have mandated GHG reduction targets, passed carbon cap-and-trade legislation, and actively sponsored renewable energy infrastructure investment. These efforts have had some effect. Between 1990 and 2008, GHG emissions in the EU-27 fell by 11 percent and per capita emissions decreased by 19 percent, while GDP increased by 44 percent, according to the European Environmental Agency. It should be noted that progress in GHG emissions reduction at the EU level masks national differences. There has been significant variation at the national level, with Germany and the United Kingdom lowering GHG emissions the most while Greece, Spain, Italy, and Portugal had sharp increases (EEA 2011). Moreover, estimates of EU emission reductions include reductions purchased in offset markets and do not account for the emissions embodied in goods imported from nonsignatory nations such as the United States and China (Jorgenson 2012).

The European Environmental Agency sees macroeconomic factors such as industry restructuring and business cycles as the major drivers of GHG emission rates, with policy changes by governments (e.g., subsidies and pollution legislation) as important but secondary. The biggest drop in emissions in the eighteen-year period from 1990 to 2008 occurred in the 1990s as a result of the restructuring of the economies of Eastern Europe after the fall of the Soviet Union. Many heavily polluting plants were closed, and the abandonment of collective agriculture led to radical transformation of agriculture, including large decreases in the number of cattle (EEA 2011). However, emissions reductions are also due to policy initiatives. The four pillars of EU climate policy are the EU ETS, renewable energy directives, carbon capture and storage, and energy efficiency for buildings (Boasson and Wettestad 2013). Progress in each of these areas reflects coordination between government and market organizations. In the next sections, we examine organizational action within coordinated market environments, focusing on European wind energy development and the EU ETS. Both policy arenas showcase how different national environments structured action by market organizations.

Wind Energy Development

Efforts to develop wind energy showcase the different relationship between the state and market organizations in coordinated versus liberal market economies. Wong (2010) compares the evolution of wind energy in Germany and the United Kingdom. The German government passed a feed-in tariff in 1991 (Electricity Feed Law), which required electric utilities to pay a fixed and higher rate for electricity generated from non-utility wind power producers, only 90 percent of which could be passed along to customers. German utility companies, under the leadership of Preusen Elektra, successfully resisted the law in both federal courts and at the EU level until 1999, when the European Court of Justice ruled in favor of the German government. According to Wong, the industry was "sufficiently consolidated and powerful, after mergers and acquisitions under European liberalization policies, to resist the federal government's pressure to develop or purchase expensive wind energy" (2010:373). In response to opposition by large utilities, the German government, committed to developing wind energy, sponsored small, local initiatives. The government's technical assistance and subsidies made Germany the world leader in wind energy by 2007, all with "backyard" plants of ten megawatts or less of generated power. Initially, these initiatives did not threaten the coal and electric power industry, but Germany began to run out of backyards. Since Germany's big energy industries blocked the growth of big onshore wind farms, the government made an end run around them and turned to offshore sites. The wind blows stronger and more dependably a few miles out in the ocean, and offshore wind farms became the favored technology. Entrepreneurs were free to develop large, very efficient offshore wind farms. But the small, dispersed German onshore wind farms lacked the capital and the technical skills for offshore farms. The coal and oil companies and the generating companies refused to invest in offshore farms. Consequently, Germany has lagged in this area, and only in the past few years has foreign capital come in to exploit the offshore opportunities. This is an instructive example of how large organizational interests shaped the renewable energy field. A strong central government was not strong enough in this case.

The United Kingdom took a different tack, as Wong carefully describes. With deregulation and weak government the ruling ethos since the Thatcher years, local communities were able to prevent wind farms from being established, although the government offered financial incentives. The government then tried to encourage onshore wind farms but was unable to overrule local objections. But the U.K. government has legal control over most of the country's seashores. Therefore, it turned to large enterprises (including Shell, which did not see a threat, in contrast to big coal in Germany) to build offshore wind farms. With subsidies and feed-in tariffs, the United Kingdom rapidly advanced in wind generation, almost all of it offshore, and now surpasses Germany in total production.

In sum, Europe is certainly doing much more with renewables than the United States. Germany in particular appears to be moving faster. In 2011, 20 percent of Germany's electrical energy came from wind or solar power, and it may meet its 2020 goal of 35 percent. There is an organizational dimension to Germany's success that is not found in the United States. Roberts (2012) argues that roughly three quarters of renewable power investments in Germany have been made by individuals, communities, farmers, and small and midsized enterprises. Germany has more than a hundred rural communities becoming 100 percent renewable. In contrast, in the United States almost every power source is a large-scale utility; only 2 percent of installed wind power, for example, is not owned by large corporations.

The European Union Emission Trading System

The launch and development of the EU ETS provides a second example of coordinated action between market organizations and government agencies. The Europeans established EU-wide emissions targets and opted to meet them by setting up the EU ETS, a cap-and-trade system that covers 45 percent of European nations' carbon dioxide emissions. Under the EU ETS, governments determine the total amount of carbon that can be emitted by energy and industrial facilities, but the cost of an emissions allowance is set by a market price; it can go up or down, as polluters need more or fewer allowances. Within each round of trading, investments in clean energy in one sector of the system subsidize continued pollution in another. In subsequent rounds, the emissions cap is tightened, slowly shifting energy production to non-fossil sources. The most recent, third phase of the system commits the EU's twenty-seven countries to reducing GHGs by 20 percent below 1990 levels by 2020 (Boasson and Wettestad 2013).

Empirical studies suggest that the EU ETS has driven some emissions reductions. While European GHG emissions grew by 2 percent in the EU ETS pilot phase, they decreased by 2 percent in the beginning of the second phase. The second phase of course coincided with the global recession, but a firm-level analysis suggests an independent emissions reduction effect attributable to the EU ETS. Based on a dataset of over 2,000 installations across the EU, Abrell, Faye, and Zachmann (2011) demonstrate that the tightening of the emissions cap between the first and second phases of the EU ETS had a significant effect on emissions reductions, even when controlling for changes in economic activity. Ellerman, Convery, and de Perthius (2010) make a similar argument: they credit the EU ETS with reducing emissions by 5 percent over three years. These estimates offer some assessment of the environmental effectiveness of the EU ETS, although they merit caveats. Not counted in such figures is the exporting of carbon pollution (i.e., Europeans consuming products imported from China, products whose carbon emissions accrue to China and not Europe). Also, many reductions were not effected at home but purchased via international offset programs. Critics were projecting that as little as 4 percent of proposed emissions reductions under the EU ETS between now and 2020 would actually be effected in Europe. In response, the European Commission mandated that at least 50 percent of the reductions in Phase 3 must be domestic (Cronin 2008).

The EU ETS creates a different market environment for organizations operating in Europe than for those in the United States. As a result, European market organizations are ahead of their U.S. peers in investing in GHG emissions reductions. Variation among European countries further underscores the role of market environments in shaping corporate responses to climate change. Engels, Knoll, and Huth (2008) and Engels (2009) analyzed the emissions trading strategies of companies in Germany, the United Kingdom, Denmark, and the Netherlands and found that a company's position, whether short or long on allowances, had no discernible effect on its trades under the EU ETS. Firms with low allowances did not buy from firms with excess credits, and the latter often did not sell their excesses. Rather, institutional context best explained trading behavior. Firms in Germany, with a regulatory style running against market-based policy instruments, resisted the scheme with legal tactics and engaged in limited trading. In contrast, firms in the United Kingdom, with early exposure to various forms of emissions trading and operating within a liberal market economy, enthusiastically traded, and London became the profit center of carbon finance and arbitrage. These patterns highlight the extent to which market action is embedded in institutional contexts.

The EU ETS also reveals some of the pitfalls of relying on market solutions to the climate problem. Midstream assessments of the EU ETS have been very critical. Many consider it a failed experiment due to numerous design and implementation problems, some of which can be traced back to industry lobbying (Betz and Sato 2006). Organizational actors seek to shape policies to suit their interests. In the case of the EU ETS, bowing to corporate preferences led to the over-allocation of permits, causing the carbon price to drop to zero and undermining the entire trading system. Such industry capture of regulatory arenas is not unique to market-based policies, but it is perhaps more likely when policy design relies heavily on market organizations. The EU ETS has also been undermined by a series of market frauds. A scathing report by the U.S. Government Accountability Office noted the poor performance of the European and the Kyoto Protocol cap-and-trade programs. Among other things, they noted the lack of monitoring abilities and verification (GAO 2008). There have been similar problems with a carbon-trading scheme in Alberta, Canada. The province's auditor-general released a highly critical report in 2010, charging lax enforcement by the government and deception by the organizations involved. It would appear that market-driven policies run that risk (Hoag 2011).

In conclusion, coordinated market economies are characterized by a more even distribution of power between market organizations and state regulators than liberal market economies. As a result, these economies have implemented a wider array of policies aimed at reducing GHG emissions and show some evidence of GHG emissions reductions. The centerpieces of GHG reduction policies in Europe are the promotion of renewable energy and the EU ETS. In both policy arenas, organizations have been able to influence regulatory outcomes. However, wind energy policy development and trading under the EU ETS also showcase the influence of market environments on organizational action. In both cases, differences in business–government relations between Germany and the United Kingdom resulted in divergent patterns of organizational wind energy investments and organizational responses to emissions trading, respectively.

Emerging Economies: Corporate Responses and Carbon Markets in the Global South

International and domestic drivers intersect to create the market environments encountered by organizations in the large emerging economies, specifically the BRICs (Brazil, Russia, India, and China). In these countries, organizations have mostly continued to operate in environments free of GHG regulation, and national governments have vociferously defended their rights to do so. The major emerging economies initially avoided obligations to reduce GHG emissions, either formally in the case of the developing countries or via loopholes in the Kyoto Protocol in the case of Russia and the former Soviet states. China and India based their international negotiations strategy on the premise that any GHG emissions reductions would interfere with their continued development and blocked attempts to negotiate a global agreement with binding reduction targets for developing countries (Held, Roger, and Nag 2013). Russia agreed to a Kyoto target that was essentially meaningless. A highly critical paper by Spash (2010) discusses how Russia refused to sign the Kyoto treaty until it received large carbon credits for its forests. In addition, Russia and the other former Soviet states were awarded excess carbon allowances based upon the industrial infrastructure of past years, much of which was no longer operative after the economic collapse in the 1990s. These excess allowances were known as "hot air" because it was meaningless in terms of carbon emissions reductions but very profitable for the former Soviet states. For example, the Czech electricity giant CEZ was allocated a third of the country's allowances, sold them when the price was high, bought them back after the price collapsed, and used the profits to invest in lucrative coal energy production.

While avoiding domestic commitments to reduce GHG emissions has been a key aim of the large emerging economies, Pulver and Benney's (2013) review of corporate responses to climate change in the Global South identifies a slow shift in perspective. For example, Brazil, India, China, Mexico, and South Africa have all enacted domestic legislation at federal and/or state levels to reduce domestic GHG emissions over the next decades; the physical risks of climate change in these countries have generated media and NGO attention, creating some public pressure for organizational action; and opportunities in international carbon markets through CDM and REDD have motivated action (Fuhr and Lederer 2009; Held, Roger, and Nag 2013). The CDM is a voluntary carbon market by which organizations in developing countries can sell emissions reductions effected at home to international buyers. REDD proposes a market solution to GHG emissions from deforestation through the addition of forest carbon credits to carbon markets.

CDM and REDD

International policy initiatives such as CDM and REDD have been central to evolving patterns of carbon governance in emerging economies. The Kyoto Protocol's CDM was launched in 2001, and by 2013 over 8,000 CDM projects had been submitted for review from the group of developing economies (UNEP Risoe Center 2013). The market was initially dominated by projects initiated in Brazil. By late 2005, India overtook Brazil as carbon market leader and remained the leader until 2007, when it was surpassed by China, which is currently host to the majority of CDM projects. CDM projects are recorded in a registry that provides an overview of climate activities by market organizations in developing countries. Organizational investments generating CDM credits vary from sector to sector and reflect investments in new green technology, such as solar power plants, and in efficiency improvements in existing industrial processes, such as waste heat recovery (Pulver and Benney 2013). REDD, a more recent program, was launched in 2008 and offers funding, via carbon credits, to keep carbon sequestered in forests. By creating financial encouragement for maintaining forests, the hope is that developing countries will no longer rely on deforestation as a means of income (Hall 2008; UN-REDD Programme 2013).

As with the EU ETS, CDM and REDD merit special attention for lessons to be learned from trying to harness markets to address climate change. The first lesson is that institutional context matters. Organizational participation in CDM varies by national context. In particular, in Brazil, Mexico, and South Africa, corporate investment in GHG reductions has been limited because although governments have been active GHG regulators, they have done relatively little to support private-sector activities (Held et al. 1999). In contrast, the Chinese government has focused on climate change–related market opportunities rather than GHG regulation. As a result, market organizations in China have the highest rates of participation in the CDM but the lowest rates of carbon emissions disclosure (Schroeder 2009). In India, corporate responses to climate change reflect an active private sector, with high levels of awareness about climate change and optimistic assessments of climate-related business opportunities (Pulver 2011b). Benecke describes India as "a case of market-dominated carbon governance taking place under a weak shadow of hierarchy and with little civil society involvement" (2009:346).

The second lesson of the CDM and REDD underscores the design challenges and potential for cheating prevalent in market systems. Unlike the EU ETS, CDM and REDD are offset markets (Bohm and Dabhis 2009; Lovell 2012). Credits purchased through the CDM and REDD are effected in countries without mandatory carbon reduction targets. The environmental integrity of these systems relies on the idea of additionality—that is, the presumption that the offset projects resulting in emissions reductions would not happen without the added income from selling those reductions. Offset systems are predisposed to cheating because proving additionality is based on a counterfactual comparison to what would have happened without the carbon market incentive. Despite such limitations, there is some evidence that the CDM has worked as intended. Pulver, Hultman, and Guimaraes' (2010) case study of CDM in the Brazilian sugar industry is illustrative. Sugar mills generate waste products that can, with significant investment and expertise, be turned into electricity. Doing so, they generate carbon offset credits, which can be bought by polluting firms. For some sugar mills, CDM helped overcome informational and technical barriers that had prevented the mills from shifting to electricity generation. Carbon consultants played a central role in this dynamic. "Consultants approached sugar mill owners, identified firm-level carbon market opportunities, created project templates by developing methodologies, convinced mill owners/managers of the benefits of carbon abatement investments, and guided projects through the bureaucratic CDM project approval process" (Pulver et al. 2010:260). However, the experiences of other sugar mills highlight the problems of the CDM. Financially, many sugar mill projects were non-additional—that is, the extra revenues from selling electricity were sufficient to motivate the investments without the added incentive of selling carbon credits. This means that some mills were paid extra for something they were planning to do anyway. This is a ubiquitous problem for the CDM across industries and countries. Wara and Victor (2008) estimate that the majority of CDM projects are non-additional. Moreover, the mechanics of the CDM pushed organizations towards non-additional projects. Because of the somewhat arbitrary CDM project approval process, organizations were not guaranteed the income stream from carbon credits when initially committing investment funds. Therefore, they gravitated toward projects that would earn sufficient return on investment, even without CDM revenues.

Certain CDM projects generated even worse outcomes. The case of hydrofluorocarbons (HFCs) illustrates how states, and the organizations they control, can "game" the system. About half of all emission reductions expected under the CDM result from reductions of HFC-23, a powerful GHG produced in connection with manufacturing HFC-22, another GHG that is used in the refrigeration and air conditioning industries. The credits given to producers who destroy HFC-23 instead of emitting it are so generous that revenue from the credits is twice the revenue from the sale of HFC-22. As a result, it is financially lucrative to build new HFC-22 production facilities only to then reduce HFC-23 emissions. This incentive structure has led to the perverse outcome of more pollution from HFC-23 since the CDM than before the program was put into place. The main beneficiaries have been organizations in China, where eleven of the nineteen registered HFC-23 projects exist (Wara 2008). The HFC case also undermines the idea that markets are the most cost-effective means of addressing climate change. One factory in China had yearly emissions of HFC-23 equivalent to a million American cars driven 12,000 miles. It would cost US$5 million to eliminate the HFC emissions, much less than the cost of avoiding the pollution of a million cars. However, foreign companies will pay the Chinese factory a hundred times this mitigation cost because the carbon price is set in a market in London. The US$500 million paid for the emissions reductions will be divided by the chemical factory's owners, a Chinese government energy fund, and bankers and consultants in London (Bradsher 2005).

Critiques of REDD also focus on perverse incentives. Environmental groups concerned with deforestation immediately criticized REDD for lacking the language to keep untouched forests standing and for lacking monitoring measures to restrict corruption and secure enforcement. The result, they said, would be more forest losses because REDD provides a subsidy mechanism for large-scale logging and establishment of plantations rather than protection of intact rainforests. Plantations have at most 20 percent of the carbon-storing capacity of native forests. Countries could get paid to raze virgin forests and grow single-species plantations in their place, such as those for palm oil (Feldman 2010). Though REDD could potentially be an effective method to reduce GHG emissions, the detailed rules will be difficult to "get right." A review published in 2009 indicated "disappointing outcomes" for REDD, with badly designed projects, narrow focuses, weak capacity, inadequate local involvement, and extensive corruption (Angelsen et al. 2009).

In conclusion, the market environments in emerging economies have generated mixed results for organizational action on climate change. In contrast to the United States and EU, where domestic politics drive climate policy, international programs have played a key role in structuring organizational responses to climate change in the Global South. With a few exceptions, emerging economies have resisted regulation of GHG emissions. Even those with ambitious domestic targets lack feasible implementation strategies. As a result, most organizational action to reduce GHG emissions has been stimulated by the market environments created by the CDM and REDD. However, both these policies showcase the challenges of market solutions, including design flaws and pervasive corruption.

WHAT IS TO BE DONE?

How can we continue to grow economies, even if only those in the developing nations, without catastrophic consequences? To limit global warming to 2 degrees Celsius—the generally accepted goal—the developed world would have to reduce its emissions by 85 percent and the developing world by 50 percent. For the developing world to catch up with the living standards of the developed world by 2050, the global carbon intensity of production, a convenient measure of GHG emissions, would have to be almost 130 times lower than it is at present. The International Energy Agency now predicts global warming will be 3.5 degrees Celsius above 1990 global average surface temperatures, and it could even be up to 6 degrees. The last time the planet saw 4 degrees of warming, some 55 million years ago at the beginning of the Eocene, the average sea level was some 75 meters higher than now (Lovell 2010). In short, an enormous amount must be done.

So, we return to where this chapter started: What insights does a sociological analysis of market organizations and market environments offer to the challenge of addressing climate change? Can organizations be expected to act independently as climate capitalists? What market environments incentivize emissions reductions and clean technology investments? How can existing market environments be reconfigured?

The first conclusion of our analysis is that market organizations will not act independently to reduce emissions at rates needed to prevent severe impacts of climate change. Reviews of organizational greening validate this skepticism. Engels (2011) carefully examines varieties of corporate greening, analyzing its impact on the bottom line (generally somewhat favorable in efficiency and public relations terms); its consequences internally (risky for careers); and its consequences for firm performance (largely unmeasured). While there are some bright spots, by and large the evaluation is negative. Moreover, Engels notes that there is a pervasive bias in this literature, and in policy debates more broadly, toward the incentives *for* greening, with little attention to how profitable is it to pollute. The IPCC assessments provide a case in point. Government panels in the United States and internationally call for strong action but avoid mentioning major polluting organizations. Yet, the industries that account for most of global GHG emissions show no signs of major reductions; major reductions would reduce profits. An analysis of oil, gas, and coal resources argues that 80 percent of the capital of energy companies would have to remain in the ground to prevent going beyond 2 degrees centigrade warming by midcentury. Imagine the stock price of a giant energy corporation falling as it abandons 80 percent of its capital (McKibben 2012). The firms that do little polluting (e.g., high-tech firms, financial and consumer product firms) are alarmed but have not challenged oil, gas, coal, and electric energy firms sufficiently.

Recognizing the limited potential of independent action by market organizations shifts attention to market environments. Here the prospects for reducing emissions are also gloomy. While there is some variation in market environments across the United States, Europe, and the BRICs, they all continue to foster increasing GHG emissions and fail to incentivize rapid deployment of mitigation technologies. Emissions from China, India, and other developing nations continue to mount. The European nations have made some progress, but at the expense of importing goods and services that create emissions elsewhere, principally in the developing nations. The second largest emitter, the United States, has failed to adopt a carbon tax or a cap-and-trade program on a national level, and the attempts at regional programs are small, slow, and politically vulnerable. Likewise, investment in carbon capture and storage, which could eliminate most of the 36 percent of U.S. carbon dioxide emissions that come from coal-fired electric power stations, has stalled (Berlin and Sussman 2007). Renewables are another option with some potential in the long term (MacKay 2009; Myhrvold and Calderia 2012) but remain a niche energy source in most markets. York's (2012) analysis notes that over the past fifty years, renewables displaced less than one tenth of fossil fuel energy in the electric generation sector.

How, then, might these market environments be reconfigured? In the United States, the primary obstacle to reducing emissions and investing in clean technology is the politically stabilized low price of energy from coal and oil, and the impasse in federal politics. In their analysis of climate politics in the United States, MacNeil and Paterson (2012) look not to greening corporations or national politics as a source of limited optimism, but to subnational governments and very slow, state-by-state, city-by-city change. They show how the United States, despite being a liberal market economy, has been able to make incremental improvements in its climate policy through subnational legislation and the executive authority of the White House. In the coordinated market economies of the EU, the focus should be on strengthening and improving carbon regulation. The political process in

Europe has created market environments that force emissions reductions. Initial experiments with cap-and-trade have generated some emissions reductions by market organizations but also problems with over-allocation, price fluctuations, leakage, and fraud. These problems need to fixed, and it remains to be seen if the political will exists to implement stringent carbon reduction targets (Newell, Pizer, and Raimi 2014). In the emerging economies, creating new market environments will have to center on subsidized technology transfer. There currently exists technology capable of reducing carbon emission by 90 percent or more (Adams and Caldeira 2008; Global CCS Institute 2011; Haszeldine 2009; Rubin 2008). However, it is expensive, and emerging economies will continue to allocate funds to development rather than climate protection (Held et al. 2013).

Such pathways have the potential for reconfiguring market environments in order to shift the choices of market organizations. However, with each pathway, there is a problem with scale. Promoting GHG regulation via local politics in the United States, stimulating emissions reductions via the EU ETS, and facilitating technology transfer through the CDM are pathways based on small-scale enterprise and local innovation and entrepreneurship. Most social scientists and some policymakers prefer such bottom-up approaches rather than large, top-down government efforts focusing on a few technologies. The issue is illustrated in an exchange between a sociologist and two economists (Mowery, Nelson, and Martin 2010; Perrow 2010b). Top-down efforts run the risk of choosing the wrong technology, drying up the resources needed for smaller projects such as wind and solar, and enriching the huge corporations that are the biggest source of emissions. However, decentralized efforts are also problematic: to date, they have remained weak, effecting only minor change. Given the scale of the climate challenge, bottom-up, local solutions are not likely to be enough.

REFERENCES

Abrell, Jan, Anta Ndoye Faye, and Georg Zachmann. 2011. "Assessing the Impact of the EU ETS using Firm-Level Data." Bruegel Working Paper 2011/8. Brussels: Bruegel.

Adams, E. Eric and Ken Caldeira. 2008. "Ocean Storage of CO_2." *Elements* 4(5):319–324.

Angelsen, Arild (Ed.). 2009. *Realising REDD+: National Strategy and Policy Options.* Bogor, Indonesia: Center for International Forestry Research (CIFOR).

Begg, Kathryn, Frans van der Woerd, and David Levy (Eds.). 2005. *The Business of Climate Change: Corporate Responses to Kyoto.* Sheffield, UK: Greenleaf Publishing.

Bell, Shannon and Richard York. 2010. "Community Economic Identity: The Coal Industry and Ideology Construction in West Virginia." *Rural Sociology* 75(1):111–143.

Benecke, Gudrun. 2009. "Varieties of Carbon Governance: Taking Stock of the Local Carbon Market in India." *Journal of Environment & Development* 18(4):346–370.

Berkhout, Frans, Julia Hertin, and David Gann. 2006. "Learning to Adapt: Organizational Adaptation to Climate Change Impacts." *Climatic Change* 78:135–156.

Berlin, Ken and Robert M. Sussman. 2007. *Global Warming and the Future of Coal: The Path to Carbon Capture and Storage.* Washington, DC: Center for American Progress.

Betz, Regina and Misato Sato. 2006. "Emissions Trading: Lessons Learnt from the 1st Phase of the EU ETS and Prospects for the 2nd Phase." *Climate Policy* 6:351–359.

Boasson, Elin and Jorgen Wettestad. 2013. *EU Climate Policy: Industry, Policy Interaction and External Environment.* Farnham, UK: Ashgate.

Bockman, Johanna. 2013. "Neoliberalism." *Contexts* 12(3):14–15.

Bohm, Steffen and Siddhartha Dabhis (Eds.). 2009. *Upsetting the Offset: The Political Economy of Carbon Markets.* London: Mayfly Books.

Bradsher, Keith. 2005. "Outsized Profits and Questions in Effort to Cut Warming Gases." *New York Times,* December 21.

Carbon Disclosure Project. 2013. *CDP Global 500 Report 2013.* London: Carbon Disclosure Project.

Catton, William Jr. and Riley Dunlap. 1978. "Environmental Sociology: A New Paradigm." *American Sociologist* 13(February):41–49.

Chandler, Alfred D. Jr. 1990. *Scale and Scope: The Dynamics of Industrial Capitalism.* Cambridge, MA: Harvard University Press.

Cogan, Douglas G. 2006. *Corporate Governance and Climate Change: Making the Connection.* Boston, MA: Ceres, Inc.

Cragg, Michael I., Yuyu Zhou, Kevin Gurney, and Matthew E. Kahn. 2013. "Carbon Geography: The Political Economy of Congressional Support for Legislation Intended to Mitigate Greenhouse Gas Production." *Economic Inquiry* 51(2):1640–1650.

Cronin, David. 2008. "From EU, 4 Percent Less Reduction Till 2020." *IPS News—Climate Change,* December 17. Retrieved July 5, 2013 (http://www.ipsnews.net/news.asp?idnews = 45147).

DiMaggio, Paul and Walter W. Powell. 1983. "The Iron Cage Revisited: Institutional Isomorphism and Collective Rationality." *American Sociological Review* 48(April):147–160.

Dobbin, Frank. 2001. "Why the Economy Reflects the Polity." Pp. 401–424 in *The Sociology of Economic Life,* edited by M. Granovetter and R. Swedberg. Boulder, CO: Westview Press.

Domhoff, William. 2002. *Who Rules America? Power and Politics.* New York: McGraw-Hill Higher Education.

Dunlap, Riley E. and Angela G. Mertig (Eds.). 1992. *American Environmentalism: The U.S. Environmental Movement, 1970–1990.* Washington, DC: Taylor and Francis New York, Inc.

Eilperin, Juliet. 2010. "Coal-state Dems Hit EPA on Climate." *New York Times,* February 19.

Ellerman, A. Denny, Frank J. Convery, and Christian de Perthuis. 2010. *Pricing Carbon: The European Union Emissions Trading Scheme.* Cambridge: Cambridge University Press.

Emergent Ventures India and The Financial Express. 2008. *How Green is Your Business—FE-EVI Green Business Survey*. New Delhi: Emergent Ventures India and The Financial Express.

Engels, Anita. 2009. "The European Emissions Trading Scheme: An Exploratory Study of How Companies Learn to Account for Carbon." *Accounting, Organizations and Society* 34:488–498.

Engels, Anita. 2011. "Corporate Greening: Greenwashing, Partial Adaptation or Transformation?" Center for Globalization and Governance Working Paper. Hamburg: University of Hamburg.

Engels, Anita, Lisa Knoll, and Martin Huth. 2008. "Preparing for the 'Real' Market: National Patterns of Institutional Learning and Company Behaviour in the European Emissions Trading Scheme (EU ETS)." *European Environment* 18:276–297.

Enkvist, Per-Anders and Helga Vanthournout. 2007. *How Companies Think About Climate Change*. New York: McKinsey & Company.

Environmental Protection Agency. 2008. "Methane to Markets." U.S. Environmental Protection Agency. Retrieved July 5, 2013 (http://www.methanetomarkets.org/).

European Environmental Agency. 2011. "Greenhouse Gas Emissions in Europe." EEA Report No 6/11. Brussels: European Environmental Agency.

Evans, Peter. 2004. "Development as Institutional Change: The Pitfalls of Monocropping and the Potentials of Deliberation." *Studies in Comparative International Development* 38(4):30–52.

Feldman, Stacy. 2010 "UN Forestry Plan Could Increase Logging, Advocates Warn." *InsideClimate News*. October 15. Retrieved July 5, 2013 (http://insideclimatenews.org/news/20101015/un-forestry-plan-could-increase-logging-advocates-warn).

Foster, John Bellamy, Brett Clark, and Richard York. 2010. *The Ecological Rift: Capitalism's War on the Earth*. New York: Monthly Review Press.

Freudenburg, William. 2005. "Privileged Access, Privileged Accounts: Toward a Socially Structured Theory of Resources and Discourses." *Social Forces* 84(1):89–114.

Fuhr, Harald and Markus Lederer. 2009. "Varieties of Carbon Governance in Newly Industrializing Countries." *Journal of Environment & Development* 18 (4):327–345.

General Accounting Office (Ed.). 2008. *International Climate Change Programs: Lessons Learned from the European Unions Emissions Trading Scheme and the Kyoto Protocol's Clean Development Mechanism*. Washington, DC: General Accounting Office.

Global CCS Institute. 2011. *The Global Status of CCS: 2010*. Canberra: Global CCS Institute.

Granovetter, Mark. 1985. "Economic Action and Social Structure: The problem of embeddedness." *American Journal of Sociology* 91:481–510.

Grant, Don and Andrew W. Jones. 2003. "Are Subsidiaries More Prone to Pollute? New Evidence from the EPA's Toxics Release Inventory." *Social Science Quarterly* 84(1):162–173.

Grant, Don, Andrew W. Jones, and Albert J. Bergesen. 2002. "Organizational Size and Pollution: The Case of the U.S. Chemical Industry." *American Sociological Review* 67:389–407.

Grant, Don, Andrew K. Jorgenson, and Wesley Longhofer. 2013. "Targeting Electricity's Extreme Polluters to Reduce Energy-Related CO_2 Emissions." *Journal of Environmental Studies and Science* 3:376–380.

Grant, Don, Mary Nell Trautner, Liam Downey, and Lisa Thiebaud. 2010. "Bringing the Polluters Back In: Environmental Inequality and the Organization of Chemical Production." *American Sociological Review* 75(4):479–504.

Hall, Anthony. 2008. "Better RED than Dead: Paying the People for Environmental Services in Amazonia." *Philosophical Transactions of the Royal Society, B* 363:1925–1932.

Hall, Peter A. and David W. Soskice. 2001. *Varieties of Capitalism: The Institutional Foundations of Comparative Advantage*. New York: Oxford University Press.

Haszeldine, R. Stuart. 2009. "Carbon Capture and Storage: How Green Can Black Be?" *Science* 325(5948):1647–1652.

Hawken, Paul, Amory Lovins, and Hunter Lovins. 1999. "The Next Industrial Revolution." Pp. 1–21 in *Natural Capitalism: Creating the Next Industrial Revolution*. Boston: Little, Brown & Co.

Held, David, Anthony McGrew, David Goldblatt, and Jonathan Perraton. 1999. *Global Transformations*. Stanford, CA: Stanford University Press.

Held, David, Charles Roger, and Eva-Maria Nag, eds. 2013. *Climate Governance in the Developing World*. Cambridge: Polity Press.

Hoag, Hannah. 2011. "The Problems With Emission Trading." *Nature News*. October 25.

Hoffman, Andrew J. 2001. *From Heresy to Dogma: An Institutional History of Corporate Environmentalism*. Palo Alto, CA: Stanford University Press.

Howarth, Robert W., Anthony Ingraffen, and Terry Engelder. 2011. "Natural Gas: Should Fracking Stop?" *Nature* 477:271–275.

Ikwue, Tony and Jim Skea. 1994. "Business and the Genesis of the EU Carbon Tax Proposal." *Business Strategy and the Environment* 3(2): 1–10.

Intergovernmental Panel on Climate Change. 2007a. "Climate Change 2007: The Physical Science Basis." In *IPCC Fourth Assessment Report*, edited by S. Solomon, D. Qin, M. Manning, Z. Chen, M. Marquis, K. B. Averyt, M. Tignor, and H. L. Miller. Cambridge and New York: Intergovernmental Panel on Climate Change.

Intergovernmental Panel on Climate Change. 2007b. "Climate Change 2007: Mitigation of Climate Change." In *IPCC Fourth Assessment Report*, edited by B. Metz, O. R. Davidson, P. R. Bosch, R. Dave, and L. A. Meyer. Cambridge and New York: Intergovernmental Panel on Climate Change.

Intergovernmental Panel on Climate Change. 2014a. "Summary for Policymakers." In *Climate Change 2014, Mitigation of Climate Change*. Contribution of Working Group III to the IPCC Fifth Assessment Report, edited by O. Edenhofer, R. Pichs-Madruga, Y. Sokona, E. Farahani, S. Kadner, K. Seyboth, A. Adler, I. Baum, S. Brunner, P. Eickemeier, B. Kriemann, J. Savolainen, S. Schlomer, C. von Stechow, T. Zwickel, and J. C. Minx. Cambridge and New York: Cambridge University Press.

Intergovernmental Panel on Climate Change. 2014b. *Climate Change 2014, Mitigation of Climate Change*. Contribution of Working Group III to the IPCC Fifth Assessment Report, edited by O. Edenhofer, R. Pichs-Madruga, Y. Sokona, E. Farahani, S. Kadner, K. Seyboth, A. Adler, I. Baum, S. Brunner, P. Eickemeier, B. Kriemann, J. Savolainen, S. Schlomer, C. von Stechow, T. Zwickel, and J. C. Minx. Cambridge and New York: Cambridge University Press.

Jacoby, H. D., A. C. Janetos, R. Birdsey, J. Buizer, K. Calvin, F. de la Chesnaye, D. Schimel, I. Sue Wing, R. Detchon, J. Edmonds, L. Russell, and J. West. 2014. "Chapter 27: Mitigation." Pp. 648–669 in *Climate Change Impacts in the United*

States: The Third National Climate Assessment, edited by J. M. Melillo, Terese (T.C.) Richmond, and G. W. Yohe. Washington, DC: U.S. Global Change Research Program.

Jones, Charles A. and David L. Levy. 2007. "North American Business Strategies Towards Climate Change." *European Management Journal* 25(6):428–440.

Jorgenson, Andrew K. 2012. "The Sociology of Ecologically Unequal Exchange and Carbon Dioxide Emissions, 1960–2005." *Social Science Research* 41:242–252.

Kanter, James and Jad Mouawad. 2008. "Money and Lobbyists Hurt Your Opinion Efforts to Curb Gases." *New York Times*, December 11.

Kinne, Beth, Michael Finewood, and David Yoxtheimer. 2014. "Making Critical Connections Through Interdisciplinary Analysis: Exploring the Impacts of Marcellus Shale Development." *Journal of Environmental Studies and Sciences* 4(1):1–6.

Kolk, Ans and Jonatan Pinkse. 2005. "Business Responses to Climate Change: Identifying Emergent Strategies." *California Management Review* 47(3):6–20.

Kolk, Ans and Jonatan Pinkse. 2007. "Multinationals' Political Activities on Climate Change." *Business & Society* 46(2):201–228.

Krugman, Paul. 2010. "Building a Green Economy." *New York Times Magazine*, April 7.

Leggett, Jeremy. 1999. *The Carbon War: Global Warming and the End of the Oil Era.* London: Penguin Books.

Levy, David and Ans Kolk. 2002. "Strategic Responses to Global Climate Change: Conflicting Pressures on Multinationals in the Oil Industry." *Business and Politics* 4(3): 275–300.

Levy, David L. and Daniel Egan. 1998. "Capital Contests: National and Transnational Channels of Corporate Influence on the Climate Negotiations." *Politics & Society* 26(3):337.

Lovell, Bryan. 2010. *Challenged by Carbon: The Oil Industry and Climate Change.* Cambridge: Cambridge University Press.

Lovell, Heather. 2012. "Governing the Carbon Offset Market." *WIREs Climate Change* 1(3):353–362.

Lowe, Ernest A. and Robert J. Harris. 1998. "Taking Climate Change Seriously: British Petroleum's Business Strategy." *Corporate Environmental Strategy* 5(2):23–31.

MacKay, David J. C. 2009. *Sustainable Energy: Without the Hot Air.* Cambridge: UIT Cambridge.

MacNeil, Robert and Matthew Paterson. 2012. "Neoliberal Climate Policy: From Market Fetishism to the Developmental State." *Environmental Politics* 21(2):230–247.

Marx, Karl. 1867 [1990]. *Capital: Critique of Political Economy.* Volume 1. London: Penguin Books.

McCright, Aaron M. and Riley E. Dunlap. 2003. "Defeating Kyoto: The Conservative Movement's Impact on US Climate Change Policy." *Social Problems* 50(3):348–373.

McKibben, Bill. 2012. "Global Warming's Terrifying New Math." *Rolling Stone*, July 12.

Mol, Arthur P. J. 1995. *The Refinement of Production: Ecological Modernization Theory and the Chemical Industry.* Utrecht: Van Arkel.

Mowery, David C., Richard R. Nelson, and Ben R. Martin. 2010. "Technology Policy and Global Warming: Why New Policy Models Are Needed (or Why Putting New Wine in Old Bottles Won't Work)." *Research Policy* 39(8):1011–1023.

Myhrvold, Nathan P. and Ken Caldeira. 2012. "Greenhouse Gases, Climate Change and the Transition From Coal to Low-carbon Electricity." *Environmental Research Letters* 7(1):8pp. doi:10.1088/1748-9326/7/1/014019.

Newell, Peter. 2000. *Climate for Change: Non-state Actors and the Global Politics of the Greenhouse*. Cambridge: Cambridge University Press.

Newell, Peter and Matthew Paterson. 2010. *Climate Capitalism: Global Warming and the Transformation of the Global Economy*. Cambridge: Cambridge University Press.

Newell, Richard, William Pizer, and Daniel Raimi. 2014. "Carbon Market Lessons and Global Policy Outlook." *Science* 343:1316–1317.

Nordhaus, William. 2009. *A Question of Balance: Weighing the Options: Global Warming Policies*. New Haven, CT: Yale University Press.

Perrow, Charles. 1986. *Complex Organizations: A Critical Essay*. New York: Random House.

Perrow, Charles. 2010a. "Organizations and Global Warming." Pp. 59–77 in *Handbook of Climate Change and Society*, edited by C. Lever-Tracy. New York: Routledge.

Perrow, Charles. 2010b. "Comment on Mowrey, Nelson and Martin." *Research Policy* 39(8):1030–1031.

Perrow, Charles. 2011. *The Next Catastrophe: Reducing Our Vulnerabilities to Natural, Industrial and Terrorist Disasters*. Princeton, NJ: Princeton University Press.

Peters, Glen P., Gregg Marland, Corinne Le Quere, Thomas Boden, Josep G. Canadell, and Michael R. Raupach. 2012. "Rapid Growth in CO_2 Emissions After the 2008–2009 Global Financial Crisis." *Nature Climate Change* 2(1):2–4.

Plumer, Brad. 2011. "Global CO_2 Emissions Rising Faster Than Worst-case Scenarios." *Washington Post*, November 4.

Polanyi, Karl. 1944 [1957]. *The Great Transformation: The Political and Economic Origins of Our Time*. Boston: Beacon Press.

Porter, Michael E. and Claas van der Linde. 1995. "Green and Competitive: Ending the stalemate" *Harvard Business Review* (September-October):120ff.

Prechel, Harland and Lu Zheng. 2012. "Corporate Characteristics, Political Embeddedness and Environmental Pollution by Large U.S. Corporations." *Social Forces* 90(3):947–970.

PriceWaterhouseCoopers. 2009. *Carbon Disclosure Project 2009: Global 500 Report*. London, UK: Carbon Disclosure Project.

Princen, Thomas. 1997. "The Shading and Distancing of Commerce: When Internalization Is Not Enough." *Ecological Economics* 20(3):235–253.

Pulver, Simone. 2002. "Organizing Business: Industry NGOs in the Climate Debates." *Greener Management International* 39(Autumn):55–67.

Pulver, Simone. 2007a. "Importing Environmentalism: Explaining Petroleos Mexicanos' Proactive Climate Policy." *Studies in Comparative International Development* 42(3/4):233–255.

Pulver, Simone. 2007b. "Making Sense of Corporate Environmentalism: An Environmental Contestation Approach to Analyzing the Causes and Consequences of the Climate Change Policy Split in the Oil Industry." *Organization & Environment* 20(1):44–83.

Pulver, Simone. 2011a. "Corporate Responses." Pp. 581–593 in *Oxford Handbook of Climate Change and Society*, edited by J. Dryzek, R. Norgaard and D. Shlosberg. Oxford: Oxford University Press.

Pulver, Simone. 2011b. "Corporate Responses to Climate Change in India." Pp. 254–265 in *Handbook of Climate Change and India: Development, Politics, and Governance*, edited by N. Dubash. New Delhi: Oxford University Press.

Pulver, Simone and Tabitha Benney. 2013. "Private-sector Responses to Climate Change in the Global South." *WIREs Climate Change* 1:1–19.

Pulver, Simone, Nathan Hultman, and Leticia Guimaraes. 2010. "Carbon Market Participation by Sugar Mills in Brazil." *Climate and Development* 2(3):248–262.

Roberts, David. 2012. "Why Germany is Phasing Out Nuclear Power." *Grist*. March 23. Retrieved July 5, 2013 (http://grist.org/renewable-energy/why-germany-is-phasing-out-nuclear-power/).

Rubin, Edward S. 2008. "CO_2 Capture and Transport." *Elements* 4(5):311–317.

Sanderson, Katherine. 2009. "Aerosols Make Methane More Potent." *Nature News*. October 29. Retrieved July 5, 2013 (http://www.nature.com/news/2009/091029/full/news.2009.1049.html).

Schiermeier, Quirin. 2008. "Europe Agrees on Emissions Deal." *Nature* 456:847.

Schmidheiny, Stephan. 1992. *Changing Course: A Global Business Perspective on Development and the Environment*. Cambridge, MA: MIT Press.

Schnaiberg, Allan and Kenneth Gould. 1994. *Environment and Society: The Enduring Conflict*. New York: St. Martin's Press.

Schneider, Malte, Volker H. Hoffmann, and Bhola R. Gurjar. 2009. "Corporate Responses to the CDM: The Indian Pulp and Paper Industry." *Climate Policy* 9(3):255–272.

Schneider, Stephen. 1997. *Laboratory Earth*. New York: Basic Books.

Schroeder, Miriam. 2009. "Varieties of Carbon Governance: Utilizing the Clean Development Mechanism for Chinese Priorities." *Journal of Environment & Development* 18(4):371–394.

Scott, Richard. 2004. "Reflections on a Half-Century of Organizational Sociology." *Annual Review of Sociology* 30(August):1–21.

Shindell, Drew T., Greg Faluvegi, Dorothy M. Koch, Gavin A. Schmidt, Nadine Unger, and Susanne E. Bauer. 2009. "Improved Attribution of Climate Forcing to Emissions." *Science* 326(5953):716–718.

Skocpol, Theda. 2012. "Naming the Problem: What It Will Take to Counter Extremism and Engage Americans in the Fight Against Global Warming." Prepared for symposium on *The Politics of America's Fight Against Global Warming*. Harvard University, Cambridge, MA, February 14, 2013. Retrieved July 5, 2013 (http://www.scholarsstrategynetwork.org/sites/default/files/skocpol_captrade_report_january_2013y.pdf).

Smith, Adam. 1776 [1991]. *The Wealth of Nations*. Amherst, NY: Prometheus Books.

Spash, Clive L. 2010. "The Brave New World of Carbon Trading." *New Political Economy* 15(2):169–195.

Stavins, Robert N. 2008. "A Meaningful U.S. Cap-and-Trade System to Address Climate Change." *Harvard Environmental Law Review* 32:293–371.

Sullivan, Rory (Ed.). 2008. *Corporate Responses to Climate Change: Achieving Emissions Reductions Through Regulation, Self-regulation, and Economic Incentives*. Sheffield, UK: Greenleaf Publishing.

Tollefson, Jeff. 2012. "Air Sampling Reveals High Emissions From Gas Field." *Nature News*. February 7. Retrieved July 5, 2013 (http://www.nature.com/news/air-sampling-reveals-high-emissions-from-gas-field-1.9982).

UNEP Risoe Center. 2013. *Approved CDM Methodologies.* Retrieved March 1, 2013 (http://www.cdmpipeline.org/cdm-methodologies.htm).

UN-REDD Programme. 2013. "About the UN-REDD Programme." Retrieved July 5, 2013 (http://www.un-redd.org/AboutUN REDDProgramme/tabid/102613/Default.aspx).

Vogel, David. 1989. *Fluctuating Fortunes: The Power of Business in America.* New York: Basic Books.

Waldermann, Anselm. 2009. "Green Energy Not Cutting Europe's Carbon." *Business Week.* February 10. Retrieved July 5, 2013 (http://www.businessweek.com/global-biz/content/feb2009/gb20090210_228781.htm).

Wara, Michael. 2008. "Measuring the Clean Development Mechanism's Performance and Potential." *UCLA Law Review* 55(6):1759–1791.

Wara, Michael and David G. Victor. 2008. "A Realistic Policy on International Carbon Offsets." Program on Energy and Sustainable Development Working Paper #74. Palo Alto, CA: Stanford University.

Wong, Shiufai. 2010. "Case Study: Wind Energy Regulation in Germany and the UK." Pp. 369–378 in *Handbook of Climate Change and Society,* edited by C. Lever-Tracy. New York: Routledge.

York, Richard. 2012. "Do Alternative Energy Sources Displace Fossil Fuels?" *Nature Climate Change* 2(6):441–443.

4

Consumption and Climate Change

Karen Ehrhardt-Martinez and Juliet B. Schor with Wokje Abrahamse, Alison Hope Alkon, Jonn Axsen, Keith Brown, Rachael L. Shwom, Dale Southerton, and Harold Wilhite

INTRODUCTION

Consumption, specifically household consumption, is a major contributor to anthropogenic climate change. Direct energy use by households for residential energy and transportation accounts for approximately 38 percent of all U.S. CO_2 emissions (Dietz et al. 2009; Gardner and Stern 2008). Indirect household effects via consumption of food, water, and goods and services further increase households' contribution. Furthermore, the contribution of residential energy to CO_2 emissions has increased by roughly 21 percent since 1990 (Environmental Protection Agency 2013:Table 2-5, authors' calculations), making this sector an important and growing component of greenhouse gas emissions.

As this chapter will show, there is now a considerable literature on how the actions of households both contribute to and mitigate climate change, and how innovative policy approaches to households, rooted in sociological findings, are gaining influence. (For a thorough discussion of mitigation see Chapter 7 in this volume). A number of studies have documented the possibilities for household-level mitigation. Some estimates suggest that total U.S. energy consumption could be reduced by 9 to 11 percent (Gardner and Stern 2008; Laitner and Ehrhardt-Martinez 2010). Under reasonable assumptions about the effectiveness of nonregulatory behavioral interventions, and without energy price increases, other researchers estimate that the volume of CO_2 emissions associated with direct energy and transportation can be reduced approximately 20 percent over a decade, representing 7.4 percent of the yearly national total (Dietz et al. 2009:18452; see also Laitner and Ehrhardt-Martinez 2010).

This potential for mitigation at the individual and household level[1] has been relatively underappreciated in the national and international policy discourse. Of course, relationships between consumption and environmental degradation have long been recognized by environmental and social scientists. In the widely cited IPAT equation, "affluence," or consumption, is identified as a key variable in determining environmental outcomes (Ehrlich and Holdren 1971):

$$\text{Environmental Impact} = \text{Population} \times \text{Affluence} \times \text{Technology}$$

However, in this formulation affluence is modeled at the macro or economy-wide level, and its drivers are also macro-level variables (see Chapter 2 in this volume). We discuss a now-extensive sociological literature studying micro, or household-level factors that provides new insights and strategies for creating more effective climate policies. Currently, economics is the only social science that is well integrated into the content of major national and international climate policy reports. For example, the National Research Council's volume *Limiting the Magnitude of Future Climate Change* (2010) proposes using market incentives such as carbon pricing or government support for technological innovation to reduce consumption-related emissions, relying exclusively on economic assessments. Furthermore, the National Research Council considers only a narrow set of activities associated with direct household energy consumption, such as residential heating/cooling and motor vehicle use, and ignores the implications of food-related consumption patterns. Similarly, although the Intergovernmental Panel on Climate Change (IPCC)'s 2007 synthesis report does note "high agreement and medium evidence that changes in lifestyle and behavior can contribute" to mitigation, detailed information on the role of households is absent from the report (IPCC 2007:59). The National Academy of Science's *America's Climate Choices* (2011) also fails to consider consumption or the role of households specifically, focusing instead on risk analysis and decision making and recommending economic policies (taxes or cap-and-trade), efficiency standards, and education and information. As illustrated in the remainder of this chapter, the limited approaches taken in these reports are inadequate to fully understand and successfully address household consumption.

That said, there *are* barriers to mitigation at the household level. Sociological research on consumption and climate change has found that households are often constrained by existing social structures and cultural norms that limit their ability to act independently and make free choices. As such, a key focus of this chapter is to explore what is known about households' "agency," or ability to act, to mitigate climate impacts, in comparison to the importance of larger social and cultural forces. The challenge is to find the middle ground in which both agency and structure play a role in shaping current, past, and future experiences. This question has been contentious not only among sociologists but also among policymakers, nonprofit organizations, and government leaders. One view holds that households are relatively passive actors in a system in which greenhouse gas (GHG) emissions are structurally driven by the decisions of business and government (Gould, Pellow, and Schnaiberg 2008). From this perspective, households bear little responsibility for, and consequently have little role in addressing, climate change. However, another school of thought suggests that the consumption choices and practices of households and individuals do matter. While we agree that production-side structural change and national and international policy interventions are crucial, the findings that we report on here reveal that there is considerable potential for mitigation at the household level through efficiency, conservation, consumption reduction, and changing patterns of food, energy, and other types of consumption.

However, achieving that potential requires recognizing that households are embedded within larger socioeconomic, socio-technical, sociocultural, and natural systems that affect the human dimensions of carbon emissions and possibilities for mitigation. Throughout this chapter, these issues are revisited as we address theories of consumption, technological innovation, and the empirical evidence of consumption patterns across various sectors. We assess the ways in which larger social contexts both shape and constrain consumption patterns and practices. We show how the dynamics of social inequality or cultural practices affect how and why households consume as they do.

Finally, this chapter also considers a broader framing of consumption-related climate issues to consider the inter-relationship between consumption and the environment but also their connections to a variety of other important issues such as quality of life, cost, equality, and health issues. This more systemic type of perspective has proven valuable to research on food production and consumption and has proven useful to many activists who are attempting to transform the food system.

The chapter begins with a brief introduction to individual and household consumption patterns in three carbon-intensive areas: energy, transportation, and food. We then turn to a discussion of major theoretical approaches that shape consumption patterns, beginning with classic sociological approaches to resource-intensive consumption. We also discuss "practice theories" that emphasize inertia in consumption patterns and the weaknesses of economic and psychological approaches that focus on values, information, or economic incentives. The fourth section presents four synopses of sector-specific insights on the consumption–climate connection: energy, transportation, food, and lifestyles. The conclusion draws lessons for reducing the climate impacts of household consumption.

GENERAL OVERVIEW OF ENERGY AND RESOURCE CONSUMPTION

Efforts to quantify and measure the impact of consumption on the Earth gained momentum in the mid-1990s with the development of ecological footprint analysis (Wackernagel and Rees 1995) and more recently with carbon footprinting. The ecological footprint is a measure of the land area and ocean resources that are necessary to produce the annual consumption of a nation, city, household, business, or any other unit with well-defined consumption. The carbon component of the ecological footprint is the subset that measures fossil fuel use, including forests necessary for sequestering the CO_2 emitted when the fuels are burned. Unlike national GHG accounting, which is territorially and production based, the carbon footprint measure accounts for exports and imports and is hence a consumption measure. From a global perspective, Americans are voracious consumers, and national consumption trends and patterns have both direct and indirect impacts on climate emissions. According to the calculations of the Global Footprint Network, the United States has one of the highest per capita carbon footprints in the world, at 5.57 hectares, more than twice the European average of 2.54. In Asia, Latin America, and Africa, footprints are far lower (0.90, 0.72, and 0.26 respectively), from five to twenty times less than in the United States (Global Footprint Network, data tables 2013).

In recent years, analyses of household climate impacts have become more sophisticated, as social scientists compute emissions from detailed consumer expenditure data that take into account the effect of imports and exports. Incorporating direct and indirect effects, households can be considered to be responsible for between 80 and 120 percent of U.S. territorial emissions (Jones and Kammen 2011). Jones and Kammen (2011) estimate that the top three areas of high impact are transportation, housing, and food, which in 2005 represented 32 percent, 28 percent, and 15 percent of the total household impact respectively. Purchases of goods represented approximately 12 percent of households' GHG impact, and services represented approximately 13 percent (Jones and Kammen 2011, authors' calculations from Figs. 4.1 and 4.2). Looking back over a number of decades, emissions from household consumption have risen dramatically, although there are a few recent countertrends.

Direct energy consumption is the largest component of the household footprint. Over the last sixty years (1950–2010), residential energy consumption nearly quadrupled from 6 to 21.9 quadrillion BTUs. By contrast, industrial energy consumption has witnessed a recent decline (U.S. Energy Information Administration 2012). Although the rate of growth of emissions in the household sector has now slowed due to the development of energy efficiency standards for new homes and technologies, other factors

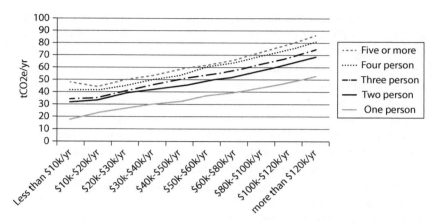

Figure 4.1 Carbon Footprints by Household Size and Income.

Source: Jones and Kammen 2011

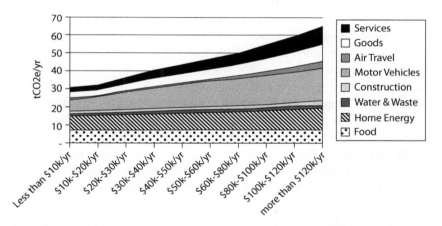

Figure 4.2 Carbon Footprints by Emission Category and Income.

Source: Jones and Kammen 2011

have worked against the gains in energy efficiency. Among the most important contributors to continued increases are the growing number of homes (due to population growth and a decline in household size); the trend toward larger homes (Dwyer 2007); the increasing prevalence of air conditioning; and the proliferation of appliances and electronics. For example, in 1980 only 27 percent of U.S. households had central air conditioning, compared to 63 percent in 2009 (U.S. Energy Information Administration 2012). Similar statistics for appliances and electronics show that the number of households with a second refrigerator grew from 14 percent in 1980 to 23 percent in 2009, while the proportion of households with three or more televisions grew from 14 percent to 44 percent in the same period (U.S. Energy Information Administration 2012). At the same time the average size of refrigerators and televisions has grown dramatically, as has the number of computers and consumer electronics (McNary and Berry 2012).

Energy consumption associated with personal transportation has witnessed similar trends. Between 1950 and 2008 the number of vehicles in operation increased 475 percent, from 43 million to 248 million. Population growth accounted for slightly more than half of the increase (251 percent). In addition, vehicles per person rose from 0.29 to 0.8, and the proportion of households without a vehicle has shrunk from 17.5 percent in 1950 to 9.1 percent in 2010. Per capita vehicle miles driven have also more than tripled, from just over 3,000 per year in 1950 to roughly 9,600 in 2010. Fuel consumption for light vehicle transportation has also increased 62 percent between 1970 and 2008, as improvements in fuel efficiency have been outweighed since 1970 by consumers' preference for larger vehicles. In recent years a number of these trends have been reversed: total miles driven peaked in 2006 (U.S. Department of Energy 2012); per capita vehicle miles driven have fallen sharply and are now at their 2004 level; car ownership has declined; and young people are much less likely to acquire driving licenses than in the past (U.S. Public Research Interest Group 2012). These developments have prompted some researchers to wonder whether we have reached the point of "Peak Car." They have also encouraged other researchers to explore the potential magnitude of savings that could be achieved through changes in consumption patterns. According to several prominent studies, such savings could make a large contribution to mitigation efforts. For example, changes in household practices associated with energy and personal transportation alone could reduce U.S. carbon emissions by an estimated 123 million metric tons (Dietz et al. 2009) and lower total U.S. energy consumption by 9 to 11 percent (Gardner and Stern 2008; Laitner and Ehrhardt-Martinez 2010).

Food is a third area with high energy and climate impacts (Carlsson-Kanyama and Gonzalez 2009; Weber and Matthews 2008). A range of food trends are contributing to higher GHG emissions, including increases in the consumption of meat and dairy products (especially red meat), highly processed foods, highly packaged foods, and bottled water. Increases in the overall volume of food consumed has also led to higher GHG emissions, as has an increase in the distance food travels from its production to consumption location, a concept known as "food miles."

According to the U.S. Department of Agriculture (2003), the average American is now consuming 500 more calories a day than in the 1970s. Meat consumption increased from an annual average of roughly 100 pounds per person during the period from 1909 to 1950 to a peak of 183 pounds in 2007. Beef consumption in 2012 represented 32 percent of total meat consumption in the United States, along with chicken, pork, and turkey (Larsen 2012). The GHG footprint of meat is three to twenty-five times that of rice depending on the type of meat (Carlsson-Kanyama and Gonzalez 2009). According to Weber and Matthews (2008), the carbon footprint of red meat is about 150 percent higher than chicken or fish. They also find that food miles contribute approximately 11 percent to the life-cycle GHG emissions.

The concept of food miles highlights the issue of production location, processing, and transportation. For U.S. households, an estimated 30 percent of emissions are generated abroad and enter the economy through imports. In the United States, imports are carbon intensive because so much trade is conducted with China, where coal is a heavily used energy source (Weber and Matthews 2007). Finally, it is important to note that averages obscure large differences in climate impacts across households; Weber and Matthews (2007) estimate a tenfold spread, with income and expenditure being the main determinants of difference. Jones and Kammen (2011, Figs. 4.1 and 4.2) find that carbon footprints roughly double as annual household income rises from less than $10,000 to more than $120,000.

THEORETICAL PERSPECTIVES ON CONSUMPTION

Within the social sciences, studies of consumption tend to focus on either the realm of individual decision making and behavior or broader cultural and social contexts that shape decisions and behaviors. The fields of economics and psychology typically take an individualist approach, while sociology and anthropology situate consumption within social and cultural contexts and study how people's consumption is influenced by others. Most economists model human behavior by characterizing individuals with a rational-actor model in which consumers are unaffected by the choices of others—what is termed "preference independence." The consumer has a fixed and unlimited set of desires (preferences) and is a self-interested, optimizing "actor" with a strong responsiveness to economic variables such as prices and incomes (Becker 1978). To explain energy consumption, this model focuses on the price of energy relative to other goods and the level of income. However, reviews of past studies reveal that estimates of the responsiveness of consumers to energy prices vary by a factor of ten, which leaves ample room for noneconomic policy approaches (Dietz et al. 2009:18453).

Psychologists similarly focus on individual behavior; however, they are less parsimonious in their modeling and more expansive in their discussion of preferences, examining environmental attitudes for example. Moreover, as we will note, there is now robust interdisciplinary collaboration between environmental psychologists and sociologists. By contrast, sociological and cultural theories of consumption analyze individuals as embedded in social contexts and explain consumption desires, choices, and practices within these contexts. Social and cultural theories often focus on consumption as a form of social communication and an expression of culture, rather than an activity that is carried out primarily to satisfy individual desires. They also focus explicitly on the role of inequality in structuring and driving consumption, whether it is inequality of class, race, or gender. These approaches have led to innovative policies aimed at changing consumption patterns for energy and other goods, as we will discuss.

Status, Class, and Culture

A key driver of carbon-intensive consumption is the desire for social esteem and position. Since Thorstein Veblen's influential *Theory of the Leisure Class* (1899), sociologists have understood that status is primarily gained by the visible or public consumption of goods and services that are, by social consensus, of high status. Because status is relative, or positional, people aim to keep up with or do better than those who comprise their social circles, or reference groups. This approach predicts a treadmill of consumption that may intensify over time as consumption competitions heat up, as a result of growing inequality, advertising, or marketing (Bell 2012:Chapter 2; Schor 1998). Veblenian consumption dynamics yield heavy climate impacts. Veblen himself noted that houses and transport are central to consumption competitions, and contemporary status dynamics have led to large increases in the expense and size of homes and vehicles. In the 1990s and 2000s, the treadmill of consumption was associated with a speed-up in the fashion cycle in goods, in which products lost their social value long before their practical usefulness was exhausted, which led in turn to more rapid discarding of products and acquisition of new ones (Schor 2010). In combination with manufacturers' attempts to plan obsolescence in products, this created a consumer regime with high and rising carbon impacts.

The status approach provides one explanation for a widely studied gap between people's attitudes to environmental issues and their actual behaviors. Consumers may simultaneously hold strong pro-environmental attitudes and live high-impact lifestyles because visible consumption is vital to their social position and core identity. There are relatively few empirical studies to date that test the Veblenian approach in the field of climate. However, Pedersen finds that consumers are more likely to opt for low-impact choices in food, energy, and other environmentally significant expenditures when they are socially visible as well as tangible (Pedersen 2000). Mau et al. find a "neighborhood effect" in which consumer preferences for hybrid vehicles are dynamic and increasing as they diffuse throughout an area (Mau et al. 2008).

Scholars have also augmented the pure Veblenian model, in which economic resources determine social status. Pierre Bourdieu (1984) argued that social status is also dependent on what he termed cultural capital, a set of dominant class dispositions, tastes, and knowledge about culture, consumption, manners, and lifestyle, the possession of which yields social status. Bourdieu argued that tastes are learned early in life; that there is a deep, nonconscious, or naturalized aspect to taste and distaste; and that in domains such as food, our bodies actually adjust to the tastes we are used to (Warde 1997; Wilk 1997). One implication of this approach is that preferences for energy, food, and transport may be ingrained and hard to change. They are also "coded" with social class meanings, often in complex and changing ways. For example, drying clothes on a line became "lower" class once middle-class households acquired clothes dryers. If energy-saving practices are stigmatized due to their association with low-income households, policy will need to address that stigma to increase their take-up.

In the United States, the traditionally high education and economic status of environmentalists, as well as a racial skew toward whites, means that green products and lifestyles tend to be coded as high-class, which can affect whether lower-income households are open to adopting them. Bennett and Williams (2011) found that "rich and elite" were associated with green products. Laidley (2013) found that across socioeconomic groups, residents of a northeastern city felt that concern for climate change is the domain of the privileged and that low-income households do not have the "luxury" of worrying about it because they are too concerned with "putting food on the table." Similarly, a study of U.S. Hummer drivers found that they characterize Prius drivers as elite and un-American (Luedicke, Thompson, and Giesler 2010). Laidley (2012) also found a significant association between education level and hybrid vehicle ownership across townships in Massachusetts. While Bourdieu himself focused primarily on class, there is a larger literature examining the role of differences such as education, gender, age, geography, and ethnicity in how people consume.

In contrast to Bourdieu, Anthony Giddens (1991) has argued that in this "late" modern era, consumers attempt to create individualized rather than class identities, exercising personal reflectiveness, or reflexivity, as a dynamic, continuous self-awareness. Individuals seek to create lifestyles or combinations of practices. The conspicuousness of some low-carbon technologies, such as the Toyota Prius, facilitates reflexivity by prompting users and observers to share and negotiate interpretations (Shove and Warde

2002). Qualitative studies of individuals attempting to live low-carbon life-styles (Connolly and Prothero 2008) have found that many engage in high levels of personal reflexivity and aim for consistent lifestyles, although this is often difficult given the complexity of environmental impacts and high levels of informational uncertainty.

Other Social Approaches to Consumption: Media and Networks

Sociologists have tied status consumption to advertising and media via a treadmill of consumption, driven by desire creation as marketers attempt to sell products in an increasingly saturated market. Advertising, media content, visual culture, and popular culture all stimulate desire and expenditure. Most of this literature is qualitative and historical, tracing the rise of consumer society, advertising, and marketing. However, Brulle and Young (2007) find that over the period from 1900 to 2000, advertising outlays were significantly related to consumption expenditures and advertising has been far more effective in spurring demand for energy-intensive luxury goods (household appliances and cars) than for necessities. Occasionally, the persuasive power of advertising is used to encourage households to engage in less environmentally destructive expenditure patterns and to reduce consumption (Peattie 2001).

A body of literature also looks at norms and social influence without the structuring variable of class. Robert Cialdini and his collaborators have found that when social norms about what other people are doing or believe are made salient, energy use is affected (Schultz et al. 2007). People are more likely to participate in hotel conservation programs when they are primed by information that others are also participating (Goldstein, Cialdini, and Griskevicus 2008), and households reduce electricity consumption after being informed that others in their neighborhood used less than they do (Schultz et al. 2007). A growing literature on social networks finds that they have large impacts on food consumption (Christakis and Fowler 2009). In addition, social ties between people in networks facilitate pro-environmental behaviors (Hopper and Nielsen 1991; Weenig and Midden 1991).

Bridging the Individual–Social Divide: Interdisciplinary Efforts

Recently, there have been attempts to create interdisciplinary collaboration between psychologists and sociologists to integrate theoretical and methodological insights from both fields (Evans and Abrahamse 2009; Nye, Whitmarsh, and Foxon 2010; Stets and Biga 2007). These cross-disciplinary literatures frequently provide better explanations of climate-related behaviors. One widely studied issue is a gap between people's attitudes to environmental behaviors and their actual behaviors. This "attitude–behavior" gap has been documented for household energy conservation (Ehrhardt-Martinez and Laitner 2009), vehicle choice (Turrentine, Kurani, and Heffner 2008), and surveys of multiple pro-environmental actions (Kollmuss and Agyeman 2002). (Pro-environmental actions include a variety of behaviors, such as reduced household energy use, transportation behavior, as well as local eating, water consumption, and recycling.) Explanations for the gap include the role of barriers, such as lack of public transportation (McKenzie-Mohr and Smith 2001) and omitted sociological factors such as comfort, convenience, identity, and status (Gifford 2011; Swim et al. 2009). Olli et al. (2001) estimated a multidisciplinary model that included social context, sociodemographic factors, and attitudes and found that social context—measured as participation in environmental networks—was as predictive as the combined effect of the other two factors.

Real-world behavior change initiatives are also yielding findings that bridge disciplines. EcoTeams, which are small neighborhood support groups where people receive advice on reducing their carbon footprints, have been studied by psychologists and sociologists. Psychologists have found EcoTeams to be effective in achieving energy savings and raising levels of environmental concern (Staats, Harland, and Wilke 2004), while sociologists have provided analyses of the processes evoked by participation and how changes are enabled by the social context (Hobson 2002; Nye and Hargreaves 2010).

Rejecting Individualism: Inconspicuous Consumption, Routines, and Theories of Practice

In the 1990s, a group of sociologists shifted their attention from conspicuous forms of consumption toward accounts of high-impact but inconspicuous and ordinary forms of consumption, such as household energy and water use (Gronow and Warde 2001; Shove and Warde 2002). This led to growing attention to the mundane "use of things" and interest in habitual and routine practices of consumption. In contrast to the standard models of economists, much human behavior appears to be performed without deliberation, and everyday actions reveal a high degree of repetition, periodicity, and predictability. Routines refer more to the unreflexive and repetitive aspects of actions, while habit tends to capture highly normative and culturally rooted actions (Warde and Southerton 2012). Eventually this literature came to adopt a focus on consumer "practices" and the difficulties of inducing climate-friendly norms once resource-intensive practices have taken root.

That conclusion came partly from case studies undertaken by British sociologists who studied the shift from non-daily bathing to daily hot showers (Hand, Shove, and Southerton 2005), the growing use of stand-alone freezers (Hand and Shove 2007) and changes in norms of household heating and cooling that have led to less seasonal variation (more heating in the winter and more air-conditioning use in the summer) (Shove 2003). In all three cases, the expansion of energy-intensive routines and practices was due to producers' actions, technological changes, and contextual dimensions of daily life, such as time pressures and changing leisure patterns. The factors of comfort, cleanliness, and convenience have been identified as particularly salient (Shove 2003) in the growth of energy use. This literature has shown how resource-intensive consumption can become normalized across a society, after which attempts to reverse course are difficult.

This line of research evolved toward what is called practice theory (Schatzki, Cetina, and von Savigny 2001), which has become an increasingly influential approach in Europe. Practice theory conceptualizes action as deriving less from a causal relationship between an individual's beliefs, desires, and conscious behavioral choices and more as being rooted in the logic of practices. Emphasis is placed on how practitioners are recruited into practices, forms of knowhow and competence that circulate regarding performance, and the institutional and material forms that particular practices take. Practices involve a combination of material objects, practical knowhow, and socially sanctioned objectives, often in the context of socio-technical systems, social and economic institutions, and modes of spatial and temporal organization (Shove, Pantzar, and Watson 2012). Consumption occurs because people perform practices, and do so for multiple reasons. To understand why practices take their current resource-intensive form requires studying dynamics between the reproduction of existing practices and the introduction of new ones (McMeekin and Southerton 2012).

Warde (2005) identifies key categories for analyzing practices. The first is social differentiation, as discussed earlier in the section on status and class. The second is that practices multiply and diversify, creating an upward ratchet effect in the use of goods and services. The growth of cultural omnivorousness (Peterson and Kern 1996), in which people engage in an ever-expanding repertoire of cultural practices, is another reason for the upward trend of consumption. Socio-technical arrangements (discussed later) also affect practices.

Practice-based approaches emphasize processes of reproduction and change and consumption is understood as the outcome of unfolding, often path-dependent, practices. Individuals matter only in that they represent the point of intersection for multiple practices and that they perform practices. However, even those intersections (the range of practices that individuals perform and the ways in which they do so) should not be understood as a function of individual discretion, but as a result of the dynamics of social relations and the multiplication and diversification of practices. Unlike economic and psychological approaches that seek to explain the discretionary decisions of millions of autonomous consumers, a practice-based approach offers a distinct set of conceptual tools for understanding how consumption is reproduced and transformed.

Practice theory has also been useful for studying reductions in the resource intensity of consumption. A study of inhabitants in Scandinavian eco-villages showed that they quickly became attuned to the norms and rules of appropriate environmental behavior and picked up new practices once they joined the community (Georg 1999). Social learning from engagement in a different "social world" is a key component of the expansion of sustainable practices. Southerton, McMeekin, and Evans (2011) reveal how behavior-change initiatives that address practices have had much greater success than those that attempt to change behavior through information or prices alone. For example, the CoolBiz program in Japan focused on changing the clothing practices of those who worked in government buildings. The lowest temperature at which air-conditioning units could be used was set to 28 degrees Celsius and a new dress code was introduced. An estimated 1.14 million ton reduction of CO_2 emissions (equivalent to the annual emissions associated with the average energy use of 85,000 U.S. homes) was achieved—although necktie sales declined by 36 percent.

Practice approaches also present the opportunity to think in broader terms about behavior and the sustainability of everyday life. Southerton, Mendez, and Warde (2012) suggest that the resource intensity of eating practices could be affected by policies that influence the timing of eating events. Compared to Spain, eating patterns in the United Kingdom are far more individualized—at no time of the day were more than 20 percent of the U.K. population eating, compared to two peaks in Spain where 40 percent of the population were eating. Individualized patterns can even out peak loads in energy consumption, but collective timings present opportunities for resource efficiencies. Radical transitions toward sustainable consumption require the reorganization of everyday practices.

Time Use, Practices, and Carbon Emissions

The time budgets of households and the temporal requirements of sustainable practices are important aspects of social responses to climate change. Practices create their own temporal demands, whether related to the frequency of performance, the need for synchronization with other practices, sequencing (e.g., washing clothes before drying them), particular durations, or coordination with the practices of others (Southerton 2006). Rush hours and peak loads in energy consumption are driven by the temporal organization of practices. Domestic appliances such as the freezer and microwave allow for the resequencing of food practices by extending the time between the acquisition and consumption of food through reheating food prepared months in advance or eating meals prepared in a factory long distances away (Hand and Shove 2007).

In societies with established patterns of time use, peak load demands on energy systems are often more pronounced, requiring more built capacity for electricity production. Peak demand is typically highest from 3 to 8 p.m., when households increase their use of appliances. Temporal patterns associated with commuting to school and work create rush hour traffic patterns that increase transportation-related energy demand as a result of traffic congestion, engine idling, and longer hours of vehicle operation. In 2011, congestion caused urban Americans to use an extra 2.9 billion gallons of fuel for a total congestion cost of $121 billion (Schrank, Eisele, and Lomax 2012). However, the weakening of common temporal rhythms also raises energy demand. In the United Kingdom, between 1937 and 2000, common temporal rhythms (e.g., Sunday dinners and Monday washdays) weakened due to new household technologies and greater flexibility of working, shopping, and eating times (Southerton 2009), which in turn led to time-deepening strategies such as doing practices faster, simultaneously, and with more temporal planning (Robinson and Godbey 1996). Time scarcity, whether caused by the rising value of time (Linder 1970), psychological factors (Robinson and Godbey 1996; Southerton 2003), or rising working hours (Schor 1992), results in time deepening. Time deepening allows for increased energy use, as people turn to "energy slaves," machines that act as the equivalent of multiple human workers to provide the means to engage in multiple tasks simultaneously (e.g., dry clothing, watch television, and cook dinner).

Time use is also connected to consumption through powerful economic dynamics related to productivity and working hours. Market economies tend to create work-and-spend cycles (Schor 1992), in which there is a structural tendency to translate labor productivity into increases in production and consumption rather than increases in leisure. A small literature on OECD countries finds that higher average hours of work have large positive effects on carbon emissions (Knight, Rosa, and Schor 2013; Schor 2005).

Social Analysis of Technology Adoption and Use

Another body of sociological work studies the adoption of new and/or less resource-intensive technologies. In contrast to practice theory, this literature studies consumers as intentional agents deciding whether to adopt and use technologies. It is often thought that new technologies (e.g., electric cars) will help shift consumption patterns and reduce GHG emissions.

Economic approaches focusing on costs and benefits have dominated the study of technological innovation. However, sociological research reveals that attention to social process provides a more complete understanding of climate-friendly technology adoption (Axsen and Kurani 2012). Two major sociological perspectives on technology adoption have emerged. The social psychological perspective, which dominates in North America, emphasizes beliefs and attitudes toward technologies and the role of social networks and social learning in technological adoption. It also focuses on behavioral aspects of technology adoption (Dietz et al. 2009; Laitner and Ehrhardt-Martinez 2010). The most prominent approach is Roger's Theory of Innovation and Diffusion, or DOI (Rogers 1962, 1995), in which technology adoption is theorized as a five-step process: awareness, persuasion, decision, implementation, and confirmation. DOI, as initially formulated, has been criticized for not adequately accounting for social processes (Axsen et al. 2011). To address these shortcomings, social scientists have studied how social variables influence each of the five diffusion steps. Axsen et al. (2011) considered the influence of social networks on the uptake of technologies, while Vishwanath and Goldhaber (2003) explored the influence of income and race on the social construction of technological attributes and necessity.

In contrast to the DOI approach, a socio-technical systems perspective integrates knowledge of both social and technical systems to explain technological transitions. Socio-technical approaches focus on institutions, techniques, and artifacts, in addition to the rules, practices, and social networks that determine how technologies are developed and used (Smith, Stirling, and Berkhout 2005). This approach analyzes the development of alternative technology niches, such as wind energy and solar installations, and the expansion of the niche through processes of social learning and networked connections with policymakers. Socio-technical regimes try to account for varying levels of human agency and institutions in shaping and using technologies.

Research on the Dutch effort to make their electric system more sustainable has explored the impact of state-led efforts to engage the public and set policies to guide energy development and has provided lessons on the incremental politics of planning and challenges in influencing corporate energy companies (Kemp and Loorbach 2006; Kemp, Loorbach, and Rotmans 2007; Verbong and Geels 2007). Other efforts have focused on automobility and housing transitions (Brown et al. 2003; Brown and Vergragt 2008; Geels et al. 2012). These studies reflect the complexity of technology adoption at the household or individual level and the ways it is embedded in a variety of contexts.

HOUSEHOLD CONSUMPTION BY SECTOR

In the foregoing, we have discussed a number of the major sociological approaches to consumption, emphasizing the unique perspectives of sociologists. In this section we will look in more detail at specific sectors and the empirical literatures with an emphasis on research that addresses reductions in carbon emissions.

Sociological Insights on Household Energy Consumption

In 2011, 21 percent of U.S. territorial CO_2 emissions were attributable to residential energy use (Environmental Protection Agency 2013; Table 2-5). While house size is an important determinant of energy use, sociologists have found that direct energy consumption varies significantly in households with similar technologies and square footage (Lutzenhiser 1993; Lutzenhiser and Hackett 1993; Wilhite et al. 2000). These findings document the importance of moving beyond traditional economic policy that is generally limited to efforts to increase the price of energy (to encourage conservation) or the provision of incentives (to adopt more efficient home technologies). Sociological studies show that cultural and lifestyle practices are equally important factors.

Sociology's first insight concerning energy use is the recognition that the household is a social unit, not a single decision-maker as traditional policy approaches assume. In the United States, 75 percent of households are multiperson ones, and the demographic characteristics of households (size, income, life stage) partially account for differences in energy usage (Hackett and Lutzenhiser 1991; Schipper and Bartlett 1989) and result in discrete patterns of energy consumption. Such patterns are not the result of any one characteristic but a combination of factors. An improved understanding of these patterns and the social variables that shape them offers the opportunity to improve interventions that have historically relied on one-size-fits-all strategies. For example, a study of energy use in northern California revealed that social factors were more important than building characteristics or environmental factors in explaining household energy use (Lutzenhiser and Bender 2008). Households with teenagers (thirteen to seventeen years old) were found to use considerably more energy, while households whose members were predominantly Latino, African American, or Asian used less energy.

The second major insight from sociologists is the importance of collective routines, as noted earlier. In most instances, conceptualizing home energy use as active and purposeful decision making is misleading. Energy is used as households carry out social routines that provide comfort, cleanliness, and convenience and create a "good home" as socially defined across various groups (Gram-Hanssen and Bech-Danielsen 2004; Lutzenhiser 1993; Shove 2003; Wilk and Wilhite 1985). These conventions are generally taken-for-granted assumptions or norms. While ignoring the role of individual choice in home energy use is too extreme, research findings increasingly suggest that much of what households do does not involve active deliberative decision making, as most economic models assume. To be successful, intervention strategies need to address the routinization and embeddedness of energy consumption practices in daily life.

The third sociological insight on home energy use is that households operate in a larger community and social context. While the motivations that drive household consumption may be visible (i.e., Westernized clothing fashions), household energy use itself has often been described as invisible (Ehrhardt-Martinez, Donnelly, and Laitner 2011; Hargreaves and Nye 2010; Shove 1997). It is largely a form of ordinary consumption and is not motivated by status or identity presentation (Gram-Hanssen 2010; Gronow and Warde 2001). With the exception of solar panels, most household energy practices are not socially visible. Energy is most often produced, transmitted, and used in ways that households don't think about. For example, in one Danish study, visible actions such as food consumption were more prevalent than reductions in home energy consumption (Pedersen 2000). Research has found some exceptions to this pattern when people are motivated by status considerations, such as the purchase of Energy Star appliances that are visible to visitors, or when they are given feedback on their energy consumption (Ehrhardt-Martinez 2010; Ehrhardt-Martinez et al. 2011). In a study of energy use in Manila (Sahakian and Steinberger 2011), status seeking was found to be important in shaping cooling habits, for example in enabling some people to adopt high-status Western fashions that are not comfortable in tropical weather, or to host family gatherings in cooled rooms. The high energy requirements of large homes and powerful cars are additional examples of how status seeking drives high energy consumption (Ehrhardt-Martinez, Laitner, and Reed 2008).

Household Transportation Practices and Patterns

The U.S. transportation sector is fueled almost exclusively by petroleum (96 percent) and accounted for 27 percent of the nation's GHG emissions in 2010. More than half of these emissions (56 percent) are from passenger vehicles and light-duty trucks (including SUVs) (U.S. Department of Energy 2012). An understanding of transportation-related consumption is rooted in vehicle purchasing decisions and the use practices associated with noncommercial vehicles.

Opportunities for GHG abatement in transportation are typically framed as either technical or behavioral solutions. Technical solutions include emitting less CO_2 per mile driven (as with hybrid vehicles) and switching to vehicles powered by potentially lower-carbon fuels such as electricity, hydrogen, ethanol, or natural gas. Behavioral solutions generally involve a reduction in vehicle miles traveled through trip consolidation, increased uptake of public transit, cycling, walking, carpooling, or new consumer norms such as eco-driving, which improves fuel efficiency (Barth and Boriboonsomsin 2009). Sociological perspectives recognize two key axes of benefits and motives—vehicles' uses and social meanings—and the divergence between private and social costs and benefits. The achievement of deep GHG abatement targets will require both behavioral and technical solutions (Williams et al. 2012).

Research on reducing transportation emissions has been dominated by engineering and economic perspectives, focusing on individual decisions motivated by private, functional benefits such as improved vehicle performance and cost minimization. Studies of market potential for low-carbon vehicles have used the rational-actor model of consumer behavior, quantifying consumer preferences regarding vehicle cost, fuel savings, and other functional attributes such as vehicle range and horsepower (Bunch et al. 1993; Hidrue et al. 2011; Potoglou and Kanaroglou 2007). These studies highlight the barriers of limited technical performance and high costs. Several rational-actor–based studies have also included social parameters such as word-of-mouth effects (Struben and Sterman 2008), information search channels (van Rijnsoever, Farla, and Dijst 2009), and social network position (Paez, Scott, and Volz 2008). However, low-carbon transportation options have strong social benefits and consumers' attitudes toward them have been shown to influence their decisions. For example, studies have revealed that preferences for environmentally preferable vehicles change as consumers are exposed to supportive peers and car buyers in a "neighbor effect" (Axsen, Mountain, and Jaccard 2009; Mau et al. 2008). Other research explores the mechanisms through which social interactions affect consumers' perceptions. Axsen and Kurani (2012) identify five processes: diffusion, conformity, dissemination, translation, and reflexivity. Diffusion involves information sharing—car buyers with higher education and higher socioeconomic status are more likely to spread information that might induce a social transition toward alternative-fueled transportation technologies (de Haan, Peters, and Mueller 2006; Santini and Vyas 2005). Conformity describes consumers' adaptation to their perceptions of others, such as assessing their own interest on the basis of how many car buyers have already bought an alternative-fuel vehicle (Eppstein et al. 2011; Heutel and Muehlegger 2010).

The social construction of technology approach (Pinch and Bijker 1984) argues that technologies are interpreted and shaped differently across social groups. Kline and Pinch (1996) explore early automobile use among rural Americans and find that initial negative interpretations of automobiles were gradually overcome by positive views that were both functional (providing stationary power assistance for farm tasks) and symbolic (reinforcing of gender roles). Automobile technology also transformed rural populations, increasing connections among rural communities and facilitating new methods of reducing agricultural labor. To achieve a transition to low-carbon mobility, the current dominant vision of automobiles as a luxurious private good (Canzler 1999) will have to be transformed to incorporate pro-social values and benefits while maintaining the functional and symbolic benefits desired by consumers.

In transportation studies, the reflexivity perspective of Giddens has been applied by electric vehicle researchers who reject the stable preference assumption of rational-choice theory (Turrentine, Sperling, and Kurani 1992). Using the concept of lifestyle sectors, simulation experiments (Kurani, Turrentine, and Sperling 1994; Turrentine and Kurani 1998) find that novel, limited-range electric vehicles can be adopted by a range of consumers, a finding that has been supported by studies of actual driver behavior (Cocron et al. 2011; Pierre, Jemelin, and Louvet 2011; Turrentine et al. 2011). However, a study by Axsen and Kurani (2012) found that among participating drivers of plug-in hybrid vehicles, only those who found positive social support were willing to commit to a lifestyle transition and come to value the pro-social attributes of the vehicle technology.

By contrast, social practice theory has a more pessimistic view of the possibilities for low-carbon mobility. From this perspective, the use of passenger vehicles has been argued to have become such an extensive, ingrained practice that vehicle use is seen as a nearly imperceptible habit that almost sustains itself (Rajan 2007). However, in the United States the downward trends in miles traveled, cars per person, and acquisition of driving licenses may be evidence of an unexpected trend away from private automobility.

Food Consumption and Alternative Food Movements

Food is a significant contributor to GHGs (Kramer et al. 1999; Weber and Matthews 2008). Scholars have attempted to quantify food's contribution to climate change throughout the food system, including production, processing, transport, and consumption. The consumption of livestock is particularly carbon intensive. The FAO estimates that livestock accounts for 14.5 percent of GHG emissions (Gerber et al. 2013), with beef production being the major contributor. In the United States, where levels of beef consumption are among the highest in the world, annual per capita consumption is equivalent to driving 1,800 miles per year (Fiala 2009). One kilogram of beef consumption is equivalent to driving 160 miles (Subak 1999). After direct energy use, food contributes more to GHG emissions than any other sector (Fiala 2009).

As in energy and transportation, trends in U.S. meat consumption followed a steady upward trajectory for much of the twentieth century. In 1950, average beef consumption was 50.2 pounds, rising to 91.5 pounds at its 1976 peak (Larsen 2012). While poultry consumption continues to rise, the trend for pork is down and per capita beef consumption has fallen to nearly its 1950 level. However, other trends in food consumption are less positive from the point of view of GHG emissions: increased reliance on globalized food services and highly processed foods (Black and Ehrhardt-Martinez 2008), the globalization of supply chains, rising food calories, and high levels of food waste all have large climate impacts. A key theme of this chapter—the need to employ a sociocultural rather than just an individualist or economistic approach—is perhaps truer in the area of food consumption than any other, because of the special role food plays in constituting culture, identity, and social connections (Douglas 1972). Sociological research on food consumption has often considered environmental dimensions alongside social, health, and other aspects, in part because consumers tend not to think about environmental impacts in isolation.

Sociologists have done extensive research on the "corporate food regime," in which the drive to realize profits has led to energy-intensive farming practices and unhealthy offerings to consumers (McMichael 2000). Sociologists have also been central to a rapidly growing body of research on alternative food movements. These movements are focused largely on local and organic foods that reduce the use of fossil fuels associated with pesticides, fertilizers, and food transportation; improve health outcomes; and yield social benefits such as stronger community ties (Bell 2004). One approach is to educate customers about "where their food comes from" by building first-hand relationships with local producers. Such connected consumers are more likely to choose organically produced, minimally processed, and locally distributed foods, even if they are more expensive, as they "vote with their dollars" in favor of environmental, economic, and social advantages (Shaw 2007). Demand for local and organic food, advocates argue, will increase opportunities for ecologically oriented farmers, eventually outcompeting the corporate food regime (Kloppenburg, Hendrickson, and Stevenson 1996). Efforts in this area have expanded dramatically as the move to consume locally has led to new practices for restaurants (e.g., farm to table), schools (e.g., farm to school), individuals (e.g., farmers' markets, community-supported agriculture), and other institutions. Others in the food movement are focusing on issues such as creating networks of alternative producers, building regional "food sheds" by using environmental and land use legislation, and food justice.

Lifestyles and New Consumer Practices

The slow pace of governmental and business response on climate change has led many consumers to attempt to reduce their own emissions through voluntary actions. In addition to energy efficiency, these include downshifting, new living arrangements such as eco-villages and co-housing, low-carbon mobility options, and the shift to products with lower embodied carbon. In contrast to previous decades, when the environmental conversation was stuck in a narrow debate about "paper" versus "plastic," innovative approaches are now proliferating, and sociologists have been studying them. For example, once "cloth" broke through the unpalatable paper/plastic tradeoff, municipalities began enacting bans on plastic bags, grocery stores offered incentives, and consumers began bringing their bags (Cherrier 2006). Lifestyle and consumption innovations represent a growing cultural trend.

The expansion of consumerism in the post–World War II period has been well documented (Cohen 2003). Homes grew larger and contained more and larger appliances and more goods overall. As a popular literature emerged to criticize overconsumption for both environmental and social effects, the message appeared to resonate with the public, especially before the 2009 recession. In surveys, Americans said they should be reducing the amount they consume (Markowitz and Bowerman 2011). The Harris Poll (2008) found that more than half the U.S. public reported they had done something to change their lifestyle "to make it more environmentally sustainable." While the public may have limited knowledge of the direct and indirect carbon embedded in various expenditures and practices, concern about the links between lifestyle and climate change is considerable. However, actions to combat climate change have been reduced since the recession (Leiserowitz et al. 2012).

There is some evidence of a trend toward downshifting (i.e., working and earning less) to get more time and less stress. In 2004, 48 percent of respondents reported having downshifted in the previous five years (Schor 2010). A small subset of downshifters, termed voluntary simplifiers, have lower measured ecological footprints than a control group of Americans (Kasser and Brown 2003). In a global online survey (Alexander and Ussher 2012) of more than 2,000 self-identified voluntary simplifiers, 46 percent said they reduce energy "at every opportunity." They are also heavily involved in local food movements (35 percent) and at least some home production of food (83 percent).

There is a growing number of qualitative studies of low-impact lifestyles and assessments of programs that help households achieve lower energy and resource use. Connolly and Prothero (2008) found that while their interviewees were committed to environmental values, they became skeptical and uncertain about the increasing number of shopping decisions they have to make to contain their environmental footprints. Stress about everyday shopping decisions is compounded by the "certification revolution," where an increasing number of third-party organizations are certifying products and companies as socially responsible (Conroy 2007). However, other research on sustainable consumption has found that when individuals believe they can make a difference in combating climate change, they are more likely to alter their lifestyles (Goldblatt 2005).

Many consumers have begun to practice "green consumption" by switching to more environmentally friendly products, such as organic foods, health and beauty aids, cleaning products, and clothing; hybrid vehicles; and also lower-impact services, such as nontoxic dry cleaning and eco-tourism. Green consumers are motivated by a number of concerns, including personal health, environmental considerations, and in some cases saving money (Black and Cherrier 2010; Lockie et al. 2002).

There is debate among sociologists about the effectiveness of these actions. Some see them as a necessary but not sufficient component of a shift to low carbon consumption on the grounds that businesses will not change without consumer support. However, Szasz (2007) has made an influential argument that shopping to save the environment undermines motivations to pursue effective collective solutions. In contrast, Willis and Schor (2012) found that in the United States, people who practice green consumption are more, not less, likely to be activists.

Perhaps the biggest barrier to green consumption is the real and perceived costs of eco-friendly products, particularly since the 2009 downturn. Black and Cherrier (2010) found that many environmentally motivated mothers in Canada and Australia did not purchase organic food because of its high price. However, the entrance of Wal-Mart into the organic foods market and increasing price competition by large chains such as Whole Foods Market have reduced the premium for many organic foods.

Another barrier is that consumers may knowingly ignore the environmental implications of products that will conflict with their value systems (Brown 2009; Ehrich and Irwin 2005). Similarly, citizens who benefit financially from the oil industry have been shown to turn a "blind eye" to initiatives designed to reduce carbon emissions (Norgaard 2006). This research suggests that quality, aesthetics, value, interests and taste often take precedence over moral attributes.

New low-impact consumer practices are emerging, based on sharing, collaboration and alternatives to the dominant consumer culture, even when it has a green hue (Schor and Thompson 2014). The "sharing economy" is largely peer to peer and includes car sharing, ride sharing, food swapping, clothing swaps, service bartering, and other practices. These initiatives have been enabled by Web 2.0 software, which is time saving and includes ratings and reputational information that addresses problems of trust that arise when strangers enter into economic relationships. Sharing is believed to reduce climate impacts because it reduces the demand for energy and durable goods (e.g., sharing cars, rides or tools); reduces the need for buildings (e.g., peer-to-peer lodging sites reduce hotel demand); or recirculates goods and materials (e.g., reuse of construction material; clothing, toy, book, and other goods swapping or resale). However, few studies of actual impact have been conducted, and second-round impacts are undoubtedly present (e.g., lodging exchanges also likely increase the volume of travel). There is also a trend toward eco-villages and co-housing, in which participants reduce the square footage of private space, share common areas, and commit to ecological lifestyles. In a global study of fourteen eco-villages, Litfin (2014) finds that they reduce ecological footprints 10 to 50 percent below national averages. While these new practices are by no means mainstream, some are growing rapidly. Their most enthusiastic adopters are often young people with high levels of education, a group that typically leads trends in consumer culture.

CONCLUSIONS

Although policymakers are often reluctant to address it, there is an indisputable relationship between consumption and climate change. From a global perspective, Americans are outsized consumers, although trends show increasing levels of consumption in most countries around the globe. These consumption trends and patterns have both a direct and an indirect impact on climate emissions. From energy to electronics to food, the size of our individual and collective climate footprint is a function of both the quantity and the characteristics of the goods and services that we consume. For example, the climate impact of electricity consumption is determined in part by the amount of electricity that households use but is also shaped by the manner in which the electricity is generated (hydroelectric, nuclear, natural gas, and/or coal) and where it is generated (distributed generation can minimize transmission losses). Energy-related consumption decisions associated with household energy demand and personal transportation alone could lower total U.S. energy consumption by between 9 and 11 percent (Gardner and Stern 2008; Laitner and Ehrhardt-Martinez 2010). Changes in food and goods and services consumption offer another significant area for emissions reductions.

While there is no doubt that consumption matters, understanding the factors that drive and shape consumption patterns has proven difficult. Sociologists have made important contributions in expanding theoretical perspectives on consumption as well as informing the efforts of practitioners who are attempting to mitigate climate impacts. Most efforts to understand consumption are rooted in rational-actor models focused on individual decision making and the relationship between attitudes and practices. However, sociologists have added to and challenged these models through the development of theories that take into account social context, status, identity, and lifestyles. Sociological insights recognize that choices are often shaped by assessments of the messages that they convey to others concerning an individual's social status or identity and that social context is important in defining the messages as well as shaping normative practices and decision-making processes. Beginning in the 1990s, some sociologists of sustainable consumption moved away from a focus on individual choices to study routine and habitual practices and their role in the growth of energy consumption.

Sociological insights have been integrated into climate mitigation strategies in different ways. Electric utilities and policymakers concerned with energy consumption still typically employ older theoretical models to understand and address consumption, particularly models that assume rational actors. These types of approaches are focused on using information and economic incentives to get consumers to make "greener" choices. However, some innovative initiatives, such as those that use social influence to reduce household energy use, have made important breakthroughs in the incorporation of social variables including social norms and social context. In addition, sociological research on the relationship among lifestyles, time use, and consumption has helped inform a range of efforts to shift cultural trends and patterns in favor of lifestyles that embrace local and organic food, reduce levels of income and consumption, and result in fewer work hours (downshifting).

The many studies discussed in this chapter present compelling evidence of the relationship between consumption and climate change and also offer insights into the nature of that relationship. They also suggest that households can be an important part of efforts to reduce climate-related emissions. Nevertheless, important questions remain. Future research on this topic should work to reveal the influence of social contexts and agency in decision making and everyday practices; to explore the effects of social inequality and cultural capital on consumption, waste, and efficiency; to assess the malleability of energy- and carbon-intensive practices and routines such as driving alone; to unravel the complexities of household interactions and their impact on energy consumption; and to help formulate policies and programs that can help guide people and organizations toward more sustainable practices and consumption patterns.

NOTE

1. Throughout this chapter we will use the terms "individual" and "household." While energy studies are largely done at the household level, studies of food consumption and transport and other types of consumption are often at the individual level. Where we are not referring to specific research results we may use the terms interchangeably, or just use one. We may also use the term "consumers." In the section on household consumption by sector, we discuss differences between households in energy consumption.

REFERENCES

Alexander, Samuel and Simon Ussher. 2012. "The Voluntary Simplicity Movement: A Multi-National Survey Analysis in Theoretical Context." *Journal of Consumer Culture* 12(1):66–86.

Axsen, Jonn and Kenneth S. Kurani. 2012. "Interpersonal Influence within Car Buyers' Social Networks: Applying Five Perspectives to Plug-in Hybrid Vehicle Drivers." *Environment and Planning A* 44(5):1047–1065.

Axsen, Jonn, Kenneth S. Kurani, Ryan McCarthy, and Christopher Yang. 2011. "Plug-in Hybrid Vehicle GHG Impacts in California: Integrating Consumer-informed Recharge Profiles with an Electricity-dispatch Model." *Energy Policy* 39(3):1617–1629.

Axsen, Jonn, Dean C. Mountain, and Mark Jaccard. 2009. "Combining Stated and Revealed Choice Research to Simulate the Neighbor Effect: The Case of Hybrid-electric Vehicles." *Resource and Energy Economics* 31:221–238.

Barth, Matthew and Kanok Boriboonsomsin. 2009. "Energy and Emissions Impacts of a Freeway-based Dynamic Eco-driving System." *Transportation Research Part D* 14:400–410.

Becker, Gary S. 1978. *The Economic Approach to Human Behavior*. Chicago: University of Chicago Press.

Bell, Michael. 2004. *Farming For Us All: Practical Agriculture and the Cultivation of Sustainability*. State College, PA: Penn State University Press.

Bell, Michael M. 2012. *An Invitation to Environmental Sociology*. Los Angeles: Sage.

Bennett, Graceann and Freya Williams. 2011. "Mainstream Green." *The Red Papers* (4). New York: Ogilvy and Mather.

Black, Iain R. and Helene Cherrier. 2010. "Anti-consumption as Part of Living a Sustainable Lifestyle: Daily Practices, Contextual Motivations and Subjective Values." *Journal of Consumer Behavior* 9:437–453.

Black, Sarah and Karen Ehrhardt-Martinez. 2008. "A High Energy Diet: The Impact of Lifestyle Changes on Energy Consumption in Food Sales and Food Service." Proceedings of the 2008 ACEEE Buildings Summer Study. Washington, DC: ACEEE.

Bourdieu, Pierre. 1984 [1979]. *Distinction: A Social Critique of the Judgement of Taste.* Translated by Richard Nice. Cambridge: Harvard University Press.

Brown, H. S. and P. J. Vergragt. 2008. "Bounded Socio-technical Experiments as Agents of Systemic Change: The Case of a Zero-energy Residential Building." *Technological Forecasting and Social Change* 75:107–130.

Brown, H. S., P. Vergragt, K. Green, and L. Berchicci. 2003. "Learning for Sustainability Transition Through Bounded Socio-technical Experiments in Personal Mobility." *Technology Analysis & Strategic Management* 15:291–315.

Brown, Keith. 2009. "The Social Dynamics and Durability of Moral Boundaries." *Sociological Forum* 24:4.

Brulle, Robert J. and Lindsay E. Young. 2007. "Advertising, Individual Consumption Levels, and the Natural Environment, 1900–2000." *Sociological Inquiry* 77(4):522–542.

Bunch, David S., Mark Bradley, Thomas F. Golob, Ryuichi Kitamura, and Gareth P. Occhiuzzo. 1993. "Demand for Clean-fuel Vehicles in California: A Discrete-choice Stated Preference Pilot Project." *Transportation Research Part A: Policy & Practice* 27:237–253.

Canzler, W. 1999. "Changing Speed? From the Private Car to CashCar Sharing." Pp. 23–31 in *Speed: A Workshop on Space, Time and Mobility*, edited by J. Beckman. Copenhagen: The Danish Transport Council.

Carlsson-Kanyama, Annika and Alejandro Gonzalez. 2009. "Potential Contribution of Food Consumption Patterns to Climate Change." *American Journal of Clinical Nutrition* 89 (suppl.):1704S–1709S.

Cherrier, Helene. 2006. "Ethical Consumption Practices: Co-Production of Self-expression and Social Recognition." *Journal of Consumer Behaviour* 6 (Sept–Oct):321–335.

Christakis, Nicholas A. and James H. Fowler. 2009. *Connected: The Surprising Power of our Social Networks and How They Shape our Lives.* Chapters 1, 6, 7. New York: Little, Brown and Company.

Cocron, P., F. Buhler, I. Neumann, T. Franke, J. Krems, M. Schwalm, and A. Keinath. 2011. "Methods of Evaluating Electric Vehicles From a User's Perspective—The MINI E Field Trial in Berlin." *IET Intelligent Transport Systems* 5:127–133.

Cohen, Lizabeth. 2003. *A Consumer's Republic: The Politics of Mass Consumption in Postwar America.* New York: Knopf.

Connolly, John and Andrea Prothero. 2008. "Green Consumption." *Journal of Consumer Culture* 8(1):117–145.

Conroy, Michael. 2007. *Branded! How the 'Certification Revolution' is Transforming Global Corporations.* Gabriola Island, BC, Canada: New Society Publishers.

de Haan, Peter, Anja Peters, and Michel Mueller. 2006. "Comparison of Buyers of Hybrid and Conventional Internal Combustion Engine Automobiles."

Transportation Research Record: Journal of the Transportation Research Board 1983:106–113.

Dietz, Thomas, Gerald T. Gardner, Jonathan Gilligan, Paul C. Stern, and Michael P. Vandenbergh. 2009. "Household Actions Can Provide a Behavioral Wedge to Rapidly Reduce US Carbon Emissions." *Proceedings of the National Academy of Sciences USA* 106(44):18452–18456.

Douglas, Mary. 1972. "Deciphering a Meal." *Daedalus* 101(1):61–81.

Dwyer, Rachel E. 2007. "Expanding Homes and Increasing Inequalities: U.S. Housing Development and the Residential Segregation of the Affluent." *Social Problems* 54(1):23–46.

Ehrhardt-Martinez, Karen. 2010. "Changing Habits, Lifestyles and Choices: The Behaviours that Drive Feedback-Induced Energy Savings." Prepared for the *ECEEE Summer Study on Energy Efficiency in Buildings*. European Council for an Energy-Efficient Economy.

Ehrhardt-Martinez, Karen, Kat Donnelly, and John A. "Skip" Laitner. 2011. "Beyond the Meter: Enabling Better Home Energy Management." Pp. 273–303 in *Energy, Sustainability and the Environment: Technology, Incentives, Behavior*, edited by F. P. Sioshans. Burlington, MA: Elsevier.

Ehrhardt-Martinez, Karen and John A. "Skip" Laitner. 2009. "Breaking out of the Economic Box: Energy Efficiency, Social Rationality, and Non-Economic Drivers of Behavioral Change." Paper prepared for the 2009 ECEEE Summer Study. Stockholm, Sweden: European Council for an Energy-Efficient Economy.

Ehrhardt-Martinez, Karen, John A. "Skip" Laitner, and Wendy Reed. 2008. "Dollars or Sense: Economic versus Social Rationality in Residential Energy Consumption." *Proceedings of the 2008 ACEEE Buildings Summer Study*. Washington, DC: ACEEE.

Ehrich, Kristine and Julie Irwin. 2005. "Willful Ignorance in the Request of Product Attribute Information." *Journal of Marketing Research* XLII:266–277.

Ehrlich, Paul and John P. Holdren. 1971. "Impact of Population Growth." *Science* 171:1212–1217.

Environmental Protection Agency. 2013. "Inventory of U.S. Greenhouse Gas Emissions and Sinks: 1990–2011." Retrieved January 7, 2014 (http://www.epa.gov/climatechange/ghgemissions/usinventoryreport.html).

Eppstein, Margaret J., David K. Grover, Jeffrey S. Marshall, and Donna M. Rizzo. 2011. "An Agent-based Model to Study Market Penetration of Plug-in Hybrid Electric Vehicles." *Energy Policy* 39:3789–3802.

Evans, David and Wokje Abrahamse. 2009. "Beyond Rhetoric: The Possibilities of and for 'Sustainable Lifestyles.'" *Environmental Politics* 18(4):486–502.

Fiala, Nathan. 2009. "The Greenhouse Hamburger." *Scientific American* 300:72–79.

Gardner, Gerald T. and Paul C. Stern. 2008. "The Short List: The Most Effective Actions US Households can take to Curb Climate Change." *Environment Magazine* [online].

Geels, F., R. Kemp, G. Dudley, and G. Lyons. 2012. *Automobility in Transition? A Socio-technical Analysis of Sustainable Transport*. London: Routledge.

Georg, Susse. 1999. "The Social Shaping of Household Consumption." *Ecological Economics* 28:455–466.

Gerber, P. J., H. Steinfeld, B. Henderson, A. Mottet, C. Opio, J. Dijkman, A. Falcucci, and G. Tempio. 2013. *Tackling Climate Change Through Livestock—A Global Assessment of Emissions and Mitigation Opportunities*. Rome: Food and Agriculture

Organization of the United Nations (FAO) (http://www.fao.org/docrep/018/i3437e/i3437e.pdf).

Giddens, Anthony. 1991. *Modernity and Self-Identity: Self and Society in the Late Modern Age*. Stanford, CA: Stanford University Press.

Gifford, Robert. 2011. "The Dragons of Inaction: Psychological Barriers that Limit Climate Change Mitigation and Adaptation." *American Psychologist* 66(4):290–302.

Global Footprint Network. 2013. "The National Footprint Accounts, 2012." Oakland, CA: Global Footprint Network (http://www.footprintnetwork.org/images/article_uploads/National_Footprint_Accounts_2012_Edition_Report.pdf).

Goldblatt, David. 2005. *Sustainable Energy Consumption and Society: Personal, Technological, or Social Change?* Norwell, MA: Springer.

Goldstein, Noah, Robert Cialdini, and Vladas Griskevicius. 2008. "A Room with a Viewpoint: Using Social Norms to Motivate Environmental Conservation in Hotels." *Journal of Consumer Research* 35:472–482.

Gould, Kenneth A., David N. Pellow, and Allan Schnaiberg. 2008. *The Treadmill of Production: Injustice and Unsustainability in the Global Economy*. Boulder, CO: Paradigm Publishers.

Gram-Hanssen, Kirsten. 2010. "Standby Consumption in Households Analyzed with a Practice Theory Approach." *Journal of Industrial Ecology* 14(1):150–165.

Gram-Hanssen, Kirsten and Claus Bech-Danielsen. 2004. "House, Home and Identity From a Consumption Perspective." *Housing, Theory and Society* 21(1):17–26.

Gronow, Jukka and Alan Warde, eds. 2001. *Ordinary Consumption*. New York: Routledge.

Hackett, Bruce and Loren Lutzenhiser. 1991. "Social Structures and Economic Conduct: Interpreting Variations in Household Energy Consumption." *Sociological Forum* 6(3):449–470.

Hand, Martin and Elizabeth Shove. 2007. "Condensing Practices: Ways of Living with a Freezer." *Journal of Consumer Culture* 7(1):79–104.

Hand, Martin, Elizabeth Shove, and Dale Southerton. 2005. "Explaining Showering: A Discussion of the Material, Conventional, and Temporal Dimensions of Practice." *Sociological Research Online* 10(2) (http://www.socresonline.org.uk/10/2/hand.html).

Hargreaves, Tom and Michael Nye. 2010. "Making Energy Visible: A Qualitative Field Study of How Householders Interact with Feedback From Smart Energy Monitors." *Energy Policy* 38(10):6111–6119.

Harris Poll. 2008. "SCA Survey Conducted by Harris Interactive Shows that Despite a Weakened Economy, U.S. Consumers Willing to Spend Green to go Green." *Harris Interactive*, April 21. Retrieved February 10, 2012 (http://www.harrisinteractive.com/vault/Client_News_SCA_2008_04.pdf).

Heutel, Garth and Erich J. Muehlegger. 2010. "Consumer Learning and Hybrid Vehicle Adoption Rates." Harvard Kennedy School Faculty Research Working Paper Series RWP10-013, April 2010.

Hidrue, Michael K., George R. Parsons, Willett Kempton, and Meryl P. Gardner. 2011. "Willingness to Pay for Electric Vehicles and Their Attributes." *Resource and Energy Economics* 33:686–705.

Hobson, K. 2002. "Competing Discourses of Sustainable Consumption: Does the 'Rationalization of Life' Make Sense?" *Environmental Politics* 11(2):95–120.

Hopper, Joseph R. and Joyce M. Nielsen. 1991. "Recycling as Altruistic Behavior. Normative and Behavioral Strategies to Expand Participation in a Community Recycling Program." *Environment and Behavior* 23:195–220.

International Panel on Climate Change. 2007. "Summary for Policymakers." *Climate Change 2007: The Physical Science Basis. Contribution of Working Group I to the Fourth Assessment Report of the Intergovernmental Panel on Climate Change.* Cambridge: Cambridge University Press.

Jones, Christopher M. and Daniel M. Kammen. 2011. "Quantifying Carbon Footprint Reduction Opportunities for U.S. Households and Communities." *Environmental Science and Technology* 45(9):4088–4095.

Kasser, T. and K. W. Brown. 2003. "On Time, Happiness, and Ecological Footprints." Pp. 107–112 in *Take Back your Time: Fighting Overwork and Time Poverty in America*, edited by J. deGraaf. San Francisco: Berrett-Koehler.

Kemp, R. and D. Loorbach. 2006. "Transition Management: A Reflexive Governance Approach." Pp. 103–130 in *Reflexive Governance*, edited by J. P. Voss, R. Kemp, and D. Bauknecht. Cheltenham: Edward Elgar.

Kemp, R., D. Loorbach, and J. Rotmans. 2007. "Transition Management as a Model for Managing Processes of Co-evolution Towards Sustainable Development." *International Journal of Sustainable Development and World Ecology* 14(1):78–91.

Kline, Ronald and Trevor Pinch. 1996. "Users as Agents of Technological Change: The Social Construction of the Automobile in the Rural United States." *Technology and Culture* 37:763–795.

Kloppenburg, Jack, Jr., John Hendrickson, and G. W. Stevenson. 1996. "Coming in to the Foodshed." *Agriculture and Human Values* 13(3):33–42.

Knight, Kyle W., Eugene A. Rosa, and Juliet B. Schor. 2013. "Could Working Less Reduce Pressures on the Environment? A Cross-National Panel Analysis of OECD Countries, 1970–2007." *Global Environmental Change* 23:691–700.

Kollmuss, Anja and Julian Agyeman. 2002. "Mind the Gap: Why do People Act Environmentally and What are the Barriers to Pro-environmental Behavior?" *Environmental Education Research* 8(3):239–260.

Kramer, Klaas Jan, Henri C. Moll, Sanderine Nonhebel, and Harry C. Wilting. 1999. "Greenhouse Gas Emissions Related to Dutch Food Consumption." *Energy Policy* 27:203–216.

Kurani, Kenneth S., Tom Turrentine, and Daniel Sperling. 1994. "Demand for Electric Vehicles in Hybrid Households: An Exploratory Analysis." *Transport Policy* 1:244–256.

Laidley, Thomas M. 2012. "The Influence of Social Class and Cultural Variables on Environmental Behaviors: Municipal-level Evidence from Massachusetts." *Environment and Behavior* 20(10):1–28.

Laidley, Thomas M. 2013. "Climate, Class, and Culture: Political Issues as Cultural Signifiers." *Sociological Review* 6(1):153–171.

Laitner, John A. "Skip," and Karen Ehrhardt-Martinez. 2010. "Examining the Scale of the Behavior Energy-Efficiency Continuum." *People-Centered Initiatives for Increasing Energy Savings*. Washington, DC: American Council for an Energy-Efficient Economy (http://www.aceee.org/people-centered-energy-savings).

Larsen, Janet. 2012. "Peak Meat: U.S. Consumption Falling." Washington, DC: Earth Policy Institute.

Leiserowitz, Anthony, Edward Maibach, Connie Roser-Renouf, Nicholas Smith, and Jay Hmielowski. 2012. "Americans' Actions to Conserve Energy, Reduce Waste, and Limit Global Warming in March 2012." Yale University and George Mason University. New Haven, CT: Yale Project on Climate Change Communication.

Linder, Staffan B. 1970. *The Harried Leisure Class*. New York: Columbia University Press.

Litfin, Karen T. 2014. *Eco-villages: Lessons for Sustainable Community*. Malden, MA: Polity Press.

Lockie, Stewart, Kristen Lyons, Geoffrey Lawrence, and Kerry Mummery. 2002. "Eating 'Green': Motivations Behind Organic Food Consumption in Australia." *Sociologia Ruralis* 42(1):23–40.

Luedicke, Marius K., Craig J. Thompson, and Markus Giesler. 2010. "Consumer Identity Work as Moral Protagonism: How Myth and Ideology Animate a Brand-Mediated Moral Conflict." *Journal of Consumer Research* 36(6):1016–1032.

Lutzenhiser, Loren. 1993. "Social and Behavioral Aspects of Energy Use." *Annual Review of Energy and the Environment* 18(1):247–289.

Lutzenhiser, Loren and Sylvia Bender. 2008. "The 'Average American' Unmasked: Social Structure and Differences in Household Energy Use and Carbon Emissions. In the *Proceedings of the ACEEE Summer Study on Energy Efficiency in Buildings*. Washington, DC: ACEEE.

Lutzenhiser, Loren and Bruce Hackett. 1993. "Social Stratification and Environmental Degradation: Understanding Household CO_2 Production." *Social Problems* 40(1):50–73.

Markowitz, Ezra M. and Tom Bowerman. 2011. "How Much is Enough? Examining the Public's Beliefs About Consumption." *Analysis of Social Issues and Public Policy* 11(1):1–23.

Mau, Paulus, Jimena Eyzaguirre, Mark Jaccard, Colleen Collins-Dodd, and Kenneth Tiedemann. 2008. "The 'Neighbor Effect': Simulating Dynamics in Consumer Preferences for New Vehicle Technologies." *Ecological Economics* 68:504–516.

McKenzie-Mohr, Doug and William Smith. 2001. *Fostering Sustainable Behavior*. Gabriola Island, BC: New Society Publishers.

McMeekin, Andrew and Dale Southerton. 2012. "Sustainability Transitions and Final Consumption: Practices and Socio-technical Systems." *Technology Analysis & Strategic Management* 24(4):345–361.

McMichael, Philip. 2000. *Development and Social Change: A Global Perspective*, 2nd ed. California: Pine Forge Press.

McNary, Bill and Chip Berry. 2012. "How Americans are Using Energy in Homes Today." *Proceedings of the ACEEE Summer Study on Energy Efficiency in Buildings*. Washington, DC: ACEEE.

National Academy of Sciences. 2011. *America's Climate Choices*. Washington, DC: National Academies Press.

National Research Council. 2010. *Limiting the Magnitude of Future Climate Change*. Washington, DC: National Academies Press.

Norgaard, Kari Marie. 2006. " 'We Don't Really Want to Know': Environmental Justice and Socially Organized Denial of Global Warming in Norway." *Organization & Environment* 19(3):347–370.

Nye, Michael and Tom Hargreaves. 2010. "Exploring the Social Dynamics of Pro-environmental Behavior Change: A Comparative Study of Intervention Processes at Home and Work." *Journal of Industrial Ecology* 14(1):137–149.

Nye, Michael, Lorraine Whitmarsh, and Timothy Foxon. 2010. "Sociopsychological Perspectives on the Active Roles of Domestic Actors in Transition to a Lower Carbon Electricity Economy." *Environment and Planning A* 42(3):697–714.

Olli, Erro, Gunnar Grendstad, and Dag Wollebaek. 2001. "Correlates of Environmental Behaviors: Bringing Back Social Context." *Environment and Behavior* 33(2):181–208.

Paez, Antonio, Darren M. Scott, and Erik Volz. 2008. "A Discrete-choice Approach to Modeling Social Influence on Individual Decision making." *Environment and Planning B: Planning and Design* 35:1055–1069.

Peattie, Ken. 2001. "Towards Sustainability: The Third Age of Green Marketing." *Marketing Review* 2(2):129–146.

Pedersen, Lene H. 2000. "The Dynamics of Green Consumption: A Matter of Visibility?" *Journal of Environmental Policy & Planning* 2(3):193–210.

Peterson, Richard and Roger M. Kern. 1996. "Changing Highbrow Taste: From Snob to Omnivore." *American Sociological Review* 61(5):900–907.

Pierre, Magali, Christophe Jemelin, and Nicolas Louvet. 2011. "Driving an Electric Vehicle: A Sociological Analysis on Pioneer Users." *Energy Efficiency* 4:511–522.

Pinch, Trevor J. and Wiebe E. Bijker. 1984. "The Social Construction of Facts and Artifacts—Or How the Sociology of Science and the Sociology of Technology Might Benefit Each Other." *Social Studies of Science* 14:399–441.

Potoglou, Dimitris and Pavlos S. Kanaroglou. 2007. "Household Demand and Willingness to Pay for Clean Vehicles." *Transportation Research Part D—Transport and Environment* 12:264–274.

Rajan, Sudhir C. 2007. "Automobility, Liberalism, and the Ethics of Driving." *Environmental Ethics* 29:77–90.

Robinson, John and Geoffrey Godbey. 1996. *Time for Life: The Surprising Ways that Americans Use Their Time.* State College: Pennsylvania State Press.

Rogers, Everett M. 1962. *Diffusion of Innovations.* New York: The Free Press.

Rogers, Everett M. 1995. *Diffusion of Innovations,* 4th ed. New York: The Free Press.

Sahakian, Marlyne D. and Julia K. Steinberger 2011. "Energy Reduction Through a Deeper Understanding of Household Consumption." *Journal of Industrial Ecology* 15(1):31–48.

Santini, Danilo J. and Anant D. Vyas. 2005. "Introduction of Hybrid and Diesel Vehicles—Status Within the Life Cycle of Technology Adoption." *Transportation Research Record: Journal of the Transportation Research Board* 1941:18–25.

Schatzki Theodore, Karin Knorr Cetina, and Eike von Savigny. 2001. *The Practice Turn in Contemporary Theory.* London: Routledge.

Schipper, Lee and Sarita Bartlett. 1989. "Linking Life-styles and Energy Use: A Matter of Time?" *Annual Review of Energy* 14(1):273–320.

Schor, Juliet. 1992. *The Overworked American: The Unexpected Decline of Leisure.* New York: Basic Books.

Schor, Juliet. 1998. *The Overspent American: Upscaling, Downshifting and the New American Consumer.* New York: Basic Books.

Schor, Juliet. 2005. "Sustainable Consumption and Worktime Reduction." *Journal of Industrial Ecology, Special Issue on Sustainable Consumption* 9(1):37–50.

Schor, Juliet B. 2010. *Plenitude: The New Economics of True Wealth.* New York: Penguin Press.

Schor, Juliet B. and Craig J. Thompson (Eds.). 2014. *Sustainable Lifestyles and the Quest for Plenitude: Case Studies of the New Economy.* New Haven, CT: Yale University Press.

Schrank, David, Bill Eisele, and Tim Lomax. 2012. "2012 Urban Mobility Report." Texas A&M Transportation Institute. College Station: Texas A&M University Press (http://d2dtl5nnlpfr0r.cloudfront.net/tti.tamu.edu/documents/mobility-report-2012.pdf).

Schultz, P. Wesley, Jessica M. Nolan, Robert B. Cialdini, Noah J. Goldstein, and Vladas Griskevicius. 2007. "The Constructive, Destructive, and Reconstructive Power of Social Norms." *Psychological Science* 18:429–434.

Shaw, Deirdre. 2007. "Consumer Voters in Imagined Communities." *International Journal of Sociology and Social Policy* 27:135–150.

Shove, Elizabeth. 1997. "Revealing the Invisible: Sociology, Energy and the Environment." P. 261 in *The International Handbook of Environmental Sociology*, edited by M. Redclift and G. Woodgate. Northampton, MA: Edward Elgar Publishing.

Shove, Elizabeth. 2003. "Converging Conventions of Comfort, Cleanliness and Convenience." *Journal of Consumer Policy* 26(4):395–418.

Shove, Elizabeth, M. Pantzar, and M. Watson. 2012. *The Dynamics of Social Practice*. London: Sage.

Shove, Elizabeth and Alan Warde. 2002. "Inconspicuous Consumption: The Sociology of Consumption, Lifestyles and the Environment." Pp. 230–251 in *Sociological Theory and the Environment: Classical Foundations, Contemporary Insights*, edited by R. Dunlap, F. Buttel, P. Dickens, and A. Gijswijt. Lanham, MD: Rowman & Littlefield Publishers.

Smith, Adrian, Andy Stirling, and Frans Berkhout. 2005. "The Governance of Sustainable Socio-technical Transitions." *Research Policy* 34:1491–1510.

Southerton, Dale. 2003. "'Squeezing Time': Allocating Practices, Coordinating Networks and Scheduling Society." *Time & Society* 12(1):5–25.

Southerton, Dale. 2006. "Analyzing the Temporal Organization of Daily Life: Social Constraints, Practices and Their Allocation." *Sociology* 40(3):435–454.

Southerton, Dale. 2009. "Temporal Rhythms: Comparing Daily Lives of 1937 with Those of 2000 in the UK." Pp. 49–63 in *Time, Consumption and Everyday Life: Practice, Materiality and Culture*, edited by E. Shove, F. Trentmann, and R. W. Berg. Oxford: Oxford University Press.

Southerton, Dale, A. McMeekin, and D. Evans. 2011. "International Review of Behaviour Change Initiatives: Climate Change Behaviors Research Programme." Scottish Government Report. Edinburgh: Scottish Government Social Research.

Southerton, Dale, C. Diaz Mendez, and Alan Warde. 2012. "Behavioral Change and the Temporal Ordering of Eating Practices: A UK–Spain Comparison." *International Journal of Sociology of Agriculture and Food* 19(1):19–36.

Staats, Henk, Paul Harland, and Henk A. M. Wilke. 2004. "Effecting Durable Change. A Team Approach to Improve Environmental Behavior in the Household." *Environment and Behavior* 36:341–367.

Stets, Jan E. and Chris F. Biga. 2007. "Bringing Identity Theory into Environmental Sociology." *Sociological Theory* 21(4):398–423.

Struben, Jeroen and John Sterman. 2008. "Transition Challenges for Alternative Fuel Vehicles and Transportation Systems." *Environment and Planning B: Planning and Design* 35:1070–1097.

Subak, S. 1999. "Global Environmental Costs of Beef Production." *Ecological Economics* 30:79–91.

Swim, Janet, Susan Clayton, Thomas Doherty, Robert Gifford, George Howard, Joseph Reser, Paul Stern, and Elke Weber. 2009. "Psychology and Global Climate Change: Addressing a Multi-faceted Phenomenon and Set of Challenges: A Report by the American Psychological Association's Task Force on the Interface Between Psychology and Global Climate Change." American Psychological Association (http://www.apa.org/science/about/publications/climate-change-booklet.pdf).

Szasz, Andrew. 2007. *Shopping Our Way to Safety: How We Changed from Protecting the Environment to Protecting Ourselves.* Minneapolis: University of Minneapolis Press.

Turrentine, Tom, Dahlia Garas, Andy Lentz, and Justin Woodjack. 2011. "The UC Davis MINI E Consumer Study." Institute of Transportation Studies. Davis: University of California, Davis.

Turrentine, Tom and Ken S. Kurani. 1998. "Adapting Interactive Stated Response Techniques to a Self-completion Survey." *Transportation* 25:207–222.

Turrentine, Tom, Ken S. Kurani, and Rusty Heffner. 2008. "Fuel Economy: What Drives Consumer Choice?" *Access Magazine* 31.

Turrentine, Tom, D. Sperling, and Ken S. Kurani. 1992. "Market Potential for Electric and Natural Gas Vehicles." Institute of Transportation Studies. Davis: University of California, Davis.

U.S. Department of Agriculture. 2003. "Profiling Food Consumption in America." *Agriculture Fact Book 2001–2002,* Chapter 2. Washington, DC: USDA Office of Communications.

U.S. Department of Energy. 2012. *Transportation Energy Data Book, Edition 31.* Prepared for the Vehicles Technology Program of the Office of Energy Efficiency and Renewable Energy, Tables 2.7, 8.1, 8.2, and 8.5. Oak Ridge, TN: Oak Ridge National Laboratory.

U.S. Energy Information Administration. "2012 Annual Energy Review 2011: Table 2.1b." Washington, DC: Department of Energy.

U.S. Public Research Interest Group. 2012. *Transportation and the New Generation.* Santa Barbara, CA: Frontier Group.

van Rijnsoever, Frank, Jacco Farla, and Martin J. Dijst. 2009. "Consumer Car Preferences and Information Search Channels." *Transportation Research Part D: Transport and Environment* 14:334–342.

Veblen, Thorstein. 1899. *The Theory of the Leisure Class.* Mineola, NY: Dover Publications.

Verbong, G. and F. Geels. 2007. "The Ongoing Energy Transition: Lessons From a Socio-technical, Multi-level Analysis of the Dutch Electricity System (1960–2004)." *Energy Policy* 35:1025–1037.

Vishwanath, A. and G. M. Goldhaber. 2003. "An Examination of the Factors Contributing to Adoption Decisions Among Late-diffused Technology Products." *New Media & Society* 5:547.

Wackernagel, Mathis and William E. Rees. 1995. *Our Ecological Footprint: Reducing Human Impact on the Earth.* Gabriola Island, Canada: New Society Publishers.

Warde, Alan. 1997. *Consumption, Food and Taste.* Thousand Oaks, CA: Sage Publications.

Warde, Alan. 2005. "Consumption and Theories of Practice." *Journal of Consumer Culture* 5(2):131–154.

Warde, Alan and Dale Southerton. 2012. "Social Sciences and Sustainability." In *The Habits of Consumption,* edited by A. Warde and D. Southerton. Helsinki: Open Access Book Series of the Helsinki Collegium of Advanced Studies.

Weber, Christopher and H. Scott Matthews. 2007. "Embodied Environmental Emissions in U.S. International Trade, 1997–2004." *Environmental Science and Technology* 41(14):4875–4881.

Weber, Christopher and H. Scott Matthews. 2008. "Food Miles and the Relative Climate Impacts of Food Choices in the United States." *Environmental Science and Technology* 42(10):3508–3513.

Weenig, Meineke W. and Cees J. Midden. 1991. "Communication Network Influences on Information Diffusion and Persuasion." *Journal of Personality and Social Psychology* 61:734–742.

Wilhite, Harold, Elizabeth Shove, Loren Lutzenhiser, and Willett Kempton. 2000. "The Legacy of Twenty Years of Energy Demand Management: We Know More About Individual Behaviour but Next to Nothing about Demand." Pp. 109–126 in *Society, Behavior, and Climate Change Mitigation,* edited by E. Jochem, J. Sathaye, and D. Bouille. Boston, MA: Kluwer Academic Publishers.

Wilk, Richard. 1997. "A Critique of Desire: Distaste and Dislike in Consumer Behavior." *Consumption Markets and Culture* 1(2):175–196.

Wilk, Richard and Harold Wilhite. 1985. "Why Don't People Weatherize Their Homes? An Ethnographic Solution." *Energy* 10(5):621–629.

Williams, James, Andrew DeBenedictis, Rebecca Ghanadan, Amber Mahone, Jack Moore, William Morrow, Snuller Price, and Margaret Torn. 2012. "The Technology Path to Deep Greenhouse Gas Emissions Cuts by 2050: The Pivotal Role of Electricity." *Science* 335:53–59.

Willis, Margaret and Juliet B. Schor. 2012. "Does Changing a Light Bulb Lead to Changing the World? Civic Engagement and the Ecologically Conscious Consumer." *Annals of the American Academy of Political and Social Science* 644(1):160–190.

5

Climate Justice and Inequality

Sharon L. Harlan, David N. Pellow,
and J. Timmons Roberts with Shannon Elizabeth Bell,
William G. Holt, and Joane Nagel

INTRODUCTION

Climate change is a justice issue for three reasons. First, there are *causes* of climate change: social inequalities drive overconsumption, a key source of unsustainable levels of greenhouse gas (GHG) emissions. Second, the *impacts* of climate change are unequally felt by the rich and poor, and disparate impacts will continue to increase in future generations. Third, *policies* designed to manage climate change have starkly unequal consequences, and the processes by which emissions reductions and climate adaptation policies are decided tend to exclude the poor and the powerless.

Climate injustice is caused by inequalities: the most politically, culturally, and economically marginalized communities and nations use vastly less fossil fuel–based energy and bear far less responsibility for creating environmental problems than do wealthier nations and people, who use far more than is needed for a decent quality of life. Overconsumption beyond ecologically sustainable levels is driven in large part by humans' desire for social status, as conspicuous consumption and leisure secure our position in society (Bell 2013; Veblen 2007). Nearly 75 percent of the world's annual carbon dioxide (CO_2) emissions come from the Global North, which makes up only 15 percent of the global population (Holdren 2007). One extremely wealthy person may emit as much carbon as 70,000 poor individuals in the world's poorest countries (Roberts and Parks 2007). The governments of wealthy nations have largely failed to address this because they perceive that cutting back on fossil fuels could reduce or eliminate the economic growth that has produced their prosperity. Yet if historic responsibility is taken into account, Global North nations have consumed more than three times their share of the atmosphere (in terms of the amount of emissions that we can safely put into the atmosphere) while the poorest 10 percent of the world's population has contributed less than 1 percent of CO_2 emissions (Roberts and Parks 2007; but see PBL Netherlands Environmental Assessment Agency 2013).

Climate injustice also refers to the fact that climate change or climate disruptions affect nations and people very differently, compounded by cascading effects of globalization that conspire to place the most vulnerable people at a cumulative disadvantage (O'Brien and Leichenko 2000). The Global South and people of color, Indigenous communities, the poor, and women and children in all nations are precisely the populations that bear the brunt of climate disruption in terms of its ecological, economic, and health burdens. Many governments of the Global South feel strongly that they have paid a heavier price for climate change, or are likely to pay as impacts grow worse, while receiving very few of the benefits (World Bank 2006). Most basically, the world's most marginalized people are suffering worst and first from the rise in climate-related disasters (Kasperson and Kasperson 2001; Roberts and Parks 2007). As Kasperson and Kasperson put it so powerfully over a dozen years ago, "Recognizing and understanding this differential vulnerability is a key to understanding the meaning of climate change" (2001:2).

Therefore, to understand climate disruption and how to address the potentially devastating problems of climate change requires understanding inequalities in human wealth, power, and privilege. Sociologists are unique among scientists in our relentless focus on inequality. Social inequality is the core of sociology, a conceptual bridge between widely divergent researchers who study forms of social behavior at the micro (individuals and families) and macro (communities, nations, and world economic systems) levels, using a range of methodologies. Many other disciplines are less attentive to inequality as a subject of scientific inquiry because they assume human differences are an inevitable result of biological inheritance or individual abilities and preferences, rather than structural causes, such as those deriving from economic change, inheritance of social class, or institutional policies. Sociologists generally dispute those scholars, policy practitioners, and even environmental activists, who assume that inequalities are receding, or will be addressed by economic growth or sustainability campaigns (Agyeman 2005).

This chapter highlights four kinds of significant contributions that sociologists are making to advance our understanding of climate injustices and how we might lessen their effects:

1. Theorizing the origins and perpetuation of social and environmental inequalities at global to local scales of human organization;
2. Understanding the unjust impacts of climate change on vulnerable populations;
3. Critically evaluating the potential of climate mitigation and adaptation policies for achieving climate justice;
4. Studying the ideas and strategies of grassroots social movements and nongovernmental organizations (NGOs) advocating climate justice.

In the next sections we review the work of some pioneering sociologists and related social scientists who are pointing out the value of bringing an understanding of human institutions, inequality, and justice into the largely natural science and technical solutions–based discussion on climate change. First we briefly review some classical theories of inequality, which have been expanded and enriched by generations of sociologists. These theories explain the social divisions and power imbalances that cause climate injustice. In the next section we review principles of social justice and pose two questions—one factual and one normative—that frame climate justice issues. The following section illustrates human vulnerabilities to the impacts of climate change with examples of unequal burdens borne by communities in "sacrifice zones" of energy extraction and among people in places affected by extreme weather. In the next section we assess how different stakeholders (e.g., international governing bodies, nations, and social movements) position themselves on climate change mitigation policies (e.g., global strategies to limit emissions through international cap-and-trade systems) and sociologists' engagement with grassroots organizations that propose more holistic adaptation policies focusing on the importance of place and the interdependence between sustainable communities and local ecosystems. We conclude by proposing ways to advance sociological research and contributions to climate change policy, and call for interdisciplinary cooperation on efforts to engage localities, nations, and the United Nations system as they move to address climate change.

THEORIES OF INEQUALITY: FOUNDATION FOR UNDERSTANDING CLIMATE INJUSTICES

Class, Gender, and Race Inequalities

Inseparable from any understanding of justice—and climate justice—is the problem of social inequality. Theorizing inequality, which dates back to the origins of our discipline, is one of sociology's major contributions to the science of climate change. What exactly is inequality and why does it matter? Inequality is a means of ordering the human and nonhuman worlds for the relative benefit of some, and to the detriment of others.

Some of the most important sociological research on inequality dates back to the late nineteenth and early twentieth centuries when scholars like Karl Marx, W. E. B. Du Bois, Jane Addams, Max Weber, and Robert Park produced groundbreaking work that would stand for generations. In his wide-ranging published works, Marx grappled with the role of the state and capital in maintaining and exacerbating social inequalities. For Marx, social inequalities are the result of the exertion of elite class dominance over the working classes, and therefore successful efforts to secure a better life for society's majority must involve deliberate confrontations with established institutions (Marx 2000). Sociologists Gould, Schnaiberg, and Weinberg (1996) and Leslie Sklair (2001) adopt this approach in their work on environmental damage, suggesting that citizen-workers must confront capitalism's core institutions to reform the system.

In the very first volume of the *American Journal of Sociology*, Jane Addams published a paper critical of the feudal-like character of domestic work and how it maintained a system of gender inequality and social isolation that disadvantaged women in particular (Addams 1896). Three years later (1899) Addams published an article in the same journal on the importance of trade unions and other social movements as critical bulwarks against the scourge of child labor and rising class inequalities (Adams 1899). That same year, W. E. B. Du Bois published *The Philadelphia Negro*, which was perhaps the first major research study of an urban African-American community (Du Bois 1899). In this seminal work Du Bois challenged numerous stereotypes and racist assumptions about African Americans while underscoring the importance of grasping the socially constructed character of race relations in the United States. This was also one of the first major sociological studies to incorporate multiple research methods, including interviews, observations, and census data.

In the three decades that followed, Robert Park and his colleagues at the University of Chicago articulated an early model of human ecology—a perspective that viewed urban centers as sites of human activity that mirrored much of what was believed to characterize nonhuman nature and ecosystems: competition for scarce resources. In the city, people competed for jobs, land, housing, and so forth, and this reflected and produced various class, racial, and spatial inequalities (Park 1915). Max Weber, a contemporary of both Du Bois and Park, explored the various consequences and implications of the most important organizational form of the modern era on inequality: bureaucracy (Weber [1922] 1978). Weber found that although bureaucracies are perhaps the most effective ways of addressing the administrative requirements of large-scale social systems in modern societies, they also require strict hierarchies and frequently dehumanize people and threaten the fundamental core of a democratic society.

The perils of bureaucracy are exacerbated by the growth and strength of what C. W. Mills (1956) called the "power elite"—that powerful minority of people who occupy positions at the top of military, political, and economic institutions. The power elite controls those institutions in ways that constitute a fundamental threat to the core values of a society that cherishes equality of opportunity. The ways in which power and inequality function in society are indeed quite complex, reflecting what Anthony Giddens (1984) termed structuration—the dynamics through which social structures limit individual behavior, but also make it possible. In other words, as highly unequal as any given society may be, even the person on the lowest rung of the social order exerts influence on society's structures while simultaneously being constrained by them. This reminds us that even the most vulnerable communities have agency and that they can organize to resist climate injustice.

Sociologists continue to pursue similar questions in a rapidly changing urban landscape marked by rising inequalities, segregation, unemployment, poverty, deindustrialization, bureaucratization, globalization, and the political and cultural terrain of the post–civil rights era. These scholars research and debate the role of industry and capital, the state, place, culture, and the family in producing and challenging persistent class, gender, racial, and spatial inequalities in urban America (e.g., Logan and Molotch 2007; Massey and Denton 1993; Patillo 2013; Sampson 2012; Wilson 2012). After more than a century of urban sociological research, one thing is clear: social inequalities persist, and efforts to reduce or eliminate them remain elusive (see, e.g., Oxfam 2014).

Sociologists and scholars across the social sciences and humanities are expanding the boundaries of the classics, taking research on inequality in new and generative directions. Marxist, feminist, racial/ethnic studies, environmental and animal studies, and disability studies scholars have articulated theories of inequality in order to examine the ways that hierarchies are produced, maintained, and challenged across the categories of class, race, gender, sexuality (Anderson and Collins 2006), culture, citizenship/nationality, nation (Bruyneel 2007; Nagel 2012), age, ability (Clare 2009), and species (Fitzgerald and Pellow 2014; Gaard 2004). It is important to understand that each of these differences is intertwined with the others, giving rise to the concept of intersectionality—the idea that we cannot fully comprehend or measure one form of advantage or disadvantage in isolation from others because they interrelate and affect each other (Collins 2000; Crenshaw [1991] 1994).

The literature on intersectionality underscores that inequality is, above all, *unnatural* in the sense that it does not "just happen"—it requires a great deal of energy, labor, and institutional effort to produce and maintain unequal societies. This point is crucial because there is also so much energy invested in making inequalities appear to be a "natural" state of affairs. As Greta Gaard (2004:36) writes, "Appeals to nature have often been used to justify social norms, to the detriment of women, nature, queers, and persons of color." Thus, inequality is not just a state of being unequal; it is frequently experienced as unearned privilege made possible by domination and injustice, and, not surprisingly, routinely resisted by those who suffer its consequences.

Inequalities can be institutionalized by systematically placing subordinate social groups in harm's way, exposing them to greater risks from hazardous conditions, such as climate change. Vulnerability, a concept with theoretical roots in the social, natural, and health sciences, is "a function of the exposure (who or what is at risk) and the sensitivity of the system (the degree to which people and places can be harmed)" (Cutter et al. 2008:599). Proximity to hazards in the natural environment leads to different degrees of risk exposure and sensitivity that are shaped by human decisions to alter the environment in ways that amplify the harmful effects of hazards for some populations more than others (Bolin and Sanford 1998; Zahran et al. 2011). Risks from objective hazards are "always mediated through social and cultural processes" (Wisner et al. 2004:19), or, more pointedly, as Kathleen Tierney (1999:215) said, "sociology emphasizes the contextual factors that structure vulnerability to hazards and the linkages that exist between vulnerability and social power." We specifically examine unequal vulnerability to climate change by race, class, and gender in the fourth section of this chapter.

Environmental Inequalities in Communities and Among Nations

Theories of social inequality laid the foundation for the field of environmental justice studies, which in turn is the forerunner of climate justice research in communities. For nearly four decades, scholars have known that exposure to pollution in the United States is distributed unevenly by race, class, and gender (Brulle and Pellow 2006; Bullard 2000; Mohai, Pellow, and Roberts 2009). Environmental justice studies underscore the historical and ongoing institutional practices that produce environmental inequalities at the local, regional, national, and transnational scales.

Communities of color and low-income, immigrant, and Indigenous communities often face disproportionately high levels of exposure to industrial pollution from power plants, manufacturing facilities, transportation corridors, and extractive industries that are sited in their communities (Crowder and Downey 2010; Pais, Crowder, and Downey 2014; Pulido 2000; Stephenson 2014; Zahran et al. 2008). Technological changes in energy and industrial production frequently produce new environmental inequalities and vast disparities between the privileged and the disadvantaged in exposure to climate-related hazards and risks associated with increasing emissions. The communities most adversely affected by emissions are commonly excluded from decision-making bodies that authorize and regulate such practices (Bullard 2000).

A plethora of studies on environmental inequality provides evidence of racial inequalities (e.g., Downey 2006), whereas other studies find that social categories, such as age, poverty, and class, matter as much or more than race, depending on the context (Mennis and Jordan 2005). Scholars have generally considered and debated three explanations for environmental inequality among communities in the United States: economic, sociopolitical, and racial discrimination. Economic explanations posit that market actors seek to maximize profit by placing noxious facilities in communities where the land is cheap. Therefore, these actors are motivated by lowering the cost of doing business, rather than an agenda of animus or discrimination (Been 1994; for a critique see Mohai and Saha 2006). Of course, many places where land is cheap are also sites where low-income people and people of color reside.

Sociopolitical explanations posit that toxic facilities are placed in communities with the least capability to mount effective political opposition to such practices (Saha and Mohai 2005). The "path of least resistance" also tends to be through low-income neighborhoods and communities of color (Saha and Mohai 2005). Facilities are sited in poor and minority neighborhoods because those communities had little-to-no political power in the establishment of industrial zoning laws or in town planning bureaucracies, and their properties were seen as less valuable to the community's economic development (Roberts and Toffolon-Weiss 1999).

Racial discrimination explanations contend that environmental inequalities are the result of historical and contemporary institutional policies that lead to the concentration of polluting facilities in low-income and people-of-color communities (Bullard 2000; Bullard and Wright 2009). For example, historical and contemporary patterns of zoning and residential segregation are primary social forces that contribute to environmental inequality for people of color and the poor (Massey and Denton 1993) because these forces limit individuals' capacity to choose where to live, thus frequently relegating these groups to highly polluted areas and restricting their mobility out of such compromised environments.

Global political economic theories, developed by sociologists and scholars in related fields, are promising avenues for understanding international climate injustices because they explain the foundations of inequality within the structures of national economies and political systems (Austin 2010). World systems theory contends that the historical economic development of core (wealthier) nations in the Global North occurred as a result of ecological degradation, social upheavals, and economic underdevelopment within the periphery (poorer) nations in the Global South (Bunker and Ciccantell 2005). This basic relationship continues in the contemporary era as core nations gain disproportionate access to capital and externalize the costs of capital accumulation onto nations of the periphery (Bunker and Ciccantell 2005). This perspective aids us in understanding why core nations are the largest emitters of GHGs and enjoy higher gross domestic product (GDP) and other quality-of-life measures (Bonds and Downey 2012).

For example, sociologist Stephen Bunker built upon social theories of unequal exchange developed in the 1970s (Amin 1974) to formulate a theory of "ecologically unequal exchange." In his theory, extended by Martinez-Alier (2003) and others, "extractive peripheries" of the developing world (which supply primarily minerals, lumber, or other raw materials) fail to gain the benefits of social development, while urban regions in core nations centralize resources and wealth from all across the planet, and thereby diversify and protect themselves from economic booms and busts (Bunker 1985). This dynamic of natural resource extraction from the developing world is compounded by the problem of additions—that is, the pollution in the form of toxic waste and other externalities of production with which wealthy core nations burden the Global South (Jorgenson, Dick, and Shandra 2011; Pellow 2007).

An intellectual offspring of unequal exchange is "ecological debt," or more specifically "climate debt," a key concept used by climate justice scholars and activists (Martinez-Alier 2003). Autonomous University of Barcelona economist Joan Martinez-Alier worked closely with the Ecuadorian NGO Acción Ecologica to publicize the idea that rich countries owed the Global South because of a "debt accumulated by Northern, industrial countries toward Third World countries on account of resource plundering, environmental damages, and the free occupation of environmental space to deposit wastes, such as greenhouse gases, from the industrial countries" (Acción Ecologica 2000). The idea of climate debt spread quickly at a major conference of developing countries in Havana in 2000, and it has become widely adopted as a central tenet of the international climate justice movement: those who have historically created the problem of global climate change bear responsibility for repaying their "climate debt" for the atmospheric space they have consumed to those who did not create the problem (Roberts and Parks 2007). In the fifth section of this chapter we discuss international policies and interest groups that seek to define solutions for reapportioning the costs of climate change among nations.

PRINCIPLES OF JUSTICE AND CLIMATE CHANGE

Social theorists and moral philosophers, among others, have probed the meanings of social justice, environmental justice, and most recently climate justice. Climate justice is a twenty-first-century idea, but its roots are firmly established in theories of social and environmental justice from the twentieth century, which sociologists have had a key role in developing along with other social scientists, philosophers, and legal scholars. Because climate justice is now an interdisciplinary term used by many different actors on the global policy stage, we provide some background on how social theory is foundational to core justice principles, particularly as the principles pertain to climate change.

Much of the scholarship on social justice is about unequal distributions of resources and opportunities across populations and geographic space (Hayward and Swanstrom 2011). John Rawls (2001) argues that achieving justice in society must involve the expansion of opportunity and reductions in resource inequalities. Rawls's work is commonly framed as a *distributional* approach to justice, as it emphasizes outcomes of the distribution of goods in society and what might be the most appropriate principles to achieve that end.

Distribution (which can also be framed as a political economic or material approach to justice) is of critical importance, although political theorists Iris Marion Young (1990) and Nancy Fraser (1996) contend that justice cannot be reduced to a focus on inequalities in material goods distribution. In their view, research on social justice should begin with the concept of unequal power relations or domination and oppression of social groups. They contend that research should focus on the social relations and processes through which injustices are created and actions should shift to eliminating injustices through processes that involve the recognition and inclusion of oppressed and marginalized populations. *Recognition* is defined as the process of acknowledging and valorizing cultural and status differences and distinctions (Fraser 2013)—whether they be gender, class, racial, sexual, national, and so forth. Justice then requires an acknowledgement "of social structures that oppress certain social groups" (Harrison 2011:15).

The concept of recognition implies that justice is impossible without procedures for *participatory parity* to allow members of society open and full access to decision-making bodies and procedures (Fraser 2013; Harrison 2011; Schlosberg 2009). Finally, following Amartya Sen (1993), justice can be realized only if there is also a sufficient degree of *capabilities*—the resources, opportunities, freedoms, and institutions required for individuals and groups to exist as full members in a given society. Examples of capabilities include "jobs, living wages, clean air and water, and affordable and accessible public transit, health care, housing, and food" (Harrison 2011:15).

Distribution, recognition, participation, and capabilities are inseparable concepts of justice because in order for people to gain access to material (distributional) social goods, they must be valued and included (recognition) through access to society's decision-making institutions (participatory parity) and society's basic institutions (capabilities) (see also Banerjee 2010; *Principles of Environmental Justice* 1991; Shue 1993). Some scholars would amend this quartet by insisting on a consideration of *restorative* justice, which emphasizes rebuilding relationships and healing rather than punishing guilty parties (e.g., Braithwaite 1999; Grasso 2010; Paterson 2001). All of these concepts of justice have direct relevance to environmental sociology and for the prospects of addressing the climate crisis.

In the great climate change debates of our time, there are multiple, overlapping, and often conflicting ways that stakeholders have defined and pursued "justice" through the lens of climate politics (Bulkeley et al. 2013). Drawing on the broad overview of justice summarized here, we apply the principles to a definition of climate justice that includes the following:

1. Equity in *distributing* the burdens and sharing the benefits of climate change in communities and among nations;
2. Social and political processes that *recognize* currently or previously marginalized groups as rightful *participants* in the governance and management of climate change;
3. Freedom of peoples to make choices that maximize their *capabilities* to survive now and in the future;
4. *Rebuilding* damaged historical relationships between parties, correcting past wrongs against humanity, and *restoring* the Earth.

Because the impacts of climate change are projected to be much more severe in the future, climate justice is concerned with society's obligation to limit GHG emissions for future generations as well as for people living today (Broome 2012; Lawrence 2014; Postner and Weisbach 2010). The central questions in this debate are as follows: (1) How does burning fossil fuels (and other ways of emitting GHGs) harm people throughout the world, and how will it affect them in the future? (2) How should we (the world community of people and governments) decide what to do or not to do about climate change (Broome 2012:99)? In the next sections, we highlight socio-logical research that addresses these two questions.

DISPARATE IMPACTS: HARMS CAUSED BY CLIMATE CHANGE

People living in poverty are exposed to persistent, intersecting, and entrenched structural inequalities, making them particularly vulnerable to harm from the hazards unleashed by climate change (Intergovernmental Panel on Climate Change [IPCC] 2014, Chapter 13). People of color, Indigenous peoples, and women are examples of social groups that are dis-proportionately vulnerable to the long-term effects of increased air pollu-tion, extreme heat, drought, food and water shortages, infectious disease, storms, and floods. They are more vulnerable to climate disruption due to discrimination, cultural expectations, and subordinate positions in social hierarchies (e.g., Kasperson and Kasperson 2001). Here we relate only two of many possible illustrations that describe harms directly linked to carbon emissions from energy production and the changes in weather events that are increasing in frequency, intensity, and destructiveness in different parts of the world.

Energy Sacrifice Zones

Human-induced climate change is caused by a "basket" of gases that includes carbon dioxide, methane, nitrogen gases, and chlorofluorocarbons. About 57 percent of the greenhouse effect is driven by carbon emissions from burning fossil fuels, and coal-fired power plants are responsible for 40 per-cent of carbon emissions in the world (PBL Netherlands Environmental Assessment Agency 2013; U.S. Environmental Protection Agency [EPA] 2014; but see Jorgenson and Birkholz 2010). In addition to its contributions to hastening climate change, the coal industry also causes other great eco-logical and social harms. Coal is responsible for a large proportion of air pol-lution and damage to human health, ecosystems, crops, and infrastructure (Bell and York 2012).

Sociological research on the effects of extractive industries on U.S. rural communities is well documented. Some of the effects of mining "booms" are increased social isolation and crime, stymied social and economic development, and deteriorating health of residents (Bell and York 2012; Freudenburg 1992; Malin 2014). Vast numbers of people around the world suffer at the front end of climate change because they live in such "energy sacrifice zones"—places that are exploited for the purpose of supplying cheap fossil fuels and electricity to power the world's growing energy demands.

Throughout the entire life cycle of coal—mining, processing, washing, and burning of coal—workers and nearby communities are endangered by industry practices. "Mountaintop removal," a particularly destructive method of coal mining, damages fragile ecosystems, endangers water supplies, and renders nearby communities vulnerable to increasingly devastating flood events (Bell 2013; Bell and York 2010). Throughout central Appalachia, this type of coal mining has led to losses in human life, structural damage to homes, contaminated well water, and a loss of access to land once used for hunting food and gathering medicinal herbs (Bell 2013). After coal is mined, the harm to local communities continues as a result of waste impounded as sludge or injected underground (Orem 2006). In comparison with the rest of the nation, coal-mining areas of Appalachia suffer higher rates of birth defects, cancer, and mortality, even after controlling for such variables as income and education (Ahern et al. 2011; Hendryx 2008).

People of color experience climate injustice at nearly every point in the climate disruption process, not only at sites of fossil fuel extraction, but also in the stages of refinement and processing, combustion, and waste dumping. They bear a disproportionate share of health burdens linked to energy produced from fossil fuels. In 2002, more than 70 percent of African Americans and 58 percent of whites lived in counties that are in violation of federal clean air laws and standards (U.S. EPA n.d.). Seventy-eight percent of African Americans, compared to 56 percent of whites, live within thirty miles of a coal-fired power plant (Congressional Black Caucus Foundation 2004). Inhalation of increased particulate matter and nitrogen oxide released from power plants is linked to many adverse birth outcomes and respiratory and cardiovascular diseases. Asthma, an illness commonly associated with air pollution from coal burning, is 36 percent more prevalent among African Americans than among whites (Black Leadership Forum et al. 2002), and climate disruption is expected to push that trend upward in the coming years as higher temperatures interact with pollutants to sharply increase smog in urban centers.

Latino Americans are also disproportionately likely to live near a range of toxic locally unwanted land uses in the United States, particularly coal-fired power plants (Quintero-Somaini and Quirindongo 2004) and high-traffic transportation routes. They are more likely to be exposed to pesticide poisoning because many are employed as farm laborers in industrial agriculture (Quintero-Somaini and Quirindongo 2004)—an industry that contributes to climate change through deforestation and massive use of petroleum products for fertilizers and transportation.

Indigenous peoples of the world are among the communities most affected by climate change (Anchorage Declaration 2009; Indigenous Environmental Network 2009). In the United States, the extraction of energy resources has run a long and often deadly course in Indian country, with a distinctly colonial flavor. Tribes have supplied access to abundant ecological materials at low prices in contracts promoted by the federal government, yet they often receive few if any benefits from such projects (Gedicks 1993; Snipp 1986). Even the most recent federal energy legislation and incentives are designed to encourage the development of tribal resources by outside corporate interests without ownership or participation of the host tribes. As a result, many tribal communities face a range of ecological, social, and economic challenges related to practices that contribute to climate change, including energy extractive industries, deforestation, and hydroelectric dam projects (National Aeronautics and Space Administration 2009). In Kenya, the United Nations Environment Programme (UNEP)-funded Mau forest conservation project made the forest "ready" for a carbon offset reforestation project by forceful and often violent eviction of its inhabitants by the Kenyan Forest Service, including the Indigenous Ogiek people, who had lived on their ancestral lands in the region for centuries (Lang 2009).

Oil production in developing nations is closely associated with human rights abuses and increasingly linked to the issue of climate justice. Examples include reports of Chevron's legal troubles linked to massive toxic waste dumping in waters and on lands affecting Indigenous peoples in Ecuador (Widener 2011); Unocal's investments in the brutal regime in Myanmar involving forced labor for oil pipeline construction (Dale 2011); Chevron and Shell's involvement with dictatorships that have executed Indigenous activists in Nigeria (Fishman 2006); and numerous foreign oil firms who stand accused of fueling mass displacement and killing of civilians living near oil fields in Sudan (Fishman 2006).

The construction of mega hydroelectric dams in Africa, Latin America, and Asia, while "cleaner" in carbon emissions terms, has frequently been accompanied by the evictions of Indigenous peoples occupying land in valleys to be inundated. In Guatemala, Brazil, India, and China, a series of huge dams have displaced hundreds of thousands of people. In response, some cases of resisters being tortured and massacred have been reported, and an international movement arose to fight for change in dam-building practices (McCully 2001). Moreover, large dams are not clean sources of electricity: studies reveal that rotting organic matter in some dam reservoirs produces significant levels of GHG emissions (Fearnside 2002; McCully 2001).

Climate and Weather Extremes

Analyzing over 4,000 climate-related disasters across two decades, Roberts and Parks (2007) found stark differences between wealthy and poor nations in lives lost and people made homeless. National wealth, social inequalities, and organization of civil society were among the significant predictors of suffering from climate disasters. "Natural" disasters are often triggered by human activities, and the impacts, aftermath, recovery, and legacy of such disasters are shaped heavily if not determined by social structures (Bullard and Wright 2009; Tierney 1999). A landmark study on U.S. natural disasters (Mileti 1999) drew on all the major studies of natural disasters in the twentieth century, concluding that "In the United States the key characteristics that seem to influence disaster vulnerability most are socioeconomic status, gender, and race or ethnicity" (see also Klinenberg 2002 and Tierney 2007).

The link between sites of natural disasters and environmental racism in the United States is clear and powerful (Bullard and Wright 2009). The level of public health risk is elevated when we consider the potential impact of a hurricane or tropical storm on hazardous facilities because anyone living nearby will be more likely to experience exposure to accidental toxic releases. The historic concentration of African Americans in the Southern states places them disproportionately at risk from hurricanes on the Gulf and Atlantic coasts, where the largest percentage of African Americans live. Most deaths during heat waves occur in cities, and people of color are twice as likely as whites to die in a heat wave and more likely to suffer from heat-related stress and illness (Kalkstein 1992). In the 1995 Chicago heat wave, communities in the historic "Black Belt" were hit hardest by heat-related deaths (Klinenberg 2002) and African Americans died at a rate 50 percent higher than whites (Whitman et al 1997). Trends in other U.S. cities are similar (O'Neill, Zanobetti, and Schwartz 2003).

Chronically hot summers in cities are also a serious climate injustice in low-income, minority neighborhoods (Harlan et al. 2006; Reid et al. 2009). Phoenix, Arizona, is one example of many U.S. cities where Latinos and African Americans predominantly live in the inner city because they have been concentrated and segregated there for more than a century (Bolin, Grineski, and Collins 2005). Urban core neighborhoods are several degrees warmer than the suburbs and outlying rural areas due to the urban heat island that is created by the vast expanses of paved parking lots, asphalt roads, and buildings that have replaced naturally vegetated land cover (Arnfield 2003).

Compared to wealthier, white neighborhoods in Phoenix, the inner city scores higher on a heat stress index, has more heat-related deaths, and has lower levels of coping resources, such as social networks, trees and parks, functioning and affordable air conditioning, and quality housing (Harlan et al. 2006, 2013). The mosaic of many different temperatures across residential neighborhoods combined with related inequalities in the built and social environments in the city is a human-made "heat riskscape"—a global problem that is projected to become even more serious due to the expansion of cities (Georgescu et al. 2014) and projected higher urban temperatures under global climate change (e.g., Li, Horton, and Kinney 2013).

Women and men occupy different spaces in economies—material (work) and moral (respectability)—that define their worth in society and position them differently, making women more vulnerable than men to many meteorological disasters related to climate change (Nagel 2012, 2015). Brody, Demetriades, and Esplen (2008a:3–4) report that climate-related exposures to water-borne diseases during floods and hyperthermia during heat waves disproportionately affect women caretakers and children and the elderly of all genders. Estimates of population displacements from climate change (sea level rise, storm surges and flooding, drought, resource conflict) range from 25 million to a billion people worldwide by 2050, with most estimates around 200 million (Laczko and Aghazarm 2009:9). These displacements are expected to affect women disproportionately because of their child-caring responsibilities, limited land rights, dependency on subsistence agriculture, and generally impoverished and disadvantaged status in many countries (United Nations WomenWatch 2009). For example, the International Federation of Red Cross and Red Crescent Societies estimated that of the approximately 140,000 killed in the 1991 cyclone in Bangladesh, 90 percent were women and children (Schmuck 2002). Bangladesh is one of the few countries in the world where men live longer than women, and Cannon (2002) argues that women's poverty and vulnerability to weather-related flooding are among the reasons why (see also Begum 1993).

Women's domestic responsibilities and cultural expectations for their modesty can make them especially vulnerable to extreme weather events, particularly in the case of hydro-meteorological disasters such as floods or storm surges (Spring 2006). A number of material and moral economic factors combined to make Bangladeshi women especially vulnerable when the waters rose in 1991. They were responsible for the home—caring for children; finding food, water, and fuel; cooking meals; growing crops; tending livestock—which tied poor women to low-lying residences. Their mobility was limited by cultural definitions of women's proper dress, demeanor, and public visibility—their long, loose clothing restricted movement; they were ashamed to seek higher ground occupied by unrelated men; they could not swim. Women's relative poverty made them less resilient—they had poor nutrition, poor health care, and limited family support since divorced and widowed women were discouraged from remarrying (Cannon 2002).

Vulnerabilities to the effects of climate change—health hazards from producing and consuming energy and from extreme weather—are evident around the globe, within Global South nations, in communities of color and Indigenous communities, among women, and even in places in the Global North where "cleaner" energy technologies are available. These trends speak to the need to understand complex social and historical forces that underlie and often maintain and exacerbate inequality, layering climate injustices on top of past wrongs in society.

POLITICS AND JUST POLICIES FOR MANAGING CLIMATE CHANGE

How should we decide what to do to address the injustices inherent in the current set of causes and impacts of climate change? What kinds of actions are equitable and might amend past wrongs, and what types of actions are likely to make the consequences of climate change worse for future generations and those with less power and wealth? In this section, we address these policy questions about climate justice.

Mitigation and Adaptation Policies Raise Concerns About Climate Justice

At the international level, in the attempt to avoid the worst impacts of climate change, treaty negotiations have focused almost exclusively on *mitigation*, or reducing GHG emissions, with very little effort until recently directed toward *adaptation* to the changing climate (Khan and Roberts 2013). The United Nations Framework Convention on Climate Change (UNFCCC) (1992) declared that parties should "avoid dangerous climate change," acting according to principles of equity and their "common but differentiated responsibility and respective capabilities," meaning that the wealthy and those who caused the problem should act first. It did not stipulate how principles of climate justice should be applied to mitigation—that is, whether nations should be assigned shares of allowable emissions based on a global emissions budget, whether remedial damages should be assessed for past emissions, whether "loss and damage" (those climate-related events that cannot be readily adapted to) should be compensated directly, or how any of these decisions should be made (Vanderheiden 2013). That is, the design and implementation of specific strategies to protect the world's vulnerable populations from climate harm were left to international negotiations, which are now into their third contentious decade (Roberts, Ciplet, and Khan 2015).

Most activists in the international climate justice movement, as well as in the United States, have advocated for equal per capita emission allowances across nations that include a nation's historical responsibility for atmospheric disruption (see, e.g., Hallstrom 2012). In particular, the "Contraction and Convergence" model was proposed by the Global Commons Institute in the early 1990s (Meyer 2000). Rather than commit to a leveling approach, however, government negotiators adopted a "grand-fathering" scheme of burden sharing that gave current emitters rights to continue at similar levels as historically, but with reductions averaging 5 percent. The main strategy for mitigation endorsed by the 1997 Kyoto Protocol, the first and only international treaty to place binding limits on emissions, is to place a cap on carbon emissions in wealthy nations. This framework allows them to trade permits with other nations and/or firms (i.e., cap-and-trade), including the damaged but still functioning Clean Development Mechanism (CDM).[1] After a few years of uncertainty, cap-and-trade systems are expanding again with their adoption across China and possible expansions in Canada and the United States, among other places (Newell, Pizer, and Raimi 2014).

There are many distributive and procedural justice issues that arise with cap-and-trade that relate directly to the core questions of this chapter. International climate justice groups in the Climate Justice Now! network have taken a hard stand against carbon trading of any sort, believing it is morally bankrupt to commodify the air we breathe and charging that trading systems have quickly become corrupt or ineffective (e.g., Clifton 2009; Lohmann 2012). Moreover, while leaders and citizens of many poor nations seek compensation for the ecological and economic harms associated with climate damage, some developing-nation NGOs have expressed concern that compensation could serve as a way for industrialized nations to "buy" the right to continue to emit and that, due to corruption, no compensation will reach the needy.[2]

After the 2009 Copenhagen negotiations at the COP 15 (Fifteenth Annual UN Conference of the Parties) failed to fully extend the Kyoto system, nations have continued to pull in different directions as they assert different interpretations of equitable allowances for emissions. India wants a "climate debt" formula for emissions targets, taking into account its historically low level of emissions, but this is adamantly opposed by the United States. China makes an argument that it has the right to develop economically and therefore a right to continue emitting GHGs (Winkler et al. 2011). The Copenhagen Accord took the modest level of binding emissions off the table in favor of a bottom-up "pledge and review" system favored by the United States and the rapidly developing countries in the BASIC (Brazil, South Africa, India, and China) negotiating group. The weakness of such negotiating positions and current agreements is that they lead directly to emissions that far exceed the 1992 agreement to "avoid dangerous climate change," which is widely accepted as limiting global mean temperature rise to 2 degrees Celsius (United Nations Environment Programme 2010). The vast inequalities between Global North and South nations with regard to information, accessibility, technical expertise, and attorneys produce further difficulties for negotiating fair treaties on sustainable climate policy based on a democratic process (e.g., Roberts and Parks 2007).

The foundering of UN COP negotiations on limiting emissions has led to proposals by various national and substate actors to intervene in the climate cycle by removing carbon from the atmosphere with geoengineering projects that aim to stop or reverse global warming by technical means. There are proposals, for example, to pump sulfate aerosols into the atmosphere to deflect incoming solar heat and cool the Earth or to seed ocean algae beds to increase their carbon absorptive capacity (Gardiner 2013; Hallstrom 2012).

Many social scientists, ethicists, and others recognize a variety of problems with these kinds of schemes. They are likely to be expensive and ineffectual, they might relegate any plans to mitigate or reduce greenhouse gases to the back burner, and they might create rather than forestall a climate disaster. Moreover, some argue that geoengineering is a unilateral, militaristic approach to the problem because they are large-scale projects potentially undertaken by one country but influencing the global environmental system. Most usefully for our review, philosopher Stephen Gardiner (2013) has raised important procedural justice questions concerning geoengineering, such as whether it encourages the current generation to harm future ones by relieving the immediate necessity to control emissions. Will it result in greater subjugation of poor nations by rich nations that have the technological means to impose these potentially dangerous solutions on all global citizens without their consent? Who has a right to control the Earth's common space? Two things that geoengineering projects have in common with mitigation strategies are that they are not necessarily decided democratically and they are expressly not designed to redress the uneven distribution of harmful impacts of climate disruption on vulnerable populations. As the number of options for climate engineering increase, however, a review by sociologist Rachael Shwom and interdisciplinary colleagues concluded that while reducing emissions is still most desirable, forest and soil management for carbon storage raised the fewest ethical concerns among engineering interventions (Cusack et al. 2014).

Some social science critics of top-down technological fixes for managing climate change are actively engaged in scholarship that provides a different set of priorities for *adapting* to climate change, or preparing to live with a destabilized climate system by reducing vulnerability to its harmful effects. Article 4 of the UNFCCC directs the wealthy nations, whose industrial activities are responsible for climate change, to help developing countries "that are particularly vulnerable to the adverse effects of climate change in meeting costs of adaptation to those adverse effects." In the last decade there has been gradual recognition that many impacts from the changing climate are now inevitable and that, therefore, adaptation plans are urgently needed (Khan and Roberts 2013).

Proposals for different types of climate adaptations have huge justice implications because the costs and benefits of individual and collective actions can be distributed in ways that amplify risks for vulnerable groups and benefit those that are socially and economically more capable of dealing with climate change (Adger, Paavoli, and Huq 2006). One of the most important and difficult issues is ensuring that mitigation strategies do not impede adaptions of vulnerable populations in developing nations or in certain communities of industrialized countries. From a justice standpoint, marginalized groups must be recognized as having *a right to participate* in negotiations and a *right to increase their capabilities* to survive and thrive in the future.

There are significant justice challenges concerning who participates in choices among alternative adaptations. Stakeholders from underrepresented groups are usually not at the negotiating table, whether the forum is an international or local assembly. For example, women are important producers of the world's staple food crops, working mostly as small farmers in the Global South (Brody, Demetriades, and Esplen 2008b), but female small-scale farmers are seldom included in fora that make policies designed to address the impact of climate change on agriculture.

Domestically, the climate disaster of Hurricane Katrina that devastated New Orleans, Louisiana, and much of the Gulf Coast Region in August 2005 is a good example of inequality in adaptive capacity between white and African-American neighborhoods, the latter being poorly protected from potential storms. Katrina's flooding was followed by a rebuilding period that excluded elements of the community from decision making based on race and class. With over 1,800 people killed, thousands displaced, and billions of dollars in damage to infrastructure and housing, most of the official plans for rebuilding the city reflected long and deep racial and class divisions and revealed the limitations of local, state, and federal governmental efforts to implement emergency response plans. According to many activists and scholars, this constituted a "second disaster" that included discriminatory and exclusionary practices directed at African Americans and low-income populations seeking financial assistance, small business loans, FEMA grants, insurance settlements, jobs, access to housing, and constitutional protection for voting rights. Conversely, the "rebuilding" plans were seen to focus on supporting elite, white, and wealthier residents and businesses in the area (Holt 2014).

As Desmond Tutu wrote in an introduction to the 2007 UN *Human Development Report*, inequality in adaptation to climate change opens a whole new realm of injustice and inequity:

> No community with a sense of justice, compassion or respect for basic human rights should accept the current pattern of adaptation. Leaving the world's poor to sink or swim with their own meager resources in the face of the threat posed by climate change is morally wrong. Unfortunately . . . this is precisely what is happening. We are drifting into a world of "adaptation apartheid." (cited in Hall and Weiss 2012).

Rethinking Climate Change Policy with Justice at the Core

Justly enhancing adaptive capacity and resiliency is an increasingly important response to climate change because it deals with the ability of affected units at multiple scales (households, neighborhoods, cities, countries, or the entire global ecosystem) to withstand and recover from the stresses and shocks of environmental change (Smit and Wandel 2006). In recognition of this complexity, some social scientists and activists are moving beyond the UNFCCC's focus on global carbon trading to new ways of understanding the interdependence of human society and the functioning of Earth systems in diverse local settings.

For example, more radical critics of mainstream approaches say scientific representations of climate change separate the scientific "facts," statistics, and models from places and time scales that humans can understand and act on. Sheila Jasanoff (2010:249) writes,

> The institutions through which climate knowledge is produced and validated (most notably, the IPCC) have operated in largely uncharted territory, in accordance with no shared, pre-articulated commitments about the right ways to interpret or act upon nature. The resulting representations of the climate have become decoupled from most modern systems of experience and understanding.

In a similar vein, sociologist Thomas Rudel (2009:130) describes how "land change" researchers identify "drivers" of land use change but produce "disembodied, ahistorical explanations" that could be improved by sociological approaches that "identify who transforms landscapes and when the transformations take place."

By making climate change an abstraction that occurs "everywhere and nowhere," it becomes a specialized area of scientific knowledge that has no cultural meaning (Jasanoff 2010). It is inaccessible to most people who need to have a voice in decision making and governmental actions to protect the environment, both global and local, as well as the freedom to decide how to live with climate threats. Patricia Romero Lankao, sociologist and co-author of the Fourth and Fifth Assessment IPCC Working Group II reports, believes bottom-up scientists (e.g., those who study communities) need to be heard on the necessity of understanding real-world priorities and constraints that limit local adaptive capacity.

Salleh (2010) argues for an integrated socio-ecological approach to the complexity of climate change and environmental justice through an understanding of how healthy local ecosystems—plants, soil, water, and temperature—function and how social justice can be achieved by democratizing control of decision making about resources. Salleh's position incorporates the procedural dimension of environmental justice by positioning local struggles as the focus of decision making about climate adaptations rather than relying on reductionist scientific models of global change that are not based on vulnerability assessments or local knowledge. It also differentiates approaches to climate vulnerability and adaptation focused on justice from those examining it from a perspective of improving "governance," as is frequently the case with political science and public policy studies of the area.

Agyeman's (2005) work on "just sustainability" attempts to forge a new paradigm by melding a definition of sustainability that is only focused on the durability of the natural environment with an environmental justice tradition that seeks fairness in the treatment of people. Sociologists McLaughlin and Dietz (2008) recommend that scientific disciplines move toward a new theoretical synthesis of vulnerability that promotes the study of social and ecological diversity, multiple paths of change, and the dynamics of adaptation in historical and future local contexts. Sociology and interpretive social sciences have the potential to make major contributions to framing climate change as a scientific problem with cultural meaning and pathways toward reducing human and environmental impacts.

Some social scientists have joined environmental and Indigenous activists in pushing for more transformative and grassroots-driven approaches to climate change. Those in environmental movements believe that the core elements of our economic system created climate change, and that solutions must come from outside of that system (Angus 2009; Clifton 2009; Rising Tide North America and Carbon Trade Watch n.d.). These calls to action, generally spearheaded by NGOs, involve many different local, national, and Indigenous organizations in global coalition building to demand recognition and participation in decision making about natural resources.[3]

Many leaders of Indigenous communities in the United States and around the world assert that their populations, cultures, traditions, and potential contributions are consistently excluded from the highest levels of climate change policymaking. They have demanded recognition of aboriginal peoples' rights under international law, their roles as stewards of ecosystems, the inherent value of traditional ecological knowledge, and their position on the front lines of climate disruption and in forging environmentally sustainable solutions. In April 2009, Indigenous representatives to the Indigenous Peoples' Global Summit on Climate Change produced the Anchorage Declaration (2009), which calls upon the UNFCCC to recognize the role of traditional ecological knowledge among Indigenous communities as a means of moving toward climate solutions. The Declaration demanded that nation states uphold existing treaty agreements in future climate mitigation strategies and land use planning.[4]

There is a growing activist literature by NGOs like the Indigenous Environmental Network (2009) that labels many of the policies reviewed in this chapter and elsewhere as "false solutions" to climate change, including large-scale dams, geoengineering techniques, "clean coal," agro-fuels (or biofuels), and tree plantations. Privatizing and marketizing ecological services through policies such as the CDM and Reducing Emissions from Deforestation and Land Degradation in Developing Countries (REDD+) are seen as empowering states and investors while disempowering local communities. Thus there is an ongoing debate among many activist groups over whether existing policy safeguards can adequately balance the highly uneven power relations among communities, nation-states, and corporate actors vying for influence across various climate change policy initiatives.

Therefore, some climate justice leaders call for a focus on the root social, ecological, political, and economic causes of the climate crisis, seeking a systemic transformation of societies (e.g., Angus 2009). For example the international anti-dam movement advocates for more sustainable, equitable, and efficient technologies and management practices for rivers, and more transparent and democratic decision-making processes for river projects. It also advocates for affordable, community-based methods of providing water and energy to the millions of people currently without access to these essentials. Some local struggles for environmental justice have forged alliances with national and international environmental organizations such as Earth First!, Rainforest Action Network, and the Sierra Club. Many coal field activists in central Appalachia express their agenda and goals as being inextricably tied to the climate justice movement. One organization whose stated mission is to end mountaintop removal coal mining calls itself "Climate Ground Zero." "Fenceline" environmental justice communities are increasingly tying their struggles to climate change (Stephenson 2014).

In April 2010, 20,000 activists from around the world met in Cochabamba, Bolivia, for the World People's Conference on Climate Change and the Rights of Mother Earth. They named capitalism as the core problem and Indigenous cultural and economic arrangements as a solution. These developments have many implications with regard to the link between environmental justice and citizenship. The environmental justice movement and other grassroots movements are expanding citizenship for marginalized people, for nonhuman animals, and for the Earth itself. For example, Ecuador announced in 2008 a revised constitution that affords the Earth and nature constitutional rights. One passage in the law says that nature "has the right to exist, persist, maintain and regenerate its vital cycles, structure, functions and its processes in evolution." The aim is to create legal systems "that change the status of ecosystems from being regarded as property under the law to being recognized as rights-bearing entities" (Revkin 2008).

Nation-state representatives at the UN COP 20 meeting in Lima, Peru, in December 2014 replicated previous tense COP negotiations about financing, technology transfer, and metrics for measuring carbon reductions. They produced a compromise agreement that encourages but does not require specific targets for national reductions and contains no guarantee of new funds to support climate adaptation in poorer countries (Jacobs 2014; UNFCCC 2014). Meanwhile, more than three dozen civil society groups articulated a vision statement for a world marked by "a transformation of energy systems, away from fossil fuels, towards access to decentralised, renewable, safe, community controlled energy systems for all" (Friends of the Earth 2014). International NGOs and civil society groups from around the globe declared that they would redouble their efforts to force nation-states and fossil fuel industries to ensure that the Global North cuts 70 percent of carbon emissions over the next decade, focusing radical mitigation efforts on "the richest several percent of the world population" without relying on ineffectual market solutions (Foran, Ellis, and Grey 2014). The tensions and divisions between those authorized to participate in climate negotiations and those outside the proceedings suggest not only a gulf in vision and resources between different stakeholders but a disconnect between those who are "seeing like a state" (Scott 1999) and those who experience the realities of climate change on the ground every day. Sociologists such as John Foran contribute to theory building and practice by shedding light on the process and outcomes of climate justice movements.

Ultimately, any effort to address the impacts of climate change must contend with the underlying problem of the growth imperative that is inherent to global capitalism. The main premise of a capitalist system of commerce rests on the assumption that an economy can achieve and sustain infinite growth (Gould et al. 1996). Of course, this is simply impossible in an ecological system with finite resources, yet our political and economic institutions behave as if this were not the case.

CONTRIBUTIONS OF SOCIOLOGY: A WAY FORWARD
IN CLIMATE CHANGE RESEARCH AND POLICY

The discipline of sociology is pivotal in studying the unequal impacts of climate change because sociology's core intellectual problems center on disparities in power, wealth, and privilege. Sociologists have made two broad contributions toward relating inequality to climate justice. One is to scholarship that defines and measures unequal contributions to the causes of climate change and its unjust impacts, including justice implications of current proposals to control and manage climate change. This work covers theoretical treatments of justice and the social structures that create and maintain inequality and injustice, empirical case studies of local communities, and international negotiations on carbon emissions.

Having studied inequalities for more than a century, sociologists are currently relating problems of climate justice to class, race, gender, and other dimensions of institutionalized and intersecting inequalities in everyday lives. These studies illuminate, among other things, how the life cycle of energy production and weather disasters affect people's health and livelihoods in all kinds of local social-environmental contexts, from Appalachia to New Orleans to Bangladesh and Kenya. Drawing upon world systems theory and the related concept of unequal ecological exchange, sociologists have associated the "drivers" or causes of climate change with historically rooted power and wealth inequalities between nations that are impediments to economic change in the Global North and barriers to attaining climate justice in the Global South. Some of these scholarly examinations have had profound impacts upon climate policymaking, especially in driving the discourse to consider justice and the needs of the most vulnerable.

Sociology's second contribution is bridging the social sciences and social movements by studying grassroots organizations advocating climate justice. This scholarship has brought a nuanced and multifaceted set of justice principles into mainstream, narrow policy discourses and stimulated new lines of thinking about how climate change might be addressed in a profoundly different way. Sociologists have decades of experience in studying the beliefs, methods, and strategies of social movements, which they often see as the only force in society capable of driving sufficient action on an issue like climate change (Roberts et al. 2015). They have investigated the social forces that produce, influence, support, oppose, destabilize, and destroy social movements. Knowledge gained from studying organizations of all sorts—international, national, state, and local bureaucracies, small and large firms, and radical and mainstream social movement organizations—has been applied to understanding the organizations that both manage and protest the new global threat of climate change. Rather than imagining that minor tweaks to institutions could improve the situation (as political scientists in the Institutionalist School suggest), sociologists frequently conclude that this issue is being malgoverned for the self-interest of powerful actors. The failure of moderate "Earth System Governance" (see, e.g., Biermann 2009) approaches to muster an adequate response to climate change brings new urgency and legitimacy to this aspect of sociology's inclination to study and herald the grassroots.

Sociologists have participated in formative events around the development of climate science *and* the concept of "climate justice"; not many disciplines can make that claim. Sociologists have done independent research, co-authored scientific assessments, acted as advisors to governments and UN negotiating groups, organized pivotal conferences on climate justice, and worked with activists and NGOs, expanding the scope and capacity of both sets of partners. As has been touched upon in this chapter, by providing research tools to examine claims of disproportionate impacts and contributing to theories of just process and outcomes, the field of sociology provides insights on why many proposals for dealing with the climate crisis are recipes for failure in achieving social justice.

Sociologists can build upon this work to make many more contributions to scholarship on climate inequalities and injustices. It is quite difficult to predict the many directions in which sociology's diverse scholars and public sociologists will take their work. Still, we see some promising areas for expansion, both scholarly and engaged:

1. More sociologists could apply theories of inequality and injustice to data on environmental change, and on climate change in particular, by using existing research tools to explain GHG emissions and unequal climate impacts as dependent variables at the individual, household, neighborhood, community, regional, national, and global levels.

2. Sociologists could press for positioning inequality and injustice at the center of environmental assessments such as the IPCC reports (Intergovernmental Panel on Climate Change 2014) and the National Climate Assessment (2014). They should also seek to serve and support social science representation on local, state, national, and international climate science and policy boards.

3. Sociologists can advance this repositioning effort by providing research that demonstrates how inequalities in consumption and production are responsible for climate injustices between and within nations and subnational groups, and how understanding these inequalities provides new routes for addressing climate change. Assessments like the IPCC can only cite research published in peer-reviewed journals, and the assessment authors' gaze is fixed on major natural science journals

such as *Nature, Science,* and *Proceedings of the National Academy of Sciences (PNAS)*. Therefore sociologists should address audiences in those outlets.

4. Sociologists can produce high-quality case studies of vulnerable populations in vulnerable places that link macro processes (like economic and global atmospheric change) with finer-scale local contextual analyses. Relatedly, more studies are needed on human agency and capability in responding to or recovering from disasters and to the socioeconomic and cultural characteristics that enhance or deter abilities to respond. Better measures and indicators for comparing the effects of disasters in terms of human life, economic loss, and foregone opportunities are needed to inform adaptation policies and interventions at the local, state, national, and international levels.

5. Building upon social movement, political, and political economy theories, sociology can contribute to theory building and practice by shedding light on when climate justice movements succeed and fail. It can identify the barriers social movements and unorganized groups face in addressing climate justice, and the key coalition partners and external supporters that prove decisive for success. And sociology can identify pathways to success in complex social conjunctures. This work should deal with the opportunities and challenges of organizing transnationally and of facing adversaries such as global fossil fuel corporations that are not accountable to a particular national constituency. The literature on the globalization of social movements offers a strong starting point here.

6. Sociologists can participate in developing just interventions and provide a vision for mitigation and adaptation programs that include voices of workers and other vulnerable communities in a "just transition" to low-carbon and resilient societies. Sociologists can engage by working with frontline community action organizations and city, state, and national agencies to identify and communicate with climate-vulnerable groups, and by helping co-develop outreach and communication strategies for incorporating their needs into public policies. Sociological research on organizations and social movements will become even more important in the future as human institutions—households, cities, nations, and intergovernmental organizations—continue to struggle with problems created and exacerbated by climate injustices.

7. Finally, sociologists can question the implicit and explicit commitment to economic growth in all policy debates surrounding climate disruption. Belief that growth will address and overcome social inequality is an article of faith in the United States and many other societies around the world, in spite of mountains of sociological research disconfirming that

hypothesis. The growth of the global economy means consumption of an ever-increasing amount of goods, which require an ever-increasing amount of energy, mineral, agricultural, and forest resources. New definitions of social development are needed that capture progress that can be "decoupled" from carbon emissions (Jorgenson and Clark 2012; Steinberger and Roberts 2010; see also York, Rosa, and Dietz 2003).

Sociologists bring an essential and unique toolkit to explore, explain, and help society address climate inequality and injustice. Our field, however, needs to initiate and foster interdisciplinary cooperation in the development of new theory, methods, and substance. Human geography, ecology, political science, communications, psychology, and economics are obvious places to begin, but finding common ground (by developing common language and collaborative projects) with the vast fields of biology, geosciences, and engineering is crucial for sociology to help move forward a global agenda to limit climate change. A major challenge is creating a global paradigm to supplant unrestrained economic growth—a new paradigm that is rooted in meeting human needs equitably while also respecting the Earth's finite capacity to sustain healthy ecosystems. To meet this challenge, we need to ask how it can be possible for the principles of climate justice to serve as the guide for policymakers as they begin seriously to weigh options for drastically decarbonizing society and adapting to the inevitable climate disruptions we face.

NOTES

1. Kyoto does not apply to the United States and Australia, which have not ratified the treaty. The CDM was to promote sustainable development in the Global South countries while allowing industrialized countries to earn emissions credits from their investments in emission-reducing projects in the South. The CDM allows Global North countries to purchase credits from Southern nations that reduce carbon emissions. Those Northern countries can then use or sell those credits in the North. According to microeconomic theory, given perfect information and fair trading, the polluters who can reduce emissions most cost-effectively should do so, and sell permits to industries whose whole ability to function in a globally competitive economy might be at risk from having to make substantial reductions in carbon emissions.

2. A recent national case study of seventy-five Kenyan environmental NGOs shows that civil society support for climate justice may be dampened due to mistrust of both the Global North and the national government (Beer 2012). Activists and scholars have argued that the CDM has removed much of the incentive to actually reduce emissions in the North, and that it has failed to deliver the promised

sustainable development benefits. Most CDM credits have been captured by Chinese chemical industries installing fairly low-cost equipment to capture CFCs and N-gases—both are GHGs that have huge multipliers in the CDM credits they generate. Large-scale projects in Brazil and India also have secured large amounts of the credits, while small nations and poor areas like Africa have been nearly entirely bypassed (Hultman et al. 2009). Developing-country governments are empowered to require sustainable development benefits of their CDM projects, but the successful host countries are those with extremely lax criteria for what that means (Cole 2009).

3. The Bali Principles of Climate Justice, using a blueprint of environmental justice principles developed at the 1991 First National People of Color Environmental Leadership Summit, was the first attempt to define climate change as a human rights and environmental justice issue (International Climate Justice Network 2002). The principles consider the causes of climate change and offer a far-reaching vision for fair solutions.

4. One of the demands articulated in this statement was that the UNFCCC adhere to the United Nations Declaration on the Rights of Indigenous People. Leaders also demanded that this declaration be fully recognized and respected in all decision-making processes and activities related to climate disruption policy at the UNFCCC.

REFERENCES

Acción Ecologica. 2000. "Trade, Climate Change and the Ecological Debt." Unpublished paper, Quito, Ecuador.

Addams, Jane. 1896. "A Belated Industry." *American Journal of Sociology* 1(5):536–550.

Addams, Jane. 1899. "Trade Unions and Public Duty." *American Journal of Sociology* 4(4):448–462.

Adger, N. W., J. Paavoli, and S. Huq (Eds.). 2006. *Fairness in Adaptation to Climate Change*. Cambridge, MA: MIT Press.

Agyeman, Julian. 2005. *Sustainable Communities and the Challenge of Environmental Justice*. New York: New York University Press.

Ahern, M. M., M. Hendryx, J. Conley, E. Fedorko, A. Ducatman, and K. J. Zullig. 2011. "The Association between Mountaintop Mining and Birth Defects among Live Births in Central Appalachia, 1996–2003." *Environmental Research* 111(6):838–846.

Amin, Samir. 1974. *Accumulation on a World Scale*, 2 vols. New York: Monthly Review Press.

Anchorage Declaration. 2009. *Indigenous People's Global Summit on Climate Change*. April 24, Anchorage, Alaska. Retrieved October 28, 2013 (http://unfccc.int/resource/docs/2009/smsn/ngo/168.pdf).

Anderson, Margaret and Patricia Hill Collins. 2006. *Race, Class, and Gender: An Anthology*. Boston, MA: Wadsworth.

Angus, Ian. 2009. *The Global Fight for Climate Justice: Anticapitalist Responses to Global Warming and Environmental Destruction*. London: Resistance Books.

Arnfield, A. J. 2003. "Two Decades of Urban Climate Research: A Review of Turbulence, Exchanges of Energy and Water, and the Urban Heat Island." *International Journal of Climatology* 23:1–26.

Austin, K. F. 2010. "Soybean Exports and Deforestation from a World-systems Perspective: A Cross-national Investigation of Comparative Disadvantage." *Sociological Quarterly* 51(3):511–536.

Banerjee, Damayanti. 2010. "Justice as Rights: Revisiting Environmental Justice Theory." Paper presented at the Annual Meetings of the American Sociological Association, August, Atlanta, GA.

Been, Vicki. 1994. "Locally Undesirable Land Uses in Minority Neighborhoods: Disparate Siting or Market Dynamics?" *Yale Law Journal* 103:1383.

Beer, Christopher Todd. 2012. "The Influence of Transnational Actors on Kenyan Environmental NGOs." Doctoral dissertation. Department of Sociology, Indiana University, Bloomington, Indiana.

Begum, Rasheda. 1993. "Women in Environmental Disasters: the 1991 Cyclone in Bangladesh." *Gender and Development* 1:34–49.

Bell, Michael M. 2013. *An Invitation to Environmental Sociology.* 4th ed. Boulder, CO: Pine Forge Press.

Bell, Shannon Elizabeth. 2013. *Our Roots Run Deep as Ironweed: Appalachian Women and the Fight for Environmental Justice.* Chicago and Urbana: University of Illinois Press.

Bell, Shannon E. and Richard York. 2010. "Community Economic Identity: The Coal Industry and Ideology Construction in West Virginia." *Rural Sociology* 75(1):111–143.

Bell, Shannon E. and Richard York. 2012. "Coal, Injustice, and Environmental Destruction: Introduction to the Special Issue on Coal and the Environment." *Organization & Environment* 25(4):359–367.

Biermann, Frank. 2009. *Earth System Governance: People, Places, and the Planet: Science and Implementation Plan of the Earth System Governance Project.* IDEP, The Earth System Governance Project.

Black Leadership Forum, The Southern Organizing Committee for Economic and Social Justice, The Georgia Coalition for the Peoples' Agenda, and Clear the Air. 2002. *Air of Injustice: African Americans and Power Plant Pollution.* Retrieved October 28, 2013 (http://www.energyjustice.net/files/coal/Air_of_Injustice.pdf).

Bolin, Bob, Sara Grineski, and Timothy Collins. 2005. "The Geography of Despair: Environmental Racism and the Making of South Phoenix, Arizona, USA." *Human Ecology Review* 12(2):156–168.

Bolin, R. and L. Sanford. 1998. *The Northridge Earthquake: Vulnerability and Disaster.* New York: Routledge.

Bonds, Eric and Liam Downey. 2012. "'Green' Technology and Ecologically Unequal Exchange: The Environmental and Social Consequences of Ecological Modernization in the World-system." *Journal of World Systems Research* 18(2):167–186.

Braithwaite, J. 1999. "Restorative Justice: Assessing Optimistic and Pessimistic Accounts." Pp. 1–127 in *Crime and Justice: A Review of Research,* edited by M. Tonry and N. Morris. Chicago: University of Chicago Press.

Brody, Alyson, Justina Demetriades, and Emily Esplen. 2008a. *Gender and Climate Change: Mapping the Linkages—A Scoping Study on Knowledge and Gaps.* Institute of Development Studies (IDS), University of Sussex. Retrieved June 26, 2011 (http://www.bridge.ids.ac.uk/reports/Climate_Change_DFID.pdf).

Brody, Alyson, Justina Demetriades, and Emily Esplen. 2008b. "Gender and Desertification: Expanding Roles for Women to Restore Dryland Areas." Institute of Development Studies (IDS), University of Sussex. Retrieved April 10, 2009 (http://www.ifad.org/pub/gender/desert/gender_desert.pdf).

Broome, John. 2012. *Climate Matters: Ethics in a Warming World*. New York: W.W. Norton.

Brulle, Robert J. and David N. Pellow. 2006. "Environmental Justice: Human Health and Environmental Inequalities." *Annual Review of Public Health* April 27:103–124.

Bruyneel, K. 2007. *The Third Space of Sovereignty: The Postcolonial Politics of U.S.– Indigenous Relations*. Minneapolis: University of Minnesota Press.

Bulkeley, H., J. Carmin, V. C. Broto, G. A. S. Edwards, and S. Fuller. 2013. "Climate Justice and Global Cities: Mapping the Emerging Discourses." *Global Environmental Change* 23(3):914–925.

Bullard, Robert D. 2000. *Dumping in Dixie: Race, Class, and Environmental Quality*. Boulder, CO: Westview Press.

Bullard, Robert D. and Beverly Wright. 2009. *Race, Place, and Environmental Justice after Hurricane Katrina*. Boulder, CO: Westview Press.

Bunker, Stephen. 1985. "Modes of Extraction, Unequal Exchange, and the Progressive Underdevelopment of an Extreme Periphery: The Brazilian Amazon 1600–1980." *American Journal of Sociology* 89(5):1017–1064.

Bunker, Stephen G. and Paul S. Ciccantell. 2005. *Globalization and the Race for Resources*. Baltimore, MD: Johns Hopkins University Press.

Cannon, Terry. 2002. "Gender and Climate Hazards in Bangladesh." *Gender and Development* 10:45–50.

Clare, E. 2009. *Exile and Pride: Disability, Queerness & Liberation*. Boston, MA: South End Press.

Clifton, Sarah-Jayne. 2009. *A Dangerous Obsession: The Evidence Against Carbon Trading and for Real Solutions to Avoid a Climate Crunch*. UK: Friends of the Earth. Retrieved October 28, 2013 (http://www.foe.co.uk/resource/reports/dangerous_obsession.pdf).

Cole, John C. 2009. "The Clean Development Mechanism (CDM) and the Legal Geographies of Climate Policy in Brazil." Unpublished Doctoral Thesis, University of Oxford, Oxford, UK.

Collins, Patricia Hill. 2000. *Black Feminist Thought: Knowledge, Consciousness, and the Politics of Empowerment*. 2nd ed. New York: Routledge.

Congressional Black Caucus Foundation, Inc. 2004. *African Americans and Climate Change: An Unequal Burden*. July 21. Retrieved October 28, 2013 (http://rprogress.org/publications/2004/CBCF_REPORT_F.pdf).

CorpWatch. 2002. Bali Principles of Climate Justice. Retrieved June 4, 2014 (http://www.corpwatch.org/article.php?id = 3748).

Crenshaw, Kimberle. [1991] 1994. "Mapping the Margins: Intersectionality, Identity Politics, and Violence against Women of Color." Pp. 93–118 in *The Public Nature of Private Violence*, edited by M. A. Fineman and R. Mykitiuk. New York: Routledge.

Crowder, Kyle and Liam Downey. 2010. "Inter-Neighborhood Migration, Race, and Environmental Hazards: Modeling Microlevel Processes of Environmental Inequality." *American Journal of Sociology* 115(4):1110–1149.

Cusack, Daniela, Jonn Axsen, Rachael Shwom, Lauren Hartzell-Nichols, Sam White, and Katherine R. M. Mackey. 2014. "An Interdisciplinary Assessment of Climate Engineering Strategies." *Frontiers in Ecology and the Environment* 12:280–287.

Cutter, S. L., L. Barnes, M. Berry, C. Burton, E. Evans, E. Tate, and J. Webb. 2008. "A Place-Based Model for Understanding Community Resilience to Natural Disasters." *Global Environmental Change* 18:598–606.

Dale, John G. 2011. *Free Burma: Transnational Legal Action and Corporate Accountability.* Minneapolis: University of Minnesota Press.

Downey, Liam. 2006. "Environmental Racial Inequality in Detroit." *Social Forces* 85(2):771–96.

Du Bois, W. E. B. 1899. *The Philadelphia Negro: A Social Study.* Philadelphia: University of Pennsylvania Press.

International Climate Justice Network. 2002. *Bali Principles of Climate Justice.* Retrieved May 11, 2013 (http://www.ejnet.org/ej/bali.pdf).

Fearnside, Philip M. 2002. "Greenhouse Gas Emissions From a Hydroelectric Reservoir (Brazil's Tucuruí Dam) and the Energy Policy Implications." *Water, Air, and Soil Pollution* 133(1–4):69–96.

Fishman, Benjamin. 2006. "Binding Corporations to Human Rights Norms through Public Law Settlement." *New York University Law Review* 81(4):1433–1468.

Fitzgerald, A. and D. N. Pellow. 2014. "Ecological Defense for Animal Liberation: A Holistic Understanding of the World." Pp. 28–48 in *Defining Critical Animal Studies: An Intersectional Social Justice Approach for Liberation,* edited by A. Nocella II, J. Sorenson, K. Socha, and A. Matsuoka. New York: Peter Lang Publishing.

Foran, John, Corrie Ellis, and Summer Gray (Eds.). 2014. *At the COP: Global Climate Justice Youth Speak Out.* E-book. Retrieved December 16, 2014 (https://climate-justiceproject.files.wordpress.com/2014/12/foran-ellis-and-gray-2014-at-the-cop.pdf).

Fraser, Nancy. 1996. *Justice Interruptus: Critical Reflections on the "Post Socialist" Condition.* New York: Routledge.

Fraser, Nancy. 2013. *Fortunes of Feminism: From State-Managed Capitalism to Neoliberal Crisis.* Brooklyn, NY: Verso.

Freudenburg, William R. 1992. "Addictive Economies: Extractive Industries and Vulnerable Localities in a Changing World Economy." *Rural Sociology* 57(3):305–332.

Friends of the Earth. 2014. "No Justice in Lima Outcome. Lima 2014 Climate Justice Statement." December 14. Retrieved December 16, 2014 (http://www.foei.org/news/no-justice-in-lima-outcome/).

Gaard, Greta. 2004. "Toward a Queer Ecofeminism." Pp. 21–44 in *New Perspectives on Environmental Justice: Gender, Sexuality, and Activism,* edited by R. Stein. New Brunswick, NJ: Rutgers University Press.

Gardiner, Stephen M. 2013. "The Desperation Argument for Geoengineering." *PS: Political Science & Politics* 46(1):28–33.

Gedicks, Al. 1993. *The New Resource Wars: Native and Environmental Struggles against Multinational Corporations.* Boston: South End Press.

Georgescu, Matei, Philip Morefield, Britta Bierwagen, and Christopher Weaver. 2014. "Urban Adaptation Can Roll Back Warming of Emerging Megapolitan Regions." *Proceedings of the National Academy of Sciences USA* 111 (8):2909–2914.

Giddens, Anthony. 1984. *The Constitution of Society: Outline of the Theory of Structuration.* Boston, MA: Polity Press.

Gould, Kenneth A., Allan Schnaiberg, and Adam S. Weinberg. 1996. *Local Environmental Struggles: Citizen Activism in the Treadmill of Production.* Cambridge: Cambridge University Press.

Grasso, Marco. 2010. "An Ethical Approach to Climate Adaptation Finance." *Global Environmental Change* 20(1):74–81.

Hall, Margaux J. and David C. Weiss. 2012. "Avoiding Adaptation Apartheid: Climate Change Adaptation and Human Rights Law." *Yale Journal of International Law* 37:309–366.

Hallstrom, Niclas (Ed.). 2012. Vol. 3. *What Next: Climate, Development and Equity.* Dag Hammarskjold Foundation and the What Next Forum.

Harlan, Sharon L., Anthony Brazel, Lela Prashad, William L. Stefanov, and Larissa Larsen. 2006. "Neighborhood Microclimates and Vulnerability to Heat Stress." *Social Science & Medicine* 63:2847–2863.

Harlan, Sharon L., Juan Declet-Barreto, William L. Stefanov, and Diana Petitti. 2013. "Neighborhood Effects on Heat Deaths: Social and Environmental Predictors of Vulnerability in Maricopa County, Arizona." *Environmental Health Perspectives* 121(2):197–204.

Harrison, J. L. 2011. *Pesticide Drift and the Pursuit of Environmental Justice.* Cambridge, MA: MIT Press.

Hayward, Clarissa Rile and Todd Swanstrom. 2011. *Justice and the American Metropolis.* Minneapolis: University of Minnesota Press.

Hendryx, M. 2008. "Mortality Rates in Appalachian Coal Mining Counties: 24 Years Behind the Nation." *Environmental Justice* 1(1):5–11.

Holdren, John. 2007. "Global Climate Disruption: What Do We Know? What Should We Do?" Presentation at Harvard University, November 6.

Holt, William. 2014. "Do You Know What it Means to Rebuild New Orleans? Cultural Sustainability After Disasters." Pp. 267–287 in *From Sustainable Cities: Global Concerns/Urban Efforts*, Vol. 15 in the Research in Urban Studies Series edited by W. Holt (R. Hutchinson, series editor). Bingley, UK: Emerald.

Hultman, Nathan E., Emily Boyd, J. Timmons Roberts, John Cole, Esteve Corbera, Johannes Ebeling, Katrina Brown, and Diana M. Liverman. 2009. "How Can the Clean Development Mechanism Better Contribute to Sustainable Development?" *AMBIO: A Journal of the Human Environment* 38 (2):120–122.

Indigenous Environmental Network. 2009. "Press Statement: Report Calls for the Rejection of REDD in Climate Treaty." Bangkok, Thailand, October 1. Retrieved October 28, 2013 (http://www.youtube.com/watch?v = RavcHZlYDBI).

Intergovernmental Panel on Climate Change (IPCC). 2014. Climate Change 2014: Impacts, Adaptation, *and Vulnerability. Part A: Global and* Sectoral Aspects. *Contribution of* Working Group *II to the* Fifth Assessment *Report of the* Intergovernmental Panel *on* Climate Change, edited by C. B. Field, V. R. Barros, D. J. Dokken, K. J. Mach, M. D. Mastrandrea, T. E. Bilir, M. Chatterjee, K. L. Ebi, Y. O. Estrada, R. C. Genova, B. Girma, E. S. Kissel, A. N. Levy, S. MacCracken, P. R. Mastrandrea, and L. L. White. Cambridge and New York: Cambridge University Press. Retrieved December 9, 2014 (http://www.ipcc.ch/pdf/assessment-report/ar5/wg2/WGIIAR5-FrontMatterA_FINAL.pdf).

Jacobs, Michael. 2014. "Lima Deal Represents a Fundamental Change in Global Climate Regime." *The Guardian*, December 15. Retrieved December 16, 2014 (http://www.theguardian.com/environment/2014/dec/15/lima-deal-represents-a-fundamental-change-in-global-climate-regime).

Jasanoff, Sheila. 2010. "A New Climate for Society." *Theory, Culture & Society* 27(2–3):233–253.

Jorgenson, Andrew K. and Ryan Birkholz. 2010. "Assessing the Causes of Anthropogenic Methane Emissions in Comparative Perspective, 1990–2005." *Ecological Economics* 69:2634–2643.

Jorgenson, Andrew K. and Brett Clark. 2012. "Are the Economy and the Environment Decoupling? A Comparative International Study, 1960–2005." *American Journal of Sociology* 118:1–44.

Jorgenson Andrew K., C. Dick, and J. M. Shandra. 2011. "World Economy, World Society, and Environmental Harm in Less Developed Countries." *Sociological Inquiry* 81(1):53–87.

Kalkstein, L. S. 1992. "Impacts of Global Warming on Human Health: Heat Stress-related Mortality." In *Global Climate Change: Implications, Challenges and Mitigating Measures*, edited by S. K. Majumdar, L. S. Kalkstein, B. Yarnal, E. W. Miller and L. M. Rosenfield. Philadelphia: Pennsylvania Academy of Science.

Kasperson, R. E. and J. X. Kasperson. 2001. *Climate Change, Vulnerability and Social Justice*. Stockholm Environment Institute, Risk and Vulnerability Programme, Stockholm, Sweden. Retrieved October 28, 2013 (http://stc.umsl.edu/essj/unit4/climate%20change%20risk.pdf).

Khan, Mizan and J. Timmons Roberts. 2013. "Towards a Binding Adaptation Regime: Three Levers and Two Instruments." In *Successful Adaptation*, edited by S. Moser and M. Boykoff. London: Routledge Publishers.

Klinenberg, Eric. 2002. *Heat Wave: A Social Autopsy of Disaster in Chicago*. Chicago: University of Chicago Press.

Laczko, Frank and Christine Aghazarm, eds. 2009. *Migration, Environment and Climate Change: Assessing the Evidence*. International Organization for Migration, Geneva, Switzerland. Retrieved June 26, 2011 (http://publications.iom.int/bookstore/free/migration_and_environment.pdf).

Lang, Chris. 2009. "Ogiek Threatened with Eviction from Mau Forest, Kenya." *REDD Monitor*. November 19. Retrieved October 28, 2013 (http://www.redd-monitor.org/2009/11/19/ogiek-threatened-with-eviction-from-mau-forest-kenya/).

Lawrence, P. 2014. *Justice for Future Generations: Climate Change and International Law*. Northampton, MA: Edward Elgar.

Li, Tiantian, Patrick Horton, and Patrick Kinney. 2013. "Projections of Seasonal Patterns in Temperature-related Deaths for Manhattan, New York." *Nature Climate Change* 3:717–721.

Logan, John and Harvey Molotch. 2007. *Urban Fortunes: The Political Economy of Place*. Berkeley: University of California Press.

Lohmann, L. 2012. "The Endless Algebra of Climate Markets." *Capitalism Nature Socialism* 22(4):93–116.

Malin, Stephanie. 2014. "There's No Real Choice But to Sign: Neoliberalization and Normalization of Hydraulic Fracturing on Pennsylvania Farmland." *Journal of Environmental Studies and Sciences* 4(1):17–27.

Martinez-Alier, Joan. 2003. "Marxism, Social Metabolism and Ecologically Unequal Exchange." Paper presented at Lund University Conference on World Systems Theory and the Environment, 19–22 September, Lund, Sweden.

Marx, Karl. 2000. *Das Kapital: A Critique of Political Economy*. Washington, DC: Regnery.

Massey, Douglas and Nancy Denton. 1993. *American Apartheid: Segregation and the Making of the American Underclass*. Cambridge, MA: Harvard University Press.

McCully, Patrick. 2001. *Silenced Rivers: The Ecology and Politics of Large Dams.* London: Zed Books.

McLaughlin, P. and T. Dietz. 2008. "Structure, Agency and Environment: Toward an Integrated Perspective on Vulnerability." *Global Environmental Change* 18:99–111.

Mennis, J. and L. Jordan. 2005. "The Distribution of Environmental Equity: Exploring Spatial Nonstationarity in Multivariate Models of Air Toxic Releases." *Annals of the Association of American Geographers* 95:249–68.

Meyer, Aubrey. 2000. *Contraction and Convergence: The Global Solution to Climate Change.* Cambridge: Green Books Ltd.

Mileti, D. 1999. *Disasters by Design: A Reassessment of Natural Hazards in the United States.* Washington, DC: Joseph Henry Press.

Mills, C. Wright. 1956. *The Power Elite.* New York: Oxford University Press.

Mohai, Paul, David N. Pellow, and J. Timmons Roberts. 2009. "Environmental Justice." *Annual Review of Environment and Resources* 34:405–430.

Mohai, Paul and R. Saha. 2006. "Reassessing Racial and Socioeconomic Disparities in Environmental Justice Research." *Demography* 43:383–399.

Nagel, Joane. 2012. "Intersecting Identities and Global Climate Change." *Identities: Global Studies in Culture and Power* 19(4):467–476.

Nagel, Joane. 2015. *Gender and Climate Change: Impacts, Science, Policy.* Boulder, CO: Paradigm Publishers.

National Aeronautics and Space Administration (NASA). 2009. "The Second Native Peoples/Native Homelands Workshop." November 18, Prior Lakes, Minnesota.

National Climate Assessment. 2014. U.S. Global Change Research Program, Washington, DC. Retrieved June 21, 2014 (http://nca2014.globalchange.gov/report).

Newell, Richard G., William A. Pizer, and Daniel Raimi. 2014. "Carbon Market Lessons and Global Policy Outlook." *Science* 343(6177):1316–1317.

O'Brien, Karen L. and Robin M. Leichenko. 2000. "Double Exposure: Assessing the Impacts of Climate Change within the Context of Economic Globalization." *Global Environmental Change* 10(3):221–232.

O'Neill, M., A. Zanobetti, and J. Schwartz. 2003. "Modifiers of the Temperature and Mortality Association in Seven U.S. Cities." *American Journal of Epidemiology* 157(12):1074–1082.

Orem, W. H. 2006. "Coal Slurry: Geochemistry and Impacts on Human Health and Environmental Quality." U.S. Geological Survey, Eastern Energy Resources Team. PowerPoint Presentation to the Coal Slurry Legislative Subcommittee of the Senate Judiciary Committee, West Virginia Legislature, November 15.

Oxfam. 2014. "Working for the Few: Political Capture and Economic Inequality." 178 Oxfam Briefing Paper. Retrieved June 24, 2014 (http://www.oxfam.org/en/policy/working-for-the-few-economic-inequality).

Pais, J., K. Crowder, and L. Downey. 2014. "Unequal Trajectories: Racial and Class Differences in Residential Exposure to Industrial Hazard." *Social Forces* 92(3):1189–1215.

Park, Robert. 1915. "The City: Suggestions for Investigation of Human Behavior in the City Environment." *American Journal of Sociology* 20(5):577–612.

Paterson, M. 2001. "Principles of Justice in the Context of Global Climate Change." Pp. 119–126 in *International Relations and Global Climate Change,* edited by Urs Luterbacher Detlef Sprinz. Cambridge, MA: MIT Press.

Patillo, Mary. 2013. *Black Picket Fences: Privilege and Peril among the Black Middle Class.* 2nd ed. Chicago: University of Chicago Press.

PBL Netherlands Environmental Assessment Agency. 2013. "Trends in Global CO2 Emissions: 2013 Report." The Hague: PBL/EC-JRC. Retrieved December 17, 2014 (http://www.pbl.nl/en/publications/trends-in-global-co2-emissions-2013-report).

Pellow, David N. 2007. *Resisting Global Toxics: Transnational Movements for Environmental Justice.* Cambridge, MA: MIT Press.

Postner, E. A. and D. Weisbach. 2010. *Climate Change Justice.* Princeton, NJ: Princeton University Press.

Principles of Environmental Justice. 1991. Environmental Justice Resource Center at Clark Atlanta University. Retrieved June 4, 2014 (http://www.ejnet.org/ej/principles.html).

Pulido, Laura. 2000. "Rethinking Environmental Racism: White Privilege and Urban Development in Southern California" *Annals of the Association of American Geographers* 90(1):12–40.

Quintero-Somaini, Adrianna and Mayra Quirindongo. 2004. *Hidden Danger: Environmental Health Threats in the Latino Community.* Natural Resources Defense Council. Retrieved October 28, 2013 (http://www.nrdc.org/health/effects/latino/english/contents.asp).

Rawls, John. 2001. *Justice as Fairness: A Restatement.* Edited by E. Kelly. Cambridge, MA: Harvard University Press.

Reid, Colleen E., Marie S. O'Neill, Carina J. Gronlund, Shannon J. Brines, Daniel G. Brown, Ana V. Diez-Roux, and Joel Schwartz. 2009. "Mapping Community Determinants of Heat Vulnerability." *Environmental Health Perspectives* 117(11):1730–1736.

Revkin, Andrew. 2008. "Ecuador Constitution Grants Rights to Nature." Dot Earth Blog. September 29. Retrieved October 28, 2013 (http://dotearth.blogs.nytimes.com/2008/09/29/ecuador-constitution-grants-nature-rights/).

Rising Tide North America and Carbon Trade Watch. n.d. *Hoodwinked in the Hothouse: False Solutions to Climate Change.* 2nd ed. Retrieved October 28, 2013 (http://risingtidenorthamerica.org/special/hoodwinkedv2_WEB.pdf).

Roberts, J. Timmons, David Ciplet, and Mizan Khan. 2015. *The New Global Politics of Climate Change.* Cambridge, MA: MIT Press.

Roberts, J. Timmons and Bradley Parks. 2007. *A Climate of Injustice: Global Inequality, North-South Politics, and Climate Policy.* Cambridge, MA: MIT Press.

Roberts, J. Timmons and Melissa M. Toffolon-Weiss. 1999. *Chronicles from the Environmental Justice Frontline.* Cambridge: Cambridge University Press.

Rudel, Thomas K. 2009. "How Do People Transform Landscapes? A Sociological Perspective on Suburban Sprawl and Tropical Deforestation." *American Journal of Sociology* 115(1):129–154.

Saha, R. and P. Mohai. 2005. "Historical Context and Hazardous Waste Facility Siting: Understanding Temporal Patterns in Michigan." *Social Problems* 52:618–648.

Salleh, Ariel. 2010. "A Sociological Reflection on the Complexities of Climate Change Research." *International Journal of Water* 5(4):285–297.

Sampson, Robert J. 2012. *Great American City: Chicago and the Enduring Neighborhood Effect.* Chicago: University of Chicago Press.

Schlosberg, David. 2009. *Defining Environmental Justice: Theories, Movements, and Nature.* Oxford: Oxford University Press.

Schmuck, Hannah. 2002. "Empowering Women in Bangladesh." International Federation of Red Cross and Red Crescent Societies, February 25. Retrieved April

4, 2009 (http://www.reliefweb.int/rw/rwb.nsf/AllDocsByUNID/570056eb0ae625 24c1256b6b00587224).

Scott, James. 1999. *Seeing Like a State: How Certain Schemes to Improve the Human Condition Have Failed*. New Haven, CT: Yale University Press.

Sen, Amartya. 1993. "Capability and Well-Being." Pp. 30–53 in *The Quality of Life*, edited by A. Sen and M. Nussbaum. Oxford: Clarendon Press.

Shue, Henry. 1993. "Subsistence Emissions and Luxury Emissions." *Law & Policy* 15(1):39–60.

Sklair, Leslie. 2001. *The Transnational Capitalist Class*. London: Blackwell Publishing.

Smit, B. and J. Wandel. 2006. "Adaptation, Adaptive Capacity and Vulnerability." *Global Environmental Change* 16:282–292.

Snipp, C. M. 1986. "American Indians and Natural Resource Development: Indigenous Peoples' Land, Now Sought After, Has Produced New Indian-White Problems." *American Journal of Economics and Sociology* 45(4):457–474.

Spring, Ursula. 2006. "Vulnerability and Resilience Building of Gender Confronted with Extreme Hydro-meterological Events." July 28. Presentation at the Regional Multidisciplinary Research Centre (CRIM), National University of Mexico, Mexico City.

Steinberger, Julia K. and J. Timmons Roberts. 2010. "From Constraint to Sufficiency: The Decoupling of Energy and Carbon from Human Needs." *Ecological Economics* 70(2):425–433.

Stephenson, Wen. 2014. "Ground Zero in the Fight for Climate Justice." *The Nation* 23/30 June:17–25.

Tierney, Kathleen J. 1999. "Toward a Critical Sociology of Risk." *Sociological Forum* 14(2):215–242.

Tierney, Kathleen J. 2007. "From the Margins to the Mainstream? Disaster Research at the Crossroads." *Annual Review of Sociology* 33:503–525.

United Nations Environment Programme (UNEP). 2010. *The Emissions Gap Report: Are the Copenhagen Accord Pledges Sufficient to Limit Global Warming to 2° C or 1.5° C?* Retrieved October 28, 2013 (http://www.unep.org/publications/ ebooks/emissionsgapreport/).

United Nations Framework Convention on Climate Change (UNFCCC). 1992. United Nations, New York. Retrieved October 28, 2013 (http://unfccc.int/resource/docs/ convkp/conveng.pdf).

United Nations Framework Convention on Climate Change (UNFCCC). 2014. *Report of the Ad Hoc Working Group on the Durban Platform for Enhanced Action*. Conference of the Parties, Twentieth Session. Lima, Peru. Retrieved December 16, 2014 (http://unfccc.int/resource/docs/2014/cop20/eng/l14.pdf).

United Nations WomenWatch. 2009. "Women, Gender Equality and Climate Change." Retrieved June 25, 2011 (http://www.un.org/womenwatch/feature/climate_change/downloads/Women_and_Climate_Change_Factsheet.pdf).

U.S. Environmental Protection Agency (EPA). n.d. *Green Book*. Data compiled by MSB EnergyAssociates (http://www.epa.gov/oar/oaqps/gbook/).

U.S. Environmental Protection Agency (EPA). 2014. "Greenhouse Gas Emissions Data." Retrieved June 26, 2014 (http://www.epa.gov/climatechange/ghgemissions/global.html).

Vanderheiden, Steve. 2013. "What Justice Theory and Climate-Change Politics Can Learn from Each Other." *PS: Political Science & Politics* 46(1):18–22.

Veblen, Thorstein. [1899] 2007. *The Theory of the Leisure Class.* New York: Oxford University Press.

Weber, Max. [1922] 1978. *Economy and Society: An Outline of Interpretive Sociology.* Berkeley: University of California Press.

Whitman, S., G. Good, E. Donoghue, N. Benbow, W. Shou, and S. Mou. 1997. "Mortality in Chicago Attributed to the July 1995 Heat Wave." *American Journal of Public Health* 87(9):1515–1518.

Widener, Patricia. 2011. *Oil Injustice: Resisting and Conceding a Pipeline in Ecuador.* Lanham, MD: Rowman & Littlefield.

Wilson, William J. 2012. *The Truly Disadvantaged: The Inner City, the Underclass, and Public Policy.* 2nd ed. Chicago: University of Chicago Press.

Winkler, H., T. Jayaraman, J. Pan, A. S. de Oliveira, Y. Zhang, G. Sant, J. D. G. Miguez, T. Letete, A. Marquard, and S. Raubenheimer. 2011. *Equitable Access to Sustainable Development: Contribution to the Body of Scientific Knowledge,* Beijing, Brasília, Cape Town, and Mumbai: BASIC Experts Group.

Wisner, B., P. Blaikie, T. Cannon, and I. Davis. 2004. *At Risk: Natural Hazards, People's Vulnerability, and Disaster.* 2nd ed. London: Routledge.

World Bank. 2006. *Where is the Wealth of Nations?* Washington, DC: Author.

York, Richard, Eugene A. Rosa, and Thomas Dietz. 2003. "A Rift in Modernity? Assessing the Anthropogenic Sources of Global Climate Change with the STIRPAT Model." *International Journal of Sociology and Social Policy* 23(10):31–51.

Young, Iris M. 1990. *Justice and the Politics of Difference.* Princeton, NJ: Princeton University Press.

Zahran, S., D. W. Hastings, and S. D. Brody. 2008. "Rationality, Inequity, and Civic Vitality: The Distribution of Treatment, Storage, and Disposal Facilities in the Southeast." *Society and Natural Resources* 21:179–196.

Zahran, S., L. Peek, J. G. Snodgrass, S. Weiler, and L. Hempel. 2011. "Economics of Disaster and Risk, Social Vulnerability, and Mental Health Resilience." *Risk Analysis* 31(7):1107–1119.

6

Adaptation to Climate Change

JoAnn Carmin, Kathleen Tierney, Eric Chu, Lori M. Hunter, J. Timmons Roberts, and Linda Shi

INTRODUCTION

Scientific predictions suggest that climate change will lead to sea level rise, increased intensity and frequency of storms, and greater variability in temperature and precipitation.[1] The consequences of these changes include losses of wetlands and fisheries, greater flooding and drought, stress on physical infrastructure and buildings, and alterations in food security, livelihoods, and human health and safety. In many instances, the places and people that will be hardest hit are the ones least able to cope. While it is essential to reduce greenhouse gas (GHG) emissions, most projections suggest that the impacts of climate change affect all nations, cities, and communities (Parry et al. 2007; Stott et al. 2010). The most recent comprehensive report from the Intergovernmental Panel on Climate Change (IPCC) concluded on the basis of increasingly strong research evidence that the consequences of anthropogenic climate change are now being felt on all continents and oceans; that impacts on water resources and crops are generally negative; and that climate-related extremes such as heat waves and wildfires reveal growing vulnerabilities and shortcomings in the capacity to prepare for such events (IPCC 2014). Many changes now under way are irreversible. Important for this chapter, the IPCC also found that such negative impacts are affecting the poor more, particularly those living in low-lying coastal areas and small island nations, and will continue to do so. On a more optimistic note, the IPCC found evidence of an increase in adaptation activities in both the Global North and the Global South (IPCC 2014). Clearly, action is imperative in order to protect economies, environmental quality, individual and collective assets, and human well-being.

In human systems, climate adaptation is "the process of adjustment to actual or expected climate and its effects, which seeks to moderate harm or exploit beneficial opportunities" (IPCC 2012, 2014). Sociological knowledge and perspectives offer a means for understanding the ways in which social dynamics shape underlying conditions; the choices that are available in different societal contexts; and the decision processes involved in individual, community, organizational, and governmental climate change adaptation efforts. Longstanding social and institutional forces have given rise to vulnerabilities and inequities at international, national, and local levels, which are compounded by the emergence of risks and impacts stemming from climate change. Sociological perspectives are important for identifying factors and forces that can promote institutional and collective action, reduce vulnerability, and facilitate equitable decisions and actions.

We begin this chapter with an introduction to climate adaptation, highlighting the ways in which social dynamics give rise to vulnerabilities, as well as how they can provide a foundation for adaptation. We then use a sociological lens to explore existing conditions and adaptation dynamics in four thematic areas: international finance, international development, disaster risk reduction, and human migration. Since adaptation research by sociologists is still in its infancy, these sections draw on works from a range of disciplines to illustrate the societal dimensions of adaptation and demonstrate links to sociological thought. At national, regional, and local levels, adaptation requires investments in engineered and technical solutions. While these are critical, in the course of this chapter, we highlight socially based strategies for adaptation and show how these approaches benefit from the knowledge generated through sociological scholarship.

CLIMATE CHANGE ADAPTATION: A SOCIAL CHALLENGE

Climate change adaptation measures aim to reduce existing and future vulnerability, commonly viewed as comprising three elements: exposure, sensitivity, and adaptive capacity (Parry et al. 2007). Vulnerability is thus a combination of the stress faced by an entity, such as a physical system, social system, organization, or individual; the extent to which the entity will be affected; and the degree to which the entity can resist, cope with, or respond to stressors (Adger 2006; Cutter, Boruff, and Shirley 2003; O'Brien et al. 2007). With respect to the social aspects of vulnerability, some individuals and groups are more vulnerable than others, owing to their limited capacity to prepare for or cope with stresses. For instance, individuals with disabilities and elderly persons may be physically incapable of undertaking protective actions in the face of hazards. Those who are socially isolated may have difficulty adjusting to the changes taking place around them, while ethnic minorities have a long history of unequal treatment in many parts of the world (Cutter et al. 2003; Paavola and Adger 2006). Those who are poor and powerless are particularly vulnerable to the impacts of both climate change and climate-related extreme events as a consequence of their lack of financial resources and limited access to basic services (Huq et al. 2007; Rosenzweig et al. 2011; Tol et al. 2004). Worldwide, poor people often concentrate in places that have high exposure to hazards, such as floodplains and hillsides that are prone to landslides, and in structures that are physically vulnerable to disasters (Pelling 2003a, 2003b). They also face difficulties during disaster recovery because they often lack the resources to rebuild their homes and reestablish livelihoods (Heltberg and Lund 2009).

In the domain of natural and technological hazards, there is longstanding agreement among social scientists that vulnerability to disasters is a product of political and economic decisions and policies, as well as collective and individual choices and actions (Mileti 1999; Tierney, Lindell, and Perry 2001). Similarly, both vulnerability and the limited preparedness for climate impacts in most parts of the world can be traced to economic, political, and policy decisions. In the United States, for instance, policies that encourage intensive growth in hazardous coastal zones that place businesses and households at risk to climate change impacts are not aligned with floodplain regulations that have stricter standards. Globally, the desire to achieve short-term economic and development gains gives rise to policies, plans, and measures in areas ranging from infrastructure to health care to ecosystem management that fail to account for predicted climate variability. At the individual and household levels, lifestyles and consumer choices are also impediments to action to combat the risks posed by climate change.

Just as institutional and social dynamics have contributed to climate change and are shaping vulnerabilities, they also are integral to preparing for climate impacts. Research on adaptation often emphasizes the importance of adaptive capacity (Nelson, Adger, and Brown 2007; Parry et al. 2007; Smit and Pilifosova 2003) and the availability of tangible resources such as money and technology (Smit and Wandel 2006; Yohe and Tol 2002). Such resources are often related to the need for investments in engineered, technological, and ecosystem-based measures. While tangible resources are critical to fostering large-scale change, the capacity to adapt also depends on less tangible resources such as policies, governance, networks, and support systems (Brooks, Adger, and Kelly 2005; Eakin and Lemos 2006; Engle 2011). A dual emphasis on tangible and intangible resources highlights how adaptation requires investments in structural options, but also the design and implementation of institutional and societal measures.

As Table 6.1 suggests, structural, institutional, and societal adaptation measures are three critical pathways for reducing vulnerability and enhancing adaptive capacity. Structural measures aim at making physical, technological, and ecological systems less vulnerable and more resilient.

Table 6.1 Pathways and Examples of Adaptation Options

Structural

Engineered	Seawalls; expanded aquifer storage; beach nourishment; redirecting rivers; infrastructure provision and upgrading
Technological	Environmental monitoring systems; green roofs; hazard mapping and monitoring systems; structural weatherproofing and upgrading
Ecosystem-based	Reestablishing wetlands and managing floodplains

Institutional

Laws and regulations	Executive orders; easements; zoning and land use regulations; coastal zone and disaster loss reduction legislation
Government policies, programs, and services	Slum upgrading programs; National Action Plans for Adaptation; National Adaptation Plans; climate, coastal, water, land use, and disaster management plans; urban upgrading programs
Economic	Economic incentives; insurance, lending and finance programs; development aid; provision of social safety nets; tax-related policies to encourage and finance adaptation

Societal

Educational	Agricultural and other types of extension programs; technical assistance and training programs; public education
Informational	Early warning systems; water quality monitoring; epidemiological monitoring; emergency alerts and warnings
Behavioral	Rainwater harvesting; reduced water consumption; storm drain clearance; individual and community gardens; retreat and migration; household disaster mitigation and evacuation planning
Social services	Food banks; vaccination programs; preventive health services
Sociodemographic	Livelihood replacement; resettlement or displacement seasonal or permanent settlement abandonment

Based on Birkmann et al. 2010; Broadleaf Capital International and Marsden Jacob Associates 2006; Burton 1996; IPCC 2014; National Research Council 2010; Parry et al. 2007; Smithers and Smit 1997.

Alternatively, institutional approaches draw on laws, regulations, and economic incentives to encourage adaptation. Societally based approaches aim to encourage adaptive behaviors and ameliorate problems that are caused or exacerbated by climate change.

Maturation in our understanding of adaptation has been accompanied by a shift in perspective. Early views on adaptation emphasized maintaining the status quo and were rooted in theories of natural systems and ecological dynamics. In contrast, current views focus on the importance of transforming existing systems and underscore the centrality of human decisions and actions, ranging from policy interventions to community-based efforts (IPCC 2014). These transitions highlight the importance of balancing structural measures with those that are based on an understanding of institutional and social dynamics.

SOCIOLOGICAL INSIGHTS ON CLIMATE ADAPTATION

Sociological research, along with scholarship from other social science fields, offers insight into the ways in which vulnerability to climate change can be reduced and equitable adaptation decisions and actions achieved. The sections that follow discuss contributions of sociological knowledge to the following thematic areas, each of which is central to both climate and general scholarship within the field of sociology: international adaptation finance, international development, disaster risk reduction, and human migration. We begin with an exploration of global adaptation finance negotiations, both because the global scale is where much of the discussion on climate adaptation began and because global perspectives and strategies influence adaptation framings and actions at other scales. At the same time, we note that while negotiations have often stalled at global and national scales, cities and communities have begun to integrate adaptation into their development initiatives, either autonomously or with support from development partners. This is particularly the case in the Global South, where infrastructure and service gaps present the greatest opportunity for climate integration (Anguelovski, Chu, and Carmin 2014), but autonomous activity is also occurring in the Global North, where cities have suffered major economic losses, have greater access to climate projections data, and are motivated to promote local resilience (Aylett 2014; Carmin, Nadkarni, and Rhie 2012). As the section on disasters risk reduction indicates, human psychology and institutional incentives favor short-term responses, leading to brief windows of mitigation and adaptation opportunity following disaster events. All too often, however, these windows quickly close. At individual and group levels of analysis, migration can become the most permanent and drastic form of adaptation. However, while migration represents poten-tially transformative adaptation, experience indicates the most vulnerable are the least likely to migrate and therefore the most likely to bear the brunt of climate impacts.

These thematic discussions point to the ways in which social and institu-tional dynamics are critical to understanding and advancing climate change adaptation. The adaptation field has tended to emphasize the identification and implementation of engineered and technological solutions, such as improved urban infrastructure and early warning systems, as a means of coping with climate change. While such measures are critical, they fail to take into account the fact that just as vulnerability to the impacts of climate change is rooted in social and institutional conditions, reducing vulnerabil-ity requires social and institutional analysis and action as well. By examin-ing the cross-cutting nature of these themes, subsequent sections highlight their effect on global, national, and local capacities to engage in various adaptation pathways, such as those shown in Table 6.1.

ADAPTATION FINANCE

Obtaining adequate funding for climate change adaptation and delivering it in appropriate, equitable ways are major challenges that are closely linked to rising global inequalities. If the financing pledged by the world's wealthy countries to the world's poorer nations to cope with climate impacts and green their economies were delivered, it would result in an unprecedented level of global wealth transfer, doubling or tripling all foreign assistance within this decade, and having a significant impact on global stratification and international political relations. Much is at stake in climate adaptation efforts, including the course of development in much of the world and the global (re)distribution of wealth.

Several sociological theories offer useful insights into both climate change and adaptation issues. These include world systems theory and broader political economy approaches, institutional analysis of international organizations and national governments, game theory approaches, and political sociology. For example, world systems theory describes an endur-ing and evolving global division of labor, investment, and trade in which wealthy "core" nations capture high-income niches and poorer "peripheral" countries sell their labor and natural resources cheaply. The theory helps explain the persistent vulnerability of peripheral societies, and largely lines up with "realist" theories in international relations that focus on how wealthy nations act in their short-term self-interest to delay meaningful action to address world poverty (e.g., Ciplet, Roberts, and Khan 2013; Roberts and

Parks 2007). Theories like these can help account for the bargaining among actors and nations over who owes whom, who should be prioritized in receiving funding, and whether climate aid should be considered charity, compensation, or mutual cooperation for economic development. We point to other issues later but first provide a broader context by discussing climate justice in the world system context.

Imbalances and Inequities of Adaptation Finance

Estimates of the cost of adaptation through climate-proofing investments in developing countries range from US$4 to $109 billion per year by 2030, with more recent estimates pushing toward and even exceeding the higher end of the scale (Flåm and Skjærseth 2009; Jones, Keen, and Strand 2012; World Bank 2010). Parry et al. (2009) contend that these figures may underestimate actual costs of climate adaptation substantially, owing to unaccounted-for sectors, decisions regarding whether to design the built environment for historical climate variability or for projected climate extremes, and the cost of residual damages that cannot be remedied through adaptation measures. In addition, the costs of climate mitigation in developing countries are esti-mated to range from US$177 to $695 billion per year by 2030 (Parry et al. 2007). These are vast amounts: foreign assistance from all nations for all issues (including health, education, agriculture, and infrastructure) totals about $140 billion per year (AidData 2013).

To address climate change mitigation and adaptation in developing coun-tries that have had very little role in creating the problem of climate change (Roberts and Parks 2007), developed countries committed to providing "new and additional resources" under the 2009 Copenhagen Accord. To show that such aid could be delivered and well managed, the Fast Start Finance period from 2010 to 2012 was initiated with a pledge of $30 billion in funding. This initial climate financing was then promised to scale up to amounts approaching $100 billion a year by 2020, primarily channeled through a new Green Climate Fund, which remains empty to date. Retrospectively apply-ing a new 2012 definition of what classifies as adaptation finance, the seven main multilateral development banks (MDBs) estimate that they provided US$3.7 billion for adaptation projects in 2011 (MDB 2012). Donor coun-tries together announced that $33 billion had been committed during the Fast Start Finance period, but much was in the form of loans, promised before Copenhagen and therefore not "new and additional" monies (Ciplet et al. 2012).

Total pledged funding therefore falls far short of the estimated costs of mitigation and adaptation in developing countries, and the gap between pledged and delivered funding may increase, especially in the face of economic crises among donor countries. Although a UNFCCC Least Developed Countries Fund targets those nations for which adaptation will be the primary climate issue, climate finance continues to prioritize mitigation, with adaptation receiving only 19 to 25 percent of the delivered Fast Start funding (Ciplet et al. 2012). Adaptation, in short, remains a "poor stepchild" to mitigation in global priority setting, apparently because it represents a local rather than a "global public good" (see, e.g., Khan and Roberts 2013).

As Chapter 5, "Climate Justice and Inequality," discusses in depth, the adaptation finance debate underscores a "profound unfairness" (Huq et al. 2007) regarding which entities have caused and continue to cause climate change and which suffer from its effects, both within and among countries. The question of who should be prioritized for funding is prominent (Ciplet et al. 2013), even as negotiations and political maneuvering have delayed action and funding disbursement to those already affected by climate change. The 2007 Bali Roadmap set down three groups of nations as "most vulnerable": Africa, the least developed countries, and small island developing states. Other nations, including Pakistan, Guatemala, and some landlocked, glacier-dependent nations, have demanded priority, also calling themselves "highly vulnerable." Numerous vulnerability rankings exist, such as those of DARA, the Stockholm Environment Institute, and Barr, Fankhauser, and Hamilton (2010), but their use of different assumptions and criteria does not resolve this wedge issue (Klein 2010). Inequality in the disbursement of funds is also affected by which entities manage the funds and by bureaucratic requirements for applying for and monitoring their use, which favor nations with more capacity to fill out paperwork and compose attractive proposals. The allocation of foreign aid also favors those same countries and is often driven by the colonial histories and geopolitical interests of donor countries (Hicks et al. 2008; Nielson and Tierney 2003). Nor does the availability of adaptation funds necessarily lead to their absorption, much less that they will translate into equitable and effective adaptation initiatives (Ayers 2009).

The Political Economy of Adaptation Finance

As suggested above, studying financing decisions for adaptation reveals power dynamics within the world system. With only a few billion U.S. dollars a year being delivered for adaptation, available financing is falling far below what is required for effective adaptation. A long struggle has gone on between wealthy and poor nations over language in climate treaties and the charter of the World Bank's Global Environment Facility. The Global Environment Facility is the core implementing agency for much early adaptation work, and its charter states that adaptation funds will pay only for the "additional costs" required for a project due to climate change, rather than for other parts of a project, including preparations for climate variability, which incorporates both baseline variation and climate change (Huq and Reid 2004). This "additionality" debate, which in some cases has led to unreasonable decisions to fund only the incremental portion of a project when there was no available funding for baseline costs (Ayers and Dodman 2010), delayed early adaptation efforts. Lack of understanding about how to mainstream climate adaptation into larger development projects and core governmental and private-sector decisions also has hampered international support.

There is great expectation in the capitals of OECD countries that the private sector and loans will fill some of the gap in climate funding, but civil society groups in developing countries view adaptation loans as a double climate injustice, arguing that direct funding for adaptation, rather than loans, is the moral responsibility of developed countries (Regmi and Bhandari 2012). In 2011, over fifty organizations in developing countries that were scheduled to receive adaptation loans submitted a letter to members of Parliament in the United Kingdom, asking that climate loans be converted to unconditional grants managed by the United Nations Adaptation Fund, which helped to create the World Bank's Pilot Program for Climate Resilience. Such controversies have crowded out equally important considerations of efficient, effective, and equitable uses of funds (Fankhauser and Burton 2011), reflecting the legacy of distrust between the Global North and South—a legacy with origins in colonial and neocolonial times that is reinforced in the frustrating "development decades" and trade pact disputes of the General Agreement on Trade and Tariffs (GATT) and the World Trade Organization (WTO).

Further debates center on whether projects funded by the MDBs should make adaptation an issue of loan compliance under the environment and social safeguard process. Members of the World Bank Group argue that they should develop flexible guidelines to incorporate climate risk without creating new "conditionalities," which are often resented by receiving countries (World Bank 2012). One concern is over the uncertainty of predicted impacts, compared to better-understood risks under current environmental and social safeguards. Despite significant international efforts to create adaptation finance arrangements, led in large part by MDBs themselves, the treadmill of production theory (Schnaiberg 1980) appears to assert itself in the end. That is, national and local interests drive decisions about the allocation of national treasury funds for adaptation, and sometimes there is significant impetus for generous aid funding only when there are business opportunities for national elites (Hicks et al. 2008).

With adaptation finance dwarfed by total private investment in urbanization worldwide, international institutions and governments are turning to the private sector for so-called win–win adaptation partnerships that generate co-benefits (Huq et al. 2007). Real estate developers' and investors' claims about "going green," rooted in the theory of ecological modernization that posits a synergy between economy and environment in both political decision making and business practices, have generated mixed results. Sometimes voluntary programs and innovative solutions are developed (Mol and Sonnenfeld 2000; Rudel, Roberts, and Carmin 2011), but more often it is government regulation, not voluntary action, that has spurred private sector innovation. Despite agglomeration economies, many global firms today are not as tied to their locations as, say, manufacturing and extractive industries, and are liable to relocate if local political and economic risks are perceived as too great (Huq et al. 2007). Competition may arise among current and aspiring global cities to make particular kinds of adaptation investments, regardless of whether they are effective or equitable, in order to attract development (Hodson and Marvin 2010). Unlike the last major period of global environmental regulatory reform, the past three to four decades have witnessed a "hollowing out" of the state (Hodson and Marvin 2010), particularly vis-à-vis regulations on the private sector, which has shifted the social pact between powerful players, as posited by political economic regulation theorists such as Aglietta (1976) and Lipietz (1992). Awareness of an adaptation imperative has begun to take off in localities around the world, with some seeking to turn it into an economic development opportunity and a chance to distinguish themselves from competing locations (Anguelovski, Chu, and Carmin 2014; Carmin, Anguelovski, and Roberts 2012).

In the United States, the drafting of the massive Waxman-Markey Climate Solutions Act of 2010 presented a revealing case of climate funding and the ways in which domestic interests drive decision making. The bill famously included a cap-and-trade system for carbon dioxide, in which an increasing proportion of the emissions permits would be auctioned off. The apportionment of those auction revenues made plain the influence of different lobbying groups over its crafters in the House of Representatives, where it narrowly passed. Larger proportions of funding were allocated to energy subsidies for the U.S. poor, to prevent deforestation, and to assist adaptation efforts in the United States, with half that funding spent on wildlife and natural resources adaptation (Sheppard 2009). Much smaller amounts (only 1 percent) would have gone for developing country adaptation through 2021, when allocations would start to scale up to 4 percent after 2027. Even this percentage for adaptation funding was only kept in the bill through an effort of church and aid groups in Washington, working together with a few environmental lobbyists. The same small group of lobbyists also managed to sustain the Obama administration's substantial increase in international adaptation aid (Kincaid and Roberts 2013).

BRIDGING CLIMATE AND DEVELOPMENT

Insights on the political economy and dynamics of development are critical for understanding issues related to climate change adaptation. With the important exceptions of China and the oil producers, developing countries not only have relatively low carbon emissions but are also projected to experience the severest climate impacts. Climate change is projected to erode infrastructure, the availability of resources and services, livelihoods, and past gains in reducing poverty (Parry et al. 2009; Hunt and Watkiss 2011), as well as have an adverse effect on societies' capacities to adapt successfully (Ahmed, Diffenbaugh, and Hertel 2009; Ayers and Dodman 2010). Consequently, adaptation has become closely associated with the development context, particularly in terms of reducing poverty, improving livelihoods, and addressing socioeconomic inequalities (Ayers and Dodman 2010; Dodman and Satterthwaite 2008).

Improving individual and societal resilience to climate impacts is critical to maintaining and advancing development achievements (Boyd et al. 2008; Brooks, Grist, and Brown 2009; Lemos et al. 2007; Someshwar 2008). Nonetheless, the connections between climate change and development often are contended. Difficult choices and tradeoffs arise when considering whether market-based approaches to growth should be the primary tool used to address both poverty and climate change (Gasper, Portocarrero, and St. Clair 2013); the extent to which developing countries should mitigate emissions or industrialize in ways similar to those followed by developed countries; and the degree to which developed countries should be held accountable for the impacts their consumption and subsequent GHG emissions actions are having on those that are less developed (Brooks et al. 2009; Cannon and Müller-Mahn 2010; Simon 2012). Competing economic priorities and social values, along with power differentials, combine to give rise to tensions in the creation of policies and the promotion of adaptation action in the development context (Wolf 2011).

Mainstreaming Adaptation into Development

Rather than climate adaptation and development being viewed as antithetical, emphasis is being placed on "mainstreaming" adaptation into development. The rationale is that adaptation can contribute to the livelihoods of people and improve their capacity to deal with changes in climate (Halsnæs and Trærup 2009). Many governments are electing to mainstream climate change into various sectors, primarily by climate-proofing infrastructure and other investments made for "adaptation plus development" (Ayers and Dodman 2010). This integration takes place at a range of governance levels, from local to global (Huq and Reid 2004). For example, at the local level, programs such as the Asian Cities Climate Change Resilience Network, the Cities and Climate Change Initiative, and Asian Cities Adapt seek to integrate planning for climate change resilience into existing urban development planning activities (Anguelovski and Carmin 2011; Brown, Dayal, and Rumbaitis Del Rio 2012; Sharma and Tomar 2010). At the national level, many countries are beginning to enact legislation that addresses the nexus of climate change adaptation and development. India's National Action Plan on Climate Change (Dubash 2012; Sharma and Tomar 2010) and China's National Twelfth Five-Year Plan both highlight the importance of adapting to sea level rise in coastal zones, maintaining resilience of natural ecosystems affected by climate change, and preventing and controlling climate disasters and environmental degradation (Li et al. 2011). At the international level, programs such as the National Action Plans for Adaptation seek to foster the integration of adaptation into national economic and sector plans (Huq and Reid 2004) and to foster city adaptation plans and awareness of the need to address climate change in infrastructural projects (Füssel, Hallegatte, and Reder 2012; Parry et al. 2009).

Given the decentralized structures of many governments, the idea of mainstreaming adaptation into development to harness the strength of public, private, and nonprofit partnerships and networks has been widely addressed in the literature and is being adopted in practice (Bulkeley et al. 2012; Tompkins and Eakin 2012). A variety of multilateral organizations working on development issues and poverty reduction have started to assess the implications of climate change for their work (Diamond and Bruch 2012; Gasper et al. 2013; Harris and Symons 2010). In addition, the expansion of mechanisms for climate financing (Rübbelke 2011) is providing further incentives for integrating climate change into existing development policies (Kok et al. 2008). These approaches have the potential to give rise to partnerships that integrate adaptation and development in order to simultaneously address disaster risk reduction, poverty reduction, and infrastructure, institutional, and community resilience (Da Silva, Kernaghan, and Luque 2012; O'Brien et al. 2008).

Despite its benefits, mainstreaming can be difficult because adaptation typically requires cross-sector and regional integration (Carmin, Dodman, and Chu 2013; Huq et al. 2007) as well as navigating the political tensions between local and national governments (Dodman and Mitlin 2014). Further, the phrase "climate-proofing development" is often used carelessly, giving a false sense that climate impacts are being managed. For instance, prior to an extreme weather event, it is very difficult to determine whether local communities are prepared and whether mainstreamed initiatives will prove to be effective or maladaptive (World Bank 2012). To better gauge preparedness, the World Bank and some regional banks are trying to understand how they can monitor the efficacy of adaptation efforts (World Bank 2012).[2] While mitigation projects can be evaluated by the estimated tons of carbon avoided per dollar invested, successful adaptation is far more difficult to assess, partly because it requires measuring what has *not* taken place, such as avoided deaths and injuries or economic losses from climate impacts (Brooks et al. 2011).

Technical assistance and funding for "climate-proofing" projects have favored engineered solutions, mainly due to the emphasis MDBs place on infrastructure. Governments typically do not receive international loans or accompanying technical assistance for nonstructural solutions, such as strengthening social capital or improving governance (see Table 6.1). However, those kinds of solutions are essential for addressing conditions that contribute to climate vulnerability (Adger 1999). Supporting social structures of mutual support and disaster response in vulnerable communities may be more effective at boosting resilience than infrastructure projects such as building seawalls. Safeguards such as cash transfer, micro-insurance, and micro-finance programs are taking root in some locations. Cash transfers build adaptive capacity by boosting long-term nutritional health and education, reducing the need to cut into these expenditures during times of instability and extreme events, and promoting flexibility in recovering from disasters (Wood 2011). In Kenya and Tanzania, for instance, households with an extra source of income were better able to withstand droughts than those without (Eriksen, Brown, and Kelly 2005). Although micro-finance is undergoing tremendous transformations, the underlying function of providing poor and "unbankable" people with a source of non-exploitative loans is potentially critical in lifting people out of poverty. Small loans can help households accumulate and manage assets, adapt their homes, or even invest in community-scale safeguards directed at climate change (Hammill, Matthew, and McCarter 2009). A number of micro-insurance pilots are under way, although the issue of affordability may limit the potential for widespread scale-up (Suarez and Linnerooth-Bayer 2010). Unfortunately, while diversifying household incomes boosts adaptive capacity more than crop switching (Agrawal 2010; Eakin 2000), income diversification is often harder to justify to donors.

Equity in Adaptation and Development

Adaptation scholarship related to social and environmental justice has focused primarily on the equitable distribution of the costs of adaptation (Beckman and Page 2008; Paavola and Adger 2006; Shepard and Corbin-Mark 2009) rather than on the specific vulnerabilities and needs experienced by those at risk (Schlosberg 2012). At the international level, governments and multilateral agencies often engage only in debates over who is responsible for providing adaptation assistance or how assistance should be directed to and divided among developing countries (Duus-Otterström and Jagers 2012; Grasso 2010; Harris and Symons 2010). As a result, prescriptions for remedying existing climate injustices remain vague about how they will go about fostering the conditions that are necessary for basic human development (Harris and Symons 2010; Schlosberg 2012).

An alternative to this "rights versus responsibilities" framework (Parks and Roberts 2010; Roberts 2009) aims to understand the ecological, social, political, and cultural conditions that contribute to the unequal distribution of risks and capacities underlying vulnerabilities (Schlosberg 2012). This framing, which highlights how the inability to gain sociopolitical recognition compounds people's vulnerability to climate impacts (Schlosberg 2012), is particularly important in areas already experiencing climate change. For example, in coastal cities facing worsening floods, governments face difficult decisions about risk, land, and adaptation infrastructure allocation, with the potential to exacerbate vulnerability among the most disenfranchised (MacCallum, Byrne, and Steele 2014). In some cases, adaptation can become a justification for governments and the private sector to push through slum clearance or longstanding megaprojects (Moser et al. 2010). While the stated goal of such actions is to reduce the vulnerability of the poor by relocating them from waterways to peripheral resettlement towns, their removal allows governments to reduce flood risks for remaining titled landowners who are too costly to relocate (Zoleta-Nantes 2000). So-called pro-poor adaptation strategies will not be successful where governments refuse to work with the poor, or worse see them as the problem (Satterthwaite et al. 2007). The distributive justice questions surrounding decisions about which solutions count as low or no regret; who receives infrastructural improvements for adaptation; whether allocations of adaptation burdens are distributed proportionally; and how preexisting social vulnerabilities potentially exacerbate climate impacts are therefore central to sociological analyses of equity in climate adaptation.

DISASTERS AND DISASTER RISK REDUCTION

For over six decades, sociologists and other social scientists have been conducting research on the societal dimensions of hazards and disasters. Many of the behavioral lessons learned from research on hazards and disasters are generalizable to the study of climate change adaptation. These lessons focus on such topics as social vulnerability to environmental threats, factors that influence hazard perceptions, barriers to undertaking adaptive actions, and institutional and policy responses to hazards and disasters. Many of the measures that societies can undertake in managing risks associated with hazards are similar to those required for adaptation to climate change, suggesting that insights from disaster research are directly transferrable. Indeed, public awareness of climate change often derives from experiences of natural hazards and disasters that are framed as climate-related.

Social Vulnerability to Environmental Hazards

Research on hazards emphasizes the extent to which populations are differentially vulnerable to environmental hazards and disasters. At the individual and household levels, axes of social inequality, including gender, race, class, ethnicity, disability, and their intersections, are associated with patterns of exposure to hazards, impact severity, and differential recovery outcomes (Cutter et al. 2003; Fothergill 2004; Phillips et al. 2009; Tierney et al. 2001). Hazard vulnerability is also driven by demographic trends such as migration. For example, studies on normalized hurricane losses in the United States indicate that increasing losses can be traced directly to the movement of populations and intensified development in vulnerable coastal areas (Pielke et al. 2008). On a global scale, vulnerability to disasters is associated with the position of nations within the world system and with conditions that are characteristic of developing countries, such as rapid urbanization and the growth of informal settlements, particularly when population growth occurs in marginal and hazardous areas (Alexander 2006; Pelling 2003b; Wisner et al. 2004; World Bank 2010). Rapid urbanization is invariably accompanied by the depletion of natural resources, including those that offer protection against disasters, further increasing vulnerability (Pelling 2003a). As Chapter 5, "Climate Justice and Inequality," elaborates, these findings offer insights into which groups are vulnerable to climate change and climate-related extreme events. In the United States, minority-group members tend to be more vulnerable to climate-related hazards than their white counterparts (Emrich and Cutter 2011). For example, extreme heat is the leading cause of hazard-related deaths in the United States (Borden and Cutter 2008), and studies conducted in Phoenix, Arizona, indicate that Mexican Americans are more likely than white residents to live in areas that experience urban heat island effects (Harlan et al. 2006).

Perceptions, Salience, and Behavioral and Policy Responses

Both perceptions and behavioral responses to hazards and disaster events are influenced by a range of factors, such as the perceived characteristics of hazards (familiarity, speed of onset, presence of identifiable cues, and other attributes); characteristics of those at risk, including income, education, race and ethnicity, social class, and gender; and knowledge of and ability to carry out recommended protective measures. Studies indicate that taking action in the face of hazards has low salience for most households, organizations, and communities. There is a strong tendency to focus on daily and persistent concerns as opposed to events that are seen as infrequent, particularly for low-income populations facing continual daily pressures. Hazard familiarity and the nature of prior disaster experiences are important influences on perception and behavior, as are gender, race, ethnicity, education, and social class more generally. Household composition is also important; for example, having younger children in the home is associated with taking action to reduce disaster losses. Regarding social psychological and psychological factors, the literature emphasizes how factors such as the development of a personalized (not generalized) sense of risk and feelings of self-efficacy in the face of hazards can encourage self-protective behavior (see Lindell and Perry 2000; Mileti and Sorensen 1987; Tierney et al. 2001).

The salience of hazards and concerns about safety tend to increase in the aftermath of disaster events, but the influence of experience is time-limited. For example, following the devastating hurricane season of 2005 in the Atlantic, there was a significant upsurge in the demand for flood insurance. However, households that purchased insurance tended to drop their coverage later and on average did not keep their coverage any longer than those who bought insurance before 2005 (Michel-Kerjan, Lemoyne de Forges, and Kunreuther 2012). In many communities, only a small proportion of property owners in flood risk areas purchase flood insurance, and even when they do, many discontinue their coverage within one to three years. This is true even in areas that have experienced recent flooding (Michel-Kerjan et al. 2012). These findings suggest that incentives are lacking at the household level for sustained efforts to reduce risk, which is relevant for climate change adaptation.

Despite the belief that knowledge deficits are a cause of inaction in dealing with hazards, research indicates that hazard knowledge does not necessarily translate into practice. For example, extensive educational efforts over a thirty-year period, as well as several damaging earthquakes, have made residents of high-earthquake-hazard parts of California quite knowledgeable about the earthquake threat. However, education and experience have not encouraged many households and businesses to plan for earthquakes beyond undertaking minimal low-cost or low-effort preparatory actions that they would take for any sort of a disaster (Kano et al. 2009).

New research that focuses specifically on climate change also suggests that large numbers of people can be aware of that hazard and understand its future impacts but still not incorporate that knowledge into their daily lives and routines. Like many studies in the hazards area, Norgaard's research on perceptions and behaviors related to climate change in a Norwegian community emphasizes the disconnect that often exists among understanding hazards, feeling able to take effective action to reduce future losses, and undertaking that action. Tellingly, Norgaard's work also reveals that there are psychological, social, and cultural pressures that appear to operate in ways that cause people to "wall off" from their everyday lives uncomfortable truths about how the climate is changing (Norgaard 2011).

Nor do hazards rank high on either local or state political agendas (Rossi, Wright, and Weber-Burdin 1982). Public officials are mainly concerned with being held accountable for events that are likely to occur during their time in office, not with events that could occur at some time in the future. Disaster events often open policy windows during which loss-reduction measures can be adopted (Birkland 1997). However, policy windows typically close rapidly and programs can face ongoing opposition when the sense of crisis has passed.

When local communities do take action to reduce hazards, they typically do so because of direct experience with disaster losses that are socially defined as unacceptable and when collective action develops around the need for change. Both local and national action for disaster risk reduction tend to be driven by elite groups and coalitions, such as scientists, engineers, and concerned lawmakers, rather than by grassroots groups (see, e.g., Stallings 1995). Laws, regulations, incentives, and guidance from states and the federal government also provide incentives for local action. For example, a number of states require local land use planning for disaster reduction, and when implemented these measures have reduced insured disaster losses (Burby 2005).

Barriers to the Adoption and Implementation of Risk Reduction Measures

As high-risk coastal storms, wildfires, droughts, and floods become more common as climate change progresses, and as sea levels rise, adaptation will necessarily require changes in land use and the built environment, the retrofitting or rebuilding of various infrastructure systems, and in some cases the abandonment of at-risk areas. However, these kinds of measures have proved difficult to implement for many hazards. Political-economic forces often stand in the way of hazard loss reduction efforts. Local "growth machine" coalitions (Logan and Molotch 1987) that champion real estate-based economic development often oppose efforts to mitigate disaster impacts and support actions that increase the risk of disaster-related losses, such as intensified development of properties in high-hazard areas (Burby 2006; Freudenburg et al. 2009; Mileti 1999). Programs to reduce losses from hazards such as floods have made little headway in the face of development pressures. For example, government efforts to buy up at-risk properties in the aftermath of the massive 1993 floods in the U.S. Midwest have been more than offset by more intensive development within those same floodplains (Pinter 2005).

Wildfire hazards provide another illustration of this point. Wildfire losses are also being driven largely by development in the wildland/urban interface (WUI). In 2000, the WUI constituted 11 percent of the land area in the contiguous United States and 38 percent of all housing units. The explosion of development in the WUI has been driven by several factors: patterns of migration to the West and Southeast; movement out of urban areas and into suburban and later exurban areas; growth in the number of second homes; and amenity migration on the part of those seeking to live in close proximity with nature (Hammer, Stewart, and Radeloff 2000). Introducing additional dwelling units into forested areas adds both additional fuels and additional sources of fire ignition, increasing vulnerability to climate- and non-climate-related wildfires.

At global, national, and subnational levels, disaster risk management and climate change adaptation depend critically on the quality of governance and the capacity to carry out risk reduction projects effectively. On a global scale, deaths and other losses from natural hazards are associated with governance problems such as corruption and lack of institutional capacity (Kahn 2005; Raschky 2008). In the United States, states and local communities show considerable variation in both their willingness and ability to manage hazards. A federal law, the Disaster Mitigation Act of 2000 (DMA2K), requires local, state, and tribal jurisdictions to develop plans for mitigating the hazards to which they are exposed. However, the law is silent on plan implementation. The few studies that have been carried out since the passage of DMA2K, which include studies of coastal states and communities that are exposed to climate-related hazards, suggest that plans have serious deficiencies, not the least of which is the failure to specify paths to implementing hazard mitigation projects (Kang, Peacock, and Husein 2010; Lyles, Berke, and Smith 2012). Deficiencies also exist in other areas, such as building code enforcement and compliance with flood plain management policies specified in the National Flood Insurance Program (Burby 2001).

Analyses of impediments and incentives for the adoption and implementation of climate adaptation measures closely mirror those discussed in the literature on disaster risk reduction strategies. Paralleling that literature, deficits in leadership, financial and human resources, and information and communication have been identified as barriers to climate change adaptation (Carmin, Dodman, and Chu 2013; Moser and Ekstrom 2011). Barriers identified with respect to climate change adaptation, such as insufficient scientific information, lack of knowledge regarding best practices and the efficacy of different measures, lack of information on the costs and benefits of alternative loss and vulnerability reduction strategies, and issues related to interorganizational and interjurisdictional coordination (Moser 2009; National Research Council 2009), also exist within the hazards arena. Incentives for climate change adaptation include experience with natural disasters, supra-local regulations, the communication of information and ideas, and the diffusion of norms through networks (Carmin et al. 2012). Also significant is the role of nonstate actors, such as nongovernmental organizations, professionally-based groups, and boundary organizations, in pressing for adaptation (Moser 2009). Improved decision support tools are also needed for both disaster loss mitigation and climate change adaptation (National Research Council 2009, 2010).

HUMAN MIGRATION AND ENVIRONMENTAL CHANGE

Human migration represents a potential adaptation to environmental shocks and stresses, including natural disasters (McLeman and Smit 2006). History provides many examples of migration in the face of environmental pressure, such as drought in central North America, hurricanes in the Caribbean, and seasonal rainfall variation in the Sahel (McLeman 2013; McLeman and Hunter 2010). Such "historical analogues" can ultimately even include settlement abandonment (McLeman 2011). In this way, migration sometimes represents a more permanent form of adaptation—a *replacement* for livelihood strategies no longer viable—as opposed to temporary relocation (Bylander 2013).

Although environmental factors are relatively new to migration scholarship, recent studies have shed light on the environmental aspects of migration streams, particularly in vulnerable regions and areas with high levels of natural resource dependence (Morrissey 2014).[3] These suggest that environmental shocks and stresses interact with existing vulnerabilities and challenges, while the specific type of migration undertaken is further shaped by socioeconomic and political realities (Black, Bennett, et al. 2011). Scholarly research also calls into question the sensationalized image of mass climate-related migrations, while echoing arguments outlined earlier regarding equity and justice with respect to the distribution of climate change impacts.

Migration as Adaptation

Several key messages about migration as an adaptive response to environmental change are emerging from recent migration–environment research, many of which have relevance for more general understandings of the potential migratory implications of climate change. First, migration-related environmental pressures take different forms—described in simple terms as "fast and slow" (Gray and Bilsborrow 2012). Although migratory responses vary worldwide, acute, short-term shocks, such as sudden natural disasters or droughts, can yield different migratory impacts than long-term, chronic strains, such as declining soil quality or climate change. Migration also ranges along a continuum of forced to voluntary (Hugo 1996) and may include entire households or individual members sent to diversify labor investments with the intention of submitting remittances (Gray and Bilsborrow 2012).

Environmental scarcity, pressures, and shocks often (but not always) act as migration push factors. For example, in regions already characterized as relatively dry, such as Burkina Faso, rainfall scarcity has been associated with higher levels of outmigration (e.g., Hunter, Leyk, et al. 2013; Hunter, Murray, and Riosmena 2013; Nawrotzki, Riosmena, and Hunter 2013). Poor agricultural and environmental conditions also drive farmers from Ghana's northern regions (van der Geest 2011). In addition, short-term rainfall deficits increase long-term migration to other rural areas but have the opposite effect on short-term moves to more distant destinations (Henry, Schoumaker, and Beauchemin 2004).

It is not only rainfall trends, per se, that shape migration patterns; rainfall ultimately affects the availability of natural resources as well as the productivity of agricultural landscapes. Some research has examined these connections more directly. In Nepal, local migration is heightened by *perceived* declines in land productivity, as well as by increased time required to gather firewood (Massey, Axinn, and Ghimire 2010). The issue of risk perceptions is important because it shapes the likelihood of residents of hazard-prone regions to migrate (Slovic 1987).

Much environmentally induced migration is short distance and/or within nations, given that international migration can be costly and therefore out of reach for the most vulnerable and impoverished households. In contrast, migration across national boundaries (e.g., Massey et al 2010) is more often associated with *higher* levels of environmental resources, as opposed to their scarcity. This finding is consistent with research on the role of land as a key form of household wealth that allows for more costly forms of livelihood diversification (Gray 2010).

Reciprocal Connections Between Vulnerability and Migration

Vulnerability and the ability to adapt through migration are not equally distributed across households. Gender, race, and class status, among other social factors, shape individual and household migratory responses, including the possibility of returning. In Nepal, due to gendered harvesting responsibilities, women are more likely to migrate in the face of shortages of grazing land and collected fodder for livestock (Hunter and David 2011; Massey et al. 2010). The 1983–1985 drought in Mali revealed similar patterns, with a dramatic increase in short-term cyclical migration, particularly among women and children (Findley 1994). At the same time, migration as adaptation also can result in intensified vulnerability in destination regions, owing to migrants' lack of connections to community and land. In India, the most vulnerable group affected by Hurricane 07B was migrant, low-caste women who were predominantly landless agricultural laborers (O'Hare 2001).

Migration as adaptation can therefore result in some populations being more likely to experience long-term displacement, as illustrated by Hurricane Katrina, where impoverished households were less likely to return home (Fussell, Curtis, and DeWaard 2014; Fussell, Sastry, and Van Landingham 2010; Groen and Polivka 2010). Gender and class distinctions also emerged in Hurricane Katrina; in New Orleans, 14 percent of working women did not own cars prior to the disaster (compared to 4.3 percent of working women nationwide). The lack of transportation constrained evacuation and migration options and inhibited subsequent searches for new employment (Laska et al. 2008).

A different type of connection among migration, adaptation, and climate change is found in hazard mitigation efforts facilitated through migrant remittances. Money sent by migrants to their origin households—especially by international migrants—has fueled improved housing construction and communication equipment purchases in Burkina Faso and Ghana. Such investments may reduce vulnerability to natural disasters (Mohapatra, Joseph, and Ratha 2012), including those associated with climate change.

Migration as adaptation can also arise from opportunities for rebuilding following natural disasters. The arrival of new migrants into devastated regions is an adaptation strategy that is grounded in shifting post-disaster employment landscapes (e.g., Pais and Elliott 2008). As an example, after Hurricane Katrina, a "rapid response labor force" arrived in New Orleans, comprising predominantly Latinos from Mexico, Central America and the Caribbean, South America, and elsewhere. Initial migrants, drawn by high construction wages, tended to be younger, with relatively few strong ties to other U.S. immigrants and less migration experience overall. Yet over time, social networks facilitated additional migration as family and friends were recruited (Fussell et al. 2010).

Social Networks and Migration

As illustrated by the Hurricane Katrina research on the emergence of a "rapid response labor force," migration–environment scholarship is rooted in knowledge of social interaction, networks, and social structure. Social networks have long been understood as central in household-level migration decision making and in the manifestation of household decisions within aggregated streams of migration (White and Lindstrom 2006). Conceptual frameworks such as "cumulative causation" focus on the development of social capital within migration streams, whereby each move reduces the cost of future moves through the building of relationships and expertise (Massey 1990).

Evidence of the importance of social networks has emerged from several recent environment–migration case studies. In rural Mexico, drought-affected households in communities with strong migration histories are more likely to send a migrant to the United States, while drought in regions lacking such social networks actually reduces the propensity to migrate (Hunter, Murray, and Riosmena 2013). Recent research from rural Ecuador also finds that both environmental factors and migration networks are important predictors of migration (Gray and Bilsborrow 2012).

Analysis of migration as one form of livelihood adaptation, including adaptation in the face of climate change, must take into account nonfinancial costs associated with migration as well as financial ones (Meze-Hausken 2000). Consider Tuvalu, a small island state in the South Pacific, whose residents are at extreme risk due to sea level rise. Residents of such small island developing states are among the most likely to become international climate refugees. The provision of land and employment in future resettlement sites cannot replace the loss of their strong cultural, lifestyle, and identity connections to their homes (Mortreux and Barnett 2009). Similar findings emerge in the highlands of Peru, where vulnerable residents also express place attachment related to lifestyle, aesthetics, and other services provided by local environments, despite the challenges of increasing aridity (Adams and Adger 2013). Cultural connections to place produce strong incentives to remain, even in vulnerable areas.

CONCLUSIONS

An improved understanding of the factors that encourage successful climate adaptation will require major contributions from sociology and related fields that are based in an understanding of the underlying causes of social vulnerabilities and inequities and the conditions that perpetuate them. Fundamental social issues include tensions between entities that have historically contributed most to climate change and those who suffer most from its effects; the ways in which economic priorities, governance and institutions, social values, and power differentials both contribute to the problem of climate change and block efforts to address it; and how exposure to hazards, including those associated with climate change, is unequally distributed, affecting the ability to prepare for and respond to climate impacts. This chapter also points to the ways in which sociological research enhances our understanding of how strategies such as mainstreaming can promote adaptation and economic development, how knowledge and perceptions can support needed behavioral and policy responses, and the ways in which social networks facilitate and support adaptive responses. Insights highlighted in this chapter demonstrate that institutional and social measures must be viewed as integral aspects of adaptation strategies.

As the discussions in this chapter suggest, analyses of the political economy of the world system and of development are important to adaptation, as are those in areas such as hazards and disasters and demography and migration. However, it is important to recognize that the topical areas reviewed here illustrate only a few ways in which the field of sociology can contribute to adaptation knowledge and practice. For example, science and technology studies can aid our understanding of societal choices surrounding new and potentially dangerous climate-related technologies, such as geo-engineering. Such research can also make contributions on important topics such as the role of science in public policy development and implementation. Political sociology and the sociology of law can lead to a better understanding of how political coalitions, networks, and regulatory regimes operate with respect to adaptation. Social-psychological research can shed light on public attitude formation and attitude–behavioral linkages that relate to various adaptation strategies. Research on race, class, and gender can make further contributions to our understanding of climate change risk perceptions, social vulnerability, and the adaptive capacities of different groups within society.

Sociological theory and research on social movements and collective action are important for understanding the origins, dynamics, and impacts of social movements that both promote and resist adaptation to climate change. In an increasingly urbanized world, characterized by rapidly expanding urban agglomerations and growing urban vulnerability, the subfield of urban sociology has much to add to our understanding of factors that facilitate and impede adaptation within the urban context. Similarly, although not discussed here, constructivist frameworks are essential for an explanation of when, where, and how climate change is recognized as an urgent public and global problem and which solutions to the problem are socially constructed as feasible and appropriate. Economic sociology can provide answers to questions on the feasibility, costs, benefits, and distributional and equity effects of different adaptation approaches.

Public participation plays a central role in helping to define impacts and prioritize adaptation responses and is integral to promoting governance and building strong institutions for adaptation. Research with a sociological orientation has already offered insights into climate change adaptation. For instance, scholars considering local knowledge related to adaptation have begun to link general insights from this field to the ways in which the effects of and responses to climate change depend on local context, as well as how adaptation options can be more effective if designed, implemented, and monitored by community residents (Ayers and Forsyth 2009; Ebi and Semenza 2008; Pringle and Conway 2012), especially those who possess extensive local knowledge (Brace and Geoghegan 2010; Chu, Anguelovski, and Carmin 2015). There also are ongoing insights emerging about the nature and role of community-based adaptation (e.g., Ebi and Semenza 2008; Ensor and Berger 2009). Promoting local resilience to climate change is important, but localities are still affected by their positions within a global architecture of climate actors and institutions (Beckman and Page 2008). Therefore, sociological research on governance and institutions at all levels is critical for a more thorough understanding of the challenges associated with adapting to our changing climate and how those challenges can be overcome.

The field of sociology already brings a wealth of insights to climate adaptation knowledge, policy, and practice. Ongoing scholarship in subfields that are already engaged in the study of climate adaptation, as well as those that are yet to be engaged, will increase our understanding of how to promote adaptation at global, national, and local levels. These findings should be taken into account in order to ensure a robust understanding of both the fundamental conditions underlying existing vulnerabilities and the choices and decision processes associated with taking meaningful adaptation action.

NOTES

1. We thank Ian Noble for comments on the table and Patricia Romero-Lankao for input on urban vulnerability.

2. A promising early direction is that a group of experts met in Edinburgh in 2011 to develop an approach for monitoring and evaluating adaptation efforts, based on setting targets such as those that are part of the Millennium Development Goals (Brooks et al. 2011).

3. Social science research on human migration and environmental conditions has burgeoned over the past several years. Recent publications include conceptual

overviews and extensions (e.g., Adams and Adger 2013; Black, Adger, et al. 2011), methodological summaries (e.g., Bilsborrow and Henry 2012), case studies (e.g., Gray and Mueller 2012a, 2012b), and policy reviews (e.g., Warner 2012). In addition, several recent book collections (e.g., Laczko and Aghazarm 2009; Piguet, Pécoud, and de Guchteneire 2011) and journal special issues have focused on the topic (e.g., Adamo and Izazola 2010). A systematic review is offered by Hunter, Luna, and Norton (2015).

REFERENCES

Adamo, Susana and Haydea Izazola. 2010. "Human Migration and the Environment." *Population and Environment* 32(2–3):105–284.

Adams, Helen and W. Neil Adger. 2013. "The Contribution of Ecosystem Services to Place Utility as a Determinant of Migration Decision-Making." *Environmental Research Letters* 8(1):1–8.

Adger, W. Neil. 1999. "Social Vulnerability to Climate Change and Extremes in Coastal Vietnam." *World Development* 27:249–269.

Adger, W. Neil. 2006. "Vulnerability." *Global Environmental Change* 16(3):268–281.

Aglietta, Michael. [1976] 2001. *A Theory of Capitalist Regulation: The US Experience.* Brooklyn, NY, and London: Verso.

Agrawal, Arun. 2010. "Local Institutions and Adaptation to Climate Change." Pp. 173–197 in *The Social Dimensions of Climate Change: Equity and Vulnerability in a Warming World*, edited by Robin Mearns and Andrew Norton. Washington, DC: World Bank Publications.

Ahmed, Syud A., Noah S. Diffenbaugh, and Thomas W. Hertel. 2009. "Climate Volatility Deepens Poverty Vulnerability in Developing Countries." *Environmental Research Letters* 4(3):1–8.

AidData. 2013. AidData database of foreign assistance. Online at AidData.org.

Alexander, David. 2006. "Globalization and Disasters: Trends, Problems and Dilemmas." *Journal of International Affairs* 59:1–22.

Anguelovski, Isabelle and JoAnn Carmin. 2011. "Something Borrowed, Everything New: Innovation and Institutionalization in Urban Climate Governance." *Current Opinion in Environmental Sustainability* 3:1–7.

Anguelovski, Isabelle, Eric Chu, and JoAnn Carmin. 2014. "Variations in Approaches to Urban Climate Adaptation: Experiences and Experimentation from the Global South." *Global Environmental Change* 27:156–167.

Ayers, Jessica. 2009. "International Funding to Support Urban Adaptation to Climate Change." *Environment and Urbanization* 21(1):225–240.

Ayers, Jessica and David Dodman. 2010. "Climate Change Adaptation and Development I: The State of the Debate." *Progress in Development Studies* 10(2):161–168.

Ayers, Jessica and Tim Forsyth. 2009. "Community-Based Adaptation to Climate Change: Strengthening Resilience through Development." *Environment* 51(4):23–31.

Aylett, Alexander. 2014. *Progress and Challenges in the Urban Governance of Climate Change: Results of a Global Survey.* Cambridge, MA: MIT.

Barr, Rhona, Samuel Fankhauser, and Kirk Hamilton. 2010. "Adaptation Investments: A Resource Allocation Framework." *Mitigation and Adaptation Strategies for Global Change* 15(8):843–858.

Beckman, Ludvig and Edward Page. 2008. "Perspectives on Justice, Democracy and Global Climate Change." *Environmental Politics* 17(4):527–535.

Bilsborrow, Richard E. and Sabine J. Henry. 2012. "The Use of Survey Data to Study Migration–Environment Relationships in Developing Countries: Alternative Approaches to Data Collection." *Population and Environment* 34(1):113–141.

Birkland, Thomas A. 1997. *After Disasters: Agenda Setting, Public Policy, and Focusing Events.* Washington DC: Georgetown University Press.

Birkmann, Jörn, Matthias Garschagen, Frauke Kraas, and Nguyen Quang. 2010. "Adaptive Urban Governance: New Challenges for the Second Generation of Urban Adaptation Strategies to Climate Change." *Sustainability Science* 5(2):185–206.

Black, Richard, W. Neil Adger, Nigel W. Arnell, Stefan Dercon, Andrew Geddes, and David Thomas. 2011. "The Effect of Environmental Change on Human Migration." *Global Environmental Change* 21:S3–S11.

Black, R., S. Bennett, S. Thomas, and J. Beddington. 2011. "Climate Change: Migration as Adaptation." *Nature* 478(7370):447–449.

Borden, Kevin A. and Susan L. Cutter. 2008. "Spatial Patterns of Natural Hazards Mortality in the United States." *International Journal of Health Geographics* 7:31.

Boyd, Emily, Henny Osbahr, Polly J. Ericksen, Emma L. Tompkins, Maria Carmen Lemos, and Fiona Miller. 2008. "Resilience and 'Climatizing' Development: Examples and Policy Implications." *Development* 51(3):390–396.

Brace, Catherine and Hilary Geoghegan. 2010. "Human Geographies of Climate Change: Landscape, Temporality, and Lay Knowledges." *Progress in Human Geography* 35(3):284–302.

Broadleaf Capital International and Marsden Jacob Associates. 2006. *Climate Change Impacts & Risk Management: A Guide for Business and Government.* Report prepared for the Australian Greenhouse Office, Department of the Environment and Heritage.

Brooks, Nick, W. Neil Adger, and P. Mick Kelly. 2005. "The Determinants of Vulnerability and Adaptive Capacity at the National Level and the Implications for Adaptation." *Global Environmental Change* 15(2):151–163.

Brooks, Nick, Simon Anderson, Jessica Ayers, Ian Burton, and Ian Tellam. 2011. *Tracking Adaptation and Measuring Development.* International Institute for Environment and Development.

Brooks, Nick, Natasha Grist, and Katrina Brown. 2009. "Development Futures in the Context of Climate Change: Challenging the Present and Learning from the Past." *Development Policy Review* 27(6):741–765.

Brown, Anna, Ashvin Dayal, and Cristina Rumbaitis Del Rio. 2012. "From Practice to Theory: Emerging Lessons from Asia for Building Urban Climate Change Resilience." *Environment and Urbanization* 24(2):531–556.

Bulkeley, Harriet, Liliana Andonova, Karin Backstrand, Michele Betsill, Daniel Compagnon, Rosaleen Duffy, Ans Kolk, Matthew Hoffmann, David Levy, Peter Newell, Tori Milledge, Matthew Paterson, Philipp Pattberg, and Stacy VanDeveer. 2012. "Governing Climate Change Transnationally: Assessing the Evidence from a Database of Sixty Initiatives." *Environment and Planning C: Government and Policy* 30(4):591–612.

Burby, Raymond J. 2001. "Flood Insurance and Flood Plain Management: The US Experience." *Environmental Hazards* 3:111–122.

Burby, Raymond J. 2005. "Have State Comprehensive Planning Mandates Reduced Insured Losses from Natural Disasters?" *Natural Hazards Review* 6(2):67–81.

Burby, Raymond J. 2006. "Hurricane Katrina and the Paradoxes of Government Disaster Policy: Bringing about Wise Governmental Decisions for Hazardous Areas." *Annals of the American Academy of Political and Social Science* 604:171–191.

Burton, Ian. 1996. "The Growth of Adaptation Capacity: Practice and Policy." Pp. 55–67 in *Adapting to Climate Change: Assessments and Issues*, edited by J. B. Smith, N. Bhatti, G. Menzhulin, R. Benioff, M. Campos, B. Jallow, F. Rijsberman, M. I. Budyko, and R. K. Dixon. New York: Springer-Verlag.

Bylander, Maryann. 2013. "Depending on the Sky: Environmental Distress, Migration, and Coping in Rural Cambodia." *International Migration* [in press].

Cannon, Terry and Detlef Müller-Mahn. 2010. "Vulnerability, Resilience and Development Discourses in Context of Climate Change." *Natural Hazards* 55(3):621–635.

Carmin, JoAnn, Isabelle Anguelovski, and Debra Roberts. 2012. "Urban Climate Adaptation in the Global South: Planning in an Emerging Policy Domain." *Journal of Planning Education and Research* 32(1):18–32.

Carmin, JoAnn, David Dodman, and Eric Chu. 2013. "Urban Climate Adaptation and Leadership: From Conceptual Understanding to Practical Action." Regional Development Working Paper 2013/26. Paris, France: Organisation for Economic Co-Operation and Development.

Carmin, JoAnn, Nikhil Nadkarni, and Chris Rhie. 2012. *Progress and Challenges in Urban Climate Adaptation Planning: Results of a Global Survey*. Cambridge, MA: DUSP/MIT.

Chu, Eric, Isabelle Anguelovski, and JoAnn Carmin. 2015. "Inclusive Approaches to Urban Climate Adaptation Planning and Implementation in the Global South." *Climate Policy* [in press].

Ciplet, David, Spencer Field, Keith Madden, Mizan Khan, and J. Timmons Roberts. 2012. "The Eight Unmet Promises of Fast-Start Climate Finance." International Institute for Environment and Development Briefing, November 2012.

Ciplet, David, J. Timmons Roberts, and Mizan Khan. 2013. "The Politics of International Climate Adaptation Funding: Justice and Divisions in the Greenhouse." *Global Environmental Politics* 13(1):49–68.

Cutter, Susan L., Bryan J. Boruff, and W. Lynn Shirley. 2003. "Social Vulnerability to Environmental Hazards." *Social Science Quarterly* 84(2):242–261.

Da Silva, Jo, Sam Kernaghan, and Andrés Luque. 2012. "A Systems Approach to Meeting the Challenges of Urban Climate Change." *International Journal of Urban Sustainable Development* 4(2):125–145.

Diamond, Jordan and Carl Bruch. 2012. "The International Architecture for Climate Change Adaptation Assistance." Pp. 291–315 in *Climate Change Adaptation and International Development: Making Development Cooperation More Effective*, edited by Ryo Fujikura and Masato Kawanishi. London and Washington, DC: Earthscan.

Dodman, David and Diana Mitlin. 2014. "The National and Local Politics of Climate Adaptation in Zimbabwe." *Climate and Development* [ahead-of-print]: 1–12.

Dodman, David and David Satterthwaite. 2008. "Institutional Capacity, Climate Change Adaptation and the Urban Poor." *IDS Bulletin* 39(4):67–74.

Dubash, Navroz K. 2012. "Climate Politics in India: Three Narratives." Pp. 197–207 in *Handbook of Climate Change and India: Development, Politics and Governance*, edited by Navroz K. Dubash. London and New York: Earthscan.

Duus-Otterström, Göran and Sverker C. Jagers. 2012. "Identifying Burdens of Coping with Climate Change: A Typology of the Duties of Climate Justice." *Global Environmental Change* 22(3):746–753.

Eakin, Hallie C. 2000. "Smallholder Maize Production and Climate Risk: A Case Study from Mexico." *Climatic Change* 45:19–36.

Eakin, Hallie C. and Maria Carmen Lemos. 2006. "Adaptation and the State: Latin America and the Challenge of Capacity-Building under Globalization." *Global Environmental Change* 16(1):7–18.

Ebi, Kristie L. and Jan C. Semenza. 2008. "Community-Based Adaptation to the Health Impacts of Climate Change." *American Journal of Preventive Medicine* 35(5):501–507.

Emrich, Christopher T. and Susan L. Cutter. 2011. "Social Vulnerability to Climate-Sensitive Hazards in the Southern United States." *Weather, Climate and Society* 3:193–208.

Engle, Nathan L. 2011. "Adaptive Capacity and its Assessment." *Global Environmental Change* 21(2):647–656.

Ensor, Jonathan and Rachel Berger. 2009. "Community-Based Adaptation and Culture in Theory and Practice." Pp. 227–239 in *Adaptation to Climate Change: Thresholds, Values, Governance*, edited by W. Neil Adger, Irene Lorenzoni, and Karen O'Brien. Cambridge and New York: Cambridge University Press.

Eriksen, Siri H., Katrina Brown, and P. Mick Kelly. 2005. "The Dynamics of Vulnerability: Locating Coping Strategies in Kenya and Tanzania." *Geographical Journal* 171(4):287–305.

Fankhauser, Samuel and Ian Burton. 2011. "Spending Adaptation Money Wisely." *Climate Policy* 11(3):1037–1049.

Findley, Sally E. 1994. "Does Drought Increase Migration: A Study of Migration from Rural Mali During the 1983–1985 Drought." *International Migration Review* 28(3):539–553.

Flåm, Karoline Haegstad, and Jon Birger Skjaerseth. 2009. "Does Adequate Financing Exist for Adaptation in Developing Countries?" *Climate Policy* 9(1):109–114.

Fothergill, Alice. 2004. *Heads Above Water: Gender, Class, and Family in the Grand Forks Flood*. Albany: SUNY Press.

Freudenburg, William R., Robert Gramling, Shirley Laska, and Kai T. Erikson 2009. *Catastrophe in the Making: The Engineering of Katrina and the Disasters of Tomorrow*. Washington, DC: Island Press.

Füssel, Hans-Martin, Stéphane Hallegatte, and Michael Reder. 2012. "International Adaptation Funding." Pp. 311–330 in *Climate Change, Justice and Sustainability: Linking Climate and Development Policy*, edited by Ottmar Edenhofer, Johannes Wallacher, and Hermann Lotze-Campen. Dordrecht: Springer.

Fussell, Elizabeth, Katherine J. Curtis, and Jack DeWaard. 2014. "Recovery Migration to the City of New Orleans after Hurricane Katrina: A Migration Systems Approach." *Population and Environment* 35(3):305–322.

Fussell, Elizabeth, Narayan Sastry, and Mark VanLandingham. 2010. "Race, Socioeconomic Status, and Return Migration to New Orleans after Hurricane Katrina." *Population and Environment* 31(1–3):20–42.

Gasper, Des, Ana Victoria Portocarrero, and Asuncion Lera St. Clair. 2013. "The Framing of Climate Change and Development: A Comparative Analysis of the Human Development Report 2007/8 and the World Development Report 2010." *Global Environmental Change* 23(1):28–39.

Grasso, Marco. 2010. "The Role of Justice in the North–South Conflict in Climate Change: The Case of Negotiations on the Adaptation Fund." *International Environmental Agreements: Politics, Law* 11(4):361–377.

Gray, Clark L. 2010. "Gender, Natural Capital, and Migration in the Southern Ecuadorian Andes." *Environment and Planning A* 42:678–696.

Gray, Clark L. and Richard Bilsborrow. 2012. "Environmental Influences on Human Migration in Rural Ecuador." *Demography* 50(4):1217–1241.

Gray, Clark L. and Valerie Mueller. 2012a. "Drought and Population Mobility in Rural Ethiopia." *World Development* 40(1):134–145.

Gray, Clark L. and Valerie Mueller. 2012b. "Natural Disasters and Population Mobility in Bangladesh." *Proceedings of the National Academy of Sciences USA* 109(16):6000–6005.

Groen, Jeffrey A. and Anne E. Polivka. 2010. "Going Home after Hurricane Katrina: Determinants of Return Migration and Changes in Affected Areas." *Demography* 47:821–844.

Halsnæs, Kirsten and Sara Traerup. 2009. "Development and Climate Change: A Mainstreaming Approach for Assessing Economic, Social, and Environmental Impacts of Adaptation Measures." *Environmental Management* 43(5):765–778.

Hammer, Roger B., Susan J. Stewart, and Volker C. Radeloff. 2000. "Demographic Trends, The Wildland-Urban Interface, and Wildfire Management." *Society and Natural Resources* 22:777–782.

Hammill, Anne, Richard Matthew, and Elissa McCarter. 2009. "Microfinance and Climate Change Adaptation." *IDS Bulletin* 39(4):113–122.

Harlan, Sharon L., Anthony J. Brazel, Lela Prashad, William L. Stefanov, and Larissa Larsen. 2006. "Neighborhood Microclimates and Vulnerability to Heat Stress." *Social Science and Medicine* 63:2847–2863.

Harris, Paul G. and Jonathan Symons. 2010. "Justice in Adaptation to Climate Change: Cosmopolitan Implications for International Institutions." *Environmental Politics* 19(4):617–636.

Heltberg, Ramos and Niels Lund. 2009. "Shocks, Coping, and Outcomes for Pakistan's Poor: Health Risks Predominate." *Development Studies* 45(9):889–910.

Henry, Sabine, Bruno Schoumaker, and Cris Beauchemin. 2004. "The Impact of Rainfall on the First Out-Migration: A Multi-Level Event-History Analysis in Burkina Faso." *Population and Environment* 25(5):423–460.

Hicks, Robert L., Bradley C. Parks, J. Timmons Roberts, and Michael J. Tierney. 2008. *Greening Aid? Understanding Environmental Assistance to Developing Countries.* Oxford and New York: Oxford University Press.

Hodson, Mike and Simon Marvin. 2010. "Can Cities Shape Socio-Technical Transitions and How Would We Know If They Were?" *Research Policy* 39:477–485.

Hugo, Graeme. 1996. "Environmental Concerns and International Migration." *International Migration Review* 30:105–131.

Hunt, Alistair and Paul Watkiss. 2011. "Climate Change Impacts and Adaptation in Cities: A Review of the Literature." *Climatic Change* 104(1):13–49.

Hunter, Lori M. and Emmanuel David. 2011. "Displacement, Climate Change and Gender." Pp. 306–330 in *Migration and Climate Change*, edited by E. Piguet, A.

Pecoud, and P. de Guchteneire. Cambridge and New York: UNESCO Publishing and Cambridge University Press.

Hunter, Lori M., Stefan Leyk, Raphael Nawrotzki, Galen Maclaurin, Wayne Twine, Mark Collinson, and Barend Erasmus. 2013. "Rural Outmigration, Natural Capital, and Livelihoods in Rural South Africa." *Population, Space and Place* 20(5):402–420.

Hunter, Lori M., Jessie Luna, and Rachel Norton. 2015. "The Environmental Dimensions of Human Migration." *Annual Review of Sociology* [in press].

Hunter, Lori M., Sheena Murray, and Fernando Riosmena. 2013. "Rainfall Variation and U.S. Migration from Rural Mexico." *International Migration Review* 47(3):874–909.

Huq, Saleemul, Sari Kovats, Hannah Reid, and David Satterthwaite. 2007. "Reducing Risks to Cities from Disasters and Climate Change." *Environment and Urbanization* 19: 3–15.

Huq, Saleemul and Hannah Reid. 2004. "Mainstreaming Adaptation in Development." *IDS Bulletin* 35(3):15–21.

IPCC. 2012. "Summary for Policymakers." Pp. 3–21 in *Intergovernmental Panel on Climate Change Special Report on Managing the Risks of Extreme Events and Disasters to Advance Climate Change Adaptation*, edited by C. B. Fields, V. Barros, T. F. Stocker, D. Qin, D. J. Dokken, K. L. Ebi, M. D. Mastrandrea, K. J. Mach, G. K. Plattner, S. K. Allen, M. Tignor, and P. M. Midgley. Cambridge and New York: Cambridge University Press.

IPCC. 2014. "Summary for Policymakers." Pp. 1–32 in *Climate Change 2014: Impacts, Adaptation, and Vulnerability. Contribution of Working Group II to the Fifth Assessment Report of the Intergovernmental Panel on Climate Change*, edited by C. B. Fields, V. Barros, D. J. Dokken, K. J. Mach, M. D. Mastrandrea, T. E. Bilir, M. Chatterjee, K. L. Ebi, Y. O. Estrada, R. C. Genova, B. Girma, E. S. Kissel, A. N. Levy, and S. MacCraken. Cambridge and New York: Cambridge University Press.

Jones, Benjamin, Michael Keen, and Jon Strand. 2012. "Fiscal Implications of Climate Change." *International Tax and Public Finance* 20(1):29–70.

Kahn, Matthew E. 2005. "The Death Toll from Natural Disasters: The Role of Income, Geography, and Institutions." *Review of Economics and Statistics* 87:271–284.

Kang, Jung Eun, Walter G. Peacock, and Rahmawati Husein. 2010. "An Assessment of Coastal Zone Hazard Mitigation Plans in Texas." *Journal of Disaster Research* 5:520–528.

Kano, M., M. Wood, M. M. Kelley, and L. B. Bourque. 2009. *California Earthquake Preparedness Survey Report: Findings and Recommendations for Strengthening Household Resiliency to Earthquakes*. Los Angeles: University of California, Los Angeles.

Khan, Mizan R. and J. Timmons Roberts. 2013. "Adaptation and International Climate Policy." *Wiley Interdisciplinary Reviews: Climate Change* 4(3):171–189.

Kincaid, Graciela and J. Timmons Roberts. 2013. "No Talk, but Some Walk: The Obama Administration's First Term Rhetoric on Climate Change and its International Climate Budget Commitments." *Global Environmental Politics* 13(4):41–60.

Klein, Richard J. T. 2010. *Which Countries Are Particularly Vulnerable? Science Doesn't Have the Answer!* Stockholm Environmental Institute.

Kok, Marcel, Bert Metz, Jan Verhagen, and Sascha Van Rooijen. 2008. "Integrating Development and Climate Policies: National and International Benefits." *Climate Policy* 8(2):103–118.

Laczko, F. and C. Aghazarm. 2009. *Migration, Environment and Climate Change: Assessing the Evidence.* Geneva, Switzerland: International Organization for Migration.

Laska, Shirley, Morrow, Betty Hearn, Willinger, Beth and Mock, Nancy. 2008. "Gender and Disasters: Theoretical Considerations." Pp. 11–21 in *Katrina and the Women of New Orleans,* edited by B. Willinger. New Orleans: Newcomb College Center for Research on Women.

Lemos, Maria Carmen, Emily Boyd, Emma L. Tompkins, Henny Osbahr, and Diana Liverman. 2007. "Developing Adaptation and Adapting Development." *Ecology and Society* 12(2):26–29.

Lindell, Michael K. and Ronald W. Perry. 2000. "Household Adjustment to Earthquake Hazard: A Review of Research." *Environment and Behavior* 32:590–630.

Lipietz, Alain. 1992. *Towards a New Economic Order: Population, Ecology and Democracy.* Oxford and New York: Oxford University Press.

Logan, John and Harvey Molotch. 1987. *Urban Fortunes: The Political Economy of Place.* Berkeley: University of California Press.

Lyles, Ward, Philip Berke, and Gavin Smith 2012. *Evaluation of Local Hazard Mitigation Plan Quality.* Chapel Hill, NC: UNC Institute for the Environment, Coastal Hazards Center.

MacCallum, Diana, Jason Byrne, and Wendy Steele. 2014. "Whither Justice? An Analysis of Local Climate Change Responses from South East Queensland, Australia." *Environment and Planning C: Government and Policy* 32(1):70–92.

Massey, Douglas S. 1990. "Social Structure, Household Strategies, and the Cumulative Causation of Migration." *Population Index* 56(1):3–26.

Massey, Douglas S., William G. Axinn, and Dormire J. Ghimire. 2010. "Environmental Change and Out-Migration: Evidence from Nepal." *Population and Environment* 32(2–3):109–136.

McLeman, Robert A. 2011. "Settlement Abandonment in the Context of Global Environmental Change." *Global Environmental Change* S11, S108–120.

McLeman, Robert A. 2013. *Climate and Human Migration: Past Experiences, Future Challenges.* Cambridge: Cambridge University Press.

McLeman, Robert A. and Lori M. Hunter. 2010. "Migration in the Context of Vulnerability and Adaptation to Climate Change: Insights from Analogues." *Climate Change* 1:450–461.

McLeman, Robert A. and Barry Smit. 2006. "Migration as an Adaptation to Climate Change." *Climatic Change* 76:31–53.

Multilateral Development Banks. 2012. *Joint MDB Report on Adaptation Finance 2011* (http://www.worldbank.org/content/dam/Worldbank/document/Joint%20 MDB%20Report%20on%20Adaptation%20Finance%202011.pdf).

Meze-Hausken, Elisabeth. 2000. "Migration Caused by Climate Change: How Vulnerable Are People in Dryland Areas." *Migration and Adaptation Strategies for Global Change* 5:379–406.

Michel-Kerjan, Erwann, Sabine Lemoyne de Forges, and Howard Kunreuther. 2012. "Policy Tenure Under the U.S. National Flood Insurance Program (NFIP)." *Risk Analysis* 32:644–658.

Mileti, Dennis S. 1999. *Disasters by Design: A Reassessment of Natural Hazards in the United States.* Washington, DC: Joseph Henry Press.

Mileti, Dennis S. and John H. Sorensen. 1987. "Why People Take Precautions Against Natural Disasters." Pp. 189–207 in *Taking Care: Why People Take Precautions,* edited by N. D. Weinstein. New York: Cambridge University Press.

Mohapatra, Sanket, George Joseph, and Dilip Ratha. 2012. "Remittances and Natural Disasters: Ex-post Response and Contribution to Ex-ante Preparedness." *Environment, Development and Sustainability* 14(3):365–387.

Mol, Arthur P. J. and David A. Sonnenfeld. 2000. "Ecological Modernization Around the World: An Introduction." *Environmental Politics* 9(1):1–14.

Morrissey, James. 2014. "Environmental Change and Human Migration in Sub-Saharan Africa." Pp. 81–109 in *People on the Move in a Changing Climate*, edited by E. Piguet and F. Laczko. Dordrecht: Springer Netherlands.

Mortreux, C. and J. Barnett. 2009. "Climate Change, Migration and Adaptation in Funafuti, Tuvalu." *Global Environmental Change* 19:105–112.

Moser, Caroline, Andrew Norton, Alfredo Stein, and Sophia Georgieva. 2010. *Pro-Poor Adaptation to Climate Change in Urban Centers: Case Studies of Vulnerability and Resilience in Kenya and Nicaragua*. Washington, DC: World Bank.

Moser, Susanne C. 2009. "Communicating Climate Change: History, Challenges, Process and Future Directions." *Climate Change* 1(1):31–53.

Moser, Susanne C. and Julia A. Ekstrom. 2011. "Taking Ownership of Climate Change: Participatory Adaptation Planning in Two Local Case Studies from California." *Journal of Environmental Studies and Sciences* 1(1):63–74.

National Research Council. 2009. *Informing Decisions in a Changing Climate*. Washington, DC: National Academies Press.

National Research Council. 2010. *America's Climate Choices: Panel on Adapting to the Impacts of Climate Change*. Washington, DC: National Academies Press.

Nawrotzki, Raphael, Fernando Riosmena, and Lori M. Hunter. 2013. "Do Rainfall Deficits Predict U.S.-bound Migration from Rural Mexico? Evidence from the Mexican Census." *Population Research and Policy Review* 32(1):129–158.

Nielson, Daniel and Michael J. Tierney. 2003. "Delegation to International Organizations: Agency Theory and World Bank Environmental Reform." *International Organization* 57(2):241–272.

Nelson, Donald R., W. Neil Adger, and Katrina Brown. 2007. "Adaptation to Environmental Change: Contributions of a Resilience Framework." *Annual Review of Environment and Resources* 32(1):395–419.

Norgaard, Kari Marie. 2011. *Living in Denial: Climate Change, Emotions, and Everyday Life*. Boston: MIT Press.

O'Brien, Geoff, Phil O'Keefe, Hubert Meena, Joanne Rose, and Leanne Wilson. 2008. "Climate Adaptation from a Poverty Perspective." *Climate Policy* 8(2):194–201.

O'Brien, Karen, Siri Eriksen, Lynn P. Nygaard, and Ane Schjolden. 2007. "Why Different Interpretations of Vulnerability Matter in Climate Change Discourses." *Climate Policy* 7(1):73–88.

O'Hare, Greg. 2001. "Hurricane 07B in the Godavari Delta, Andhra Pradesh, India: Vulnerability, Mitigation and the Spatial Impact." *Geographical Journal* 167(1):23–38.

Paavola, Jouni and W. Neil Adger. 2006. "Fair Adaptation to Climate Change." *Ecological Economics* 56(4):594–609.

Pais, Jeremy F. and James R. Elliott. 2008. "Places as Recovery Machines: Vulnerability and Neighborhood Change After Major Hurricanes." *Social Forces* 86(4):1415–1453.

Parks, Bradley C. and J. Timmons Roberts. 2010. "Climate Change, Social Theory and Justice." *Theory, Culture and Society* 27(2–3):134–166.

Parry, Martin, Nigel Arnell, Pam Berry, David Dodman, Samuel Fankhauser, Chris Hope, Sari Kovats, Robert Nicholls, David Satterthwaite, Richard Tiffin, and Tim Wheeler. 2009. *Assessing the Costs of Adaptation to Climate Change: A Review of the UNFCCC and Other Recent Estimates*. London: International Institute for Environment and Development.

Parry, Martin, Osvaldo F. Canziani, Jean P. Palutikof, Paul J. van der Linden, and Claire E. Hanson. 2007. *Climate Change 2007: Impacts, Adaptation and Vulnerability. Contribution of Working Group II to the Fourth Assessment Report of the Intergovernmental Panel on Climate Change. Summary for Policymakers*. Bonn, Germany: Intergovernmental Panel on Climate Change (IPCC).

Pelling, Mark. 2003a. *The Vulnerability of Cities: Natural Disasters and Social Resilience*. London: Earthscan.

Pelling, Mark. 2003b. *Natural Disaster and Development in a Globalizing World*. London: Routledge.

Phillips, Brenda, Deborah S. K. Thomas, Alice Fothergill, and Lynn Blinn-Pike. 2009. *Social Vulnerability to Disasters*. Boca Raton, FL: CRC Press.

Pielke, R. A., J. Gratz, C. W. Landsea, D. Collins, M. A. Saunders, and R. Musulin. 2008. "Normalized Hurricane Damage in the United States: 1900-2005." *Natural Hazards Review* 9:29–42.

Piguet, Étienne, Antoine Pécoud, and Paul de Guchteneire. 2011. *Migration and Climate Change*. Cambridge: UNESCO Publishing and Cambridge University Press.

Pinter, Nicholas. 2005. "One Step Forward, Two Steps Back on U.S. Floodplains." *Science* 308(5719):207–208.

Pringle, Patrick and Declan Conway. 2012. "Voices from the Frontline: The Role of Community-Generated Information in Delivering Climate Adaptation and Development Objectives at Project Level." *Climate and Development* 4(2):104–113.

Raschky, Paul A. 2008. "Institutions and the Losses from Natural Disasters." *Natural Hazards Earth System Science* 8:627–634.

Regmi, Bimal Raj and Dinanath Bhandari. 2012. "Climate Change Governance and Funding Dilemma in Nepal." *TMC Academic Journal* 7(1):40–55.

Roberts, J. Timmons. 2009. "The International Dimension of Climate Justice and the Need for International Adaptation Funding." *Environmental Justice* 2(4):185–190.

Roberts, J. Timmons and Bradley C. Parks. 2007. "Fueling Injustice: Globalization, Ecologically Unequal Exchange and Climate Change." *Globalizations* 4(2):193–210.

Rosenzweig, Cynthia, William D. Solecki, Reginald Blake, Malcolm Bowman, Craig Faris, Vivien Gornitz, Radley Horton, Klaus Jacob, Alice LeBlanc, Robin Leichenko, Megan Linkin, David Major, Megan O'Grady, Lesley Patrick, Edna Sussman, Gary Yohe, and Rae Zimmerman. 2011. "Developing Coastal Adaptation to Climate Change in the New York City Infrastructure-Shed: Process, Approach, Tools, and Strategies." *Climatic Change* 106(1):93–127.

Rossi, Peter H., James D. Wright, and Eleanor Weber-Burdin. 1982. *Natural Hazards and Public Choice: The State and Local Politics of Hazard Mitigation*. New York: Academic Press.

Rübbelke, Dirk T. G. 2011. "International Support of Climate Change Policies in Developing Countries: Strategic, Moral and Fairness Aspects." *Ecological Economics* 70(8):1470–1480.

Rudel, Thomas K., J. Timmons Roberts, and JoAnn Carmin. 2011. "Political Economy of the Environment." *Annual Review of Sociology* 37(1):221–238.

Satterthwaite, David, Saleemul Huq, Hannah Reid, Mark Pelling, and Patricia Romero Lankao. 2007. *Adapting to Climate Change in Urban Areas: The Possibilities and Constraints in Low and Middle Income Countries*. London: International Institute for Environment and Development.

Schlosberg, David. 2012. "Climate Justice and Capabilities: A Framework for Adaptation Policy." *Ethics and International Affairs* 26(4):445–461.

Schnaiberg, Allan. 1980. *Environment: From Surplus to Scarcity*. Oxford and New York: Oxford University Press.

Sharma, Divya and Sanjay Tomar. 2010. "Mainstreaming Climate Change Adaptation in Indian Cities." *Environment and Urbanization* 22(2):451–465.

Shepard, Peggy M. and Cecil Corbin-Mark. 2009. "Climate Justice." *Environmental Justice* 2(4):163–166.

Sheppard, Kate. 2009. "Everything You Always Wanted to Know about the Waxman-Markey Energy/Climate Bill in Bullet Points." June 4, 2009. Retrieved April 23, 2013 (http://grist.org/article/2009-06-03-waxman-markey-bill-breakdown/).

Simon, David. 2012. "Reconciling Development with the Challenges of Climate Change: Business as Usual or a New Paradigm?" Pp. 195–217 in *The Political Economy of Environment and Development in a Globalised World*, edited by D. J. Kjosavik and P. Vedeld. Oslo, Norway: Tapir Akademisk Forlag.

Slovic, Paul. 1987. "Perception of Risk." *Science* 236(4799):280–285.

Smit, Barry and Olga Pilifosova. 2003. "From Adaptation to Adaptive Capacity and Vulnerability Reduction." Pp. 9–28 in *Climate Change, Adaptive Capacity and Development*, edited by J. B. Smith, R. J. T. Klein, and S. Huq. London: Imperial College Press.

Smit, Barry and Johanna Wandel. 2006. "Adaptation, Adaptive Capacity and Vulnerability." *Global Environmental Change* 16(3):282–292.

Smithers, John and Barry Smit. 1997. "Human Adaptation to Climatic Variability and Change." *Global Environmental Change* 7(2):129–146.

Someshwar, Shiv. 2008. "Adaptation as 'Climate-Smart' Development." *Development* 51(3):366–374.

Stallings, Robert A. 1995. *Promoting Risk: Constructing the Earthquake Threat*. Piscataway, NJ: Transaction Publishers.

Stott, Peter A., Nathan P. Gillett, Gabriele C. Hegeri, David J. Karoly, Daithi A. Stone, Xuebin Zhang, and Francis Zwiers. 2010. "Detection and Attribution of Climate Change: A Regional Perspective." *Climate Change* 1(2):192–211.

Suarez, Pablo and Joanne Linnerooth-Bayer. 2010. "Micro-Insurance for Local Adaptation." *Climate Change* 1(2):271–278.

Tierney, Kathleen J., Michael K. Lindell, and Ronald W. Perry. 2001. *Facing the Unexpected: Disaster Preparedness and Response in the United States*. Washington, DC: Joseph Henry Press.

Tol, Richard S. J., Thomas E. Downing, Onno J. Kuik, and Joel B. Smith. 2004. "Distributional Aspects of Climate Change Impacts." *Global Environmental Change* 14(3):259–272.

Tompkins, Emma L. and Hallie Eakin. 2012. "Managing Private and Public Adaptation to Climate Change." *Global Environmental Change* 22(1):3–11.

Van Der Geest, Kees. 2011. "North-South Migration in Ghana: What Role for the Environment?" *International Migration* 49(s1):e69–e94.

Warner, Koko. 2012. "Human Migration and Displacement in the Context of Adaptation to Climate Change: The Cancun Adaptation Framework and Potential for Future Action." *Environment and Planning C: Government and Policy* 30(6):1061–1077.

White, Michael J. and David P. Lindstrom. 2006. "Internal Migration." Pp. 311–346 in *Handbook of Population*, edited by D. Poston and M. Micklin. New York: Kluwer Academic Publishers.

Wisner, Ben, Piers Blaikie, Terry Cannon, and Ian Davis. 2004. *At Risk: Natural Hazards, People's Vulnerability and Disasters*. London: Routledge.

Wolf, Johanna. 2011. "Climate Change Adaptation as a Social Process." Pp. 21–32 in *Climate Change Adaptation in Developed Nations: From Theory to Practice*, edited by J. D. Ford and L. Berrang-Ford. London and New York: Springer.

Wood, Rachel Godfrey. 2011. "Is There a Role for Cash Transfers in Climate Change Adaptation?" *IDS Bulletin* 42(6):79–85.

World Bank. 2010. *Natural Hazards, Unnatural Disasters: The Economics of Effective Prevention*. Washington, DC: World Bank.

World Bank. 2012. *Adapting to Climate Change: Assessing the World Bank Group Experience, Phase III*. Washington, DC: World Bank.

Yohe, Gary and Richard S. J. Tol. 2002. "Indicators for Social and Economic Coping Capacity: Moving Toward a Working Definition of Adaptive Capacity." *Global Environmental Change* 12(1):25–40.

Zoleta-Nantes, Doracie B. 2000. "Flood Hazards in Metro Manila: Recognizing Commonalities, Differences, and Courses of Action." *Social Science Diliman* 1(1):60–105.

7

Mitigating Climate Change

Karen Ehrhardt-Martinez, Thomas K. Rudel,
Kari Marie Norgaard, and Jeffrey Broadbent

INTRODUCTION

The anticipated changes in the Earth's global climate and their impact on people and ecosystems can potentially be minimized through two courses of action: mitigation and adaptation. Mitigation efforts are concerned with reducing the emissions of greenhouse gases (GHGs) while adaptation efforts are focused on enhancing our ability to live with the changes that will occur. This chapter focuses on mitigation efforts and how sociological insights can help us to better understand mitigation strategies and processes.

While a variety of mitigation efforts are already under way in many countries and cities around the globe, the Intergovernmental Panel on Climate Change (IPCC)'s Fourth Assessment Reports concluded that current efforts are far from sufficient to forestall the continued growth in atmospheric concentrations of GHGs and the continued warming of the planet (IPCC Summary Report 2007c). The IPCC's Fifth Assessment Report reiterates this observation (IPCC 2014). Achieving meaningful levels of mitigation is difficult because most developed countries (and the modern global economy) rely heavily on fossil fuel energy sources. As such, GHG emissions are firmly rooted in the current organization of economic and social systems, and meaningful mitigation efforts are likely to require fundamental changes in these systems. Given the limited scope of current climate change mitigation policies and sustainable development practices, the IPCC estimates that global GHG emissions will increase by 25 to 90 percent between 2000 and 2030, leading to an increase in average global temperatures of 1.8 to 4 degrees Celsius by 2090 (compared to averages during 1980–1999).[1]

The IPCC mitigation report (2007a) forecasts an increase in global GHG emissions of 8 to 36 Gt of CO_2 equivalent by 2030. The same report estimates the global mitigation opportunity at roughly 5 to 31 Gt of CO_2 equivalent by 2030. In other words, the IPCC studies suggest mitigation efforts could offset most of the projected growth in emissions or even reduce emissions below current levels, but only if "adequate policies are in place and barriers removed" (IPCC 2007b:58).

These IPCC estimates of mitigation opportunities focus on the potential impact of new technologies, failing to adequately account for the potential influence of nontechnological factors such as lifestyle changes or the effects of social forces and shifts in social structures. Given the well-documented ability of human societies to organize and reorganize themselves in response to both environmental conditions and economic and political forces, current IPCC estimates are incomplete and likely to be conservative in their conclusions. IPCC estimates of mitigation potential indicate that the largest global mitigation opportunities are associated with enhanced energy efficiencies in buildings (both residential and commercial), industry, energy supply, and agricultural end uses. Additional, albeit less substantial, mitigation potential is attributed to technological improvements in forestry, transportation, and waste.

While the IPCC reports have a global focus, a series of reports by the National Academy of Sciences (NAS) known as *America's Climate Choices* (ACC) considers climate change and mitigation opportunities for the United States in particular (NAS 2010). Like the IPCC reports, the ACC studies are predominantly concerned with technological approaches to climate change mitigation. However, unlike the IPCC reports, the ACC discussions of mitigation opportunities focus most heavily on the role of energy systems and technologies. For example, the ACC report on "Limiting the Magnitude of Climate Change" notes that in the United States "CO_2 emissions from fossil fuel combustion in the energy system accounts for approximately 82% of total U.S. GHG emissions" (NAS 2010:38). The determinants of fossil fuel consumption are identified as (1) the overall demand for goods and services that require energy to produce or deliver, (2) the efficiency with which the energy is used to provide goods and services, and (3) the extent to which energy comes from fossil fuels. As such, the predominant focus of the mitigation opportunities discussed in the ACC reports is on energy efficiency and renewable energy technologies. Similar to the IPCC global assessment, the ACC studies find that the building sector offers large opportunities for mitigating GHG emissions through enhancements in energy efficiency. Additional mitigation opportunities are associated with industry, transport, the carbon intensity of energy, and renewable and nuclear power.

Compared to the IPCC report, the ACC report on mitigation gives some-what more weight to social and cultural factors. However, its discussion is limited to two areas: (1) efforts to understand the variation in consumer demand for energy-intensive and energy-efficient goods and services (and to potentially influence consumer preferences and choices) and (2) possible equity implications of mitigation strategies (especially for disadvantaged populations). In understanding consumer demand, the ACC report recognizes that "Consumer choices among market offerings in different societies shape demand for everything from living space and electrical appliances to dietary choices" (NAS 2010:39) and that "long-term sustained changes will be driven by the interactions of technology markets, the policy environment and consumer choices" (NAS 2010:41). Despite the range of insights that a sociological perspective might offer, this analysis reduces the social aspect mainly to market decisions. Among the report's conclusions, though, the authors note that "the adoption of many energy efficient technologies and practices requires significant changes in human behavior, lifestyle, and consumer spending practices" (NAS 2010:66). The report also stresses that more social science research is needed to understand how social and behavioral dynamics interact with technology.

An overarching commonality between the IPCC and ACC assessments is their common characterization of climate change and mitigation strategies as *technological* hurdles, generally ignoring the possibilities of *social and cultural change and neglecting to acknowledge* the limited effectiveness of ongoing, technology-focused strategies to date. For example, within the IPCC reports' discussion of mitigation limits, the human dimensions of climate change are mostly relegated to the chapter on *Framing Issues* in which the authors explore the linkages between climate change and sustainable development.

Notably, however, neither the IPCC nor the ACC reports consider the importance of the myriad other aspects of social organization and culture: governance, power structures, political activism, labor policies, the countless drivers of consumption, the force of social routines and expectations, systems of global production, cultural values, and a range of other sociological factors that shape and constrain mitigation opportunities apart from technologically focused solutions.

In contrast, in the present chapter we suggest that there are important synergies and tradeoffs among climate change, the capacity to mitigate climate change, and particular configurations of social organization and cultural practices. Social organization and culture produce variation in values among stakeholders and decision makers, variation in the perception of risks and uncertainties, differences in costs and benefits, and variation in the capacity of decision makers to implement mitigation policies (Beck 1992; Fisher 2006; Roberts and Parks 2007).

UNDERSTANDING CLIMATE CHANGE MITIGATION FROM A SOCIOLOGICAL PERSPECTIVE

From a sociological standpoint our present output of GHG emissions stems from the current organization of social, political, and economic systems and the social and cultural practices that reinforce those systems. Therefore, efforts to mitigate GHG emissions will require a better understanding of the working of these systems and the various components that make them up (organizations, cultural practices, economic policies and regulations, technologies, individuals, and networks) as well as an understanding of the relationships among these components.

A particular strength of the sociological perspective is its foundation in approaches that recognize the nested nature of social systems from individuals and households to organizations, cities, states, and nations to global systems. The nested nature of social systems requires the acknowledgment that nations comprise numerous subunits, including states (or provinces), organizations, communities, households, and individuals. Similar to the work of ecologists, the work of social scientists conceptualizes "the systems they study as hierarchical but with complex embeddings" (Dietz, Rosa, and York 2010:84). As such, a sociological perspective can help shed light on the interplay between actions at different levels, from individuals and households at the micro level, to organizations and cities at the meso level, to national and international governmental organizations at the macro level. By considering the nested nature of social systems, such assessments can uncover the effects of agency, culture, social structure, institutions, power, inequality, and spatial characteristics and the roles that they play in shaping and constraining our efforts to reduce climate emissions. These insights reveal that while the actions of individuals and households can have a significant impact on reducing GHG emissions and may even induce policy change, in isolation such actions are unlikely to be sufficient for meeting the challenges of reducing emissions to the levels needed to avert major shifts in the Earth's climate. The enormity of the mitigation challenge requires a "both/and" solution that engages actors at all levels to change both individual and household practices as well as implement organizational, municipal, state, national, and international policies. How can we instigate, support, encourage, and catalyze the changes that are needed? Sociological insights can help to answer this question, but first we need to break free from traditional policy approaches that focus exclusively on economic factors and technological fixes.

While this chapter is far from comprehensive in its discussion of the range of sociological approaches to climate change mitigation, it does highlight some of the sociocultural factors that influence the effectiveness of mitigation efforts while also calling into question many of the assumptions and biases that underlie technology-centric approaches (like those used by the IPCC and ACC assessments). As a discipline, sociology offers a wealth of resources, including a range of theories and methods that acknowledge and address the influence of agency, culture, power, diversity, social inequality, global systems, social movements, and other social factors that shape and constrain mitigation efforts. As demonstrated in this chapter, the use of a sociological lens offers the means to expand and enhance our understanding of mitigation challenges by—for example—acknowledging the social construction of consumer demand (Wilhite et al. 2000), understanding the ways in which individuals and organizations make sense of climate challenges in the contexts of their everyday lives (Norgaard 2011), and revealing cross-national differences in discourse about climate change mitigation and networks of advocacy coalitions, and their implications for the success of national mitigation policies (Broadbent et al. 2013). Through the lens of sociology, policymakers can gain the ability to recognize the oddities, discrepancies, omissions, and inconsistencies of standard techno-economic definitions of mitigation challenges and prescriptions for change. They can also become more cognizant of the range of sociocultural factors that are equally—if not more—important in facilitating or impeding mitigation efforts.

The remainder of this chapter provides a selection of sociological insights on factors affecting climate change mitigation. The chapter is organized in parallel with the nested nature of social structures, beginning with a discussion of sociological insights at the micro level (individuals and households) followed by meso-level insights (organizations, companies, industries, networks, and states) and concluding with macro-level insights (nations, international organizations, and global systems). While each of these three sections has a *primary* focus on one of the three levels, the divisions that separate these realms of sociological thinking represent more of a disciplinary convention rather than firm boundaries not to be crossed. In fact, because the essence of a sociological perspective is the study of complex social systems, the failure to recognize the interconnectedness of levels would be antithetical to the task at hand.[2] For these reasons, the discussions in each section try to identify some of the ways in which the social processes operating at one level influence and interact with processes operating at other scales. The chapter ends with a brief discussion of the interplay between levels and a summary of many of the sociological insights highlighted in this chapter.

MICRO: INDIVIDUALS AND HOUSEHOLDS

How can the agency[3] of individuals be leveraged toward emissions reductions? Climate emissions are produced by individuals directly through the consumption of energy in the household and personal travel, and indirectly through the use of commercial goods and services. In the United States, the activities of individual consumers directly account for around 38 percent of total carbon emissions (Dietz et al. 2009). However, while the ability of any one individual to effect change is limited, sociological work on the agency of individuals to carry out cultural practices and internalize cultural beliefs emphasizes how social structure is both enacted and contested by individuals in social interactions. Thus, as sociologists we recognize that the micro-level agency of individuals can be mobilized in a variety of ways to leverage lower emissions. Here we describe sociological insights regarding the role of individuals and households in the mitigation of direct emissions, and the mitigation potential of individuals as agents of democratic renewal and social change. We begin with a discussion of the role of micro-level actors in determining GHG emissions and the associated opportunities for mitigation. The second subsection discusses the opportunities for individuals to shape mitigation efforts through indirect processes and social movements.

Direct Emissions

Scholars in economics, psychology, and science communication have placed the human response to climate change more centrally in their research agendas than have sociologists, with the overall result that within the larger interdisciplinary conversation about mitigation (1) micro-level approaches have dominated the general policy and scientific discourse; (2) within this micro focus the potential role for individuals in mitigation has largely been construed in terms of their ability to reduce individual and household consumption; and (3) explanations for consumer behavior have undertheorized the role of social context in shaping consumption behavior. (See Chapter 4 on the sociology of consumption for a more in-depth discussion of this topic.) Taken together, the resulting emphasis on individual consumption choices in the absence of social context overrepresents the importance of individual decision making and the potential power of individuals as the actors most responsible for climate emissions (Maniates 2002; Szasz 2011; Webb 2012).

Direct emissions by individuals in the "household sector" do of course matter. A key sociological insight is that individual behaviors and choices regarding direct contributions to household and transportation emissions take place in a context of social opportunities and constraints that include everything from the availability of low-carbon transportation options and the price of energy to the social construction of beliefs, values, and knowledge. So while information about the seriousness of climate change and the relation of fossil fuel use to individual behaviors may be a necessary condition for people to respond, normative constructions of time and space beyond the individual shape whether climate change is perceived to be "near" or "far" (Norgaard 2011) or the responsibility of people as citizens or a consumption choice (Maniates 2002; Szasz 2007). Furthermore, cultural constructions of space affect the visibility of problems and shape an individual's sense of personal responsibility for emissions (Frantz and Mayer 2009; Norgaard 2011; Ungar 2003). No doubt a central challenge in our attempt to grapple with climate change is its connection to qualities outlined in Ulrich Beck's work on the "risk society" (1992). In particular, rather than a problem we can touch and see for ourselves, climate change is a threat that must be interpreted for us through scientific expertise, using complex instrumentation. The reliance on experts opens the door to the politicization of what is known and what needs to be done.

In addition to relying on expert knowledge, however, people also have the ability to look to those around them for social cues in order to assess normative behavior and determine their own course of action. And in fact, social norms are often an important source of influence in shaping individual patterns of behavior. Notably, however, their influence appears to be higher under conditions of uncertainty and matter most when activities are visible (Cialdini and Goldstein 2004; Schultz et al. 2007). The public nature of curbside recycling, for example, is well suited to the development of such programs (Schultz 2002). On the other hand, most household energy practices are performed behind closed doors, making the use of social norms more difficult. Given the proven effectiveness of social norms in changing behavior, however, many program designers are looking for innovative ways to induce normative changes in emissions-intensive activities like heating homes. (A more in-depth discussion of household consumption and social norms is provided in Chapter 4.)

Many efforts to enhance environmental sustainability rely on value-driven campaigns aimed at changing attitudes and mobilizing individuals with strong environmental values to engage in new practices or to support particular policies. However, research has found that the removal of structural constraints is often a more effective means of changing individual behaviors. For example, in a study of recycling behaviors, researchers found that the existence of curbside recycling programs was a more reliable indicator of whether households recycled than environmental orientation (Derksen and Gartrell 1993). And while strong environmental values were associated with the push for environmental legislation in the past, Szasz (2007) suggests that today, people are more likely to express their environmental concerns in individualized acts of self-protection that he characterizes as an *inverted quarantine*. Instead of engaging in political mobilization, people buy bottled water, sunscreen, and organic foods to protect themselves from the environmental risks of modern life. Szasz outlines several social conditions that would be necessary for consumers to aggressively switch to low-carbon lifestyles, including a widespread sense that climate change is real and urgent, affordable and attractive low-carbon alternatives, motivation (i.e., from an organized consumer movement), and trust. He concludes that because most of these conditions do not currently exist, targeting the consumption of individual consumers should not be a priority.

It is important to note, however, that social science researchers disagree about the potential importance of individual behavior for mitigation efforts. As noted earlier, some research suggests that the decisions and practices of individuals and households represent a "behavioral wedge" that could, through policy inducements, reduce national carbon emissions in the short term by as much as 7.4 percent (Dietz et al. 2009; Vandenbergh et al. 2010). Moreover, this philosophy is being embraced by a growing number of cities and organizations that are working to mitigate GHG emissions through local programs that work collaboratively with individuals, households, and organizations. Such efforts include the development of energy feedback programs focused on providing electricity and natural gas consumers (both residential and commercial) with information about their energy consumption in ways that incorporate social norms research and other social psychological research to effectively motivate action, shift energy use practices, and encourage conservation (Alcott and Mullainathan 2010; Nolan et al. 2008). As documented by Ehrhardt-Martinez, Donnelly, and Laitner (2010), energy feedback programs have been effective in reducing average household energy consumption by 4 to 12 percent, depending on the type of feedback provided and the degree to which feedback programs successfully integrate social science insights. Indirect forms of feedback, such as Opower's use of monthly statements that incorporate both descriptive and injunctive norms to compare household electricity use to neighborhood averages, were shown to reduce household electricity use by 2 to 4 percent on average. If this program were scaled to the national level, it could reduce U.S. emissions by an estimated 12.7 million metric tons of CO_2 annually, or roughly 1 percent of U.S. carbon emissions from electric power (Alcott and Mullainathan 2010). Energy savings among high energy users were shown to be even more dramatic, and when social norms were used in conjunction with goal setting, average savings reached 8 percent. It is important to note that most of the savings from feedback-related interventions are generated through low-cost or no-cost types of activities such as simple shifts in everyday practices rather than high-cost investment activities (Ehrhardt-Martinez et al. 2010). These findings suggest that shifts in the everyday behaviors of individuals and households can make a measurable difference in reducing carbon emissions.

Research findings such as these have sparked the interest of a growing number of electric utilities and other types of organizations who have begun studying and applying social science insights in innovative new programs aimed at reducing household energy consumption. For example, public utility commissions in a growing number of states are requiring utilities to invest in enhancing energy efficiency before they invest in new power-generation facilities (although not always with the goal of addressing climate concerns). These types of utility policies and the growing interest in household energy consumption recently prompted the California Public Utilities Commission to sponsor a set of nine white papers on energy efficiency and behavior (California Institute for Energy and Environment 2009). Similarly, the New York State Energy Research Administration has recently funded social science research to test the effectiveness of various behavioral mechanisms for reducing energy use in commercial buildings and transportation.

In New Zealand, an interdisciplinary group of researchers is working in collaboration with the national government to gain a better understanding of the social factors that shape the existing energy culture and the opportunities to reshape it. Their approach is rooted in a sociological perspective, drawing both from Bourdieu's (1990) concept of "habitus" (the persistent patterns of thought, perceptions, and actions within which the individual exists) and Lutzenhiser's research on the cultural models of household energy consumption (Stephenson et al. 2010). In combination, these ideas have helped inform an interdisciplinary approach to understanding modern energy and climate mitigation challenges that focuses on the culture of consumption. At its core, the energy cultures model "suggests that consumer energy behavior can be understood at its most fundamental level as the interactions between cognitive norms (e.g. beliefs, understandings), material culture (e.g. technologies, building form) and energy practices (e.g. activities, processes)" (Stephenson et al. 2010:6123–6124). Similar to other studies, their research concludes that people often behave in ways that are inconsistent with their attitudes or values, but there is value in exploring and understanding the rationalizations that people provide and mechanisms for overcoming them (Lawson, Mirosa, and Gnoth 2011; Mirosa et al. 2011).

Indirect Processes

While the ability of individuals to effect change through their role as *consumers* may be limited, sociological research also acknowledges the role that individuals play in maintaining or changing established social orders that in turn shape GHG emissions. Changes in fertility patterns demonstrate how shifts in practices can have powerful indirect effects on the mitigation of GHGs.

Micro-level changes in fertility decisions are important because they add up, producing large aggregate impacts on national and global climate emissions. In the developed world, declines in fertility rates over the past century are expected to persist into the future, resulting in significant GHG emissions reductions during the twenty-first century (all else equal). While these estimates are encouraging, it is important to note that even when fertility rates decline to below replacement levels, they do not immediately result in negative population growth (or emissions reductions) because populations have momentum. In fact, it takes roughly twenty years after fertility rates fall below replacement levels to generate actual declines in population size—only after absolute declines occur in the size of cohorts entering their reproductive years. Acknowledging this dynamic, one recent calculation suggests that population declines in high-GHG-emitting societies during the twenty-first century could generate anywhere from a 15 percent to a 29 percent decline in global GHG emissions (O'Neill et al. 2010).

Globally, however, births continue to exceed deaths, and estimates suggest we will reach roughly 9 to 10 billion people by the late twenty-first century. Most of this population growth will occur in less-developed countries and is often described as a driver of increased GHG emissions. This type of global-scale analysis can be misleading because it fails to differentiate between the effects of population growth in high-emitting versus low-emitting societies. The distinction is important because per capita emissions levels in affluent societies are many times the levels found in less-affluent societies, where fertility rates are the highest. Moreover, economic development often results in lower fertility rates. As countries become more developed and skills become more specialized, the status of children shifts from being an economic asset to being an economic liability. Shifts in gender roles and opportunities are also common. Lower fertility rates and enhanced economic opportunities are associated with growth in women's access to education and work as well as income equality, legal protections, and social safety nets (Oppenheim Mason 1987). And as opportunities for women continue to expand, fertility rates fall. Such findings suggest that the relationship between household fertility and GHG emissions is complex but that changes in women's social status may play an important mediating role between economic development and population-driven trends in GHG emissions.

Individual Involvement in Social Movements

Sociologists emphasize the need for individuals to work collectively to challenge the social structures that shape unsustainable behaviors by engaging in social movements and participating in a renewal of democracy more broadly (Brulle 2010; Kent 2009). Viewed from this perspective, there are numerous micro-level mitigation opportunities to contest and reshape social norms and social structures as people participate in a variety of activities that affect both indirect consumption and large-scale climate policy. Individuals may choose to join with others to live more simply, adopt a vegan or vegetarian diet, or eat locally grown food. Individuals also volunteer with community groups working on climate change through churches, schools, watershed councils, and local nonprofits. Individuals engage in changing normative ideas about the world through their involvement in more formal political work through voting, letter writing, volunteering, involvement in local political parties and campaigns, testifying before city council, running for office, and marching in the streets. Collective activities pertaining to climate change provide an opportunity to reshape taken-for-granted practices and may also reshape individual personalities. According to McAdam (2012), individuals who participate in social movements sometimes experience a process of "cognitive liberation." This term describes the realization gained by social actors that society can be changed and that engaging in such change is worthwhile (McAdam 2012). Liberation occurs when individuals recognize the importance of their actions and feel empowered to change social orders.

According to Gamson (1992), movement mobilization also requires an integration of information from three different sources in order to gain traction—media, direct experience, and personal networks. As people gather information about the large-scale changes in the world around them, this information shapes their personal motivations to accept the status quo or take action (Roy 1994). Although the American public is currently divided into different camps on climate change (Leiserowitz et al. 2011), the growing number of climate-related disasters and the growth in scientific knowledge is likely to shift public opinion in the future. As the objective damages of climate change intensify (e.g., fires, droughts, heat waves, floods), support for mitigation policies is likely to increase. When combined with measures of self-efficacy, sociological insights on movement mobilization and cognitive liberation could inform future mitigation efforts.

MESO: STATES, CITIES, ORGANIZATIONS, AND NETWORKS

At a more aggregate level of analysis, sociologists study the functioning of states, cities, organizations, and networks to better understand the effect of organizational dynamics and structures on GHG emissions. These studies focus on the ways in which meso-level entities shape human responses to climate change. Meso-level actors intent on reducing emissions include nongovernmental organizations (NGOs), social movements, the larger private enterprises, cities, and state governments. Their impacts on emissions are discussed in the following order: the impact of focusing events as catalysts for new regulatory efforts, the influence of political-economic contexts in shaping mitigation efforts, the influence of spatial proximity in the spread of mitigating practices, the roles of city and state governments in promoting mitigation, and, finally, the influence of intraorganizational coalitions and business cycles on the establishment of new norms and standards.

Focusing Events and Punctuated Equilibria

The effectiveness of organizations is always at least partially determined by the social context in which they operate. For social movement organizations, effectiveness has been linked to measures of political opportunity that suggest that the most far-reaching reforms occur in the aftermath of events that transform social structures. The *Exxon Valdez* spill is a good example of such an event. In the two years after the spill, the federal government and more than thirty state governments adopted more restrictive laws regarding oil spills (Faass 2009). A pattern of punctuated equilibria in policymaking (Repetto 2006) emerges with alternating periods of political stasis and event-triggered surges in legislative activity. In times of environmental crisis, more progressive agencies within government can find themselves in charge of clean-ups and of making new regulations (as with the National Oceanic and Atmospheric Administration in the BP Gulf oil spill case). Perhaps surprisingly, the new regulations often come from proposals put together in preceding, quieter political periods by movement activists. In the case of proposals for state laws, government officials frequently use legislation passed in adjacent jurisdictions as models for their own legislation (DiMaggio and Powell 1983). For this reason, prior mobilization by movement activists around an issue is important because it increases the likelihood that activists will be able to take advantage of the political moment in the aftermath of galvanizing events when political change typically occurs (Boudet 2011).

Economic Opportunities, Power Structures,
and Ecological Modernization

Economic conditions play an important role in determining whether or not and when enterprises invest in sustainable practices that reduce their environmental impact. Certainly time-sensitive government incentives such as tax breaks have been shown to spur businesses to substitute newer, cleaner-burning technologies for older, more emissions-intensive technologies. Proximate factors frequently play an important role in decisions to modernize. For example, it's not uncommon for companies to make a decision to overhaul their manufacturing processes based on an assessment of aging machinery that has little to do with the environmental impact of old versus new machinery. But once the company executives have embarked on the overhaul, research findings suggest that they may be more likely to decide to purchase "greener" machinery (Florida 1996). Such decisions may also be shaped by voluntary standards programs, such as ISO 14001, which provide the firm with positive publicity in a concerned consumer market. If the adopting firm is an industry leader, other firms are more likely to adopt similar technologies in order to stay competitive. Such opportunities help shape a process such that a decision by one company to build a new, greener technology initiates a cascade of similar decisions throughout an industry.

A somewhat similar set of proximate circumstances influenced organizational participation in the Kyoto Protocol's clean development opportunities in Brazil and India. In this case, companies that had worked with a particular set of consulting firms were faced with a unique opportunity to benefit from the consultants' expertise and familiarity with a new clean development program. These consulting firms, which had previously given advice to cement mill and sugarcane mill owners about their production processes, became the conduit for further overhauls in production processes in order to receive carbon credits, which the mill owners could then sell in carbon markets (Pulver, Hultman, and Guimaraes 2010). In effect, the consultants became agents of diffusion for emission-reducing changes, but the effects of their activities were largely limited to the particular subset of organizations with whom they had previously worked.

Mitigation opportunities are also influenced by political ideology and the power of economic elites—although not always in ways that we would expect. In a comparative study of state-level emissions in the United States, Dietz et al. (2012) found that measures of conventional liberal/conservative politics mattered little in explaining state-by-state variations in emissions. The power of economic elites was, however, significant. When economic elites were politically strong, states exhibited higher levels of emissions. In states where economic elites were less powerful, reforms were more likely, resulting in lower emissions levels. This assessment builds on Shwom's (2011) research on appliance efficiency standards in the United States that concluded that societal power relations (including the power of economic elites and the strength of the state) play an important role in determining when treadmill of production theories are more applicable or, alternatively, when ecological modernization theories are likely to hold true. Based on her findings, Shwom emphasizes the importance of regulatory actions in shaping appliance standards that, in turn, shape the range of options available to consumers. Shwom concludes that, unless the strength of the state can counterbalance the interests of industry, progress will be slow at best and that "a purely voluntary approach without regulation or the threat of regulation is unlikely to lead to action on climate change" (2011:724). Similarly, Dietz et al. (2012:20) suggest that when business interests are especially powerful, "liberals will have little ability to promote environmental reforms and may not consider the environment a matter of central concern. When capital is less strong, a bipartisan consensus may form around such reform."

The Effect of Spatial Spillovers on Sustainable Practices

Spatial proximity has also been shown to influence changes in organizational behaviors and norms. Sometimes these patterns result from proximity to a focusing event, and in other cases they result from simple observation facilitated by spatial proximity or from geographically defined access to enthusiastic service providers. After the *Exxon Valdez* accident, for example, contiguous states adopted similar legislation regulating the transport of oil (Faass 2009). Spatial proximity also facilitates learning through observation. This type of spillover is exemplified by neophyte organic farmers who benefited from their proximity to more experienced organic farmers because they were able to track the progress of various agricultural experiments in their neighbors' fields. As a result, the farmers were able to learn best practices more quickly (Risgaard, Frederiksen, and Kaltoft 2007). In some cases, as with geographic patterns of no-till agriculture, adoption follows the spatial lines of jurisdictions. Counties with extension agents who were enthusiastic about conservation agriculture (no-till or minimum-till) had more farmers who adopted the new techniques than nearby counties with less enthusiastic extension agents (Kane et al. 2011). In addition, spatial proximity also facilitated the development of local farmers' networks focused on the sharing of relevant information. In this case, farmers from the same locations who attended the first "farm day demonstrations" of no-till techniques in the 1960s were able to reach out to each other and later formed networks of no-till farmers in their counties or states (Coughenour and Chamala 2000).

Proximity also has played a role in the creation of agricultural markets that link urban consumers and peri-urban farmers. For example, movements to establish farmers' markets and community-supported agriculture in the United States has resulted in a surge in the number of farmers' markets to more than four thousand across the United States and an increase in the numbers of consumers who consider themselves to be (emissions-reducing) locavores. If successful, these types of social movements reduce emissions and offer the opportunity to begin to bridge the metabolic rift between city and countryside that developed with urbanization during the nineteenth century (Foster 1999; Marx 1976).

Challenging the Status Quo Through Interorganizational Coalitions

Reducing the impact of organizations and businesses on the environment generally requires changing organizational norms and industry practices. The development of interorganizational coalitions can facilitate this process through the development of common goals and common "frames" for defining environmental problems and solutions. By working collectively, networks of NGOs institutions can aggregate their power and influence. For example, coalitions have developed among churches that espouse eco-theology and promote ideas like ecological stewardship (Ellingson, Woodley, and Paik 2012). These coalitions frequently bring different kinds of organizations together. For example, the U.S. Green Building Council brings together builders, urban planners, engineers, energy professionals, and environmental researchers and advocates to promote more energy-efficient building construction practices and the energy-efficient operation of existing buildings. By bringing together diverse interests, these coalitions have the opportunity to weave together the varied interests of disparate groups through the use of smartly crafted "issue frames." These types of coalitions can be effective because they are able to leverage common values across organizations. In the case of real estate interests and energy efficiency, for example, effective coalition organizers will recognize that while most real estate professionals are not likely to be motivated by concerns over climate change, they are often motivated by the status associated with LEED certification, the use of state-of-the-art technologies, the cost savings of efficient building operations, and the need to conform to industry norms concerning building performance. By tapping into these common values, coalitions offer the opportunity to shift industry norms and standards, particularly as building performance standards, publicized through city, state, and national programs, become more widely known. Similar coalitions have been formed by activists, sympathetic government officials, and national environmental NGOs to block the further expansion of energy-intensive sprawl in suburbs (Rudel 2009).

Cities, States, and Regions as Mitigators

Over the past two decades, cities and states have slowly developed programs that help residents shift their behaviors in ways that reduce both direct and indirect sources of emissions. Such programs work with households to lower thermostat settings, hang laundry to dry, car pool, reduce bottled water consumption, and compost biodegradable waste. Since 2005 more than a thousand cities have signed on to the U.S. Conference of Mayors Climate Protection Agreement, which calls on cities to (1) meet or beat the Kyoto Protocol targets in their own communities; (2) urge state and federal governments to enact policies to do the same; and (3) urge Congress to pass GHG reduction legislation that would establish a national emissions trading system. So far, more than six hundred cities have crafted their own climate action plans.

State and provincial governments have also made efforts to reduce GHG emissions. The most notable effort in the United States has come in California, which has established a cap-and-trade system for reducing GHG emissions. It also has mandated reductions in the carbon content of fuels and promoted the use of zero-emission vehicles for travel. In 2013 the scope of this political initiative to mitigate GHGs expanded to the region when the governors of British Columbia, California, Oregon, and Washington agreed to coordinate their efforts. Similarly, fuel efficiency standards adopted in California were later adopted at the national level by the U.S. Environmental Protection Agency (Wines 2013). These adoption–diffusion dynamics that accompany regulatory and technological innovations have long been studied by sociologists, but studies about the spread of GHG-reducing innovations across enterprises and jurisdictions are only just beginning.

Business Cycles, Exogenous Events, and New Industrial Norms

Finally, business cycles also matter in decisions to "go green." During periods of economic expansion, enterprises are enriched, enhancing the likelihood that they will invest in more environmentally efficient technologies. Research on the association between economic expansion and business investments in energy efficiency explains the paradoxical relationship between the growth in energy efficiency investments and the simultaneous growth in total energy consumption. As Jevons first noted, instead of decreasing overall energy use, new efficiencies in energy use often occur alongside increases in the absolute amounts of energy used (York 2006).

Apart from cycles of expansion and contraction, several additional factors have been shown to contribute to shifts away from engrained industrial norms and toward more sustainable practices. In particular, sociological research suggests that the perturbations caused by system-transforming events, when coupled with an abundance of capital, can sometimes bring a sudden end to preexisting communities of practice that organizations adhere to in order to produce a commodity with as little uncertainty as possible (Biggart and Beamish 2003). Exogenous events disturb these routines and increase the opportunities for change as organizations cast around for a new set of cultural tools or conventions around which to organize their daily operations. People open up to new schemata, and in these confused situations ideologies often become stronger because they provide ready-made explanations for events that do not fit well into the preexisting culture (Swidler 1986).

MACRO APPROACHES: NATIONS AND GLOBAL REGIMES

At the most aggregated level, researchers study the ways in which emergent characteristics and processes of entire societies influence global mitigation strategies and outcomes. In particular, the application of cross-societal comparison can provide unique insights concerning the range of national mitigation responses and the social conditions that either facilitate or impede mitigation policies, practices, and outcomes. Studies at the macro level are particularly appropriate given that climate change is a global phenomenon that can only be countered through strong national policies and an unprecedented level of international cooperation.

Macro-level studies generally focus on global (and international) social movements, global (and international) policy regimes, or cross-national patterns of mitigation. Research on global social movements studies the actions of NGOs, their strategies and tactics, and the roles that they play in reducing climate emissions. Studies of global policy regimes and international cooperation are concerned with the ways in which global (and international) policies are shaped, their implications for various participants, and their effectiveness in changing outcomes. Finally, national and cross-national studies look at the ways in which differences in social structures, history, and culture result in different mitigation outcomes. The research discussed in this section will focus on the latter two of these approaches. A discussion of the influence of social movements can be found in Chapter 8.

This section begins with an overview of the Kyoto Protocol and a short discussion of global mitigation policy efforts. The remainder of the section highlights some of the key sociological contributions to understanding the development and ratification of a global climate policy, the ineffectiveness of international climate agreements, and the lack of meaningful climate legislation in the United States.

The Kyoto Protocol and International Efforts to Mitigate Climate Change

The Kyoto Protocol, adopted in Kyoto, Japan, in 1997, attempts to mitigate climate change. It is an international agreement that commits developed countries to internationally binding emissions reduction targets of 5 percent below 1990 emissions levels during the first commitment period (2008–2012) and 18 percent below 1990 emissions levels during the second commitment period (2013–2020). The detailed rules for the implementation of the protocol were adopted in 2001, and the protocol entered into force in February 2005.

Countries first expressed their concern over climate change through the adoption of the United Nations Framework Convention on Climate Change in 1992. The treaty committed the signatories to "cooperatively consider what they could do to limit average global temperature increases and the resulting climate change, and to cope with whatever impacts were, by then, inevitable" (United Nations Framework Convention on Climate Change n.d.). By 1995, it had become clear that the emissions reductions provisions in the convention would not be sufficient to address climate change, sparking the development of more stringent approach through the Kyoto Protocol. Currently there are 195 parties to the convention and 192 parties to the Kyoto Protocol. Notably, the United States has failed to ratify the protocol, and Canada denounced the convention effective December 2012 and ceased to be a member from that time.

The lack of U.S. participation in the protocol has generated high levels of international concern given the critical role of the United States in both producing emissions and establishing international standards. The United States was the single largest producer of GHG emissions in the world until 2007, when it was overtaken by China. Moreover, the United States plays an important leadership role in the world both economically and politically. Social scientists have sought to better understand why some countries agreed to sign the protocol while others did not and why many of the signatories have failed to cut their emissions. The failure of the United States to sign the protocol and the lack of any significant national-level legislation on climate change in the United States have also drawn the scrutiny of social scientists. These topics are discussed in more detail in the following three sections.

Understanding Participation in International Climate Policies and Mitigation Efforts

How can we best understand why some countries agree to ratify a climate policy treaty while others do not? What conditions increase the likelihood of treaty ratification? This section explores three sociological approaches to these questions. The first considers the impacts of ecological modernization on the preparedness of nation-states to battle climate change and their willingness to ratify the Kyoto Protocol. The second considers the role of a global set of environmental values and norms in shaping the policies of nation-states and their willingness to ratify the protocol. Finally, a hybrid approach considers how national and global forces work together to shape national policies.

By definition, global climate change is among those environmental problems that are truly international in scope. Because a solution requires international cooperation and no supranational authority has the authority to compel compliance, social scientists often seek to explain cooperation by looking at the compatibility of national interests with international policy or structural differences in country-specific characteristics (e.g., education levels or the strength of civil society organizations). International regime theory and ecological modernization theory are two common approaches to understanding how and when cooperation among nation-states is achieved through international treaties or regimes.

Regime Theories

Regimes can be thought of as "implicit or explicit principles, norms, rules and decision-making procedures around which actors' expectations converge in a specific issue area" (Hasenclever, Mayer, and Rittberger 1997:9). Formal regimes generally include explicit agreements while less formal regimes may be identified through observations of state actions or through the reactions of nation-states to noncompliance. Classical (international regime) theory posits that the actions of nation-states are driven by national interests and rational decisions. Generally speaking, international regime theory predicts that countries will support international agreements only when they serve domestically determined national interests. Since dominant national interests are often focused on immediate problems, a traditional international regime analysis would be pessimistic about the possibility of developing effective international mitigation agreements.

Ecological Modernization and Global Environmental Systems

In contrast to international regime theory, approaches based on ecological modernization theory consider the structural differences between countries that serve to "circumscribe or enable their ability to reduce GHG emissions" (Zahran et al. 2007:39). Such studies are concerned with measures of countries' structural and strategic readiness to commit to and comply with the Kyoto Protocol (Fisher 2003; Zahran et al. 2007). The expectation is that through the process of ecological modernization, societies design economic, political, and cultural institutions that regulate human interactions with nature in a way that offers greater environmental protections. To do so, the process of modernization must become more reflexive and adaptable, allowing for the reconsideration, reorganization, and reform of past ways of doing things (Mol 2003). Economic changes center around the restructuring of production practices to increase energy efficiency and decrease environmental externalities. Politically, ecologically modernized countries are more likely to combine market-based approaches with command-and-control approaches to overcome production-side environmental problems (Spaargaren 2000), and they are more likely to cooperate internationally and recognize the value of multilateral relationships. Culturally, popular value systems tend to value aesthetics, identity, and self-actualization over economic growth. Such "postmaterialist" values are linked to higher levels of membership in environmental groups, donations for environmental protection efforts, and environmentally friendly behaviors (Kidd and Lee 1997). Studies of the prevalence of ecologically modernized institutions across countries found notable differences (Sonnenfeld and Mol 2002; Spaargaren and Mol 1992), and these differences were found to provide a partial explanation of the likelihood of Kyoto Protocol ratification (Zahran et al. 2007). In particular, Zahran's research found that "societies characterized by extensive civil liberties and political rights, high energy efficiency, low carbon dioxide emissions per capita, high education levels, and records of international cooperation on other transboundary environmental issues are significantly more likely to commit to the Protocol" (50). Governments have a range of domestic policy instruments at their disposal to enhance energy efficiency, including utility incentives, tax credits, energy-sensitive building codes, equipment efficiency standards, and regulation of alternative energy markets. By engaging in these actions, governments set the stage for participation in binding international agreements.

World Society Theory and an International Culture of Environmentalism

In contrast to ecological modernization theory and classic international regime theory, world society theory provides an alternative perspective that is rooted in the evolution of global institutions and global culture. More specifically, world society theory conceives of international policies and common patterns of national development as an outgrowth of emerging supranational (often global) institutions and the common set of norms, values, and cultural rules that they share. Rather than viewing nation-states as rational actors pursuing their own unique interests, neoinstitutional theory suggests that nation-states are guided by the emergence of a commonly held set of ideas about particular issues such as approaches to governance or environmental practices. The formation of a dominant set of global norms results in the institutionalization of cultural models that come to define what a "normal" or appropriate nation-state looks like (Meyer et al. 1997). Through the process of institutionalization, there is an enhanced likelihood that similar policies and practices will be transmitted and adopted across and within a broad spectrum of nation-states. In fact, sociological research has found that state environmentalism is measurable and is a viable concept for describing the propensity of a nation-state to take political action in support of the environment (Dietz and Kalof 1992; Frank 1999). Such research suggests that an international culture of environmentalism is likely to be an important factor in shaping the likelihood that nations will ratify a common international policy despite competing interests at the national level. As noted by Frank (1999:538), "Environmentalization often appears to conflict with other national interests, especially economic development, generating intense resistance from nation states." Nevertheless, the same study found that those countries that are most embedded in the world social system are also most likely to ratify international environmental treaties despite competing domestic interests.

Hypotheses rooted in a world society perspective would generally be more optimistic than international regime theory about the possibility of developing a collaborative path toward international climate mitigation policies through an international learning process about climate change. To understand the development of international regimes, world society theory often looks to the role of international NGOs in both establishing and reinforcing cultural norms (e.g., Schofer and Hironaka 2005). According to Meyer and his colleagues (1997), international environmental NGOs and other types of civil society organizations are the carriers of world culture that is subsequently adopted by local actors. International environmental NGOs help shape the language of international treaties, assist in setting standards, develop codes of conduct, and create technical guidelines; they may even monitor compliance by nations in the absence of formal enforcement mechanisms (Jorgenson, Dunlap, and Clark 2013).

With regard to climate change mitigation policy, recent sociological research has explored the creation of this type of learning process through the facilitation of the United Nations. By actively bringing countries together, the UN is acting as an agent of change with the goal of persuading self-interested actors to see the longer-term view and to act in accordance with the collective good. As described above, these processes are consistent with the principles of neoinstitutional theory and the tenet that new norms can diffuse on a global scale, changing the very definition of national interests toward new goals while also institutionalizing new forms of international cooperation (Busch and Jörgens 2005; Frank, Hironaka, and Schofer 2000).

The UN-sponsored regime seeks to engage societies and governments to work toward the development of (1) increasingly consensual and certain scientific information disseminated in the reports of the IPCC; (2) cogent moral injunctions to reduce emissions as part of the United Nations Framework Convention on Climate Change; and (3) specification of specific policy mechanisms for reducing GHG emissions.

A recent study considered the role of major national newspapers in diffusing scientific knowledge and mitigation norms. Consistent with neo-institutional theory, Broadbent et al. (2013) found that the prevalence of climate information in newspapers was more pronounced in UN member countries than in non-UN member countries. The findings suggest that such processes may prove to be a mechanism for enhancing the emergence of a global mitigation culture and agreement on basic scientific knowledge, norms of action, and a fair distribution of responsibilities for action.

Global mitigation norms are also being communicated directly by the UN through its new Climate Performance Index in an effort to increase pressure for action among nation-states (Lee et al. 2010). The new UN index scores two measures of climate action, *climate accountability* (effectiveness of programs and policies) and *climate performance* (actual emissions reductions). The index clearly communicates the values and expectations that countries should aspire to. In the period following the 2009 Copenhagen meetings, the best combination of overall accountability and performance was achieved by Sweden, Denmark, Germany, Japan, and France, while Germany, China, and the Republic of Korea had achieved the largest improvements on both scores. During the same period, measureable improvements in climate accountability were achieved by India, Indonesia, Kenya, Mexico, the Philippines, and Rwanda. Notably, the climate performance of Switzerland and Austria was rated as good, while the United Kingdom and the United States were found to have strong accountability but not strong performance.

Theory of Global Environmental Systems

Research by Fisher (2003, 2004) suggests a third approach for understanding protocol ratification that incorporates both measures of nation-state characteristics and the influence of global environmental norms and expectations and the interactions between these sources of influence. Given the findings of prior research (Fisher and Freudenburg 2004), Fisher's work is focused exclusively on the environmental regimes of advanced capitalist countries with the goal of identifying factors that can help explain the variation in the environmental policies between those countries.

Like the ecological modernization approach described above, Fisher's research includes an assessment of nation-state characteristics as potential sources of variation. However, Fisher's domestic measures are focused on the roles of four sets of domestic actors and their participation in the formation of a domestic climate change regime: the state, science, the market, and civil society. More specifically, the study considers why responses to the Kyoto Protocol have been so different among advanced capitalist nation-states by assessing the interrelationships between state strength, the centrality of science to policymaking, the material composition of the economic market and the role it plays with the state, and the level of civil society participation in domestic decision-making processes.

Through a comparative assessment of three countries—Japan, the Netherlands, and the United States—Fisher's findings suggest that when strong states work with collaborative market actors in an environment where scientists are highly and collaboratively engaged in the policymaking processes, treaty ratification and favorable environmental outcomes are more likely. Politically, a culture of collaboration was also found to benefit treaty ratification.

Explaining the Failure of International Climate Agreements

A notable group of researchers have suggested that while treaty ratification might be the first step toward finding a global solution to climate change, the success or failure of mitigation efforts is closely tied to issues of international inequality (Parks and Roberts 2008, others). Global inequalities exist on a variety of climate-related issues, including who will suffer the effect of climate change the most, which countries hold the most responsibility for the problem, and who is willing and able to address climate problems.

As noted by Parks and Roberts (2008:623–624):

> With only four percent of the world's population, the US is responsible for over 20 percent of all global emissions. That can be compared to 136 developing countries that together are only responsible for 24 percent of global emissions (Roberts and Parks 2007). Poor countries therefore remain far behind wealthy countries in terms of emissions per person. Overall, the richest 20 percent of the world's population is responsible for over 60 percent of its current emissions of greenhouse gasses. That figure surpasses 80 percent if past contributions to the problem are considered, and they probably should be, since carbon dioxide (the main contributor to the greenhouse effect) remains in the atmosphere for over one hundred years.
>
> A casual observer might think that the best way to resolve the issue of responsibility for climate change would be to give all humans equal

atmospheric rights and assign responsibility to individuals based on how much "environmental space" they use. This is a basic rule of civil justice and kindergarten ethics: those who created a mess should be responsible for cleaning up their fair share. But in international politics things are not so simple.

Despite the Kyoto Protocol's recognition of historically rooted inequalities and its attempt to address them, tensions over issues of inequality, justice, and fairness have continued to plague international climate discussions and impede the success of international mitigation efforts (Roberts and Parks 2007).

In reference to the vast disparities described above, four distinct proposals about the roles that different types of countries should be obligated to play in the cleaning up the atmosphere have emerged as part of climate negotiations: grandfathering, carbon intensity, contraction and convergence to a global per capita norm, and historical responsibility. The grandfathering approach would require the world's wealthier nations to reduce their emissions relative to a particular baseline year rather than requiring the establishment of a global emissions average or a much earlier historical baseline. The carbon intensity approach would require voluntary changes in the *efficiency* of emissions to allow for economic growth alongside higher emissions standards. Both of these approaches favor the short-term economic interests of industrial nations while producing only incremental mitigation benefits. The third option, the historical responsibility approach, would require an accounting of the amount of damage done as a result of nations' past emissions. And finally, the per capita contraction and convergence approach would require that countries work toward a global average per capita emissions target that would allow nations with low levels of emissions per capita to increase and require high emitters to decrease emissions. In addition, low emitters could trade their carbon emissions credits in exchange for the capital they need for economic development. These last two approaches involved more benefits for less-developed countries.

Ultimately, the Kyoto Protocol was developed based on grandfathering, allowing high-emitting nations to make voluntary reductions using 1990 base year targets but only requiring commitments from the more-developed countries and allowing voluntary emissions reductions from less-developed ones. This "resolution" of the inequality problem has been blamed for ongoing political tensions and inaction on the part of many less-developed countries and the United States alike (Parks and Roberts 2008; Roberts and Parks 2007; Roberts, Parks, and Vasquez 2004).

Many sociological studies of the impact of inequality have emphasized the importance of climate justice in shaping both treaty ratification and the likelihood of successful climate mitigation efforts. Among such studies,

Roberts and Parks (2007) specifically consider how the globalization of production systems has further complicated issues of climate justice. Their study suggests that while the globalization of production has created the *illusion* that the economies of more-developed countries are dematerializing, a closer examination reveals evidence indicating that declines in resource consumption in these countries is achieved through their ability to export material-intensive production and its environmental consequences to less-developed nations. Such relationships create situations of ecological indebtedness and ecologically unequal exchange in which "poor nations export large quantities of under-priced products whose value does not include the environmental (and social) costs of their extraction, processing or shipping" (Roberts and Parks 2007:196). Sociologists note that these types of relationships are historically rooted in colonial and neocolonial power relations. The ongoing impacts of such relationships can be measured using a materials flow accounting methodology. Such assessments suggest that more-developed nations are in fact exhausting the ecological capacity of extractive economies by importing resource-intensive products and have shifted environmental burdens to less-developed ones through the export of waste (Andersson and Lindroth 2001). Similarly, statistical research suggests that when less-developed nations trade more products, it results in higher emissions for poor nations and lower emissions for more-developed ones (Heil and Selden 2001; Roberts and Parks 2007).

To address climate justice dilemmas, Roberts and Parks suggest the adoption of one of several potential hybrid approaches that take into account factors such as historical emission levels, per capita emissions levels, geography, climate, energy supply, and domestic economic structures.

In an extension of such research, a more recent sociological study (Roberts 2011) considers how changes in global economic and political systems and the international balance of power have reshaped climate coalitions, strategies, and negotiations. The article suggests that over time, discussions have moved further away from the principles and practices of climate justice due to the fragmentation of less-developed nations' perspectives and continued resistance on the part of the United States. Importantly, the article attributes growing U.S. stubbornness to its increasing insecurity about its ability to provide jobs for its workers in the context of job outsourcing (to China and India). At the same time, China, India, and other rapidly developing nations must deal with their own concerns that negotiations might dampen growth in their countries and defer their aspirations for higher levels of prosperity.

A sociological evaluation of climate mitigation policies underscores the importance of taking a systems approach that also recognizes the historical influences on current international relations and that questions widespread assumptions about the development trajectories of today's less-developed nations. By recognizing the importance of inequality in the global system, its historical roots in colonialism and neocolonialism, and its perpetuation in today's system of global production, sociologists have revealed the mechanisms that have resulted in the ecological indebtedness of developed countries like the United States. Such studies suggest that the success of global mitigation efforts is likely to depend on hybrid policy strategies that adequately take such factors into account.

Assessments of National Mitigation Legislation in the United States

The lack of any large-scale climate strategy in United States and the country's failure to ratify the Kyoto Protocol have led many sociologists to focus their attention on explanations of what some refer to as the U.S. climate "non-policy" (Lutzenhiser 2001). Such studies have pointed to a variety of potential causes, including climate change denial and the lack of popular policy support (McCright and Dunlap 2011a; Norgaard 2011), the politicization of climate change (McCright and Dunlap 2011b), the influence of powerful special interests (Brulle 2013), the role of natural resource interests (Fisher 2006), and the interrelations among state, market, civil society, and the scientific community (Fisher 2003).

According to Norgaard, the public's inability or unwillingness to recognize the climate imperative and support policies that address it may be rooted in emotional responses to climate dilemmas as well as measures of political economy. Norgaard's research suggests that while people experience deep fears regarding climate change, they often "normalize their inaction through a variety of cultural tools and narratives" (2011:177). However, in the United States, there is a literal denial of climate change among many Americans that is linked to political efforts to divert action by suggesting that the problem needs further study or by questioning the quality of the science. U.S. climate skepticism is further fueled by corporate-funded campaigns that have actively worked to instill doubt and question the legitimacy of science in the public sphere. As noted by Norgaard, the effectiveness of such efforts is at least partially rooted in the anti-intellectualism that pervades American culture and is complicated by America's cultural reliance on fossil fuels and comparatively high levels of GHG emissions. Finally, Norgaard's research also suggests that in the United States, political alienation and a culture of individualism contribute greatly to disengagement on climate change issues. In general, Americans don't trust that government institutions can respond to modern problems, while a culture of individualism leaves Americans at a loss in terms of what to do about climate change. Under these conditions, it isn't surprising that the American public feels both disempowered and ineffective in the face of global climate change (Macnaghten 2003:77). The prevalence of climate denialism was found to be particularly prominent among conservative white males in the United States (McCright and Dunlap 2011a).

In addition to the broad disempowerment felt by many Americans, sociologists have also attributed the lack of U.S. climate policy to the politicization of the issue. Beginning in the early 1990s

> a coordinated anti-environmental countermovement, spearheaded by conservative foundations, think tanks, and politicians, emerged in response to the rise of global environmentalism—symbolized by the 1992 Rio "Earth Summit"—and its perceived threat to the spread of neoliberal economic policies worldwide (McCright and Dunlap 2011b:158).

As a result of these efforts and the rightward shift in U.S. political culture, conservative interests were able to shift the climate perspectives of politically conservative Americans, resulting in a "sizable political divide" between liberals/Democrats and conservatives/Republicans (McCright and Dunlap 2011b:178). In general, the views of liberals were more likely to be consistent with scientific consensus. Importantly, this political divide became increasingly prominent between 2001 and 2010. Such findings are important because they call into question the assumption that advanced, modern societies will necessarily adopt more environmentally friendly policies when such policies are supported by sound science. Instead, this work suggests that forces of antireflexivity must also be accounted for.

As noted earlier, part of the cause of U.S. inaction on climate change mitigation can also be attributed to the active funding of a climate change countermovement that Brulle (2013) describes as a cultural contestation between efforts to restrict carbon emissions and those opposed to the restrictions. The countermovement comprises a number of conservative think tanks, trade associations, and advocacy organizations focused on confounding public understanding of climate science and delaying government mitigation policies. According to Brulle, these efforts justify the unlimited use of fossil fuels by delegitimizing climate science and efforts to impose mandatory limits on carbon emissions. Such research is important because it documents the role of political and philanthropic interests in actively manipulating and misleading the public over the threat posed by climate change and delaying the ability of U.S. policymakers to take action on the issue. Similarly, an assessment of Senate voting records on climate-related policies reveals the influence of partisan politics (Fisher 2006). Fisher's work is unique in that it also highlights the influence of states' dependency on fossil fuel resources (oil and coal) as a key determinant of climate policy voting patterns. In fact, her research suggests that it is the resource dependence of the state that matters most in determining both national policy and the implementation of state-level GHG emission targets.

Finally, in a comparative assessment of nation-state action on climate change mitigation policies, Fisher (2003) attributes U.S. nonaction to some of the unique characteristics of the U.S. climate change regime. According to Fisher's assessment, the relative power of four key influencers (the state, science, the market, and civil society) determines the opportunities for policy action. In the United States, the federal government has relatively little autonomy relative to other social actors, limiting its ability to act in the absence of strong support from other sectors. In addition, the heavy dependence of the U.S. economy on fossil fuels and the peripheral role of science and environmental organizations in climate policy processes have stacked the deck against the likelihood of climate legislation.

THE INTERPLAY OF LEVELS AND FUTURE RESEARCH

Efforts to mitigate climate change are happening throughout the social system, from the micro-level efforts of individuals and households to the macro-level efforts of nations and intergovernmental bodies. A sociological perspective helps shed light on the interplay among actions at the micro, meso, and macro levels, revealing the effects of agency, culture, social structure, institutions, power, inequality, and spatial characteristics in shaping and constraining our success in reducing climate emissions. These insights reveal that while the actions of individuals and households can have a significant impact in reducing GHG emissions, these actions alone will be insufficient for meeting the challenges of reducing emissions to the levels needed to avert major shifts in the Earth's climate. The enormity of the mitigation challenge requires a "both/and" solution that engages actors at all levels to change both individual and household practices and organizational, municipal, state, national, and international policies. How can we instigate, support, encourage, and catalyze the changes that are needed? A sociological perspective can contribute greatly in answering this question as well.

In a social system, change can begin anywhere, and isn't necessarily a linear process. Social psychology and social movement theory tend to highlight the importance of individual agency in a social context and how individual choices and actions can reshape policies through shifts in everyday practices, involvement in political processes, and the establishment of broad social movements that encourage or demand shifts in national and international policies. Organizational sociologists can shed light on the dynamics and actions of businesses, nonprofits, and other groups that operate in the shadow of national policy but offer the ability to scale up change quickly through organizational networks. Economic sociology and political sociology are often concerned with the roles that national governments and institutions play in determining national and international policy directly.

As described in this chapter, efforts are being made at all levels, but their success at reducing GHG emissions has been limited. At the micro level, the challenges of climate change have largely been met with denial and consumerist-type responses. As Norgaard (2011) points out, the scale of the problem and its perceived distance in time have allowed people to find ways to continue to live their lives as if the problem didn't exist. And for those who have been mobilized to action, the actions have often been limited to consumerist and individualized forms of action (Szasz 2007). At the macro level, much progress has been made in documenting the threat of climate change, in the formation of some international agreements, and even some actual carbon reductions in a subset of countries.

Nevertheless, decades of efforts to establish a global agreement or strong U.S. policy have largely failed in the face of a variety of political and economic challenges. More encouraging progress may be found at the meso level, where cities and organizations have begun to step forward *en masse*. Similarly, organizations such as the U.S. Green Building Council and the Urban Land Institute have begun engaging their networks of builders, building owners, and building managers in actions to reduce energy consumption and carbon emissions.

Whether or not we, as a global community, will be successful in mitigating the worst impacts of climate change remains to be seen. History provides ready examples of the factors that have led to social collapses in the past (Diamond 2004). Given the global scale of the climate problem, it is unlikely that efforts to address the problem at the micro or meso levels alone will avoid some type of collapse of global proportions in the future. In other words, shifts in national and international policy are required to avoid the catastrophic consequences of climate change that are expected to occur unless emissions are dramatically reduced. Perhaps micro- and meso-level initiatives will help lay the groundwork necessary for achieving the national and international policies that are needed.

Research Recommendations

The predominant programmatic emphasis on *individual* consumption choices (in the absence of social context, organizational behavior, and national and international policies) has resulted in an unbalanced level of attention on the power of individuals and households to change consumption patterns and achieve much-needed changes in emissions. Instead, a more balanced research agenda should look more broadly at the social dynamics influencing climate mitigation efforts in each of the three levels of the social system (micro, meso, and macro) as well as the interplay among the levels with the goal of revealing constraints, opportunities, and policies that are likely to facilitate action across all three levels. Such efforts need to explore more fully the effects of agency, culture, social structure, institutions, power, inequality, and spatial characteristics as they shape and constrain our success in reducing climate emissions. More research is needed in three broad categories:

1. Which factors most constrain and/or facilitate the implementation of national and state policies? When, to what degree, and in what ways do the actions and interests of different political actors shape policy outcomes? What are the contexts in which power elites are successful in shaping political agendas? What is the role of social movements, advocacy organizations, and voters in influencing outcomes, and when do their efforts matter most? And how can cities, states, utilities, and other meso-level actors influence both climate-related practices and national policy?
2. What can we learn from cross-societal comparisons that offer unique insights concerning the range of national mitigation responses and the social conditions that either facilitate or impede mitigation policies, practices, and outcomes?
3. Under what conditions does social, political, economic, and cultural change occur? How might policy efforts such as a carbon tax leverage momentum across spatial and temporal scales? How do state-level or city-level initiatives influence broader levels of change? How can shifts in industry norms reshape social debate, cultural beliefs, and the range of viable policy options? How can shifts in the social organization of energy technologies reshape public perceptions and policy options? Under what circumstances does social movement mobilization occur? How can sociological insights on movement mobilization and cognitive liberation leverage efforts at larger scales (e.g., state policy)?

NOTES

1. These estimates are based on scenarios in which fossil fuels maintain their dominant position in the global energy mix to 2030 and beyond, such that global emissions from energy use are expected to grow 40 to 110 percent between 2000 and 2030. The IPCC scenario estimates also assume a world of rapid economic growth, a global population that peaks in midcentury, and the rapid introduction of new and more efficient technologies. The most conservative GHG emissions estimates

assume important shifts in global economic structures toward a service and information economy (IPCC 2007b).

2. Environmental sociology also expands beyond the social systems to include their relationship to the encompassing ecology: the biosphere, geosphere, and other natural phenomena.

3. *Agency* refers to the ability of individuals to act independently and make their own free choices.

REFERENCES

Alcott, Hunt and Sendhil Mullainathan. 2010. "Behavior and Energy Policy." *Science Magazine* 327.

Andersson, Jan Otto, and Mattias Lindroth. 2001. "Ecologically Unsustainable Trade." *Ecological Economics* 37:13–122.

Beck, Ulrich. 1992. *Risk Society: Towards a New Modernity*. Translated by Mark Ritter. London: Sage.

Biggart, Nicole W. and Thomas D. Beamish. 2003. "The Economic Sociology of Conventions: Habit, Custom, Practice, and Routine in Market Order." *Annual Review of Sociology* 29:443–464.

Boudet, Hilary S. 2011. "From NIMBY to NIABY: Regional Mobilization against Liquefied Natural Gas Facility Siting in the U.S." *Environmental Politics* 20:786–806.

Bourdieu, Pierre. 1990. *The Logic of Practice*. Translated by Richard Nice. Palo Alto, CA: Stanford University Press.

Broadbent, Jeffrey, Sun-Jin Yun, Dowan Ku, Kazuhiro Ikeda, Keiichi Satoh, Sony Pellissery, Pradip Swarnarkar, Tze-Luen Lin, Ho-Ching Lee, and Jun Jin. 2013. "Asian Societies and Climate Change: Global Events and Domestic Discourse." *Globality Studies Journal* 32, July 26, 2013 (http://globality.cc.stonybrook.edu/wp-content/uploads/2013/07/032JBroadbent.pdf).

Brulle, Robert J. 2010. "From Environmental Campaigns to Advancing the Public Dialogue: Environmental Communication for Civic Engagement." *Environmental Communication: A Journal of Nature and Culture* 4(1 March):82–98.

Brulle, Robert J. 2013. "Institutionalizing Delay: Foundation Funding and the Creation of U.S. Climate Change Counter-Movement Organizations." *Climate Change* 122(4):681–694.

Busch, Per-olof and Helge Jörgens. 2005. "The International Sources of Policy Convergence: Explaining the Spread of Environmental Policy Innovations." *Journal of European Public Policy* 12:860–884.

California Institute for Energy and Environment. 2009. "Behavior and Decision Making." White papers available at http://uc-ciee.org/behavior-decision-making/overview.

Cialdini, Robert B. and Noah J. Goldstein. 2004. "Social Influence: Compliance and Conformity." *Annual Review of Psychology* 55:591–621.

Coughenour, C. Milton and Shankariah Chamala. 2000. *Conservation Tillage and Cropping Innovation. Constructing the New Culture of Agriculture*. Ames, IA: Iowa State University Press.

Derksen, Linda and John Gartrell. 1993. "The Social Context of Recycling." *American Sociological Review* 58:434–442.

Diamond, Jared. 2004. *Collapse: How Societies Choose to Fail or Succeed.* New York: Penguin.

Dietz, Thomas, Gerald T. Gardner, Jonathan Gilligan, Paul C. Stern, and Michael P. Vandenbergh. 2009. "Household Actions Can Provide a Behavioral Wedge to Rapidly Reduce U.S. Carbon Emissions." *Proceedings of the National Academy of Sciences USA* 106(44):18452–18456.

Dietz, Thomas and Linda Kalof. 1992. "Environmentalism among Nation-States." *Social Indicators Research* 26:353–366.

Dietz, Thomas, Eugene A. Rosa, and Richard York. 2010. "Human Driving Forces of Global Change: Dominant Perspectives." Pp. 83–134 in *Human Footprints on the Global Environment: Threats to Sustainability,* edited by Eugene A. Rosa, Andreas Diekmann, Thomas Dietz, and Carlo C. Jaeger. Cambridge, MA: MIT Press.

Dietz, Thomas, Cameron T. Whitley, Jennifer Kelly, and Rachel Kelly. 2012. "Treadmill of Production or Ecological Modernization: The Political Economy of Greenhouse Gas Emissions in the United States." Proceedings of the American Sociological Associations Annual Meeting, Denver, Colorado.

DiMaggio, Paul J. and Walter W. Powell. 1983. "The Iron Cage Revisited: Institutional Isomorphism and Collective Rationality in Organizational Fields." *American Sociological Review* 48:147–160.

Ehrhardt-Martinez, Karen, Kat Donnelly, and John A. "Skip" Laitner. 2010. "Advanced Metering Initiatives and Residential Feedback Programs: A Meta-Review for Household Electricity-Saving Opportunities." Washington, DC: ACEEE.

Ellingson, Stephen, Vernon A. Woodley, and Anthony Paik. 2012. "The Structure of Religious Environmentalism: Movement Organizations, Interorganizational Networks, and Collective Action." *Journal for the Scientific Study of Religion* 51(2):266–285.

Faass, Josephine. 2009. "Mission Accomplished or Mission Impossible: Current Practices, Common Challenges, and Innovative Solutions in State-level Oil Pollution Regulation." PhD dissertation, Rutgers University.

Fisher, Dana. 2003. "Global and Domestic Actors Within the Global Climate Change Regime: Toward a Theory of the Global Environmental System." *International Journal of Sociology and Social Policy* 23:5–30.

Fisher, Dana. 2004. *National Governance and the Global Climate Change Regime.* Lanham, MD: Rowman & Littlefield Publishers.

Fisher, Dana. 2006. "Bringing the Material Back In: Understanding the United States Position on Climate Change." *Sociological Forum* 21:467–494.

Fisher, Dana and William R. Freudenburg. 2004. "Postindustrialization and Environmental Quality: An Empirical Analysis of the Environmental State." *Social Forces* 83:157–188.

Florida, Richard. 1996. "Lean and Green: the Move to Environmentally Conscious Manufacturing." *California Management Review* 39:80–105.

Foster, John Bellamy. 1999. "Marx's Theory of Metabolic Rift." *American Journal of Sociology* 105(2):366–405.

Frank, David J. 1999. "The Social Bases of Environmental Treaty Ratification, 1900–1990." *Sociological Inquiry* 69(4):523–550.

Frank, David John, Ann Hironaka, and Evan Schofer. 2000. "The Nation-State and the Natural Environment over the Twentieth Century." *American Sociological Review* 65:96–116.

Frantz, Cynthia M. and F. Stephan Mayer. 2009. "The Emergency of Climate Change: Why Are We Failing to Take Action?" *Analyses of Social Issues and Public Policy* 9:205–222.

Gamson, William A. 1992. *Talking Politics*. New York: Cambridge University Press.

Hasenclever, Andreas, Peter Mayer, and Volker Rittberger. 1997. *Theories of International Regimes*. New York: Cambridge University Press.

Heil, Mark T. and Thomas M. Selden. 2001. "International Trade Intensity and Carbon Emissions: A Cross-Country Econometric Analysis." *Journal of Environment and Development* 10:35–49.

IPCC. 2007a. "Climate Change 2007: Mitigation." Contribution of Working Group III to the Fourth Assessment Report of the Intergovernmental Panel on Climate Change, edited by B. Metz, O. R. Davidson, P. R. Bosch, R. Dave, and L. A. Meyer. Cambridge and New York: Cambridge University Press.

IPCC. 2007b. "Climate Change 2007: Synthesis Report." Contribution of Working Groups I, II, and III to the Fourth Assessment Report of the Intergovernmental Panel on Climate Change, edited by R. K. Pachauri and A. Reisinger. Geneva, Switzerland: IPCC.

IPCC. 2007c. "Summary for Policymakers." Pp. 7–22 in *Climate Change 2007: Impacts, Adaptation and Vulnerability*. Contribution of Working Group II to the Fourth Assessment Report of the Intergovernmental Panel on Climate Change, edited by M. L. Parry, O. F. Canziani, J. P. Palutikof, P. J. van der Linden, and C. E. Hanson. Cambridge: Cambridge University Press.

IPCC. 2014. "Climate Change 2014: Mitigation." Contribution of Working Group III to the Fifth Assessment Report of the Intergovernmental Panel on Climate Change, edited by O. Edenhofer, R. Madruga, and Y. Sokona. Retrieved December 2, 2014 (https://www.ipcc.ch/report/ar5/).

Jorgenson, Andrew K., Riley E. Dunlap, and Brett Clark. 2014. "Ecology and Environment." In *Concise Encyclopedia of Comparative Sociology*, edited by M. Sasaki, J. Goldstone, E. Zimmermann, and S. K. Sanderson. Leiden and Boston: Brill Academic Publishers.

Kane, Stephanie, Penelope Dieble, J. D. Wulfhorst, Barbara Foltz, and Douglas Young. 2011. "Socio-economic Factors Affecting Tillage Practices in Northwest Dryland Agriculture." Proceedings of the Rural Sociological Society Meetings, July 30—August 2, Boise, Idaho.

Kent, Jennifer. 2009. "Individualized Responsibility and Climate Change: If Climate Protection Becomes Everyone's Responsibility, Does It End Up Being No-one's?" *Cosmopolitan Civil Societies Journal* 3(1):132–149.

Kidd, Quentin and Aie-Rie Lee. 1997. "Postmaterialist Values and the Environment: A Critique and Reappraisal." *Social Science Quarterly* 78(1):1–15.

Lawson, Rob, Miranda Mirosa, and Daniel Gnoth. 2011. "Linking Personal Values to Energy-Efficient Behaviours." *Environment and Behavior* (December) (http://eab. sagepub.com/content/early/2011/12/26/0013916511432332).

Leiserowitz, Anthony, Edward Maibach, Connie Roser-Renouf, and Nicholas Smith. 2011. "Americans' Actions to Conserve Energy, Reduce Waste and Limit Global Warming in November 2011." Yale Project on Climate Communications.

Lee, Hee Ryung, Alex MacGillivray, Paul Begley, and Elena Zayakova. 2010. *The Climate Competitiveness Index 2010*. AccountAbility.

Lutzenhiser, Loren. 2001. "The Contours of U.S. Climate Non-Policy." *Society and Natural Resources* 14:511–523.

Macnaghten, Phil. 2003. "Embodying the Environment in Everyday Life Practices." *Sociological Review* 51(1):63–84.

Maniates, Michael. 2002. "Individualization: Buy a Bike, Plant a Tree, Save the World?" Pp. 43–66 in *Confronting Consumption,* edited by T. Princeton, M. Maniates, and K. Conca. Cambridge, MA: MIT Press.

Marx, Karl. [1867] 1976. *Capital,* Vol. 1. New York: Vintage.

McAdam, Doug. 2012. "Cognitive Liberation." In *The Blackwell Encyclopedia of Social and Political Movements,* edited by D. Snow, D. della Porta, B. Klandermans, and D. McAdam. Malden, MA, and Oxford: Blackwell Publishing.

McCright, Aaron M. and Riley E. Dunlap. 2011a. "Cool Dudes: The Denial of Climate Change among Conservative White Males in the United States." *Global Environmental Change* 21(4):1163–1172.

McCright, Aaron M. and Riley E. Dunlap. 2011b. "The Politicization of Climate Change and Polarization in the American Public's Views of Global Warming, 2001–2010." *The Sociological Quarterly* 52:155–194.

Meyer, John W., David John Frank, Ann Hironaka, Evan Schofer, and Nancy Brandon Tuma. 1997. "The Structuring of a World Environmental Regime, 1870–1990." *International Organization* 51:623–651.

Mirosa, Miranda, Daniel Gnoth, Rob Lawson, and Janet Stephenson. 2011. "Rationalising Energy-Related Behaviour in the Home: Insights From a Value-Laddering Approach." European Council for an Energy-Efficiency Economy Summary Study, France.

Mol, Arthur P. J. 2003. *Globalization and Environmental Reform: The Ecological Modernization of the Global Economy.* Cambridge, MA: MIT Press.

National Academy of Sciences. 2010. "Limiting the Magnitude of Future Climate Change." America's Climate Choices: Panel on Limiting the Magnitude of Future Climate Change. Washington, DC: National Academies Press.

Nolan, Jessica M., P. Wesley Schultz, Robert B. Cialdini, Noah J. Goldstein, and Vladas Griskevicius. 2008. "Normative Social Influence is Underdetected." *Personality and Social Psychology Bulletin* 34(7):913–923.

Norgaard, Kari. 2011. *Living in Denial: Climate Change, Emotions and Everyday Life.* Cambridge, MA: MIT Press.

O'Neill, Brian, Michael Dalton, Regina Fuchs, Leiwen Jiang, Shonali Pachauri, and Katarina Zigova. 2010. "Global Demographic Trends and Future Carbon Emissions." *Proceedings of the National Academies of Science USA* 107(41):17521–17526. doi: 10.1073/pnas.1004581107.

Oppenheim Mason, K. 1987. "The Impact of Women's Social Position on Fertility in Developing Countries." *Sociological Forum* 2(4):718–745.

Parks, Bradley C. and J. Timmons Roberts. 2008. "Inequality and the Global Climate Regime: Breaking the North-South Impasse." *Cambridge Review of International Affairs* 21(4):621–648.

Pulver, Simone, Nathan Hultman, and Leticia Guimaraes. 2010. "Carbon Market Participation by Sugar Mills in Brazil." *Climate and Development* 2:248–262.

Repetto, Robert (Ed.). 2006. *Punctuated Equilibrium and the Dynamics of U.S. Environmental Policy.* New Haven, CT: Yale University Press.

Risgaard, Marie-Louise, Pia Frederiksen, and Pernille Kaltoft. 2007. "Socio-cultural Processes behind the Differential Distribution of Organic Farming in Denmark: A Case Study." *Agriculture and Human Values* 24:445–459.

Roberts, J. Timmons. 2011. "Multipolarity and the New World dis(Order): US Hegemonic Decline and the Fragmentation of the Global Climate Regime." *Global Environmental Change* 21(3):776–784.

Roberts, J. Timmons and Bradley C. Parks. 2007. *A Climate of Injustice: Global Inequality, North-South Politics, and Climate Policy*. Cambridge, MA: MIT Press.

Roberts, J. Timmons, Bradley C. Parks, and Alexis A. Vasquez. 2004. "Who Ratifies Environmental Treaties and Why? Institutionalism, Structuralism and Participation by 192 Nations in 22 Treaties." *Global Environmental Politics* 4(3):22–64.

Roy, Beth. 1994. *Some Trouble with Cows*. Berkeley: University of California Press.

Rudel, Thomas K. 2009. "How Do People Transform Landscapes?: A Sociological Perspective on Suburban Sprawl and Tropical Deforestation." *American Journal of Sociology* 115(1):129–154.

Schofer, Evan and Ann Hironaka. 2005. "The Effects of World Society on Environmental Outcomes." *Social Forces* 84(1):25–47.

Schultz, P. Wesley. 2002. "Knowledge, Information, and Household Recycling: Examining the Knowledge-Deficit Model of Behavior Change." Pp. 67–82 in *New Tools for Environmental Protection: Education, Information, and Voluntary Measures*, edited by T. Dietz and P. C. Stern. Washington, DC: National Academy Press.

Schultz, P. Wesley, Jessica M. Nolan, Robert B. Cialdini, Noah J. Goldstein, and Vladas Griskevicius. 2007. "The Constructive, Destructive, and Reconstructive Power of Social Norms." *Psychological Science* 18(5):429–434.

Shwom, Rachael. 2011. "A Middle Range Theory of Energy Politics: The U.S. Struggle for Energy Efficient Appliances." *Environmental Politics* 20(5):705–726.

Sonnenfeld, David A. and Arthur P. J. Mol. 2002. "Globalization and the Transformation of Environmental Governance: An Introduction." *American Behavioral Scientist* 45(9):1311–1339.

Spaargaren, Gert. 2000. "Ecological Modernization Theory and Domestic Consumption." *Journal of Environmental Policy and Planning* 2(4):323–335.

Spaargaren, Gert and Arthur P. J. Mol. 1992. "Sociology, Environment, and Modernity: Ecological Modernization as a Theory of Social Change." *Society Natural Resources* 5(4):323–344.

Stephenson, Janet, Barry Barton, Gerry Carrington, Daniel Gnoth, Rob Lawson, and Paul Thorsnes. 2010. "Energy Cultures: A Framework for Understanding Energy Behaviours." *Energy Policy* 38:10.

Swidler, Ann. 1986. "Culture in Action: Symbols and Strategies." *American Sociological Review* 51(2):273–286.

Szasz, Andrew. 2007. *Shopping Our Way to Safety: How We Changed From Protecting the Environment to Protecting Ourselves*. Minneapolis: University of Minnesota Press.

Szasz, Andrew. 2011. "Is Green Consumption Part of the Solution?" Pp. 594–608 in *The Oxford Handbook of Climate Change and Society*, edited by J. S. Dryzek, R. B. Norgaard, and D. Schlosberg. New York: Oxford University Press.

United Nations Framework Convention on Climate Change. n.d. "Background on the UNFCCC: The International Response to Climate Change" (https://unfccc.int/essential_background/items/6031.php).

Ungar, Sheldon. 2003. "Misplaced Metaphor: A Critical Analysis of the "Knowledge Society." *Canadian Review of Sociology* 40(3):331–347.

Vandenbergh, Michael, Paul Stern, Gerald Gardner, Thomas Dietz, and Jonathan Gilligan. 2010. "Implementing the Behavioral Wedge: Designing and Adopting Effective Carbon Emissions Reduction Programs." *Environmental Law Review* 40:10547–10554.

Webb, Janette. 2012. "Climate Change and Society: The Chimera of Behaviour Change Techniques." *Sociology* 46(1):109–125.

Wilhite, Harold, Elizabeth Shove, Loren Lutzenhiser, and Willett Kempton. 2000. "The Legacy of Twenty Years of Energy Demand Management: We Know More About Individual Behaviour but Next to Nothing about Demand." Pp. 109–126 in *Society, Behaviour, and Climate Change Mitigation*, edited by E. Jochem, J. Sathaye, and D. Bouille. Dordrecht, The Netherlands: Kluwer Academic Publishers.

Wines, Michael. 2013. "Climate Change Pact is Signed by Three States and a Partner." *The New York Times*, October 29.

York, Richard. 2006. "Ecological Paradoxes: William Stanley Jevons and the Paperless Office." *Human Ecology Review* 13(2):143–148.

Zahran, Sammy, Eunyi Kim, Xi Chen, and Mark Lubell. 2007. "Ecological Development and Global Climate Change: A Cross-National Study of Kyoto Protocol Ratification." *Society and Natural Resources* 20:37–55.

8

Civil Society, Social Movements, and Climate Change

Beth Schaefer Caniglia, Robert J. Brulle, and Andrew Szasz

INTRODUCTION

Over the past decade, concern over climate change has manifested itself in a number of organized efforts across the globe that advocate for action to address this issue (Beddoe et al. 2009; Cahn 1995; Fischer et al. 2007). This mobilization has engaged a large number of organizations that range in size from small neighborhood groups to large, formal international organizations with budgets that are in the hundreds of millions of dollars, and millions of members. The aims and orientations of these different groups vary widely, including groups advocating the adoption of technological measures, various types of international accords, economic measures, or wholesale transformation of capitalism (Walstrom, Wennerhag, and Rootes, 2013). They also vary regarding the level of government policy they aim to influence, from local authorities and municipalities to national and international institutions, as well as specific corporate actors (Garrelts and Dietz 2014; Soule 2009). However, despite this diversity, a clearly identifiable climate movement in the form of a loose network structure advocating for action to address climate change can be identified. As a result of this mobilization, climate change has become a major political issue across the globe (Caniglia 2010; Dietz and Garrelts 2014).

This chapter applies sociological perspectives to develop an understanding of this diverse and widespread movement. Sociology has developed an extensive and robust literature on the process of social change driven by citizen mobilization, including the development and advocacy of alternative cultural and policy perspectives, the creation of new organizations, and how these organizations can affect both corporate actions and public policy. The first part of the essay provides a brief overview of why action by social movements is seen to be a critical component of facilitating social change. After this introduction, climate change movements at the international, national, and cross-national levels are examined through the lens of sociological theory. Special attention is paid to recent scholarship on the role of religion and climate change in the United States. The chapter concludes with a discussion of the actual and potential role played by the climate change movement in addressing climate change and what further research is needed to advance our understanding in this area.

CIVIL SOCIETY, SOCIAL CHANGE, AND SOCIAL MOVEMENTS

Sociology has long studied the mechanisms by which large-scale cultural and social change is produced. Sociological scholarship points to the critical role played by the institutions of civil society for the origination of social change through citizen mobilization (Calhoun 1993; Gamson 1975; McAdam 1988; McCarthy and Zald 1977; Skocpol 2003; Sztompka 1993; Tarrow 1998). Barry Commoner (1971:300) recognized, over forty years ago, that the environmental crisis is fundamentally a social challenge and therefore its resolution lies in social and institutional change brought about through collective action by mobilized citizens. The threat of a changing climate is no different. The importance of civil society has been recognized by the Intergovernmental Panel on Climate Change (IPCC). The Fourth IPCC Assessment points out that climate policy emerges from the joint work of a number of different institutions, including nongovernmental actors and civil society (IPCC 2007:708). The IPCC defines civil society as "the arena of uncoerced collective action around shared interests, purposes and values" (IPCC 2007:713). The IPCC also points out that the climate change movement influences climate change policy through three actions: policy advocacy, providing policy research, and opening political spaces for new political reforms (IPCC 2007:713–714).

Civil society is constituted by interactions that take place outside of either market or government interactions. It comprises a number of both formal and informal organizations, including "clearly defined group interests; through associations and cultural establishments; up to public interest groups and churches or charitable organizations" (Habermas 1996:355). The importance of civil society as a site for the origination of social change is based on its structural location within society. As the modern social order developed, the institutions of society formed into three distinct spheres: the market, the state, and civil society. The actions of both the market and the state institutions center on meeting key imperatives. For market institutions, it is imperative that they act to maximize return on investment. For state institutions, the key imperatives entail providing security through the military and the law, ensuring economic growth, and maintaining political legitimacy (Alexander 2010; Habermas 1996). Accordingly, policies that have an adverse impact on these imperatives will not be fostered by market or state institutions. However, civil society is not constrained in the same manner. Through voluntary association with one another, citizens can identify and advocate for their collective interests. This independence forms the key to the capacity of civil society to serve as a site for the generation of efforts that aim at social change (Gamson 1975; McAdam, McCarthy, and Zald 1988; McCarthy and Zald 1977; Skocpol 2003). Thus within sociology, *social movements*, originating in civil society, are seen as the central actors in fostering social change. Scholars in sociology seek to specify how social movements emerge and organize themselves and how they influence people and policies (Olsen 1965). Scholars have also sought to understand the conditions that favor movements' success—or failure—to achieve their goals.

Basic Social Movement Theory

The most basic way that movements change the social landscape is through *framing* grievances in ways that resonate with members of civil society (Brulle and Benford 2012; McAdam, McCarthy, and Zald 1996; Snow and Benford 1988; Tarrow 1998). Social movements focus members of civil society on particular dimensions of social problems and provide their publics with clear definitions of those problems, along with arguments regarding who is at fault and what options exist for solving their social grievances. These framing efforts have the diffuse effect of changing the hearts and minds of civil society (Poletta 2002), but they also result in concrete mobilization efforts that increase organization memberships and participation in campaign activities, like protests, boycotts, and letter writing. Frames that resonate within civil society are associated with mobilization of key resources, including financial capital, human volunteers, and the support of elites within the target political agencies (Hewitt and McCammon 2004).

Framing efforts alone are ineffective on a large scale without mobilization of the human and financial capital needed to mount campaigns that result in broad social change (e.g., altered policy outcomes, the creation of new institutions and/or new international regimes). The *resource mobilization* dimension of movements is examined in a rigorous literature that highlights the role of networks of formal social movement organizations and other mobilizing structures like churches and universities in providing the central infrastructure needed to coordinate mass movements in effective ways (McCarthy and Zald 1977; Staggenborg 1988; Tarrow 1998) and the role of foundations and other income sources in the creation and maintenance of social movement organizations (Carmichael, Jenkins, and Brulle 2012).

A movement's message—its framing of the issue—and the resources it can bring to bear are two of the raw materials for developing and mobilizing its *strategies and tactics*. Here the question is: How does the movement try to make its case, build its influence, and become effective enough to change what it seeks to change?

One important distinction identified by social movement scholars is the strategic choice between *mobilization from below* and *mobilization from above*. In the case of mobilization from below, collective issues are first experienced as individual burdens on everyday life (Habermas 1996:365). These issues can then lead to discussions among individuals and through these discussions can rise to become common concerns within a collective. As communities further define the issues, they develop and advocate for the adoption of specific actions to address their issues of concern and advocate for them in public forums, including the media, academic venues, and professional organizations. For the mobilization-from-below model, the ability of society to learn and respond to changed conditions depends on the recognition of a shared social problem, the development and advocacy of specific policy proposals, and the application of this knowledge to restructure practices and institutions.

The second approach to the transformation of society, mobilization from above, involves an organized, top-down effort toward social change, which is typically led by established institutions (including social movement organizations). The concentration of economic wealth has created an enormous inequality regarding the ability of different groups to participate in public decision making (Leicht and Fitzgerald 2014). As a result, the public space has become increasingly dominated by powerful and wealthy organized groups, including industry interest groups, wealthy foundations, and even some powerful individuals (Caniglia 2001; Habermas 1989:141–180; Magan 2006:31–32). Public policy is produced not as the result of an open public debate to ascertain the common interest, but rather as the result of institutional interventions to secure a political and cultural advantage through the use of political marketing (Sievers 2010:136; Walker 2012). This involves "the sophisticated manipulation of public opinion by powerful actors" (Magan 2006:32). That is, it involves selling the public on a specific issue through representing the particular interests of a particular organization or sector as being in the general interest. The goal is not to engender critical reflection and debate but rather to generate good will and a prestige for a particular policy position, thus strengthening public support for the position without ever making it a matter of real public inquiry. This allows those organizations to set the terms of the debate and disadvantages participation by community organizations or those with dissenting positions (Greenberg, Knight, and Westersund 2011:69).

The most resonant frames and high levels of mobilization make little difference if there are no openings for social change. To understand the political environment in which movements operate, social movement scholars have examined the *political opportunity structure*. The political opportunity structure consists of a number of dimensions, including institutional provisions for participation, stability of political alignments, elite access and alliances, elite conflict, and level of repression (Caniglia and Carmin 2005; McAdam 1988; Tarrow 1998). While resource mobilization and framing perspectives tend to focus on the characteristics of movements at given points in time, scholars of the political opportunity structure examine movements over longer periods of time and specifically theorize cycles of activism and movement decline. Resource mobilization and political opportunity structure scholars emphasize the importance of movement organizations. Social movement organizations facilitate mobilization, cultivate leaders, and serve as clearinghouses of movement information. However, political opportunity structure scholars emphasize the sociopolitical contexts in which movements operate and the conditions that enable and constrain their success. These scholars are especially interested in how cycles of movement protests reflect responses to conditions external to the movement itself (Meyer 2006; Meyer and Imig 1993).

In the remaining pages, we review sociological findings in studies of the global climate movement and its U.S. counterpart, providing an overview of the framing, resources, strategies, and political opportunity contexts of each. We find strong overlaps between these two movements, particularly in the ways their members frame the climate change problem and its potential solutions. Unfortunately, when it comes to producing social and policy change on the national and international levels, we find that the climate change movement appears to be failing despite its ability to mobilize an impressive array of social movement organizations, coalitions, and protest actions.

GLOBAL ENVIRONMENTAL MOVEMENTS

International movements are generally facilitated by the creation of international institutions designed to govern the issues those movements care about. The global climate change movement is no exception. For that reason, our discussion of this movement strongly centers around the United Nations Framework Convention on Climate Change (UNFCCC) and the loose network of organizations and individuals who focus their mobilization and framing efforts toward this intergovernmental forum.

The International Political Opportunity Structure: The UNFCCC

The UNFCCC was initiated at the Rio Earth Summit in 1992. The goal was to create a binding treaty for the reduction of greenhouse gas (GHG) emissions in the face of mounting evidence that they were responsible for increasingly erratic weather patterns, including global warming. At the summit, the UNFCCC articulated a shared set of principles that included acknowledgment that climate change required immediate cooperative action. The UNFCCC asked nations to "protect the climate system . . . on the basis of equity and in accordance with their *common but differentiated responsibilities and respective capabilities*" (as quoted in Roberts and Parks 2007:3). However, North–South politics, along with a poor political environment in the United States, signaled that an effective treaty would be difficult to achieve (see Chapter 5 in this volume and Fisher 2004; Pulver 2004; Roberts and Parks 2007).

Because the Protocol was ratified so late in its first term, many nations, nongovernmental organization (NGOs), and social movement groups alike have treated the Protocol like a trial to create the institutional and economic mechanisms needed to implement a more rigorous treaty after 2012. The two Conference of Parties (COP) meetings prior to COP 15 in Copenhagen (2010) were particularly forward-looking and began sketching out potential modifications of the Kyoto Protocol that would lead to a more effective and truly transformative treaty. A surge of optimism accompanied the election of Barack Obama to the U.S. presidency in 2008 because he promised to rejoin the UNFCCC process and develop a carbon cap-and-trading system in the United States. Australia experienced a change in leadership and became a signatory, indicating that momentum was gaining.

Unfortunately, this momentum seems to have stalled if not reversed at the Copenhagen meeting, which failed to produce the binding treaty it promised and, according to many, lost ground compared to previous COP meetings (Athanasiou 2010; Institute for Sustainable Development 2009). The UN process, which is typically marked by evolving consensus over agreement language, devolved during the Copenhagen meeting and has followed the same lackluster trajectory in meetings since.

Analyzing social movement activities surrounding the UNFCCC provides revealing insights into typical UN treaty organization structures and their usefulness as sites of movement mobilization. Treaties are binding agreements between or among nation-states and distinctly privilege government officials in their negotiations. Systematic barriers restrict the extent and avenues of NGO and other nonstate actors' participation, and while these barriers disadvantage all nonstate actors, they often particularly disadvantage groups from the developing world (Caniglia 2010; Roberts and Parks 2007; Willetts 1996). The UNFCCC has been cited as being more favorable to NGO participation than some other treaty secretariats (Pulver 2004), but barriers to NGO participation are still present; they limit NGO effectiveness and structure the repertories of strategies available for non-state actors' influence.

Accreditation to attend UNFCCC meetings is more difficult to acquire than attendance at the meetings of other UN environmental agencies, such as the United Nations Environment Programme (UNEP) or the High Level Political Forum on Sustainable Development, which replaced the UN Commission on Sustainable Development after Rio +20. In part, this is because the UNFCCC is a focused area, while the High Level Political Forum encompasses the entire scope of sustainable development. In addition, the UNFCCC is less likely to accredit smaller, local organizations because the secretariat emphasizes the international character of those who apply. Just under 70 NGOs attended the 1995 negotiations, but in 1997 over 230 NGOs were represented by over 3,500 people (Pulver 2004), and in Copenhagen the number surged to approximately 40,000 accredited representatives of NGOs (Caniglia 2010; Walstrom et al. 2013). Early on, NGO access to the negotiating floor was restricted by the secretariat because of the heavy influence of business and industry NGOs; the secretariat wanted to restrict the influence of business and industry NGOs and chose to banish all nonstate actors from the negotiating floor to achieve this result.

Nonstate actors are given one formal intervention opportunity at each negotiating meeting of the UNFCCC—at the plenary session, each sector of nonstate actors is able to present opening comments to governments. Otherwise, interventions by NGOs must follow available *informal* channels, and the NGO community—at the UNFCCC and at other hard law forums—has found creative ways to exercise limited influence through these channels (see Dodds and Strauss 2004). These circumstances require NGOs and movement groups to build personal ties of trust with secretariat officials and members of government delegations, which can be mobilized at critical moments to increase NGO responsiveness and influence (Caniglia 2001; Dodds and Strauss 2004; Hemmati 2002). Table 8.1 highlights the distinct benefits that civil society groups receive from formal consultative status and informal ties with intergovernmental personnel.

Mobilization and Mobilizing Structures

To strengthen their position at the UNFCCC, NGOs formed a loose self-organizing committee, which meets daily and has historically been coordinated by Climate Action Network. Climate Action Network is a massive network of NGOs from around the world and carries a great deal of

Table 8.1 Benefits of Consultative Status and Informal Ties with IGOs

Benefits of Consultative Status	Benefits of Informal Relationships
• Receipt of UN documents • Distribution of position papers • Access to UN meetings • Guidelines for agency accreditation & meeting participation	• Interpretation of UN documents • Feedback on position papers • Insight into country delegation positions & political context • Insight into other UN agencies & personnel • Consideration when opportunities arise • Information & introduction re: other organizations with similar interests • Co-sponsorship of events & projects • Funding opportunities • Inside, "privileged" information

Caniglia 2001.

scientific and moral authority as a representative of civil society interests at the climate change talks. It also publishes *ECO*, a daily NGO newsletter that reviews and comments upon the negotiations that are expected to take place that day. Members of government delegations have been known to approach the Network to have statements published in *ECO*, which is one indicator of how successful this publication has been at elevating NGO viewpoints inside of the UNFCCC (Caniglia 2010; Pulver 2004). Throughout the negotiating sessions, the NGOs arrange meetings with government delegations and lobby those delegates in hallways, bathrooms, and just about anywhere they can find them.

In general, NGO influence is restricted to these informal avenues of access. As Newell (2008) points out, NGOs at the UNFCCC are restricted to traditional lobbying, shame-and-blame strategies, and watchdog roles. This does not suggest that nonstate actors cannot influence the process; clear evidence exists that they have and will continue to press governments toward more equitable and creative solutions to climate change (Pulver 2004; Roberts and Parks 2007). Nonetheless, because governments are accountable to the obligations made under treaties like the Kyoto Protocol, those governments tend to restrict the extent to which nonstate actors' voices are included at the table. NGOs face similar limitations across the UN system (Dodds 2002; Emadi-Coffin 2002) and have honed a toolkit of strategies to nudge governments in the direction of their interests. Most of these strategies require multiple years of participation in UN meetings to leverage them effectively, because they require informal relationships of trust with intergovernmental and governmental representatives to operate most effectively (Caniglia 2001; Dodds 2002). Nonetheless, numerous UN-savvy environmental NGOs have joined forces to target UNFCCC talks. Pulver's (2004) examination of how NGO positions gained favor over those of opposing industry and business NGOs illustrates that environmental justice advocates can indeed have an impact on negotiations in this and other hard law venues.

Cultural Dimensions and Framing

As described at the beginning of the chapter, one of the most powerful tactics social movements bring to the table is their ability to frame social problems—to the general public, to their current and potential members, and to the decision makers they target. The international climate change movement, like the environmental movement in general, is not a completely unified movement and, therefore, uses an array of discourses to frame the problems, perpetrators, and solutions to global climate change.

The application of frame analysis to climate change first appeared in a discussion of the international climate change movement. There are three major analyses of the different frames that operate within this movement. The first analysis of the discursive frames of the climate change movement appeared in Newell (2006:98–101). He argues that because of differential resources, expertise, and access to key government officials, as well as different strategies to address climate change, the international community of environmental organizations can be divided into three different groups. First are the *Inside-Insiders*. This group of environmental organizations has substantial access to international climate deliberations. The second group, *Inside-Outsiders*, is involved in the international deliberations but does so from a peripheral position and adopts a confrontational style. Finally, the third group, *Outside-Outsiders*, has only marginal access to the international climate negotiations and works primarily outside of the formal institutionalized processes.

Each group of organizations has a different ideological perspective. As discussed in Table 8.2, the Inside-Insiders generally accept both the primacy of market economies and the possibility of enacting significant reforms through government action. The Inside-Outsiders adopt a more confrontational approach but strive to work through the existing international governance mechanisms. Finally, the Outside-Outsiders reject the existing institutions of both the market and the international governance arrangements as being capable of significantly altering the global society toward a sustainable trajectory. Rather, they advocate a number of large-scale revisions to the existing social order. As the table also shows, the extent of acceptance within the societal discussion on climate change is tied to its

Table 8.2 Corresponding Movements and Movement Organizations

	Inside-Insiders	Inside-Outsiders	Outside-Outsiders
Examples	• World Wildlife Fund • Environmental Defense • Natural Resources Defense Council • FIELD	• Friends of the Earth • Greenpeace • Sinkswatch	• Climate Justice Movement • Indigenous People Activists • Rising Tide
Aims	• To advance action on climate change within existing frameworks • To gain access to government decision making • To directly influence the negotiations	• To advance more drastic action on climate change • To question more fundamentally how the issue is being addressed	• To question the current framing of the climate change debate • To raise popular awareness about the impact of climate change on the poor
Strategies	• Access to delegations • Research • Provision of legal advice • Support to like-minded delegations • Diplomatic lobbying	• Research for public audiences • Use of media • More confrontational styles of lobbying and exposure	• Protest, demonstrations • Parallel actions and side events • Cross-movement mobilization • Litigation • Popular education
Focus of Influence	• Governments • Regional and international institutions • Private sector (collaboratively)	• Governments • Regional and international institutions • Private sector (critical approach)	• Governments • The public • Other movements (antiglobalization movement)
Ideologies	• Generally benign view of the market • Critical view of command and control approaches but faith in governments and international institutions to respond effectively to the issue	• Critical view of market mechanisms • Residual faith in international and regional institutions to deliver action and belief in the primacy of legal-based regulation	• Failure to act on climate change seen as part of broader failing of globalization • Critical view of the willingness or ability of governments and international institutions to deliver environmental justice because of their ties to the corporate sector

Adapted from Table 3.1 in Newell 2006:116.

ideological nature. The less radical frames are more accepted and accorded an insider status, whereas the more radical frames are marginalized.

Bäckstrand and Lövbrand (2007) provide a more recent analysis with alternative frame categories. According to their study, the first frame is defined as *green governmentality* (Bäckstrand and Lövbrand 2007:126–129). This discursive frame defines a process based on a scientific analysis of climate change and the development of global governmental initiatives to address this issue. The solution to climate change in the discourse of green governmentality is the implementation of a strong system of governance of the economy, natural resource use, and individual behavior informed by the natural sciences that is developed at an international level. In this approach, natural scientists play a key role in global environmental management (Glover 2006:3–6) by mediating between science and politics and providing information to the public and decision-makers in government and industry of the need for governmental action. This places scientists in the key role of defining the nature of the problems and proposing mechanisms for their resolution. This approach underlies the many of the existing international treaty frameworks in which science-based resource management plays a central role. It also informs actions aimed at the proximate causes of environmental degradation, such as creating parks or land trusts to preserve ecosystems, or developing new technologies that can provide low carbon energy.

The second major discursive worldview is known as *ecological modernization* (Bäckstrand and Lövbrand 2007:129–131). This discursive frame focuses on the role of technological development, economic expansion, and the growth of environmental governance in creating and also mitigating environmental problems. In this perspective, economic development and shifts in technology lead to the initial generation of increased CO_2 emissions. However, further economic development can also mitigate these emissions by shifting to renewable energy technologies and energy efficiency (Cantor and Yohe 1998). This technological shift can thus result in a decrease in carbon emissions and a decoupling of economic growth and energy production with carbon emissions (Murphy 2000). Thus, in this perspective economic growth can result in an absolute decline in levels of environmental pollution (Mol 2001). At the core of ecological modernization theory is that the existing social, economic, and governmental institutions can effectively deal with environmental issues, and there is no need for radical structural changes in industrial society (Carolan 2004; York, Rosa, and Dietz 2003). This leads to a neoliberal market approach to the resolution of climate change.

The ecological modernization approach takes two forms (Christoff 1996). The first form, weak ecological modernization, focuses on technological development and energy efficiency and includes the use of market-based user fees for pollution, tax incentives, increases in energy efficiency, or the shifting of production toward "green" products (Glover 2006:4–6). The second version, strong ecological modernization, focuses on embedding environmental and ecological concerns in society by reconfiguring existing political and economic institutions. This includes adjusting economic systems to include the value of natural capital into production decisions, modifying the existing political system toward more democratic participation, and including developing countries' social justice and equity concerns into global environmental governance (Berger et al. 2001). Of these two approaches, the "weak" form is generally considered to be the dominant of the two. Over the past decade, the discursive frames of green governmentality and ecological modernization have tended toward a merger. Since the viability of ecological modernization mechanisms, such as a global price on carbon, depends on the institution of an effective consensus global accord, the two approaches to address climate change are seen as mutually constitutive. Together, they form the dominant climate change discursive frame.

Distinct from green governmentality and ecological modernization is the discourse of *civic environmentalism* (Bäckstrand and Lövbrand 2007:132). Civic environmentalism offers a counternarrative to the dominant climate change discourses. There are two related approaches within this discursive frame. The first perspective is defined by Bäckstrand and Lövbrand (2007:132) as *radical resistance*. From this discursive perspective, both ecological modernization and green governmentality are seen as favoring the interests of the existing power elites and the dominant industrialized countries, resulting in the marginalization of poor people and the governments of less-developed countries. It challenges the neoliberal approach embedded in ecological modernization and calls for the radical democratization of global governance and economic processes. It aims at "a fundamental transformation of consumption patterns and existing institutions to realize a more eco-centric and equitable world order" (Bäckstrand and Lövbrand 2007:132). It focuses on the notion of global climate justice, emphasizing the responsibilities of the developed countries to dramatically reduce their carbon emissions, and the equitable sharing of technology and capital to enable the poorest nations to address global climate change. It also presses for the democratic reform of large multilateral institutions, such as the International Monetary Fund and the World Bank (Dawson 2010).

The second perspective within civic environmentalism is reformist, or *participatory multilateralism* (Bäckstrand and Lövbrand 2007:134). This version of civic environmentalism focuses on opening up climate change treaty negotiations to wider participation by representatives of civil society. The argument is that becoming more inclusive will generate greater legitimacy for the negotiated agreements and increase the implementation of actions that maximize the benefits from neoliberal approaches.

Although they originated at different times and use different labels, the analyses by Newell (2006) and Bäckstrand and Lövbrand (2007) parallel each other. The combination of weak ecological modernization and green governmentality corresponds to Newell's Insider-Insider category. Strong ecological modernization corresponds to the Insider-Outsider category, and civil environmentalism corresponds to the Outsider-Outsider category.

In their recent piece, Walstrom, Wennerhag, and Rootes (2013) examine the prognostic frames of movement protestors from three European protests. A prognostic frame focuses on how to solve the problem of interest, in this case climate change. We focus here on their examination of the frames espoused by participants in the December 12 protest at the UNFCCC COP 15 meeting in Copenhagen. This protest was organized by a coalition of social movement organizations, trade unions, and religious organizations that spanned 67 countries and included over 500 organizations. While there was tremendous agreement among the organizers and participants that urgent action was needed to create strong resolutions at COP 15, exactly how governments, industry, social movements, and individuals should change to ameliorate climate change varied in informative ways. The six prognostic frames identified from participant interviews were *system change* (24.2 percent), *global justice* (17.2 percent), *change individual behavior/raise awareness* (35.2 percent), *technological change/investments* (22.4 percent), *legislation/policy change* (43.5 percent), and *change industry/production* (11.2 percent). Because these data focus on participants in one protest, we hesitate to generalize the conclusions, but these data provide ground-truthed insights into the motivations of climate change activists.

THE U.S. CLIMATE CHANGE MOVEMENT

The U.S. climate change movement has developed over the past three decades from a small concern of environmental groups involved in air pollution issues to a large-scale social movement with components that focus at all levels of government. Climate change first entered into the U.S. political arena in 1977. In a congressional hearing on the environmental implications of the Carter administration's energy plans, the representative of the National Wildlife Federation noted that the development of the proposed synthetic fuels program would lead to massive releases of carbon dioxide and destabilize the planet's climate. Participation in congressional hearings on climate change was initially limited to a few environmental organizations that had been involved in other air pollution issues, most importantly acid rain and ozone depletion (Brulle 2014). After the dramatic testimony by Dr. James Hansen in the summer of 1988, political interest in the issue expanded, and the number of groups testifying before Congress increased. Another rapid increase followed the release of Al Gore's movie *An Inconvenient Truth* in 2006. By 2010, 123 environmental movement organizations had testified before Congress in hearings on climate change. This growth at the congressional level was paralleled by participation by these organizations in the meetings associated with the procedures defined by the UNFCCC (Brulle 2014; Caniglia 2010). Starting with the initial discussions of the makeup of this treaty in 1991, there was a steady annual increase in participation by environmental movement organizations in the annual COP meetings. Participation in the UNFCCC meetings underwent a dramatic increase in 2009 due to the increased importance attached to COP 15, held in Copenhagen. Taken together, the number of organizations involved either in providing congressional testimony regarding climate change or participating in the UNFCCC process increased from 7 in 1980 to 240 in 2010 (Figure 8.1).

Another dimension of the historic growth of environmental movement organizations' advocacy on the issue of climate change is the growth of coalitions that are specifically focused on that issue. Coalitions facilitate formalized and regular patterns of cooperation in the development of collective action, so they can be seen as providing an empirical indicator of the extent and boundaries of a social movement industry (Adams, Jochum, and Kriesi 2008; Beamish and Luebbers 2009; Lichterman 1995; Murphy 2005;

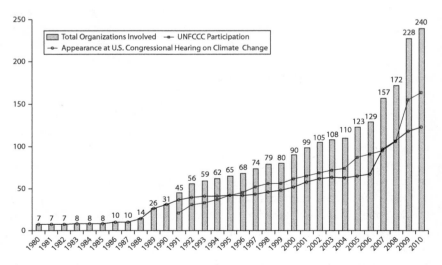

Figure 8.1 U.S. Environmental Organization Participation in Climate Change Forums

Poloni-Staudinger 2009; Van Dyke 2003). A listing of these coalitions, in order of founding year, is provided in Table 8.3 (Brulle 2014). As this table shows, the first coalition, the U.S. Climate Action Network, was founded in 1989. The Sustainable Energy Coalition followed next in 1992, along with Via Campesina. There was then a six-year break in the mid-1990s until two additional coalitions were formed in 1998. However, as this table illustrates, over half of the coalitions were formed between 2006 and 2009. This corresponds to the increased growth of organizations participating in the UNFCCC process and appearing before Congress.

Overall, the pattern of growth of U.S. environmental movement organizations involved in climate change advocacy at the national or international level was one of gradual involvement up until about 2005. From 2006 to 2009, there was a rapid increase in the number of organizations participating in formal government arenas or in coalition actions. At the end of 2010, 467 unique organizations had been identified as part of the national climate change movement (Brulle 2014).

These 467 NGOs define the core of the national U.S. climate change movement. To visualize the relationships across an entire movement,

Table 8.3 Climate Change Movement Coalitions

Coalition Name	Year Founded
U.S. Climate Action Network	1989
Sustainable Energy Coalition	1992
Via Campesina North American Region	1992
Chesapeake Climate Action Network	1998
Interfaith Power & Light Network	1998
Rising Tide North America	2000
Apollo Alliance	2001
Interwest Energy Alliance	2002
Climate Crisis Coalition	2004
Energy Action Coalition	2004
Blue Green Alliance	2006
Catholic Coalition on Climate Change	2006
Interfaith Climate Change Network	2006
350.org/1 Sky	2007
Rocky Mountain Climate Organization	2007
Season End	2007
U.S. Climate Action Partnership	2007
Mobilization for Climate Justice	2008
TckTckTck/Global Campaign for Climate Action	2008
Clean Energy Works	2009
Climate SOS	2009

Brulle 2014.

sociology uses network analysis to map the social ties between different groups. It is predicated on the powerful influence that social ties exert over organizational activities (Knoke 1990). By channeling resources, communications, influence, and legitimacy, social networks create shared identities and collective interests and thus promote the acceptance of a common discursive frame within a social movement (Knoke and Yang 2008:6). As the exchange of information increases, organizations form stable relationships with other organizations based on their knowledge of the specific competencies and reliability of one another. These relationships solidify over time, and future behavioral actions become regularized and routinized, forming a stable social network (Gulati and Gargiulo 1999:1440). Thus network analysis can provide a means to view the components of a social movement as well as how these different components interact.

A network analysis of the twenty-one climate change coalitions previously identified resulted in the overall image of the U.S. climate change movement for 2010 (Brulle 2014) shown in Figure 8.2. This diagram shows shared memberships in different coalitions by organizations. There is a great deal of diversity in the types of organizations that make up the different coalitions. The largest coalition, 350.org, is primarily composed of NGOs. It also has links to nine other coalitions. The structures of the next two largest generalist coalitions, TckTckTck, and the U.S. Climate Action Network, are quite similar, possibly due to their common ties to the global environmental movement. There are three coalitions that are primarily composed of for-profit corporations: the Interwest Energy Alliance, the Sustainable Energy Coalition, and the U.S. Climate Action Partnership. In addition, both the Apollo Alliance and the Blue Green Alliance are composed primarily of labor unions and NGOs. An examination of the centrality scores clearly shows the prominence of six coalitions that constitute the core of the U.S. national climate change movement. This network analysis shows that the core coalitions are first, 350.org, followed by the U.S. Climate Action Network and TckTckTck. These three coalitions are highly central to the overall network structure. The Apollo Alliance, Clean Energy Works, and Energy Action Coalition constitute the remaining members of the core of the network. All of the remaining coalitions are integrated into the overall network but occupy more peripheral roles.

Framing

In examining the discursive frames adopted by the different coalitions, the dominant overall viewpoint is ecological modernization (Brulle 2014; Schlosberg and Rinfret 2008). A few coalitions have adopted the discursive frame of civic environmentalism. Thus, the discursive frames of the

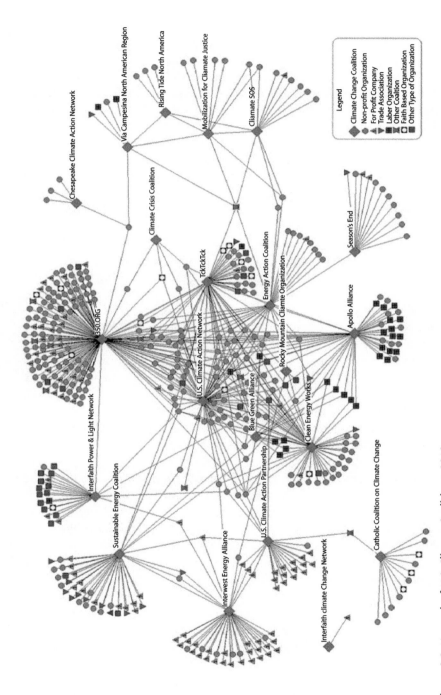

Figure 8.2 Network of U.S. Climate Coalitions, 2010

Legend
◆ Climate Change Coalition
● Non-profit Organization
◀ For Profit Company
▶ Trade Association
◢◣ Labor Organization
◥◤ Other Coalition
□ Faith Based Organization
▣ Other Type of Organization

Chesapeake Climate Action Network
Via Campesina North American Region
Rising Tide North America
Mobilization for Climate Justice
Climate SOS
Climate Crisis Coalition
TckTckTck
Energy Action Coalition
Season's End
350.ORG
U.S. Climate Action Network
Rocky Mountain Climate Organization
Apollo Alliance
Interfaith Power & Light Network
Blue Green Alliance
Clean Energy Works
Sustainable Energy Coalition
U.S. Climate Action Partnership
Catholic Coalition on Climate Change
Interwest Energy Alliance
Interfaith climate Change Network

U.S. climate change movement generally parallel the global discursive frames described by Bäckstrand and Lövbrand (2007). Also, an analysis of the financial resources available to any specific coalition shows that the coalitions that have embraced a discursive frame of ecological modernization control virtually all of the economic resources of the entire national climate change movement. Coalitions that embrace the discursive frame of civic environmentalism have only a tiny fraction of the overall economic resources available to the overall climate movement.

The dominance of ecological modernization in the network structure is also replicated in the selection of organizations who either appear in the media or testify before Congress. One of the critical areas for movement success has been the diffusion of its perspectives throughout the issue public. Appearances at congressional hearings and coverage in the mass media are some of the primary means through which a movement transmits its ideas to the larger public and to influential elites. However, there are significant barriers to the movement's access to either the media or congressional forums due to a number of factors, including media reporting norms, competition for media space, and corporate hegemony (Gamson et al. 1992).

Only recently have sociologists turned to an analysis of the participation of climate change movement organizations in either the media or congressional hearings. An analysis of the statements presented at U.S. Congress climate hearings during the 109th and 110th Congresses (Fisher, Leifeld, and Iwaki 2012) shows a general increase in the representation of pro–climate change arguments and witnesses between the 2005–2006 and the 2007–2008 Congressional sessions. Jenkins (2011) shows that the levels of media coverage of organizations that favor action on climate change was substantially higher in the time period 1986 to 1994, and that over the time period 1996 to 2003, the representation of NGOs that support or oppose action on climate change has been in rough parity. Finally, Brulle (2013, 2014) shows that there were vast differentials between the discourse of the environmental organizations that appeared before Congress or in the media. His analysis shows that virtually all of the organizations that appeared espoused ecological modernization, and that the more radical discourses that are present within the climate change movement in the United States were absent.

Overall, the U.S. climate change movement is overwhelmingly dominated by the perspective of ecological modernization. Although it has a number of different unique coalitions, the coalitions that have the most financial resources, access to public forums, and media coverage all espouse this specific framing of climate change. Civic environmentalism is an extreme minority position and has only a marginal role in the orientation of this movement.

Political Opportunity Structure

The U.S. climate change movement has confronted a political opportunity structure that limits its ability to effect change, despite impressive mobilization and the existence of sympathetic leadership (Caniglia 2010). In the United States, evidence highlights that pro-environmental reforms generally advance during Democratic administrations, while they lapse or even fall behind during Republican administrations. Even when a supportive administration is in place, social movements can fight the good fight and still fail. How the Obama administration handled climate change legislation during its first term is illustrative. During the 2009 campaign, Republicans and Democrats advocated for the creation of a cap-and-trade system to regulate GHG emissions in response to climate change. In 2003, McCain co-authored climate change legislation to advocate cap-and-trade; he is also quoted as saying, "It is time for the United States government to do its part to address this global problem, and a discussion of mandatory reductions is the form of leadership that is required" (Revkin et al. 2012). Shortly after taking office, on January 26, 2009, President Obama gave a speech about the economic crisis, energy, and climate change, in which he said, "These urgent dangers to our national and economic security are compounded by the long-term threat of climate change, which if left unchecked could result in violent conflict, terrible storms, shrinking coastlines and irreversible catastrophe" (Phillips 2009). Several social movements geared up to support Obama's commitment to climate change legislation and the need for an international treaty to curtail climate change. Rallies were held in Washington, DC; bills were introduced, debated, and passed out of the House of Representatives. Yet, by the midterm elections in 2010, the Copenhagen Conference on Climate Change had come and gone and the United States still had no policy to reduce GHG emissions.

While there are several narratives that attempt to explain why the movement to enact climate legislation by the midterm elections in 2010 failed, some illustrative lessons can be learned from this example. Regardless of consensus on cap-and-trade, America and the rest of the world were in a terrible recession at the beginning of the Obama administration; therefore, an economic stimulus was the first order of business and climate change legislation was delayed. Further complications arose when political alliances in Washington shifted into three political factions: the Republicans, the mainstream Democrats, and the so-called Blue-dog Democrats. Although Democrats held majorities in both the House and the Senate, many Blue-dog Democrats (who are primarily from an increasingly conservative South) represented constituents whose livelihoods depended upon the fossil fuel industries. And, despite the agreement that characterized the presidential campaign trail between McCain and Obama on the need for climate change legislation, the Obama administration faced a Republican minority that objected to almost every piece of legislation, regardless of whether there were points of agreement across the aisles. As Republicans geared up for the midterm elections knowing they could regain important seats given the bad economy, the odds of a climate bill passing grew slimmer and ultimately disappeared. The political opportunity structure, made up of the alignments, access points, and priorities of the political arena, simply prioritized the economy over the environment and faced a deeply divided Congress unable to acquiesce to movement pressures. Scholars are uncertain regarding what changes will be required to create openings for national action on climate change in the United States. New research, therefore, has focused on the role of local initiatives like those taken by California, Oregon, Chicago, and other local authorities.

CROSS-NATIONAL COMPARISONS
OF CLIMATE/ENVIRONMENTAL MOVEMENTS

Until recently, the vast majority of scholarship on the climate change movement focused on the U.S., European, and global climate movements (Doyle and MacGregor 2013). However, Doyle and MacGregor's new two-volume series on *Environmental Movements Around The World* has brought into focus cross-cutting themes that help to explain the ways activists are enabled and constrained by their national contexts and their positions within historical and contemporary world systems. The series features chapters written by participants and scholarly observers of environmental movements in their own countries. Rather than try to fit our interpretation of these chapters into the framework of mobilizing structures, frames, and political opportunity structures, we will highlight the frameworks offered by Doyle and MacGregor. These included North–South geographies, Left–Right political orientations, and the country's character in reference to postmaterialism, postindustrialism, and postcolonialism.

The dichotomy of North–South has been a feature of most writing on the global climate change movement and the climate justice movement (see Chapter 5 in this volume and Roberts and Parks 2007). Scholars have universally argued that social movements from the Global South face considerable barriers to effective participation in global movements. Southern NGOs often lack the financial resources to travel to the sites of international meetings (Willetts 1996); those who can participate often suffer from slow learning curves, since their governments don't typically mirror the parliamentarian structure of the UN, and English-language dominance at UN meetings presents challenges that the industrialized North movements do not have to face. Doyle and MacGregor (2013) acknowledge the usefulness of this scholarship but expand it to include the fact that people in the Global South often orient themselves toward their national governments and international corporations quite differently than those from the Northern industrialized nations. While many activists from the Global North align their movements toward ecological modernization (as we have illustrated), "nearly all environmental movements in the global South still seem to cling to ideas from the Left . . . [and] tend to be framed as 'the people' versus either a large transnational corporation or the state" (Doyle and MacGregor 2013:2–3). This tends to place activists from the Global South on the margins of the global climate movement and results in the development of "radical flank" tactics like those highlighted by Newell's (2006) Outsider-Outsider category. Reformist approaches aren't easily embraced by activists in the Global South, especially when their own governments are authoritarian or corrupt; in fact, it is a blatant sign of defeat in some parts of the Global South to work in cooperation with the state or large corporations, particularly movements in Latin America. Postcolonial viewpoints, like those of many African movements, also tend to place social distance between the framing and tactics chosen by movements in the Global South and the industrialized North and global movements.

The case studies in Doyle and MacGregor's volumes (2013) highlight that residents in the Global South do not experience their lives as occurring after the Industrial Revolution; rather, industrialization is either marching forward with severe environmental impacts or is a noticeably absent option for upward mobility and improved quality of life. They argue that climate change has been largely missing from the discourse of environmental movements in the Global South, although recently this terrain has opened somewhat. In part, the lack of resonance of climate change to residents of the Global South relates to their different experience of industrialization, but it also stems from the fact that many environmental movements in the Global South tend to emphasize preservation, conservation, and animal rights because of their spiritual orientations. In chapters focused on Africa, India, Sri Lanka, and South Korea, the authors argue that religious leaders and organizations and spiritual philosophies serve as mobilizing structures and master frames for local movements. Only recently have some of these movements placed their localized environmental concerns in the context of global climate change.

Several scholars have turned to the Cochabamba movement in Bolivia as an illustrative case of grassroots environmental movements in the Global South (Bernauer 2013; Jasanoff 2014; Martinez-Torres and Rosset 2010; Olivera and Lewis 2004; Rootes, Zito, and Barry 2012). This movement of peasants, environmental groups, and local leaders fought against the privatization of water services in the Cochabamba region of Bolivia. Beginning as a grassroots, local movement, movement participants took their cause to international meetings like the World Social Forum, where they mobilized global environmental movement leaders like Maude Barlow to take up their cause. The central argument in this fight was that natural resources like water and air are common goods that belong to no one and everyone at the same time. By mobilizing international networks from the global democracy and international environmental movements, residents of Cochabamba maneuvered complex political institutions to become a central rallying point for indigenous, peasant, and grassroots environmental movements around the world (Olivera 2004). The movement became so central to the identity of Bolivia that Evo Morales, a leader of the Cochabamba movement, was elected as president of Bolivia in 2005. Morales's leadership has solidified the Cochabamba movement as a battle cry for indigenous farmers around the world who feel their access to natural resources is being threatened by corporate interests.

Two potential areas for building ties between activists in the Global South and the industrialized North are religion and climate justice. Because climate justice is the primary focus of Chapter 5 in this volume, we will simply provide an evaluation of the role of religion in the U.S. climate change movement to illustrate this potential unifier.

RELIGION AND CLIMATE CHANGE IN THE UNITED STATES

Religious communities have been actively engaged in environmental issues for the past thirty years (Brulle 2000; Kearns 1996). These efforts started to include a climate change component in the early 2000s with the creation of the Interfaith Climate Change Network in 2002 and the Evangelical Environmental network's "What Would Jesus Drive?" campaign in 2006.

Potential Versus Actual Impacts

We first consider possibility that communities of religious belief can become a significant force as civil society mobilizes on the issue of climate change; then we will assess what empirical studies show about the actual impacts to date.

Several observers have suggested a number of reasons why the world's religions could play an important role in how civil society responds to climate change (see especially Gardner 2003; Gerten and Bergmann 2012; Kearns and Keller 2007; Posas 2007; Reder 2012; Tucker and Grim 2001; Wolf and Gjerris 2009). There are four components to this argument. First, these scholars argue that religions can encourage a response to climate change via their influence on believers' worldviews or cosmologies. These narratives provide meaning and purpose and explain the individual's and group's place in the world, thereby creating the context for ethical deliberation. Such narratives also establish what is sacred, which in turn is often set aside to be preserved, respected, and protected. Considerable research shows that religions shape adherents' perceptions of and behavior toward the natural environment (Sponsel 2005), so the idea that this influence extends to climate change seems plausible.

Second, advocates for this viewpoint maintain that religions are able to engage a broad audience, many of whom accept and respect their moral authority and leadership. When one considers the global presence of religious groups, the number of people who could be engaged is impressively large: some estimate that up to 84 percent of the world population identifies with one of the major world religions. Organized religion's influence may have declined in some parts of the world, such as Western Europe, but elsewhere it continues to have profound influence. As a pervasive and powerful force in the lives of the many of the world's people, religions are, at least in theory, well positioned to mobilize millions of people on the issue of climate change.

Third, religions have significant institutional and economic resources at their disposal. They run schools and social agencies and are, collectively, the third largest category of investors in the world. In some countries they also wield considerable political power. At the institutional level this influence enables them to reach a broad audience not only through their own networks but also via interfaith coalitions such as the Parliament of the World's Religions, ecumenical groups such as the World Council of Churches, or faith-based transnational organizations such as World Vision. The international recognition and respect of leaders such as the Dalai Lama and Pope Francis are also symbolic resources that can be deployed to some effect on issues of their choosing.

Finally, religions have the potential to provide connectivity (e.g., in the form of social capital) that fosters achievement of collective goals (e.g., Smidt 2003; Swart 2006). Church affiliation is one of the most common forms of association, reaching more people in some countries (including the United States) than political affiliation or other voluntary associations (Smidt 2003). Local faith communities are often among the first to respond in disaster situations (Wisner 2010). Worldwide, faith groups are hubs for the distribution of goods, services, and even emotional support to those who are at risk; this plays a role both locally and internationally via missionary and charity outreach activities (Clarke 2006). Such fostering of trust and strengthening of social ties position religious associations as potentially effective agents in response to climate change.

Those are the arguments for faith communities' *potential* capacities to have an impact. Only recently have we begun to see sociological, empirical assessments of *actual* engagement by faith communities (Veldman et al. 2012, 2014a, 2014b). To date, though, most analyses focus largely on three communities— mainline Protestant congregations, the Catholic Church, and conservative Protestant Christianity—and they suggest, as a whole, that actual impacts typically fall short of the kinds of potential impacts we discussed earlier.

Mainline Protestant Christianity

Mainline Protestant churches, such as Lutheran, Anglicans, Presbyterians, and Methodists, tend to be liberal in both theological orientation and political engagements. Generally, these faiths acknowledge that climate change is real and serious. They accept the scientific explanation of climate change and they all agree that climate change is clearly a religious matter, since they are deeply committed to social justice and to addressing the problems of the world's poor. Since climate change is predicted to affect the world's poor first and worst, these faiths' commitment to the poor is a large part of the explanation for why they have been able to so readily define climate change as a legitimately religious matter. They have made pronouncements about what ought to be done, by the faithful in their churches and by the institutions and political authorities of their societies. When we turn from "what they say" to "what they do," it is clear that mainline Protestants are trying to do a lot. These churches educate their membership about climate change. They encourage their adherents to green their local church's building/grounds, and they challenge their adherents to begin to individually change their consuming habits, to become more ethical consumers, and to lower their carbon footprints. They try to influence policy at the level of the nation-state. Some have used the power of investing to support development projects that increase resilience in underdeveloped nations. They travel to international treaty conferences and participate in them, trying to influence the negotiations. In underdeveloped societies they work to enhance local capacity to adapt to climate change. Although their actual accomplishments, to date, fall somewhat short of what they had hoped to accomplish, these churches have real potential to affect public opinion about climate change as the situation continues to deteriorate.

Roman Catholic Church

With about 1.1 billion adherents, the Roman Catholic Church is the largest Christian church on Earth. About one fourth of Americans belong to the Catholic Church. From popes to American bishops, the Roman Catholic Church describes climate change as a major moral issue. Both Pope John

Paul II and Benedict XVI advocated for "climate justice." Benedict did so repeatedly and with increasing passion in the years leading up to his retirement. In the United States, the National Conference of Catholic Bishops appointed twenty-four "ambassadors" to convey to individual parishes what the pope teaches about climate, and the Catholic Climate Covenant urges the faithful to "take the St. Francis pledge . . . your first step in the right direction to care for creation and the poor." Actual implementation has been uneven (Agliardo 2014). The situation is similar to what we find among mainline Protestants: understanding, caring, and commitment at the level of formal rhetoric, but modest and uneven progress at the level of concrete action. Nonetheless, with a billion adherents worldwide, the Catholic Church's potential influence on civil society cannot be underestimated.

Conservative Protestant Christianity

Conservative Christians wield considerable political power, and the alliance of some parts of the conservative Christian establishment with militant climate deniers has contributed to a policy stalemate on the issue; however, there are minorities within evangelicalism who argue that Christians should embrace "stewardship" and "creation care," which for some includes climate activism. These arguments are made out of a Christian concern for the poor and because the Earth is God's creation, with a corresponding human duty of care. Still, although the greens were gaining ground in the first decade of the twenty-first century, their "insurgency" seem to have been contained more recently (Kearns 2012). As Zaleha and Szasz (2014) point out, though, "creation care" and "stewardship" have become acceptable terminology, even among climate change–denying conservative Christians. The presence of a passionate green minority means that we cannot rule out the possibility of significant change as climate conditions worsen and a new generation of younger, politically more moderate Evangelicals comes of age.

Varieties of Religious Framing

Most studies suggest that although faith communities have the potential to make a substantial contribution to climate activism, the level of actual engagement has, to date, lagged behind. Nonetheless, that potential continues to be significant, in part because several powerful discursive frames are available to drive religious engagement with climate change.

In her analysis, Kearns (1996, 2012) identifies three distinct religious frames. The first is Christian stewardship, which focuses on an evangelical interpretation based on a biblical mandate to care for God's creation. The second is creation spirituality, which is based on a synthesis of religion and

Table 8.4 Three Christian Religious Discursive Frames

Discourse	Description
Conservational Stewardship	Creation has been created "good." This "garden of God" should be preserved, as it was created, as well as possible. Technology and development are possible threats.
Developmental Stewardship	We are called to fill and subdue the Earth and turn the wilderness into a garden, as it should become. Technology and development are a necessity for this task.
Developmental Preservation	Creation is "good" and changing: progress and preservation should be combined. God has granted us the creativity to find solutions. Technology and development can present challenges as well as help us in this task.

Adapted from Wardekker, Petersen, and van der Sluijs 2009, Table 1, p. 515.

science that seeks to inform the formation of an ecological ethic based on the interconnections between human life and the global ecosystem. The third perspective is eco-justice, which draws on and extends religious ethics regarding the treatment of others to encompass the treatment of nature. Within the religious community, these frames are interwoven to create a complex multitude of different perspectives (Kearns 2012:135).

Another analysis focuses specifically on Christian religious discourses in the U.S. climate debate. In their analysis, Wardekker, Petersen, and van der Sluijs (2009) identify three distinct religious discursive frames (Table 8.4).

There has been no unification of these disparate literatures into a comprehensive analysis of the discursive frames, organizational structure, or field practices of these different communities. Further research is needed to fill in the empirical understanding of the activities of the different faith communities.

DISCUSSION AND CONCLUSION

This review of the literature on civil society and climate change reveals a number of insights that are not commonly understood in the larger climate community. First, the discursive analysis shows that there are a wide variety of voices in the climate movement and that a simplistic dichotomy between climate advocates and climate deniers misses critical information about the range of different perspectives. The resource mobilization analysis shows that the climate change movement has experienced an accelerating rate of growth since 2003 and that numerous coalitions are engaging this issue. These coalitions span an enormous range of viewpoints and, as the network analysis shows, are composed of not only nonprofit organizations but also trade unions, for-profit organizations, trade associations, and faith-based organizations. Finally, the analysis of the political opportunity structure that the climate change movement responds to is highly limited and curtails access to government deliberations and media coverage to only a few organizations based in weak ecological modernization. It also shows that the disaster-driven transition to a low-carbon economy is unrealistic without a strong social movement that can capitalize on these occasional openings in the political opportunity structure.

However, the analysis of civil society and climate change is in its infancy. There are only a handful of empirical studies that attempt to measure and analyze the discursive frames of this movement. Unexamined is the role these different frames have in the selection of movement tactics, resource mobilization sources, or access to government and media. The resource mobilization analysis lacks any examination of the role of foundations in the creation, maintenance, and strategy formation of the climate change movement. Finally, the analysis presented is highly limited to the global climate movement and the United States. In part, that decision was strategic, given the focus of this volume, but it also reflects a lack of attention to cross-national comparisons of climate movements. The structure, aims, and influence of developing-world social movements that focus on climate change have not yet been developed. We have reported central themes that have arisen in this emerging field, but admittedly the analysis presented in this chapter is only a beginning. It can serve as an initial description of the layout of this area of inquiry at present and guide the development of further research.

The analysis of the social movements advocating for climate change is extremely limited. There are only a handful of empirical analyses of this movement. There is a need for research in a number of areas:

1. *Organizational structure*—There is a need to look at the institutional structures across the different climate change movements, and how these structures influence the strategy, tactics, and aims of the different organizations. Some questions in this area include: What is the organizational structure of different organizations—are they participatory or oligarchic in structure? What is the role of foundations and outside funders in the governance of these institutions as opposed to their membership? What are the connections between specific groups and external organizations, such as corporations, government, or foundations? How do these connections affect the actions of the organizations?

2. *Strategy and Tactics*—There is an absence of analysis of the actions that climate change groups engage in to achieve their goals. Questions in this area include: What is the repertoire of actions of climate change groups? How do these actions connect to the organization's overall strategy? How do the groups use the Internet to recruit and mobilize members? What is the relative efficacy of different actions (e.g., demonstrations, boycotts, providing information, lobbying)?

3. *Resource Mobilization*—A critical factor is to understand how the climate change movement builds and maintains its organizations. What are the sources of funding of the climate change movement? What institutional mechanisms facilitate or inhibit organizations' growth and stability? How do the resource mobilization factors influence the structure and actions of organizations?

4. *Framing Climate Change*—The analysis of the different social movement organization frames is probably the most developed component of research on the climate change movement. However, further analysis is needed to understand the links among mobilization, organizational structure, and resource mobilization. Does the framing of climate change by a particular group affect its funding and institutional support? How do different framings affect the levels of membership mobilization? How do different framings of climate change develop and change?

5. *Impacts of Political Opportunity Structure*—One of the major factors influencing movement outcomes is the external political opportunity structure. Understanding the interactions between movement outcomes and the external political opportunity structure is a critical area of research regarding movement effects. For example, what is the impact of the polarization over climate change in the United States on groups' strategies and tactics? What is the dynamic between movement and countermovement efforts? How do external events affect the political opportunity structure? Is the "radical flank" a useful pressure point, given it is largely made up of organizations from the Global South?

REFERENCES

Adams, S., M. Jochum, and H. Kriesi. 2008. "Coalition Structures in National Policy Networks: The Domestic Context of European Politics." Pp. 193–217 in *Civil Society and Governance in Europe: From National to International Linkages*, edited by W. A. Maloney and J. W. Deth. Northampton, MA: Edward Elgar.

Agliardo, Michael. 2014. "The U.S. Catholic Response to Climate Change." Pp. 174–192 in *How the World's Religions are Responding to Climate Change: Social*

Scientific Investigations, edited by R. G. Veldman, A. Szasz and R. Haluza-DeLays. New York: Routledge.

Alexander, J. C. 2010. "Power, Politics, and the Civil Sphere." Pp. 111–126 in *Handbook of Politics*, edited by K. Leicht and C. Jenkins. New York: Springer.

Athanasiou, Tom. 2010. "After Copenhagen: On Being Sadder but Wiser, China, and Justice as the Way Forward" (http://www.ecoequity.org/2010/01/after-copenhagen/).

Bäckstrand, K. and E. Lövbrand. 2007. "Climate Governance Beyond 2012: Competing Discourses of Green Governmentality, Ecological Modernization and Civic Environmentalism." Pp. 123–148 in *The Social Construction of Climate Change: Power, Knowledge, Norms, Discourses*, edited by M. E. Pettenger. Hampshire, UK: Ashgate.

Beamish, T. and A. Luebbers. 2009. "Alliance Building Across Social Movements: Bridging Difference in a Peace and Justice Coalition." *Social Problems* 56(4):647–676.

Beddoe, Rachael, Robert Costanza, Joshua Farley, Eric Garza, Jennifer Kent, Ida Kubiszewski, Luz Martinez, Tracy McCowen, Kathleen Murphy, Norman Myers, Zach Ogden, Kevin Stapleton, and John Woodward. 2009. "Overcoming Systemic Roadblocks to Sustainability: The Evolutionary Redesign of Worldviews, Institutions, and Technologies." *Proceedings of the National Academies of Science USA* 106(8):2483–2489.

Berger, G., A. Flynn, F. Hines, and R. Johns. 2001. "Ecological Modernization as a Basis for Environmental Policy: Current Environmental Discourse and Policy and the Implications on Environmental Supply Chain Management." *Innovation* 14(1):55–72.

Bernauer, Thomas. 2013. "Climate Change Politics." *Annual Review of Political Science* 16:421–448.

Brulle, Robert J. 2000. *Agency, Democracy, and Nature: The U.S. Environmental Movement from a Critical Theory Perspective*. Cambridge, MA: MIT.

Brulle, Robert J. 2013. "The U.S. National Climate Change Movement." Pp. 146–170 in *Climate Change Politics: U.S. Policies and Civic Action*, edited by Yael Wolinsky-Nahmias. Washington, DC: Congressional Quarterly Press.

Brulle, Robert J. 2014. "Environmentalisms in the United States." Pp. 163–191 in *Global Perspectives on Environmentalism*, edited by T. Doyle and S. MacGregor. New York: Praeger.

Brulle, Robert J. and Rob Benford. 2012. "From Game Protection to Wildlife Management: Frame Shifts, Organizational Development, and Field Practices." *Rural Sociology* 77(1):62–88.

Cahn, Matthew Alan. 1995. *Environmental Deceptions: The Tension between Liberalism and Environmental Policymaking in the United States*. Albany: SUNY Press.

Calhoun, Craig. 1993. "Nationalism and Civil Society: Democracy, Diversity and Self-Determination." *International Sociology* 8(4):387–411.

Caniglia, Beth Schaefer. 2001. "Informal Alliances vs. Institutional Ties: The Effects of Elite Alliances on Environmental TSMO Network Positions." *Mobilization* 6(1):37–54.

Caniglia, Beth Schaefer. 2010. "Global Environmental Governance & Pathways for the Achievement of Environmental Justice." Pp. 129–155 in *Environmental Injustice Beyond Borders: Local Perspectives on Global Inequalities*, edited by J. Agyeman and J. Carmin. Cambridge, MA: MIT Press.

Caniglia, Beth Schaefer and JoAnn Carmin. 2005. "Scholarship on Social Movement Organizations: Classic Views and Emerging Trends." *Mobilization: An International Journal* 10(2):201–212.

Cantor, Robin and Gary Yohe. 1998. "Economic Analysis: Chapter One, 3:1–104." In *Human Choice and Climate Change*, edited by S. Rayner and E. Malone. Columbus, OH: Battelle Press.

Carmichael, J., J. C. Jenkins, and Robert J. Brulle. 2012. "Building Environmentalism: The Founding of Environmental Movement Organizations in the United States, 1900–2000." *Sociological Quarterly* 53:422–453.

Carolan, Michael S. 2004. "Ecological Modernization Theory: What About Consumption?" *Society and Natural Resources* 17:247–260.

Christoff, P. 1996. "Ecological Modernisation, Ecological Modernities." *Environmental Politics* 5(3):476–500.

Clarke, Gerard. 2006. "Faith Matters: Faith-Based Organisations, Civil Society and International Development." *Journal of International Development* 18(6):835–848.

Commoner, Barry. 1971. *The Closing Circle*. New York: Alfred A. Knopf.

Dawson, Ashley. 2010. "Climate Justice: The Emerging Movement against Green Capitalism." *South Atlantic Quarterly* 109(2):313–338.

Dietz, Matthias and Heiko Garrelts (Eds.). 2014. *Routledge Handbook of the Climate Change Movement*. New York: Routledge.

Dodds, Felix. 2002. "Reforming the International Institutions." Pp. 319–340 in *Earth Summit 2002: A New Deal*, edited by F. Dodds. New York: Earthscan.

Dodds, Felix and Michael Strauss. 2004. *How to Lobby at Intergovernmental Meetings: Mine's a Caffe Latte*. New York: Earthscan.

Doyle, T. and S. MacGregor, eds. 2013. *Environmental Movements Around the World: Shades of Green in Politics and Culture*. Santa Barbara, CA: Praeger Press.

Emadi-Coffin, Barbara. 2002. *Rethinking International Organization: Deregulation and Global Governance*. New York: Routledge.

Fischer, Joern, Adrian D. Manning, Will Steffen, Deborah B. Rose, Katherine Daniell, Adam Felton, Stephen Garnett, Ben Gilna, Rob Heinsohn, David B. Lindenmayer, Ben Macdonald, Frank Mills, Barry Newell, Reid Julian, Libby Robin, Kate Sherren, and Alan Wade. 2007. "Mind the Sustainability Gap." *Trends in Ecology & Evolution* 22(12):621–624.

Fisher, Dana. 2004. *National Governance and the Global Climate Change Regime*. Lanham, MD: Rowman & Littlefield.

Fisher, Dana R., Philip Leifeld, and Yoko Iwaki. 2012. "Mapping the Ideological Networks of American Climate Politics." *Climatic Change* 116(3):1–23.

Gamson, William A. 1975. *The Strategy of Social Protest*. Homewood, IL: Dorsey Press.

Gamson, William A., David Croteau, William Hoynes, and Theodore Sasson. 1992. "Media Images and the Social Construction of Reality." *Annual Review of Sociology* 18(1):373–393.

Gardner, Gary. 2003. "Engaging Religion in the Quest for a Sustainable World." Pp. 152–175 in *State of the World 2003*, edited by L. Starke. New York: W. W. Norton & Co.

Garrelts, H. and M. Dietz. 2014. "Contours of the International Climate Movement: Conception and Contents of the Handbook." In *Routledge Handbook of the Climate Change Movement*, edited by M. Dietz and H. Garrelts. New York: Routledge.

Gerten, Dieter and Sigurd Bergmann. 2012. "Facing the Human Faces of Climate Change." Pp. 3–15 in *Religion in Environmental and Climate Change: Suffering,*

Values, Lifestyles, edited by D. Gerten and S. Bergmann. London: Continuum International.

Glover, Leigh. 2006. "From Love-ins to Logos: Charting the Demise of Renewable Energy as a Social Movement." Pp. 249–270 in *Transforming Power: Energy, Environment, and Society in Conflict*, edited by J. Byrne, N. Toly, and L. Glover. New Brunswick, NJ: Transaction Publishers.

Greenberg, J., G. Knight, and E. Westersund. 2011. "Spinning Climate Change: Corporate and NGO Public Relations Strategies in Canada and the United States." *International Communication Gazette* 73(1–2):65–82.

Gulati, R. and M. Gargiulo. 1999. "Where Do Interorganizational Networks Come From?" *American Journal of Sociology* 104(5):1439–1493.

Habermas, Jürgen. 1989. "The Public Sphere: An Encyclopedia Article." Pp. 102–107 in *Critical Theory and Society*, edited by E. Bronner and D. Kellner. New York: Routledge.

Habermas, Jürgen. 1996. *Between Facts and Norms: Contributions to a Discourse Theory of Law and Democracy*. Cambridge, MA: MIT Press.

Hemmati, Minu. 2002. *Multi-stakeholder Processes for Governance and Sustainability: Beyond Deadlock and Conflict*. London: Earthscan Publications Ltd.

Hewitt, Lyndi and Holly J. McCammon. 2004. "Explaining Suffrage Mobilization: Balance, Neutralization, and Range in Collective Action Frames, 1892–1919." *Mobilization* 9:149–166.

Institute for Sustainable Development. 2009. *Earth Negotiations Bulletin* 12(459) (http://www.iisd.ca/climate/cop15/).

IPCC. 2007. Contribution of Working Group II to the Fourth Assessment Report of the Intergovernmental Panel on Climate Change. Edited by M. L. Parry, O. F. Canziani, J. P. Palutikof, P. J. van der Linden, and C. E. Hanson. Cambridge and New York: Cambridge University Press.

Jasanoff, Sheila. 2014. "Weathering the Climate Crisis." *Current History* 113(759):12–15.

Jenkins, David G. 2011. "Biology, Movement, and Impacts of *Gambusia holbrooki* in its Native Range." Pp. 11–14 in *Gambusia Forum 2011: Small Fish . . . Big Problem!*, edited by P. Jackson and H. Bamford. Canberra City, Australia: Murray-Darling Basin Authority.

Kearns, Laurel. 1996. "Saving the Creation: Christian Environmentalism in the United States." *Sociology of Religion* 57:55–70.

Kearns, Laurel. 2012. "Religious Climate Activism in the United States." Pp. 132–151 in *Religion in Environmental and Climate Change: Suffering, Values, Lifestyles*, edited by D. Gerten and S. Bergmann. New York: Continuum.

Kearns, Laurel, and Catherine Keller, eds. 2007. *Ecospirit: Religions and Philosophies for the Earth*. New York: Fordham University Press.

Knoke, D. 1990. *Political Networks: The Structural Perspective*. Cambridge: Cambridge University Press.

Knoke, D., and S. Yang. 2008. *Social Network Analysis*. Los Angeles: Sage.

Leicht, Kevin T. and Scott T. Fitzgerald. 2014. *Middle Class Meltdown in America: Causes, Consequences, and Remedies*. 2nd ed. New York: Routledge.

Lichterman, P. 1995. "Piecing Together Multicultural Community: Cultural Differences in Community Building Among Grass-Roots Environmentalists." *Social Problems* 42(4):513–534.

Magan, Andre. 2006. "Refeudalizing the Public Sphere: "Manipulated Publicity in the Canadian Debate on GM Foods." *Canadian Journal of Sociology* 31(1):25–53.

Martinez-Torres, Maria Elena, and Peter M. Rosset. 2010. "La Via Campesina: The Birth and Evolution of a Transnational Social Movement." *Journal of Peasant Studies* 37(1):149–175.

McAdam, Doug. 1988. *Freedom Summer*. New York: Oxford University Press.

McAdam, Doug, John D. McCarthy, and Mayer N. Zald. 1988. "Social Movements." Pp. 695–737 in *Handbook of Sociology*, edited by N. J. Smelser. Thousand Oaks, CA: Sage.

McAdam, Doug, John D. McCarthy, and Mayer Zald. 1996. *Comparative Perspectives on Social Movements*. Cambridge: Cambridge University Press.

McCarthy, John D. and Mayer N. Zald. 1977. "Resource Mobilization and Social Movements: A Partial Theory." *American Journal of Sociology* 82(6):1212–1241.

Meyer, David S. 2006. *The Politics of Protest: Social Movements in America*. New York: Oxford University Press.

Meyer, David S. and Douglas R. Imig. 1993. "Political Opportunity and the Rise and Decline of Interest Group Sectors." *Social Science Journal* 30(3):253–270.

Mol, Arthur. 2001. *Globalization and Environmental Reform: The Ecological Modernization of the Global Economy*. Cambridge, MA: MIT Press.

Murphy, G. 2005. "Coalitions and the Development of the Global Environmental Movement: A Double-Edged Sword." *Mobilization* 10(2):235–250.

Murphy, Joseph. 2000. "Ecological Modernisation." *Geoforum* 31(1):1–8.

Newell, P. 2006. "Climate for Change? Civil Society and the Politics of Global Warming." Pp. 90–119 in *Global Civil Society 2005/2006*, edited by M. Glasius, M. Kaldor, and H. Anheier. Thousand Oaks, CA: Sage.

Newell, P. 2008. "Civil Society, Corporate Accountability and the Politics of Climate Change." *Global Environmental Politics* 8(3):122–153.

Olivera, Oscar and Tom Lewis. 2004. ¡*Cochabamba! Water War in Bolivia*. New York: South End Press.

Olsen, Mancur. 1965. *The Logic of Collective Action*. Cambridge: Harvard University Press.

Phillips, Macon. 2009. "From Peril to Progress (Update 1: Full Remarks)" (http://www.whitehouse.gov/blog_post/Fromperiltoprogress/).

Poletta, Francesca. 2002. *Freedom Is an Endless Meeting: Democracy in American Social Movements*. Chicago: University of Chicago Press.

Poloni-Staudinger, I. 2009. "Why Cooperate? Cooperation Among Environmental Groups in the United Kingdom, France, and Germany." *Mobilization* 14(3):375–396.

Posas, Paul J. 2007. "Roles of Religion and Ethics in Addressing Climate Change." *Ethics in Science and Environmental Politics* 2007:31–49.

Pulver, S. 2004. "Power in the Public Sphere: The Battles Between Oil Companies and Environmental Groups in the UN Climate Change Negotiations, 1991–2003." Unpublished PhD dissertation, University of California at Berkeley.

Reder, Michael. 2012. "Religion in the Public Sphere: The Social Function of Religion in the Context of Climate and Development Policy." Pp. 32–45 in *Religion in Environmental and Climate Change: Suffering, Values, Lifestyles*, edited by D. Gerten and S. Bergmann. London: Continuum.

Revkin, Andrew C., Shan Carter, Jonathan Ellis, Farhanna Hossain, and Alan Mclean. 2012. "On the Issues: Climate Change." *New York Times*, May 23 (http://elections.nytimes.com/2008/president/issues/climate.html).

Roberts, J. Timmons and Bradley C. Parks. 2007. *A Climate of Injustice: Global Inequality, North-South Politics, and Climate Policy*. Cambridge, MA: MIT Press.

Rootes, Christopher, Anthony Zito, and John Barry. 2012. "Climate Change, National Politics and Grassroots Action: An Introduction." *Environmental Politics* 21(5):677–690.

Schlosberg, David and Sara Rinfret. 2008. "Ecological Modernisation: American Style." *Environmental Politics* 17(2):254–275.

Sievers, Bruce. 2010. *Civil Society, Philanthropy, and the Fate of the Commons*. Medford, MA: Tufts University Press.

Skocpol, T. 2003. *Diminished Democracy: From Membership to Management in American Civic Life*. Norman: Oklahoma University Press.

Smidt, Corwin E. 2003. *Religion as Social Capital: Producing the Common Good*. Waco, TX: Baylor University Press.

Snow, David A. and Robert D. Benford. 1988. "Ideology, Frame Resonance, and Participant Mobilization." *International Social Movement Research* 1:197–218.

Soule, Sarah. 2009. *Contention and Corporate Social Responsibility*. New York: Cambridge University Press.

Sponsel, Leslie E. 2005. "Anthropologists." Pp. 94–96 in *The Encyclopedia of Religion and Nature*, edited by B. R. Taylor. London: Thoemmes Continuum.

Staggenborg, Suzanne. 1988. "The Consequences of Professionalization and Formalization in the Pro-Choice Movement." *American Sociological Review* 53:585–605.

Swart, Ignatius. 2006. "Churches as a Stock of Social Capital for Promoting Social Development in Western Cape Communities." *Journal of Religion in Africa* 36(3–4):346–378.

Sztompka, Piotr. 1993. *The Sociology of Social Change*. Cambridge, MA: Blackwell.

Tarrow, Sidney. 1998. *Power in Movement*. 2nd ed. New York: Cambridge University Press.

Tucker, Mary Evelyn and John A. Grim. 2001. "Introduction: The Emerging Alliance of World Religions and Ecology." *Daedalus* 130(4):1–22.

Veldman, Robin Globus, Andrew Szasz, and Randolph Haluza-Delay. 2012. "Introduction: Climate Change and Religion – A Review of Existing Literature." *Journal for the Study of Religion, Nature and Culture* 6(3):255–275.

Veldman, Robin Globus, Andrew Szasz, and Randolph Haluza-Delay. 2014a. "Social science, religions, and cimate change." Chapter 1 in *How the World's Religions Are Responding to Climate Change: Social Scientific Investigations*, edited by R. G. Veldman, A. Szasz, and R. Haluza-DeLay. New York: Routledge.

Veldman, Robin Globus, Andrew Szasz, and Randolph Haluza-Delay. 2014b. "Climate change and religion as global phenomena: Summing up and directions for further research." Chapter 19 in *How the World's Religions Are Responding to Climate Change: Social Scientific Investigations*, edited by R. G. Veldman, A. Szasz, and R. Haluza-DeLay. New York: Routledge.

Van Dyke, Nella. 2003. "Crossing Movement Boundaries: Factors That Facilitate Coalition Protest by American College Students, 1930–1990." *Social Problems* 50(2):226–250.

Walker, E. 2012. "Putting a Face on the Issue: Corporate Stakeholder Mobilization in Professional Grassroots Lobbying Campaigns." *Business and Society* 51(4): 619–659.

Walstrom, Mattias, Magnus Wennerhag, and Christopher Rootes. 2013. "Framing 'The Climate Issue': Patterns of Participation and Prognostic Frames among Climate Summit Protesters." *Global Environmental Politics* 13(4):101–122.

Wardekker, J. A., A. C. Petersen, and J. van der Sluijs. 2009. "Ethics and Public Perception of Climate Change: Exploring the Christian Voices in the US Public Debate." *Global Environmental Change* 19:512–521.

Willetts, Peter. 1996. "From Stockholm to Rio and Beyond: The Impact of the Environmental Movement on the United Nations Consultative Arrangements for NGOs." *Review of International Studies* 22(1):57–80.

Wisner, Ben. 2010. "Untapped Potential of the World's Religious Communities for Disaster Reduction in an Age of Accelerated Climate Change: An Epilogue and Prologue." *Religion* 40(2):128–131.

Wolf, Jakob and Mickey Gjerris. 2009. "A Religious Perspective on Climate Change." *Studia Theologica* 63(2):119–139.

York, Richard, Eugene A. Rosa, and Thomas Dietz. 2003. "Footprints on the Earth: The Environmental Consequences of Modernity." *American Sociological Review* 68(2):279–300.

Zaleha, Bernard and Andrew Szasz. 2014. "Keep Christianity Brown! Climate Denial on the Christian Right in the United States." Chapter 14 in *How the World's Religions Are Responding to Climate Change: Social Scientific Investigations*, edited by R. G. Veldman, A. Szasz, and R. Haluza-DeLay. New York: Routledge.

9

Public Opinion on Climate Change

Rachael L. Shwom, Aaron M. McCright,
and Steven R. Brechin with Riley E. Dunlap,
Sandra T. Marquart-Pyatt, and Lawrence C. Hamilton

INTRODUCTION

Since public support is widely recognized as an important factor shaping societal response to climate change (U.S. National Research Council 2010), this chapter provides a sociological overview of public opinion on climate change.[1] There is a significant body of research aimed at documenting and explaining citizens' views on climate change, and it has produced many valuable insights. However, much of the popular discourse on climate change public opinion has a strong psychological perspective, regularly emphasizing what individuals think about climate change as if an atomistic mind produces opinions on climate change in an asocial vacuum. This chapter extends beyond the psychological dimensions of climate change, such as those reported in the American Psychology Association's Task Force report on climate change (Swim et al. 2011), to highlight the more sociological dimensions of public opinion on climate change. Sociological research attempts to examine and understand the larger social, economic, cultural, political, and environmental factors that influence the formation and shifting of public opinion on climate change.

Public opinion on climate change is multidimensional, dynamic, and differentiated. The multiple dimensions covered under the umbrella term "public opinion on climate change" include, among others, beliefs about anthropogenic climate change, perceptions of climate change risks, concern about its seriousness, and thoughts on what, if anything, should be done to address it. Public opinion is dynamic, changing over time due to personal, social, political, economic, and environmental factors. There has been limited research on dynamics of climate change political opinion, because there are few time series datasets available. The surveys and studies discussed here often provide snapshots of a few specific dimensions of opinion at a particular point in time in a particular place. Finally, public opinion is differentiated. A range of sociodemographic, political, cultural, economic, and environmental factors predict variation in climate change public opinion.

We use the term "public opinion" to discuss beliefs, attitudes, policy support, and behavioral intentions of people and groups within a particular geographic location.[2] Traditional public opinion research focuses on the aggregate of individual attitudes and beliefs or "mass opinion" (Zaller 1992). Many political scientists have argued that *the collective preferences* of the mass public are important to understanding the political dynamics of an issue (Page and Shapiro 1992; Stimson, MacKuen, and Erikson 1995). However, Berinsky (1999) argues that research focusing on individuals' attitudes— "the microfoundations of public opinion"—provides important insights on the individual expressions of preferences that aggregate to mass opinion. In this chapter, we review the majority of survey research[3] on both mass opinion and individuals' views.

First, we provide an overview of mass opinion on climate change transnationally. We then focus on mass opinion in the United States, where many surveys have been administered in recent years. From there we move on to discuss the social, political, cultural, economic, and environmental factors that explain variation in individuals' climate change views. These topics elaborate on the useful sociological contributions found in *American Climate Choices* (especially Chapter 3, which focuses on the complexity of social systems, public understanding, and risk assessment regarding climate change). After summarizing our major findings, we identify four fruitful avenues for future research that focus on examining the extent to which public opinion influences household behavior change, political or collective action participation, policy outcomes, and corporate outcomes. Finally, we identify the contributions that the study of climate change public opinion have made to the broader field sociology of public opinion.

PUBLIC OPINION ON CLIMATE CHANGE AROUND THE WORLD

Until the early 1990s, little was known of the public's environmental views (referred to as "environmental concern") outside of Western industrialized countries. Scholars assumed that citizens of poorer developing countries could not afford to be concerned with environmental issues until they became wealthier economically (see Guha and Martinez-Alier 1997). This belief was promoted by Ronald Inglehart's (1997) postmaterialist values thesis, wherein concern for the environment is expected to emerge only after basic material needs and political and emotional security have been met. In short, public concern for the environment was framed essentially as a "luxury" good that would be "purchased" individually and nationally when sufficient numbers of the public were wealthy and feeling secure enough to do so. This theory has received generally weak empirical support (Dietz, Fitzgerald, and Shwom 2005), and when environmental concern is operationalized as views about climate change, this thesis receives mixed support at best.

Pioneering findings on general environmental concern from Gallup's 1992 Health of the Planet Survey (Dunlap, Gallup Jr., and Gallup 1993) were among the first to challenge the assumption that developing nation citizens cannot afford to be concerned about the environment. This study of twenty-four countries provides the first-ever national probability survey of respondents from richer developed and poorer developing countries. Many studies since this survey have also found high levels of environmental concern regardless of economic wealth (Brechin 1999; Brechin and Kempton 1994; Dunlap and York 2008; Marquart-Pyatt 2007, 2008, 2012; Marquart-Pyatt et al. 2011). A multilevel analysis of data from the 2010 International Social Survey Program finds that national gross domestic product (GDP) per capita is positively correlated with individual-level environmental concern (Franzen and Vogl 2013), while a study of 2005–2008 data from the World Values Survey finds that individual-level postmaterialist/materialist value orientations influence environmental concern in all nations, but with a stronger effect in industrialized nations (Dorsch 2014). This body of research suggests that while some poor nations do have high levels of environment concern, a nation's economy can nevertheless influence environmental concern levels among citizens.

The last two decades have yielded a number of cross-national studies specifically exploring climate change public opinion. Early efforts, including those by Dunlap (1998) and Bord, Fisher, and O'Connor (1998), suffered from limited data from only a few countries. Still, they found that the publics in these countries expressed concern for climate change, but at levels lower than concern for other environmental problems. The publics had very limited understanding of climate change, particularly in identifying its anthropogenic causes. Analyzing data from more countries nearly a decade later, Brechin (2003) finds slightly improved public understanding of the anthropogenic causes of climate change and also growing levels of concern. However, public concern for climate change continued to lag behind other environmental problems, especially water and air pollution.[4]

In 2007 and 2008, Gallup conducted a survey on nationally representative samples in over 150 countries. Gallup asked, "How much do you know about global warming or climate change?" and gave respondents three choices: "I have never heard of it," "I know something about it," or "I know a great deal about it." The results suggest that many individuals, approximately 39 percent globally, are unaware of the term "global warming or climate change." Nearly all of the publics of developed nations have at least heard of the term "climate change," from 99 percent in Japan to 96 percent in the United States. Yet, approximately 75 percent of Egyptians, 71 percent of Bangladeshis, 65 percent of South Africans, 65 percent of Indians, and 63 percent of Indonesians *have not even heard* of climate change (Pugliese and Ray 2009a). Regionally, those most likely to have at least heard of climate change are in Europe and the Americas (which includes North, South, and Central America). The regions least likely to have heard of it include the Middle East/North Africa, Asia, and sub-Saharan Africa regions (Pugliese and Ray 2009b).

More recent research, however, suggests that while many poor rural people in developing counties are unfamiliar with the term "climate change," they have observed changes in weather patterns that are affecting their crops and have already made changes in their farming strategies (Moghariya and Smardon 2014). The Gallup 2007–2008 poll followed up the question about awareness of climate change by asking respondents how serious a threat ("not at all" to "very") climate change was to them and their family. Globally, of the 61 percent who are aware of climate change to some extent, 41 percent believe climate change is a serious threat, while 18 percent see it as not very or not at all serious. Adults in Asia are least likely to see climate change as a serious threat, whereas adults in North and South America are most likely to see it as a threat (Pugliese and Ray 2009b).

While we know more about developing countries than ever before, there is much more to learn. There are still few quality representative surveys of developing nations. In addition, a key question remains regarding how to explain cross-national variation in climate change public opinion.

Kvaløy, Finseraas, and Listhaug (2012) analyze climate change public opinion in the forty-seven nations of the 2005–2009 World Values Survey, offering one of the more nuanced comparative analyses to date. This study predicts climate change public opinion with two models: one based on post-materialist and religious values and another based on objective conditions (i.e., levels of CO_2, exposure to natural disasters, and GDP per capita). They find little support for the objective conditions model but modest support for the values model. Perceived seriousness of climate change was positively associated with level of education, postmaterialist values, Left-leaning political identification, and religiosity. In an analysis of an AC Nielsen survey of forty-six nations, Sandvik (2008) also finds that public concern for climate change is negatively related to national levels of GDP per capita and CO_2 production per capita.

Kim and Wolinsky-Nahmias (2014) pool data across multiple cross-national surveys conducted by many organizations to examine the variation in the intensity of public concern about climate change and degree of support for various climate policies. They find that citizens in developed nations report less concern about climate change and less support for climate policies than citizens in developing nations. Further, while a measure of environmental vulnerability (theoretically an indicator of the extent or magnitude of harms posed by climate change in that nation) does not explain cross-national variation in level of concern, it does correlate with higher commitment to climate policies, including greater willingness to pay[5] and stronger support for pro-climate energy policies (Kim and Wolinsky-Nahmias 2014:18).

Using 2008 Eurobarometer data from nationally representative samples in twenty-seven European Union (EU) countries, McCright, Dunlap, and Marquart-Pyatt (2014) analyze key individual-level and national-level predictors of five dimensions of climate change public opinion: perceived seriousness of climate change, acceptance of anthropogenic climate change, beliefs about fighting climate change, personal willingness to pay to fight climate change, and support for EU greenhouse gas (GHG) emission reduction policies. Using a combination of pooled regression analysis and multilevel modeling, they find that more highly educated people and women express stronger belief in climate change and willingness to support actions to deal with it than their less educated and male counterparts. Consistent with the antireflexivity thesis (McCright and Dunlap 2010), citizens on the Left report stronger beliefs in climate change and support for climate policies than do citizens on the Right in Western Europe. However, there is no such ideological divide in former communist countries, where the Left–Right divide takes a different meaning. Due to minimal cross-national variation, national-level variables such as GDP per capita and CO_2 emissions per capita explain very little variation in climate change public opinion.

Running (2013) employs a social psychological approach to build on past research on cooperation and social dilemmas that suggests individuals around the world are more likely to cooperate with others they see in their group. Extending this tradition, she focuses on the geographic scale of a citizens' social group identification (world citizen, national citizen, local community member, and autonomous individual) and its correlation with concern about climate change. Using the data for 128 nations from the 2005–2008 wave of the World Values Survey, Running (2013) finds that those who strongly identify as being both a world citizen and an autonomous individual are 1.26 times more likely to find climate change to be a serious problem. She argues that this may be related to the nature of climate change as both an individual and a communal risk.

In reviews of cross-national public opinion on climate change, public concern for climate change varies significantly across nations, but certain generalizations can be made (Brechin 2010; Brechin and Bhandari 2011). There tend to be high levels of concern about the environment in general across most nations. This finding has been fairly consistent for several decades. However, these high levels of concern (as is the case for many issues) do not directly translate to willingness to pay to combat climate change. In both developed and developing nations, willingness to pay is found to be lower than levels of concern. Two related factors likely explain this phenomenon. First, worldwide there are serious competing environmental concerns, especially the desire for clean water and air. Second, publics perceive the impacts of climate change to be less severe and further off than the more immediate impacts of these other environmental problems. That being said, educational attainment is positively correlated with awareness of, belief in, and support for taking action on climate change (Kvaløy et al. 2012; Leiserowitz 2009; McCright, Dunlap, and Marquart-Pyatt 2014).

The studies we have discussed continue to show increasing public understanding of the human causes of climate change, concern for the problem, and general support for a range of policy alternatives to address it. However, accumulation of scientific knowledge and consensus on the nature of public opinion is difficult given that studies ask somewhat different questions in different nations over time. Puzzling findings have arisen that have yet to be explained. For example, while concern about climate change seems to be increasing in the countries surveyed globally, the publics in the United States, Great Britain, Russia, and China remain among the least concerned of wealthy nations. This is particularly noteworthy given that these four countries are among the largest GHG emitters globally. While the Chinese public is among the least concerned about climate change, it is among the most supportive of policies to address the problem. Similarly, while India's public is among the most concerned, they are among the least willing to support policy alternatives (see Brechin 2010).

MASS OPINION ON CLIMATE CHANGE IN THE UNITED STATES

A modest number of surveys have documented public opinion on climate change in the United States in the last two decades. Nationally, Gallup has collected data intermittently in the 1980s and 1990s and then annually since 2001 (Dunlap and McCright 2008; McCright 2010; McCright and Dunlap 2011a, 2011b; McCright, Dunlap, and Xiao 2013, 2014a; McCright, Dunlap, and Marquart-Pyatt 2014). The General Social Survey included questions on climate change and polar region concerns in 2006 and 2010 (Hamilton 2008).

The Pew Charitable Trust as well conducts regular surveys on public opinion on climate change in the United States and occasionally internationally.

Several other U.S. nationally representative surveys have emerged over the last five years, including the "Climate Change in the American Mind" survey at the Yale Project on Climate Change (Leiserowitz et al. 2012), the National Survey of American Public Opinion on Climate Change at Muhlenberg College (Borick and Rabe 2010), the Woods Institute for the Environment surveys on energy and climate at Stanford University (Malka, Krosnick, and Langer 2009), and the National Community and Environment in Rural America survey by the Carsey Institute at the University of New Hampshire (Hamilton 2012; Hamilton and Keim 2009). There also have been regional and statewide surveys at various points in time (Dietz, Dan, and Shwom 2007; Hamilton 2011; Hamilton and Keim 2009; Hamilton and Stampone 2013a, 2013b; Villar and Krosnick 2011).

Several scholars have examined these different surveys, documenting general patterns and trends in mass opinion on climate change in the United States (Brewer 2005; Marquart-Pyatt et al. 2011; Nisbet and Myers 2007). In this section, we summarize these general patterns and trends, using the framework of topics provided by Nisbet and Myers (2007): (1) awareness of climate change as a problem; (2) knowledge of climate change; (3) belief in the reality of climate change and scientific consensus; (4) perceived immediacy of climate change impacts; (5) relative concern about climate change compared to other issues; and (6) support for climate change policies.

First, we deal with the issue of survey wording and whether asking about climate change or global warming makes a difference in the survey results. As mentioned earlier, journalists often use the earlier term "global warming" and "climate change" interchangeably. Several studies have examined whether the use of "global warming" or "climate change" in surveys influences public opinion, and the results to date are mixed. A few studies find no substantial effect on public opinion overall (Eurobarometer 2009; Lorenzoni et al. 2006), while a couple find that word choice does matter (Schuldt, Konrath, and Schwarz 2011; Whitmarsh 2009). Further, two studies find that political orientation moderates the relationship between word choice and public opinion; Republicans are more likely to perceive climate change than global warming as real and serious, while either the reverse is true among Democrats (Villar and Krosnick 2011) or Democrats are not influenced by question wording (Schuldt et al. 2011).

In more recent analysis of two 2014 Gallup surveys, Dunlap (2014) reports that people react to both terms similarly and that the degree of partisan and ideological divergence on the two was also very similar. Unlike previous studies that find Republicans to be significantly more skeptical about global warming than climate change, the two Gallup polls find only modest differences. These recent results suggest that the term "climate change" has become as politically polarizing as "global warming."

Public awareness of climate change has been increasing since the 1980s in the United States. The percentage of Americans who are aware of climate change was 39 percent in 1986, rose to 58 percent in 1988, and is currently around 96 percent. Since public awareness has been defined as having heard of or read about climate change, it is not surprising that public awareness has increased with media coverage. It saw its first bump in 1988 after a record U.S. heat wave—which some linked to climate change—received much media coverage. Climate change awareness has almost fully saturated the United States, and although people may not know much about its causes or impacts, almost all have at least heard of it.

Public knowledge of climate change has been increasing, though not as quickly as general awareness. Self-assessed knowledge is measured by asking people to report how much they know about or how well they understand climate change. When asking people how well they understand climate change, Gallup documents that the percentage saying "very well" or "fairly well" increased from 69 percent in 2001, to 74 percent in 2006, and to 80 percent in 2011 (Jones 2011).

Objectively assessed knowledge of climate change is measured by asking people about the causes and consequences of climate change. Studies find that Americans often score low on these objective measures and express a fair amount of confusion about the causes and consequences of climate change. Early studies in the mid-1990s show that Americans often conflated climate change with ozone depletion and with air and water pollution in general (therefore seeing pollutants like pesticides as causing climate change) (Bostrom et al. 1994; Read et al. 1994). When asked to describe the process of climate change, approximately 73 percent of interviewees in one study invoked a mental model of "holes in the ozone layer" causing global warming by letting more heat in (Bostrom et al. 1994). A 2009 replication of a 1994 survey (Read et al. 1994) finds that the prevalence of the hole in the ozone layer mental model dropped to 3 percent of respondents; the percentage of people believing using aerosol spray cans causes climate change dropped from 22 percent in 1994 to 10 percent in 2009 (Reynolds et al. 2010). These research studies also highlight that individuals continue to confuse short-term weather patterns with long-term climate trends, a persistent challenge to public understanding of climate change.

While almost all Americans are aware of climate change, levels of belief in the reality and human causation of climate change have fluctuated across the last few decades, with a recent downturn from 2008 to 2010 and a rebound from 2010 to 2013 (Leiserowitz et al. 2012; Saad 2013). While nearly two thirds of Americans believe that climate change is occurring, only about half believe that human activities are causing it (Hamilton 2012; McCright and Dunlap 2011b). Unlike most other industrialized nations, the United States consistently has had a segment of the population either deny the reality or anthropogenic cause of climate change. The percentage of Americans holding multiple climate change denial beliefs was approximately 10 percent in the late 2000s (McCright and Dunlap 2011a), though others put that percentage slightly higher, at around 14 percent (Leiserowitz et al. 2012).

Whether measured as how much citizens worry about climate change or whether climate change poses a threat to them or their way of life, only about one third of Americans report a high level of concern about climate change (Dunlap and McCright 2008; McCright 2009; McCright and Dunlap 2011b). Compared to a wide range of other issues, relative concern about climate change is consistently low. For instance, concern about climate change ranks consistently lower than concern about other issues such as "the economy" or "national security" (Nisbet and Myers 2007). The Pew Research Center (2014) has been tracking the public's prioritization of climate change since 2007 and consistently finds it to fall last or second to last in a list of twenty issues (often only ranking above global trade).

More importantly perhaps, Americans worry less about climate change than about a range of local and/or more immediate environmental problems like air pollution, water pollution, and toxic waste contamination (Saad 2009). One reason that climate change may be given such low priority, particularly in comparison to other environmental problems, is that many Americans perceive it to be a risk only in the distant future. When asked when the effects of climate change will begin to happen, 27 percent of Americans report that they will only begin in the future (while another 15 percent state they will not happen at all). Additionally, 64 percent believe that climate change does not pose a serious threat to them or their way of life in their lifetime (Jones 2014).

Despite the low levels of relative concern, consistently more than 50 percent of Americans support a variety of energy and climate change policies (Brewer 2005; Krosnick and MacInnis 2011; Leiserowitz et al. 2012; Shwom, Dan, and Dietz 2008). However, this support varies. For example, Krosnick and MacInnis (2011) report an average 5 percent drop in support from 2010 to 2011 for a number of policies. Overall, support tends to be higher for "carrots" like subsidies for renewables or tax rebates for people who buy energy-efficient appliances than for "sticks" like taxes or regulations on specific industries. Gasoline or straight carbon taxes have the lowest level of support (Leiserowitz et al. 2014; Shwom et al. 2010).

Integrating many of these dimensions of climate change public opinion, Maibach et al. (2011) have developed a segmented profile of U.S. public opinion on climate change. The segmentations are based on dimensions of climate change concern, awareness, beliefs, and levels of engagement. They characterize the groups as alarmed, concerned, cautious, disengaged, doubtful, and dismissive. The segmentation is developed to guide targeted climate change communications and behavioral campaigns.

The studies we have cited reveal general patterns and trends for mass opinion on climate change in the United States. Nearly all Americans are aware of climate change, though actual knowledge of climate change remains relatively low. In general, climate change falls near the bottom of Americans' list of worries, and it is not considered as pressing as other environmental problems. Climate change is often viewed as far away in time and space. About 10 percent of the American public consistently denies the reality and seriousness of climate change, while approximately a third of the public is uncertain about whether it is caused by humans. Disaggregating the data provides the opportunity to examine theoretically relevant insights on the predictors of public opinion on climate change. We turn to this in the next section.

PREDICTORS OF INDIVIDUALS' VIEWS ON CLIMATE CHANGE

As we have mentioned, public opinion on climate change is differentiated. A range of sociodemographic, political, cultural, economic, and environmental factors help explain variation in climate change views. Research on these predictors is not driven by an overarching theoretical perspective. Rather, scholars working in this area draw upon a range of theories and hypotheses from across the social sciences, such as gender socialization theory, party sorting theory, the elite cues hypothesis, information processing theory, values beliefs norms theory and cultural cognition theory.

Much sociological research on predictors of climate change views has evolved from earlier work on the social bases of environmental concern (e.g., Jones and Dunlap 1992; Van Liere and Dunlap 1980). Not surprisingly, some of the main findings of these two literatures are quite similar. For instance, most of those sociodemographic factors that are not consistent predictors of environmental concern also are not consistent predictors of climate change views: income, race, religiosity, and place of residence. In other words, variation in these characteristics is not regularly associated with variation in climate change views. Two key predictors of environmental concern (gender and political orientation) are quite consistent predictors of climate change views, while two others (age and education) are less consistent predictors.

Gender socialization theory explains how individuals in any society learn about gender expectations and norms (i.e., what masculinity and femininity mean) in their society's dominant culture (e.g., Chodorow 1978; Gilligan 1982). In the U.S. context, boys learn that masculinity means being competitive, independent, and unemotional and entails objectively exerting mastery and control over other people and things; girls learn that femininity means being compassionate, cooperative, and empathetic and entails connecting with other people and expressing concern about their well-being. In short, a masculine identity emphasizes detachment, control, and mastery, while a feminine identity stresses attachment, empathy, and care (Keller 1995; Merchant 1980). Consistent with gender socialization theory, women express slightly more concern about climate change than do men (Brody et al. 2008; Hamilton 2008; Leiserowitz 2006; Malka et al. 2009; McCright 2010; O'Connor, Bord, and Fisher 1999). Further, women report beliefs more consistent with the scientific consensus on climate change than do men (McCright 2010; McCright, Dunlap and Xiao 2013, 2014a).

Postmaterialist values theory posits that as countries make the transition to postmodern nations, increasing cohorts of citizens will continue to experience shifts in personal values such as more support for environmental protection, including greater concern over climate change. While younger adults are more likely to report beliefs consistent with the scientific consensus than are older adults (McCright 2010), the effect of age on concern about climate change is inconsistent across several studies (Kellstedt, Zahran, and Vedlitz 2008; Malka et al. 2009; Wood and Vedlitz 2007). Also, the effect of education on climate change views is inconsistent not only across studies but also within studies. Greater education increases the likelihood of believing that global warming has already begun, but it has no effect on Americans' belief about the primary cause of recent warming (McCright and Dunlap 2011b)—the opposite of results reported by Hindman (2009). More studies find that education is negatively associated with concern about global warming (Malka et al. 2009; Marquart-Pyatt et al. 2011; McCright and Dunlap 2011a, 2011b; O'Connor et al. 1999; Wood and Vedlitz 2007) than find it is positively associated with such concern (e.g., Hamilton 2008). Others report education to have a positive effect on policy support (O'Connor et al. 1999, 2002; Zahran et al. 2006).

Paralleling its strength as a predictor of general environmental concern in the United States, by far the strongest predictor of climate change views is political orientation: political ideology and party identification. Four decades of social science research on environmental values, beliefs, and attitudes confirms that self-identified liberals and Democrats are more pro-environmental than self-identified conservatives and Republicans in the American public (Dunlap, Xiao, and McCright 2001 and the sources cited within). The pursuit of environmental protection often entails governmental intervention into markets and restrictions on property rights, challenging conservative and Republican values, but is consistent with liberals' and Democrats' view that protecting collective welfare is a proper role of government. More broadly, conservatives and Republicans are more likely than are liberals and Democrats to support the dominant social paradigm (Dunlap and Van Liere 1984), our society's prevailing worldview that includes core elements of conservative ideology but also faith in science and technology, support for economic growth, faith in material abundance, and faith in future prosperity. Support for the dominant social paradigm undermines support for environmental protection (Dunlap and Van Liere 1984).

The characterization of climate change as a major problem and the possible ratification of an internationally binding treaty to curb carbon dioxide emissions are seen as a direct threat to sustained economic growth, the free market, national sovereignty, and the continued abolition of governmental regulations—key goals supported by conservatives and Republicans. Indeed, the challenge that climate change poses to conservative ideology has partially driven the emergence and mobilization of the climate change denial countermovement (see Chapter 10 in this volume). Thus, it is no surprise that self-identified conservatives and Republicans report beliefs about climate change less consistent with the scientific consensus, express lesser personal concern about global warming, and report lesser support for climate policy proposals than do their liberal and Democratic counterparts (e.g., Borick and Rabe 2010; Dietz et al. 2007; Dunlap and McCright 2008; Hamilton 2008, 2011; Hamilton and Keim 2009; Krosnick, Visser, and Holbrook 1998; Krosnick, Holbrook, and Visser 2000; Leiserowitz 2006; McCright 2009, 2010; McCright and Dunlap 2011a, 2011b; McCright, Dunlap, and Xiao 2013, 2014a; O'Connor et al. 2002; Wood and Vedlitz 2007).

Perhaps more important, this political divide on climate change views in the American public has increased over the years in a manner consistent with party sorting theory—the prevailing theoretical explanation for political polarization in the general public (e.g., Dunlap and McCright 2008; McCright and Dunlap 2011b; McCright, Dunlap, and Xiao 2014a). In recent decades, ideological and party elites in the United States have become polarized on the issue of climate change (e.g., McCright and Dunlap 2003, 2010). According to party sorting theory, ideological and party activists drive polarization among political elites, and this elite polarization provides cues to citizens about the evolution of party positions (Baldassarri and Gelman 2008; Fiorina and Abrams 2008; Layman, Carsey, and Horowitz 2006). As such, the divide between conservatives and Republicans on one side and liberals and Democrats on the other side on beliefs about climate science and concern about global warming was significantly greater in 2010 than in 2001 (McCright and Dunlap 2011b; see also McCright, Dunlap, and Xiao 2014a).

Related to this is what has been termed the political moderator effect, whereby political orientation moderates the relationship between educational attainment and self-reported understanding on one side and climate change views on the other. Greater educational attainment and greater self-reported understanding of global warming have differing effects on climate change views for conservatives and Republicans than for liberals and Democrats (Dunlap and McCright 2008; Hamilton 2008, 2011; Hamilton and Keim 2009; Krosnick et al. 2000; Malka et al. 2009; McCright 2009, 2011; McCright and Dunlap 2011b). That is, the effects of educational attainment and self-reported understanding on climate change views are *positive* for liberals and Democrats but *weaker or negative* for conservatives and Republicans. Several theories or hypotheses—for example, the elite cues hypothesis (Krosnick et al. 2000) and information processing theory (e.g., Wood and Vedlitz 2007), among others—have been invoked to explain this relationship. What they all share in common is the idea that people selectively assimilate information that reinforces their existing political beliefs.

Recent work in this area examines the influence of political orientation on different aspects of climate change denial. For instance, McCright, Dunlap, and Xiao (2013, 2014a) find that self-identified conservatives and Republicans are more likely than their liberal and Democratic counterparts to believe there is no scientific agreement on the reality of human-caused climate change, which then has a significant negative effect on support for government action. Further, several studies find that political conservatives are more likely than liberals to espouse explicitly denialist views about climate change (McCright and Dunlap 2011a; Poortinga et al. 2011; Whitmarsh 2011). More specifically, McCright and Dunlap (2011a) document what they refer to as the "conservative white male" effect, whereby conservative white males are more likely to deny the reality and seriousness of climate change than others in the general public.

The most comprehensive attempt at a theoretical explanation of climate change views is the values-beliefs-norms theory of environmental concern and behavior. This is a broad theory of environmental concern that suggests that values influence general beliefs about the environment, which in turn shape beliefs about the consequences of environmental change on what is valued. These specific beliefs about the threat to objects valued then affect perceptions about the ability to reduce those threats, which in turn influences norms about taking action (Stern et al. 1999, 1995). A few studies provide empirical support for part of this model. For instance, environmental values and identity have a positive effect on climate change views (e.g., Brody et al. 2008; Kellstedt et al. 2008; Leiserowitz 2006; McCright and Dunlap 2011b), and beliefs about climate change consequences have a positive effect on support for climate policy proposals (Brody et al. 2008; McCright 2009; O'Connor et al. 2002; Zahran et al. 2006). An experimental survey of British Canadian citizens found that provision of additional information on the effectiveness of local climate change policies did not change the respondents' levels of climate change policy support, further supporting that factors other than knowledge (e.g., values and trust) are driving climate change views (Rhodes et al. 2014). Dietz et al. (2007) offer the most encompassing test of values-beliefs-norms theory vis-à-vis climate change views. They find that personal values—more specifically altruistic and traditional values—influence various worldviews (e.g., new ecological paradigm), beliefs (e.g., trust in different institutions), and realization of the consequences of climate change, which are then associated with support for policies to reduce GHG emissions.

A group of scholars from outside of sociology draw from the cultural theory of risk (e.g., Douglas and Wildavsky 1983) to explain how cultural factors influence climate change views. They argue that where individuals lie on core value dimensions of the hierachicalism–egalitarianism and individualism–communitarianism spectrums drives their formation of risk perceptions—and, by extension, climate change views. Referring to this as cultural cognition theory, Kahan and colleagues (Kahan, Jenkins-Smith, and Braman 2011; Kahan et al. 2012) document that individuals form views about climate change in ways structured by their cultural values.

These social psychological theories suggest that we see differentiation in public opinion because of individual-level characteristics that are socially influenced and defined. However, spatial variation in public opinion and shifts in public opinion over time may be driven more by shifts in other factors such as economic conditions, media discourse, new scientific findings, or even climate change itself. (For more detailed description, we refer readers to Chapter 12 of this volume on methodological challenges for climate change research.)

Only a few studies attempt to examine the influence of economic factors on climate change views, each using aggregated public opinion data. Analyzing ten years of polling data combined from several survey organizations, Brulle, Carmichael, and Jenkins (2012) find that the unemployment rate is negatively related and GDP is positively related to aggregated public concern about climate change. A second study provides empirical support for the negative effect of the unemployment rate on climate change views, yet its analyses are only based on data from 2008 and 2009 (Scruggs and Benegal 2012). A third study, analyzing public opinion data for all EU countries in 2008 and 2009, finds that the GDP per capita and quarterly economic growth both are positively related to the percentage of respondents in an EU country who are very worried about climate change (Shum 2012).

The original call for environmental sociology was to study human–environment interaction (Catton Jr. and Dunlap 1978). This has come to fruition in those studies of climate change views that investigate the influence of weather phenomena and climatic patterns in recent years. We divide up the relevant studies into three distinct groups, arranged by their general approach.

One group of studies uses survey items on perceptions of short-term temperature changes (Capstick and Pidgeon 2014; Krosnick et al. 2006; Semenza et al. 2008a), self-reported experiences with climate-related phenomena like air pollution and flooding (Spence et al. 2011; Whitmarsh 2008), or self-reported personal experience with global warming (Akerlof et al. 2013; Myers et al. 2013) and link those to climate change views. They demonstrate that perceived indicators of climate/weather phenomena are only modestly related to climate change views. Further, several of these studies have limited generalizability, with samples only of specific cities (Semenza et al. 2008a) or counties (Akerlof et al. 2013; Whitmarsh 2008).

People who perceive an increase in local temperatures in recent years are more likely to believe that global warming exists (Krosnick et al. 2006), but those self-reporting having experienced global warming do not perceive a greater local global warming risk (Akerlof et al. 2013). In a study examining data from two cities, those who perceive greater heat on the previous day express greater concern about climate change in the Southern city but not in one in the Pacific Northwest (Semenza et al. 2008a). In a survey of a nationally representative sample in the United Kingdom, three times as many people report that the pattern of recent cold winters is evidence for rather than against the existence of climate change (Capstick and Pidgeon 2014). Reporting that air pollution has affected your health increases the likelihood of believing climate change is real, seeing it as personally important, and perceiving that it will affect you personally, but it has no effect on whether or not you regularly take any action out of concern for climate change (Whitmarsh 2008). Experiencing flooding in the last five years increases the likelihood of seeing climate change as personally important, but it has no effect on the likelihood of taking any action out of concern for climate change (Whitmarsh 2008) or the preparedness to reduce energy use (Spence et al. 2011). While one study finds that having personally experienced a local flood increases your concern about climate change, your certainty that climate change is happening, and the perceived vulnerability of your local area (Spence et al. 2011), another reports no such influence on the likelihood of believing that climate change is real, seeing it as personally important, or regularly taking action out of concern for climate change (Whitmarsh 2008).

A second group of studies, which analyzes the relationship between weather and climate phenomena and aggregated public opinion data, yields mixed results (Brulle et al. 2012; Donner and McDaniels 2013; Scruggs and Benegal 2012; Shum 2012). Two find a significant influence in the expected direction (Donner and McDaniels 2013; Scruggs and Benegal 2012), while two do not (Brulle et al. 2012; Shum 2012). Nevertheless, in aggregating individual-level survey data to averages either for an entire survey (Brulle et al. 2012; Donner and McDaniels 2013; Scruggs and Benegal 2012) or country (Shum 2012), important variation at the individual level—which has long formed the core of public opinion research on climate change views—is lost (Raudenbush and Bryk 2002; Snijders 2011). Thus, this group of studies cannot help us understand the relationship between climate phenomena and individuals' climate change views. But these studies can help us more fully understand the political and cultural context that can illuminate public opinion trends at the national level and reflect "policy moods" that likely shape national policy.

The Brulle et al. (2012) study discussed earlier finds that political and economic factors—but not climate phenomena (as measured by the National Oceanic and Atmospheric Administration [NOAA]'s Climate Extremes Index)—affect public concern about climate change. A second study finds partial support for the effects of seasonal and global temperature anomalies on the belief that global warming is happening (Scruggs and Benegal 2012). Yet, as we mentioned, its analyses are only based on data from 2008 and 2009. A third study, analyzing public opinion data for all EU countries in 2008 and 2009, finds that the mean temperature of the previous August has no influence on the percentage of respondents in an EU country who are very worried about climate change (Shum 2012). A fourth study reports that the mean temperature anomaly over the previous year correlates with the belief that the climate is warming and whether people worry about climate change (Donner and McDaniels 2013).

A third group of studies use statewide, regionally, or nationally representative survey data to examine how actual short-term or long-term temperature or climatic patterns relate to citizens' climate change views. Some of these studies integrate data for several physical context variables such as local heat wave exposure (Egan and Mullin 2012); drought severity and precipitation (Goebbert et al. 2012); sea level rise/inundation risk, flooding vulnerability, and natural hazard casualties (Brody et al. 2008; Zahran et al. 2006;); and variation in climate extremes, as comprehensively measured by NOAA's Climate Extremes Index (Marquart-Pyatt et al. 2014). Most use long-term temperature trend data (Brody et al. 2008; Egan and Mullin 2012; Goebbert et al. 2012; Hamilton and Keim 2009; Hamilton and Stampone 2013a; McCright, Dunlap, and Xiao 2014b; Scruggs and Benegal 2012; Shao et al. 2014; Zahran et al. 2006).

While vulnerability to sea level rise has an (unexpected) negative effect on climate policy support, the reported injuries and fatalities from natural hazards in a locality have no influence on climate policy support (Zahran et al. 2006). Also, vulnerability to sea level rise and proximity to a coast increase climate change risk perceptions (Brody et al. 2008). Recent exposure to a heat wave increases the likelihood of believing there is solid evidence that the Earth is getting warmer (Egan and Mullin 2012). As local drought severity increases, the likelihood of perceiving that local droughts have been more frequent also increases; further, as precipitation and soil moisture increase, the likelihood of perceiving that local floods have been more frequent also increases (Goebbert et al. 2012). Exposure to regional climate extremes does not influence perceptions of the timing of climate change and has only a negligible effect on perceptions about the seriousness of climate change (Marquart-Pyatt et al. 2014).

Of those studies using long-term temperature trend indicators, three find no significant influence on climate change views (Brody et al. 2008; Goebbert et al. 2012; McCright, Dunlap, and Xiao 2014b), while seven do (Egan and Mullin 2012; Hamilton and Keim 2009; Hamilton and Stampone 2013a; Howe et al. 2013; Scruggs and Benegal 2012; Shao et al. 2014; Zahran et al. 2006)—including three that use a multilevel modeling approach (Egan and Mullin 2012; Hamilton and Keim 2009; Howe et al. 2013).

Part of the challenge in making sense of these results is the variation in how scholars have created these long-term temperature trend indicators. While most use a baseline of thirty years to establish a "normal" local temperature, there seems to be no standard for comparison. Scholars have used the following as their indicators:

- whether or not survey respondents live in a NOAA climatic division with a statistically significant correlation ($p < 0.05$) between year and the number of days exceeding average temperature from 1948 to 2005 (Brody et al. 2008; Zahran et al. 2006);
- thirty-eight year (1970–2007) winter temperature trend (Hamilton and Keim 2009);
- deviation of normal daily local temperature—averaged over the week prior to the date of the survey interview—from the average local temperature for the survey date calculated between 1971 and 2000 (Egan and Mullin 2012; Scruggs and Benegal 2012);
- deviation of local temperatures in the past few years (prior three-year average) from the prior thirty-year average (Goebbert et al. 2012);
- average local temperature anomaly for the day of the interview and the prior day relative to the 1981–2010 average (Hamilton and Stampone 2013a);

- standard deviation of the mean temperature/precipitation for the month prior to the survey by the monthly mean temperature/precipitation from 1981 to 2010 (Shao et al. 2014); and
- deviation of recent average temperature (prior twelve-month average for the twelve months leading up to the survey date) from the 1961–1990 average temperature (Howe et al. 2013).

A further complicating factor is that while all of the studies in this third group use some sort of geo-coding to spatially locate survey respondents, only four use multilevel modeling to appropriately model the clustering of respondents within larger contextual units (Egan and Mullin 2012; Hamilton and Keim 2009; Howe et al. 2013; McCright, Dunlap, and Xiao 2014b). Further, the choice of contextual units employed varies. That is, studies spatially locate their survey respondents differently, by ZIP code (Egan and Mullin 2012; Goebbert et al. 2012; Scruggs and Benegal 2012), county (Hamilton and Keim 2009; Shao et al. 2014), NOAA climatic division (Brody et al. 2008; Zahran et al. 2006), state (McCright, Dunlap, and Xiao 2014b), and nation (Howe et al. 2012; Shum 2012).

Finally, several studies in this third group do not control for what are regularly the most powerful individual-level predictors: political ideology and party identification (Brody et al. 2008; Howe et al. 2013; Zahran et al. 2006). Among those that include an indicator of political orientation, all find a statistically significant effect of either political ideology (Egan and Mullin 2012; Goebbert et al. 2012; Scruggs and Benegal 2012; Shao et al. 2014) or party identification (Egan and Mullin 2012; Hamilton and Keim 2009; Hamilton and Stampone 2013a; Scruggs and Benegal 2012; Shao et al. 2014) or both (McCright, Dunlap, and Xiao 2014b). Further, two studies find that political orientation moderates the influence that recent local warming has on climate change perceptions (Egan and Mullin 2012; Hamilton and Stampone 2013a). Briefly, strongly partisan respondents are least influenced by local temperature increases. The relationship between recent local warming and the belief that global warming is real is stronger for people who are less partisan (e.g., "leaning" Democrats or Republicans) than it is for their more partisan counterparts (e.g., "strong" Democrats or Republicans) (Egan and Mullin 2012). Also, while Independents' belief that recent global warming is caused mainly by human activities is positively related to recent temperature anomalies, such recent temperature variation has little effect on such belief of Democrats and Republicans (Hamilton and Stampone 2013a).

Those studies that examine the larger contextual factors that influence climate change views are a welcome addition to the studies that consider only individual-level predictors. However, the results raise a critical question: Are the factors similar or different in other countries? We do not know.

In fact, we know very little about how the socio-political-cultural context of specific nation-states influences the climate change views of their citizens (Brechin 2010). From numerous cross-national surveys we have detailed records of climate change views, but we know much less about what is shaping those opinions.

SUGGESTIONS FOR A FUTURE RESEARCH AGENDA

This chapter has provided an overview of the state of the knowledge of international and domestic public opinion on climate change. For the most part, we have focused on climate change views as an outcome of various forces and predictors. We now turn to consider climate change public opinion as a potential predictor of other outcomes—in other words, how public opinion matters. While scholars of public opinion have studied this in general, very little sociological research specifically has examined how climate change public opinion matters. We identify four research questions to drive this area of scholarship in the future.

Our first question asks how individuals' climate change views translate into household actions to mitigate climate change. Dietz et al. (2009) suggest targeting the public directly to change citizens' household behaviors to mitigate climate change. The question of the extent to which climate change views influence household behaviors is a specific manifestation of a focus on the attitude–behavior gap—the finding that a significant portion of people who have high levels of environmental concern often will continue to behave in a manner detrimental to the environment (Kollmuss and Agyeman 2002; Vermeir and Verbeke 2006). Scholars explain this gap by highlighting constraints on time and money, the difficulty in breaking habits, and the fact that many behaviors are locked in by previous decisions (for a review see Shwom and Lorenzen 2012 and Chapter 4 in this volume).

Semenza and colleagues (2008a, 2008b) find that concern about climate change is positively related to reporting having actually changed behavior to mitigate climate change. Further, they argue that various impediments to behavioral changes need to be addressed by governments and industry. More research in this area can advance our understanding of those dimensions of climate change views (i.e., awareness of climate change, belief in the reality and human cause of climate change, level of worry or concern about climate change, awareness of the consequences of climate change, support for climate policy) that are necessary (though not sufficient) for enabling behavioral change in households.

The second important future research question concerns the influence that climate change views have on individuals' participation in political activities or collective action to address climate change. The qualitative work of Lorenzoni, Nicholson-Cole, and Whitmarsh (2007) suggests that many people perceive climate change to be a collective problem that is not effectively addressed by the behavioral changes of individuals. At the same time, their work also suggests that many people are hesitant to spend much time politically organizing to act collectively because of the belief that the government is mostly unwilling to act. Some sociological research examines the factors that facilitate or inhibit individuals' participation in social movements (DeMartini 1983; Klandermans 1993, 1997; Snow, Zurcher Jr., and Ekland-Olson 1980; Snow et al. 1986). Some of this work focuses on the environmental movement (e.g., Stern et al. 1999), but no major research efforts to date focus on the roles of climate change views in influencing participation in collective action or political involvement vis-à-vis climate change.

Our third suggested question for future research concerns the extent to which climate change public opinion influences public policy through our democratic processes. Currently, there is a debate between scholars who believe that the general relationship between public opinion and public policy is strong, reflecting a robust belief in democratic values and processes, and those who say the effect is weak, reflecting an opposite view where public policies are created through less democratic forces (see Burstein 2010; Burstein and Linton 2002; Erikson, MacKuen, and Stimson 2002; Manza, Cook, and Page 2002). Existing research seems to support the "strong effect" assertion (see Erikson, Wright, and McIver 1993; Stimson et al. 1995).

Yet, little work has been done to understand how public opinion on climate change influences climate policy. In looking at environmental agreement ratification cross-nationally, Roberts, Parks, and Vásquez (2004) find that civil society (numbers of domestic nongovernmental organizations) and the extent that civil society can voice and act on their opinions on government elections and through media (measured by the Voice and Accountability index) are significant predictors of environmental treaty ratification. Analyzing data from the International Social Survey Program's 2000 Environment module for twenty-six nations (most of which were developed nations), Tjernström and Tietenberg (2008:19) conclude that the percentage of a nation's population perceiving "a rise in the world's temperature caused by the 'greenhouse effect' as either extremely or very dangerous for the environment" (which they treat as a measure of national public concern about climate change) is correlated with decreases in national GHG emissions from 1997 to 2000, but they did not examine policy adoption. In the case of climate change, it is likely important to address other key forces that influence policymaking, such as state–business relations—in particular, the size, structure, and cohesion of a country's industrial sector (Levy and Egan 1998; Shwom 2011).

Our fourth question for future research concerns the extent to which climate change public opinion can influence corporate actions on climate change directly. While sociologists typically have paid attention to how public opinion matters for policy outcomes, business researchers have begun to identify how public opinion influences corporate behavior without state intervention. Studies to date focus on corporate reputation and its relationship to financial performance (De la Fuente Sabat and De Quevedo Puente 2003; Kolk and Levy 2001). Some scholars (e.g., Hoffman 2005; Kolk and Levy 2001) argue that some corporations are responding to public concerns about climate change and voluntarily addressing climate change issues, at least in part because of concern for their corporate reputation.

Interviews with upper management in U.S. and European car manufacturing companies reveal how the European car manufacturing branches rejected the strategy of disputing climate science, citing that the European public was more likely to trust and accept climate science than was the American public (Levy and Rothenberg 2002). The extent to which climate change public opinion influences corporate actions, and the types of corporate or industry characteristics that may shape such responsiveness, are further questions worth attention.

Knowledge of the sociological dimensions of climate change public opinion has accumulated over the past three decades, and this body of research has provided important insights for the broader sociology of public opinion. First, the study of climate change public opinion highlights how multidimensional public opinion has become in the information age. That is, it highlights the extent of the contested nature of knowledge and science for issues that have varying levels of certainty. The belief that a phenomenon exists and is a problem is a primary dimension of public opinion along with actual public support for specific public policies to remedy the problem. Second, cross-national studies demonstrate how organized political interests and media coverage have real impacts on public knowledge and belief in climate science. Third, studies of climate change public opinion highlight that while political ideology and identity seem to be increasingly important factors in politics, the material world matters. This is demonstrated both by the role of economic factors (e.g., changes in GDP or unemployment) in moving mass opinion and the significance of people's experiences with weather and natural disasters in influencing their climate change views.

To advance and refine the sociology of public opinion on climate change, researchers need to be able to analyze variations in climate change public opinion over time and space with high-quality measures for all factors believed to influence public opinion. Coordination between researchers and funders to attain high-quality longitudinal surveys with large representative samples from a wide range of nations would be a logical next step in advancing the state of knowledge on climate change public opinion. In addition, data to study the theorized impacts of public opinion should be collected. For example, studies should include survey measures on household behaviors and political activities. To examine mass opinion outcomes, public opinion survey data should be paired with complementary data sets like corporate adoption rates of voluntary GHG emissions reporting for companies headquartered in different nations or policy outcomes at various scales. These steps would enable a robust future for climate change public opinion research and continued advances in the field of sociology of public opinion.

NOTES

1. While "global warming" and "climate change" are different, but related, scientific phenomena, many scholars and most journalists and policymakers use the terms interchangeably. We also use the terms interchangeably in this chapter.

2. Since little of the literature on public opinion on climate change invokes key conceptual dimensions of public opinion (e.g., attitude strength/intensity, centrality/salience), we abstain from using such technical concepts in this chapter (see Dunlap 1995; Schuman and Presser 1981).

3. For a comprehensive review of insights into climate change knowledge, perceptions, and engagement from smaller sample studies and alternative approaches (e.g., qualitative studies and focus groups), see Wolf and Moser (2011).

4. See Brechin (2010), Brechin and Bhandari (2011), and Leiserowitz (2009).

5. Willingness to pay for an item represents a "trade-off" question designed to explore the strength of public opinion.

REFERENCES

Akerlof, Karen, Edward W. Maibach, Dennis Fitzgerald, Andrew Y. Cedeno, and Amanda Neuman. 2013. "Do People 'Personally Experience' Global Warming, and If So How, and Does It Matter?" *Global Environmental Change* 23(1):81–91.

Baldassarri, Delia and Andrew Gelman. 2008. "Partisans without Constraint: Political Polarization and Trends in American Public Opinion." *American Journal of Sociology* 114(2):408–446.

Berinsky, Adam. 1999. "The Two Faces of Public Opinion." *American Journal of Political Science* 43(4):1209–1230.

Bord, Richard J., Ann Fisher, and Robert E. O'Connor. 1998. "Public Perceptions of Global Warming: United States and International Perspectives." *Climate Research* 11(1):75–84.

Borick, Christopher P. and Barry G. Rabe. 2010. "A Reason to Believe: Examining the Factors that Determine Individual Views on Global Warming." *Social Science Quarterly* 91:777–800.

Bostrom, Ann., M. Granger Morgan, Baruch Fischhoff, and Daniel Read. 1994. "What Do People Know About Global Climate Change? 1. Mental Models." *Risk Analysis* 14(6):959–970.

Brechin, Steven R. 1999. "Objective Problems, Subjective Values, and Global Environmentalism: Evaluating the Postmaterialist Argument and Challenging a New Explanation." *Social Science Quarterly* 80(4):793–809.

Brechin, Steven R. 2003. "Comparative Public Opinion and Knowledge on Global Climatic Change and the Kyoto Protocol: The US Versus the World?" *International Journal of Sociology and Social Policy* 23(10):106–134.

Brechin, Steven R. 2010. "Public Opinion: A Cross-National View." Pp. 179 in *Routledge Handbook of Climate Change and Society*, edited by C. Lever-Tracy. New York: Routledge.

Brechin, Steven R. and Medani Bhandari. 2011. "Perceptions of Climate Change Worldwide." *Wiley Interdisciplinary Reviews: Climate Change* 2(6):871–885.

Brechin, Steven R. and Willett Kempton. 1994. "Global Environmentalism: A Challenge to the Postmaterialism Thesis?" *Social Science Quarterly* 75(2):245–369.

Brewer, Thomas L. 2005. "US Public Opinion on Climate Change Issues: Implications for Consensus-Building and Policymaking." *Climate Policy* 4(4):359–376.

Brody, Samuel D., Sammy Zahran, Arnold Vedlitz, and Himanshu Grover. 2008. "Examining the Relationship between Physical Vulnerability and Public Perceptions of Global Climate Change in the United States." *Environment and Behavior* 40(1):72–95.

Brulle, Robert. J., Jason Carmichael, and J. Craig Jenkins. 2012. "Shifting Public Opinion on Climate Change: An Empirical Assessment of Factors Influencing Concern over Climate Change in the US, 2002–2010." *Climatic Change* 114(2):169–188.

Burstein, Paul. 2010. "Public Opinion, Public Policy, and Democracy." Pp. 63–79 in *Handbook of Politics: State and Society in Global Perspectives*, edited by K. T. Leicht and J. Craig Jenkins. New York: Springer.

Burstein, Paul and April Linton. 2002. "The Impact of Political Parties, Interest Groups, and Social Movement Organizations on Public Policy: Some Recent Evidence and Theoretical Concerns." *Social Forces* 81(2):380–408.

Capstick, Bryce Stuart and Nicholas Frank Pidgeon. 2014. "Public Perception of Cold Weather Events as Evidence for and against Climate Change." *Climatic Change* 122:695–708.

Catton Jr., William R. and Riley E. Dunlap. 1978. "Environmental Sociology: A New Paradigm." *American Sociologist* 13(1):41–49.

Chodorow, Nancy. 1978. *The Reproduction of Mothering*. Berkeley: University of California Press.

De la Fuente Sabat, Juan Manuel, and Esther de Quevedo Puente. 2003. "Empirical Analysis of the Relationship between Corporate Reputation and Financial Performance: A Survey of the Literature." *Corporate Reputation Review* 6(2):161–177.

DeMartini, Joseph R. 1983. "Social Movement Participation: Political Socialization, Generational Consciousness, and Lasting Effects." *Youth and Society* 15(2):195–223.

Dietz, Thomas, Amy Dan, and Rachael Shwom. 2007. "Support for Climate Change Policy: Social Psychological and Social Structural Influences." *Rural Sociology* 72(2):185–214.

Dietz, Thomas, Amy Fitzgerald, and Rachael Shwom. 2005. "Environmental Values." *Annual Review of Environment & Resources* 30(12):1–38.

Dietz, Thomas, Gerald T. Gardner, Jonathan Gilligan, Paul C. Stern, and Michael P. Vandenbergh. 2009. "Household Actions Can Provide a Behavioral Wedge to Rapidly Reduce US Carbon Emissions." *Proceedings of the National Academy of Sciences USA* 106(44):18452–18456.

Donner, Simon D. and Jeremy McDaniels. 2013. "The Influence of National Temperature Fluctuations on Opinions About Climate Change in the US since 1990." *Climatic Change* 118(3–4):537–550.

Dorsch, Michael T. 2014. "Economic Development and Determinants of Environmental Concern." *Social Science Quarterly* doi: 10.1111/ssqu.12071.

Douglas, Mary and Aaron Wildavsky. 1983. *Risk and Culture: An Essay on the Selection of Technical and Environmental Dangers.* Berkeley: University of California Press.

Dunlap, Riley E. 1995. "Public Opinion and Environmental Policy." Pp. 63–114 in *Environmental Politics and Policy: Theories and Evidence*, edited by J. Lester. Durham, NC: Duke University Press.

Dunlap, Riley E. 1998. "Lay Perceptions of Global Risk Public Views of Global Warming in Cross-National Context." *International Sociology* 13(4):473–498.

Dunlap, Riley E. 2014. "Global Warming or Climate Change: Is There a Difference?" Gallup Politics. Retrieved April 22, 2014 (http://www.gallup.com/poll/168617/global-warming-climate-change difference.aspx?utm_source = WWW&utm_medium = csm&utm _campaign = syndication).

Dunlap, Riley E., George H. Gallup Jr., and Alec M. Gallup. 1993. "Of Global Concern: Results of the Health of the Planet Survey." *Environment: Science and Policy for Sustainable Development* 35(9):7–39.

Dunlap, Riley E. and A. M. McCright. 2008. "A Widening Gap: Republican and Democratic Views on Climate Change." *Environment: Science and Policy for Sustainable Development* 50(5):26–35.

Dunlap, Riley E. and Kent D. Van Liere. 1984. "Commitment to the Dominant Social Paradigm and Concern for Environmental Quality." *Social Science Quarterly* 65(4):1013–1028.

Dunlap, Riley E., Chenyang Xiao, and Aaron M. McCright. 2001. "Politics and Environment in America: Partisan and Ideological Cleavages in Public Support for Environmentalism." *Environmental Politics* 10(4):23–48.

Dunlap, Riley E. and Richard York. 2008. "The Globalization of Environmental Concern and the Limits of the Postmaterialist Values Explanation: Evidence from Four Multinational Surveys." *Sociological Quarterly* 49(3):529–563.

Egan, Patrick J. and Megan Mullin. 2012. "Turning Personal Experience into Political Attitudes: The Effect of Local Weather on Americans' Perceptions About Global Warming." *Journal of Politics* 74(03):796–809.

Erikson, Robert S., Michael B. MacKuen, and James A. Stimson. 2002. "Public Opinion and Policy: Causal Flow in a Macro System Model." Pp. 33–53 in *Navigating Public Opinion: Polls, Policy, and the Future of American Democracy*, edited by J. Manza, F. L. Cook, and B. I. Page. New York: Oxford University Press.

Erikson, Robert S., Gerald C. Wright, and John P. McIver. 1993. *Statehouse Democracy: Public Opinion and Policy in the American States.* New York: Cambridge University Press.

Eurobarometer. 2009. "Europeans Attitudes toward Climate Change." Brussels, Belgium: European Commission. Retrieved August 31, 2011 (http://ec.europa.eu/public_opinion/archives/ebs/ebs_300_full_en.pdf).

Fiorina, Morris P. and Samuel J. Abrams. 2008. "Political Polarization in the American Public." *Annual Review of Political Science* 11:563–588.

Franzen, Axel and Dominikus Vogl. 2013. "Two Decades of Measuring Environmental Attitudes: A Comparative Analysis of 33 Countries." *Global Environmental Change* 23(5):1001–1008.

Gilligan, Carol. 1982. *In a Different Voice: Psychological Theory and Women's Development.* Cambridge, MA: Harvard University Press.

Goebbert, Kevin, Hank C. Jenkins-Smith, Kim Klockow, Matthew C. Nowlin, and Carol L. Silva. 2012. "Weather, Climate, and Worldviews: The Sources and Consequences of Public Perceptions of Changes in Local Weather Patterns." *Weather, Climate, and Society* 4(2):132–144.

Guha, Ramachandra and Juan Martinez-Alier. 1997. *Varieties of Environmentalism: Essays North and South.* New York: Earthscan Publications Ltd.

Hamilton, Lawrence C. 2008. "Who Cares About Polar Regions? Results from a Survey of US Public Opinion." *Arctic, Antarctic, and Alpine Research* 40(4):671–678.

Hamilton, Lawrence C. 2011. "Education, Politics and Opinions About Climate Change Evidence for Interaction Effects." *Climatic Change* 104(2):231–242.

Hamilton, Lawrence C. 2012. "Did the Arctic Ice Recover? Demographics of True and False Climate Facts." *Weather, Climate, and Society* 4(4):236–249.

Hamilton, Lawrence C. and Barry D. Keim. 2009. "Regional Variation in Perceptions About Climate Change." *International Journal of Climatology* 29(15):2348–2352.

Hamilton, Lawrence C. and Mary D. Stampone. 2013a. "Blowin' in the Wind: Short-Term Weather and Belief in Anthropogenic Climate Change." *Weather, Climate, & Society* 5:112–119.

Hamilton, Lawrence C. and Mary D. Stampone. 2013b. "Arctic Warming and Your Weather: Public Belief in the Connection." *International Journal of Climatology* doi: 10.1002/jov.3796.

Hindman, Douglas Blanks. 2009. "Mass Media Flow and Differential Distribution of Politically Disputed Beliefs: The Belief Gap Hypothesis." *Journalism & Mass Communication Quarterly* 86(4):790–808.

Hoffman, Andrew J. 2005. "Climate Change Strategy: The Business Logic behind Voluntary Greenhouse Gas Reductions." *California Management Review* 47(3):21–46.

Howe, Peter D., Ezra M. Markowitz, Tien Ming Lee, Chia-Ying Ko, and Anthony Leiserowitz. 2013. "Global Perceptions of Local Temperature Change." *Nature Climate Change* 3(4):352–356.

Inglehart, Ronald. 1997. *Modernization and Post-modernization: Cultural, Economic, and Political Change in 43 Societies.* Princeton, NJ: Princeton University Press.

Jones, Jeffrey. 2011. "In U.S., Concerns About Global Warming Stable at Lower Levels." *Gallup Politics*: Gallup. Retrieved August 1, 2013 (http://www.gallup.com/poll/146606/concerns-global-warming-stable-lower-levels.aspx).

Jones, Jeffrey. 2014. "In U.S., Most Do Not See Global Warming as Serious Threat." *Gallup Politics:* Gallup. Retrieved March 20, 2014 (http://www.gallup.com/poll/167879/not-global-warming-serious-threat.aspx).

Jones, Robert Emmet and Riley E. Dunlap. 1992. "The Social Bases of Environmental Concern: Have They Changed over Time?" *Rural Sociology* 57(1):28–47.

Kahan, Dan M., Hank Jenkins-Smith, and Donald Braman. 2011. "Cultural Cognition of Scientific Consensus." *Journal of Risk Research* 14(2):147–174.

Kahan, Dan M., Ellen Peters, Maggie Wittlin, Paul Slovic, Lisa Larrimore Ouellette, Donald Braman, and Gregory Mandel. 2012. "The Polarizing Impact of Science Literacy and Numeracy on Perceived Climate Change Risks." *Nature Climate Change* 2(10):732–735.

Keller, Evelyn Fox. 1995. *Reflections on Gender and Science*. New Haven, CT: Yale University Press.

Kellstedt, Paul M., Sammy Zahran, and Arnold Vedlitz. 2008. "Personal Efficacy, the Information Environment, and Attitudes toward Global Warming and Climate Change in the United States." *Risk Analysis* 28(1):113–126.

Kim, So Young and Yael Wolinsky-Nahmias. 2014. "Cross-National Public Opinion on Climate Change: The Effects of Affluence and Vulnerability." *Global Environmental Politics* 14(1):79–106.

Klandermans, Bert. 1993. "A Theoretical Framework for Comparisons of Social Movement Participation." *Sociological Forum* 8(3):383–402.

Klandermans, Bert. 1997. *The Social Psychology of Protest*. Oxford: Blackwell Publishing.

Kolk, Ans and David Levy. 2001. "Winds of Change: Corporate Strategy, Climate Change and Oil Multinationals." *European Management Journal* 19(5):501–509.

Kollmuss, Anja and Julian Agyeman. 2002. "Mind the Gap: Why Do People Act Environmentally and What Are the Barriers to Pro-Environmental Behavior?" *Environmental Education Research* 8(3):239–260.

Krosnick, J. A., Allyson L. Holbrook, Laura Lowe, and Penny S. Visser. 2006. "The Origins and Consequences of Democratic Citizens' Policy Agendas: A Study of Popular Concern About Global Warming." *Climatic Change* 77(1):7–43.

Krosnick, Jon A., Allyson L. Holbrook, and Penny S. Visser. 2000. "The Impact of the Fall 1997 Debate about Global Warming on American Public Opinion." *Public Understanding of Science* 9:239–260.

Krosnick, Jon A. and Bo MacInnis. 2011. "National Survey of American Public Opinion on Global Warming." Stanford, CA: Stanford University with Ipsos and Reuters.

Krosnick, Jon A., Penny S. Visser, and Allyson L. Holbrook. 1998. "American Opinion on Global Warming." *Resources* 133:5–9.

Kvaløy, Berit, Henning Finseraas, and Ola Listhaug. 2012. "The Publics' Concern for Global Warming: A Cross-National Study of 47 Countries." *Journal of Peace Research* 49(1):11–22.

Layman, Geoffrey C., Thomas M. Carsey, and Juliana Menasce Horowitz. 2006. "Party Polarization in American Politics: Characteristics, Causes, and Consequences." *Annual Review of Political Science* 9:83–110.

Leiserowitz, Anthony. 2006. "Climate Change Risk Perception and Policy Preferences: The Role of Affect, Imagery, and Values." *Climatic Change* 77(1):45–72.

Leiserowitz, Anthony. 2009. "Have You Ever Heard of Climate Change?" Yale Project on Climate Change Communication. New Haven, CT: Yale University. Retrieved July 23, 2012 (http://environment.yale.edu/climate/item/have-you-ever-heard-of-climate-change).

Leiserowitz, Anthony, Edward W. Maibach, Connie Roser-Renouf, Geoff Feinberg, and Seth Rosenthal. 2014. *Public Support for Climate and Energy Policies in November 2013* (http://ssrn.com/abstract = 2410649).

Leiserowitz, Anthony, Edward W. Maibach, Connie Roser-Renouf, and Jay D. Hmielowski. 2012. "Climate Change in the American Mind: Public Support for Climate & Energy Policies in March 2012." *Yale Project on Climate Change Communication.* New Haven, CT: Yale University.

Levy, David L. and Daniel Egan. 1998. "Capital Contests: National and Transnational Channels of Corporate Influence on the Climate Change Negotiations." *Politics and Society* 26(3):337–362.

Levy, David L. and S. Rothenberg. 2002. *Heterogeneity and Change in Environmental Strategy: Technological and Political Responses to Climate Change in the Global Automobile Industry.* Stanford, CA: Stanford University Press.

Lorenzoni, Irene, Anthony Leiserowitz, Miguel De Franca Doria, Wouter Poortinga, and Nick F. Pidgeon. 2006. "Cross-National Comparisons of Image Associations with 'Global Warming' and 'Climate Change' Among Laypeople in the United States of America and Great Britain." *Journal of Risk Research* 9(03):265–281.

Lorenzoni, Irene, Sophie Nicholson-Cole, and Lorraine Whitmarsh. 2007. "Barriers Perceived to Engaging with Climate Change among the UK Public and Their Policy Implications." *Global Environmental Change* 17(3):445–459.

Maibach, Edward W., Anthony Leiserowitz, Connie Roser-Renouf, and C. K. Mertz. 2011. "Identifying Like-Minded Audiences for Global Warming Public Engagement Campaigns: An Audience Segmentation Analysis and Tool Development." *PLoS One* 6(3):e17571.

Malka, Ariel, Jon A. Krosnick, and Gary Langer. 2009. "The Association of Knowledge with Concern About Global Warming: Trusted Information Sources Shape Public Thinking." *Risk Analysis* 29(5):633–647.

Manza, J., F. L. Cook, and B. I. Page. 2002. *Navigating Public Opinion: Polls, Policy, and the Future of American Democracy.* New York: Oxford University Press, USA.

Marquart-Pyatt, Sandra T. 2007. "Concern for the Environment among General Publics: A Cross-National Study." *Society and Natural Resources* 20(10):883–898.

Marquart-Pyatt, Sandra T. 2008. "Are There Similar Sources of Environmental Concern? Comparing Industrialized Countries." *Social Science Quarterly* 89(5):1312–1335.

Marquart-Pyatt, Sandra T. 2012. "Contextual Influences on Environmental Concerns Cross-Nationally: A Multilevel Investigation." *Social Science Research* 41(5):1085–1099.

Marquart-Pyatt, Sandra T., Aaron M. McCright, Thomas Dietz, and Riley E. Dunlap. 2014. "Political Orientation Eclipses Climate Extremes for Climate Change Perception." *Global Environmental Change* 29:246–257.

Marquart-Pyatt, Sandra T., Rachael Shwom, Thomas Dietz, Riley E. Dunlap, Stan A. Kaplowitz, Aaron M. McCright, and Steven Zahran. 2011. "Understanding Public Opinion on Climate Change: A Call for Research." *Environment* 53(4):38–42.

McCright, Aaron M. 2009. "The Social Bases of Climate Change Knowledge, Concern, and Policy Support in the US General Public." *Hofstra Law Review* 37(4):1017–1047.

McCright, Aaron M. 2010. "The Effects of Gender on Climate Change Knowledge and Concern in the American Public." *Population and Environment* 32:66–87.

McCright, Aaron M. 2011. "Political Orientation Moderates Americans' Beliefs and Concern about Climate Change." *Climatic Change* 104(2):243–253.

McCright, Aaron M. and Riley E. Dunlap. 2003. "Defeating Kyoto: The Conservative Movement's Impact on U.S. Climate Change Policy." *Social Problems* 50(3):348–373.

McCright, Aaron M. and Riley E. Dunlap. 2010. "Anti-Reflexivity: The American Conservative Movement's Success in Undermining Climate Science and Policy." *Theory, Culture, and Society* 27(2–3):100–133.

McCright, Aaron M. and Riley E. Dunlap. 2011a. "Cool Dudes: The Denial of Climate Change among Conservative White Males in the United States." *Global Environmental Change* 21:1163–1172.

McCright, Aaron M. and Riley E. Dunlap. 2011b. "The Politicization of Climate Change and Polarization in the American Public's Views of Global Warming, 2001–2010." *Sociological Quarterly* 52:155–194.

McCright, Aaron M., Riley E. Dunlap, and Sandra T. Marquart-Pyatt. 2014. "Climate Change and Political Ideology in the European Union." Under Review.

McCright, Aaron M., Riley E. Dunlap, and Chenyang Xiao. 2013. "Perceived Scientific Agreement and Support for Government Action on Climate Change in the USA." *Climatic Change* 119:511–518.

McCright, Aaron M., Riley E. Dunlap, and Chenyang Xiao. 2014a. "Increasing Influence of Party Identification on Perceived Scientific Agreement and Support for Government Action on Climate Change in the USA, 2006–2012." *Weather, Climate, and Society* 6:194–201.

McCright, Aaron M., Riley E. Dunlap, and Chenyang Xiao. 2014b. "Predicting Perceived Winter Warming in the USA." *Nature Climate Change* 4:1077–1081.

Merchant, Carolyn. 1980. *The Death of Nature: Women, Ecology and the Scientific Revolution*. New York: HarperCollins.

Moghariya, Dineshkumar P. and Richard Smardon. 2014. "Rural Perspectives of Climate Change: A Study from Saurastra and Kutch of Western India." *Public Understanding of Science* 23(6):660–667.

Myers, Teresa A., Edward W. Maibach, Connie Roser-Renouf, Karen Akerlof, and Anthony A. Leiserowitz. 2013. "The Relationship between Personal Experience and Belief in the Reality of Global Warming." *Nature Climate Change* 3(4):343–347.

Nisbet, Matthew C. and Teresa Myers. 2007. "The Polls-Trends: Twenty Years of Public Opinion About Global Warming." *Public Opinion Quarterly* 21(3):444–470.

O'Connor, Robert E., Richard J. Bord, and Ann Fisher. 1999. "Risk Perceptions, General Environmental Beliefs, and Willingness to Address Climate Change." *Risk Analysis* 19(3):461–471.

O'Connor, Robert E., Richard J. Bord, Brent Yarnal, and Nancy Wiefek. 2002. "Who Wants to Reduce Greenhouse Gas Emissions?" *Social Science Quarterly* 83(1):1–17.

Page, Benjamin I. and Robert Y. Shapiro. 1992. *The Rational Public: Fifty Years of Trends in Americans' Policy Preferences*. Chicago, IL: University of Chicago Press.

Pew Research Center. 2014. "Deficit Reduction Declines as Policy Priority." Retrieved March 20, 2014 (http://www.people-press.org/files/legacy-pdf/01-27-14%20 Policy%20Priorities%20Release.pdf).

Poortinga, Wouter, Alexa Spence, Lorraine Whitmarsh, Stuart Capstick, and Nick F. Pidgeon. 2011. "Uncertain Climate: An Investigation into Public Scepticism About Anthropogenic Climate Change." *Global Environmental Change* 21(3):1015–1024.

Pugliese, Anita and Julie Ray. 2009a. "Awareness of Climate Change and Threat Vary by Region: Adults in Americas, Europe Most Likely to Be Aware, Perceive Threat." Gallup. Retrieved March 13, 2014 (http://www.gallup.com/poll/124652/ Awareness-Climate-Change-Threat-Vary-Region.aspx).

Pugliese, Anita and Julie Ray. 2009b. "A Heated Debate: Global Attitudes toward Climate Change." *Harvard International Review* 31(3):64–68.

Raudenbush, Stephen W. and Anthony S. Bryk. 2002. *Hierarchical Linear Models: Applications and Data Analysis Methods.* Thousand Oaks, CA: Sage Publications.

Read, Daniel, Ann Bostrom, M. Granger Morgan, Baruch Fischhoff, and Tom Smuts. 1994. "What Do People Know About Global Climate Change? II. Survey Studies of Educated Laypeople." *Risk Analysis* 14(6):971–982.

Reynolds, Travis William, Ann Bostrom, Daniel Read, and M. Granger Morgan. 2010. "Now What Do People Know About Global Climate Change? Survey Studies of Educated Laypeople." *Risk Analysis* 30(10):1520–1538.

Rhodes, Ekaterina, Jonn Axsen, and Mark Jaccard. 2014. "Does Effective Climate Policy Require Well-Informed Citizen Support?" *Global Environmental Change* 29:92–104.

Roberts, J. Timmons, Bradley C. Parks, and Alexis A. Vásquez. 2004. "Who Ratifies Environmental Treaties and Why? Institutionalism, Structuralism and Participation by 192 Nations in 22 Treaties." *Global Environmental Politics* 4(3):22–64.

Running, Katrina. 2013. "World Citizenship and Concern for Global Warming: Building the Case for a Strong International Civil Society." *Social Forces* 92(1):377–399.

Saad, Lydia. 2009. "Water Pollution Americans' Top Green Concern." Gallup. Retrieved February 24, 2014 (http://www.gallup.com/poll/117079/water-pollution-americans-top-green-concern.aspx).

Saad, Lydia. 2013. "Americans' Concerns About Global Warming on the Rise." Gallup. Retrieved April 22, 2014 (http://www.gallup.com/poll/161645/americans-concerns-global-warming-rise.aspx).

Sandvik, Hanno. 2008. "Public Concern over Global Warming Correlates Negatively with National Wealth." *Climatic Change* 90(3):333–341.

Schuldt, Jonathan P., Sara H. Konrath, and Norbert Schwarz. 2011. "'Global Warming' or 'Climate Change'?: Whether the Planet Is Warming Depends on Question Wording." *Public Opinion Quarterly* 75(1):115–124.

Schuman, Howard and Stanley Presser. 1981. *Questions and Answers in Attitude Surveys: Experiments on Question Form, Wording, and Context.* San Diego, CA: Academic Press.

Scruggs, Lyle and Salil Benegal. 2012. "Declining Public Concern About Climate Change: Can We Blame the Great Recession?" *Global Environmental Change* 22(2):505–515.

Semenza, Jan C., David E. Hall, Daniel J. Wilson, Brian D. Bontempo, David J. Sailor, and Linda A. George. 2008a. "Public Perception of Climate Change: Voluntary Mitigation and Barriers to Behavior Change." *American Journal of Preventive Medicine* 35(5):479–487.

Semenza, Jan C., Daniel J. Wilson, Jeremy Parra, Brian D. Bontempo, Melissa Hart, David J. Sailor, and Linda A. George. 2008b. "Public Perception and Behavior Change in Relationship to Hot Weather and Air Pollution." *Environmental Research* 107(3):401–411.

Shao Wanyun, Barry D. Keim, James C. Garand, and Larry C. Hamilton. 2014. "Weather, Climate, and the Economy: Explaining Risk Perceptions of Global Warming, 2001–2010." *Weather, Climate, and Society* 6:119–134.

Shum, Robert Y. 2012. "Effects of Economic Recession and Local Weather on Climate Change Attitudes." *Climate Policy* 12(1):38–49.

Shwom, Rachael L. 2011. "A Middle Range Theorization of Energy Politics: The Struggle for Energy-Efficient Appliances." *Environmental Politics* 20(5):705–726.

Shwom, Rachael, David Bidwell, Amy Dan, and Thomas Dietz. 2010. "Understanding US Public Support for Domestic Climate Change Policies." *Global Environmental Change* 20(3):472–482.

Shwom, Rachael, Amy Dan, and Thomas Dietz. 2008. "The Effects of Information and State of Residence on Climate Change Policy Preferences." *Climatic Change* 90(4):343–358.

Shwom, Rachael and Janet A. Lorenzen. 2012. "Changing Household Consumption to Address Climate Change: Social Scientific Insights and Challenges." *Wiley Interdisciplinary Reviews: Climate Change* 3(5):379–395.

Snijders, Tom A. B. 2011. *Multilevel Analysis: An Introduction to Basic and Advanced Multilevel Modeling*, 2nd ed. Thousand Oaks, CA: Sage.

Snow, David A., E. Burke Rochford Jr., Steven K. Worden, and Robert D. Benford. 1986. "Frame Alignment Processes, Micromobilization, and Movement Participation." *American Sociological Review* 51(4):464–481.

Snow, David A., Louis A. Zurcher Jr., and Sheldon Ekland-Olson. 1980. "Social Networks and Social Movements: A Microstructural Approach to Differential Recruitment." *American Sociological Review* 45(5):787–801.

Spence, Alexa, Wouter Poortinga, Catherine Butler, and Nicholas Frank Pidgeon. 2011. "Perceptions of Climate Change and Willingness to Save Energy Related to Flood Experience." *Nature Climate Change* 1(1):46–49.

Stern, Paul C. 2000. "New Environmental Theories: Toward a Coherent Theory of Environmentally Significant Behavior." *Journal of Social Issues* 56(3):407–424.

Stern, Paul C., Thomas Dietz, Troy Abel, Gregory A. Guagnano, and Linda Kalof. 1999. "A Value-Belief-Norm Theory of Support for Social Movements: The Case of Environmentalism." *Human Ecology Review* 6(2):81–98.

Stern, Paul C., Thomas Dietz, and Linda Kalof. 1993. "Value Orientations, Gender, and Environmental Concern." *Environment and Behavior* 25(5):322–348.

Stern, Paul C., Thomas Dietz, Linda Kalof, and Gregory A. Guagnano. 1995. "Values, Beliefs, and Proenvironmental Action: Attitude Formation toward Emergent Attitude Objects." *Journal of Applied Social Psychology* 25(18):1611–1636.

Stimson, James A., Michael B. MacKuen, and Robert S. Erikson. 1995. "Dynamic Representation." *American Political Science Review* 89(3):543–565.

Swim, Janet K., Paul C. Stern, Thomas J. Doherty, Susan Clayton, Joseph P. Reser, Elke U. Weber, Robert Gifford, and George S. Howard. 2011. "Psychology's Contributions to Understanding and Addressing Global Climate Change." *American Psychologist* 66(4):241–250.

Tjernström, Emilia and Thomas Tietenberg. 2008. "Do Differences in Attitudes Explain Differences in National Climate Change Policies?" *Ecological Economics* 65(2):315–324.

U.S. National Research Council. 2010. *Advancing the Science of Climate Change: America's Climate Choices*. Washington, DC: National Academies Press.

Van Liere, Kent D. and Riley E. Dunlap. 1980. "The Social Bases of Environmental Concern: A Review of Hypotheses, Explanations and Empirical Evidence." *Public Opinion Quarterly* 44(2):181–197.

Vermeir, Iris and Wim Verbeke. 2006. "Sustainable Food Consumption: Exploring the Consumer 'Attitude–Behavioral Intention' Gap." *Journal of Agricultural and Environmental Ethics* 19(2):169–194.

Villar, Ana and Jon A. Krosnick. 2011. "Global Warming Vs. Climate Change, Taxes Vs. Prices: Does Word Choice Matter?" *Climatic Change* 105(1):1–12.

Whitmarsh, Lorraine. 2008. "Are Flood Victims More Concerned About Climate Change Than Other People? The Role of Direct Experience in Risk Perception and Behavioural Response." *Journal of Risk Research* 11(3):351–374.

Whitmarsh, Lorraine. 2009. "What's in a Name? Commonalities and Differences in Public Understanding of 'Climate Change' and 'Global Warming'." *Public Understanding of Science* 18(4):401.

Whitmarsh, Lorraine. 2011. "Scepticism and Uncertainty about Climate Change: Dimensions, Determinants and Change Over Time." *Global Environmental Change* 21(2):690–700.

Wolf, Johanna and Susanne C. Moser. 2011. "Individual Understandings, Perceptions, and Engagement with Climate Change: Insights from in-Depth Studies across the World." *Wiley Interdisciplinary Reviews: Climate Change* 2(4):547–569.

Wood, B. Dan and Arnold Vedlitz. 2007. "Issue Definition, Information Processing, and the Politics of Global Warming." *American Journal of Political Science* 51(3):552–568.

Zahran, Sammy, Samuel D. Brody, Himanshu Grover, and Arnold Vedlitz. 2006. "Climate Change Vulnerability and Policy Support." *Society and Natural Resources* 19(9):771–789.

Zaller, John R. 1992. *The Nature and Origins of Mass Opinion*. New York: Cambridge University Press.

10

Challenging Climate Change

The Denial Countermovement

Riley E. Dunlap and Aaron M. McCright

INTRODUCTION

Shortly after James Hansen's June 1988 Senate testimony placed anthropogenic global warming on the public agenda in the United States, organized efforts to deny the reality and significance of the phenomenon began, reflected by formation the following year of the Global Climate Coalition (an industry-led front group formed to call global warming into question). These efforts to deny global warming—and human-caused climate change more generally[1]—have continued over the ensuing quarter-century, involving an ever-growing array of actors, and often cresting when domestic or international action (e.g., the 1997 Kyoto Protocol) aimed at reducing greenhouse gas (GHG) emissions seems imminent. Organized denial reached an unprecedented level in 2009 when the newly elected Obama administration and a Democratic-controlled Congress increased the likelihood of U.S. action to reduce GHG emissions and the December 2009 Copenhagen Conference of the United Nations Framework Convention on Climate Change was approaching. The efforts have continued relatively unabated since then, cresting whenever climate change policymaking becomes salient on the U.S. or international agendas.

The fruits of this campaign are readily apparent in the U.S. Congress, where Republican majorities in the Senate and House of Representatives are attempting to prevent the Environmental Protection Agency from taking steps to control GHG emissions and doing their best to cut funding for federal programs dealing with climate change. Further, climate change denial[2] has become a virtual "litmus test" for Republican politicians, strongly enforced by elements of the conservative movement. Clearly the organized climate change denial campaign has achieved a great deal of success. It has delegitimated to a considerable degree global warming as a widely agreed-upon societal problem, a status many thought had been achieved in the early 1990s (McCright and Dunlap 2000), and has generated near-hegemonic acceptance of denial among staunch political conservatives—especially elites and activists—as well as widespread skepticism within the U.S. public, particularly among self-identified conservatives and Republicans (McCright and Dunlap 2011b).

In emphasizing the impact of organized climate change denial, this chapter will depart from major assessment reports that ignore the topic, such as the U.S. Global Change Research Program's *National Climate Assessment* volumes and the first four assessment reports of the Intergovernmental Panel on Climate Change (IPCC), and others that give denial short shrift. For example, a chapter in the volume produced by Working Group III (Mitigation of Climate Change) in the latest IPCC report (Assessment Report 5) on "Sustainable Development and Equity" devotes only a single sentence to denial (IPCC 2014:300–301). While the National Research Council's Panel on America's Climate Choices pays more attention to climate change denial in its 2010 volume, *Informing an Effective Response to Climate Change*, it still provides quite limited treatment by focusing mainly on the denial campaign's impact on public understanding of climate change (U.S. National Research Council 2010:25, 30, 208, 214).

Specifically, this chapter will first outline historical and cultural conditions that have provided fertile soil for denial—ranging from the dominant Western worldview that developed over the past few centuries to the more recent emergence of neoliberalism as hegemonic ideology—and help explain why powerful interests are so opposed to recognizing and dealing with climate change. We then analyze the key strategies and tactics employed by the denial campaign, noting that climate change policy advocates' reliance on climate science makes their efforts particularly vulnerable to denial campaigns. We next identify the major actors involved in the denial campaign in the United States, giving overviews of each one. This is followed by a brief description of the diffusion of denial to other nations, suggesting that organized denial has evolved into a global advocacy network (Farquharson 2003) designed to combat the IPCC and international climate change policy advocacy network. Lastly, we provide a short conclusion and identify key priorities for future research.

In covering these issues, three caveats are necessary. First, we draw upon a much broader range of research and analyses of climate change denial than that conducted within sociology. Nevertheless, we draw heavily upon sociological perspectives and concepts when analyzing climate change denial. Second, early sections on the historical and cultural context in which climate change denial emerged follow a more narrative structure than do later sections, where large numbers of empirical studies are available for review. Third, neither we nor the work we review claims that the failure of the United States to endorse international climate treaties or to pass domestic climate change legislation can be attributed solely or even primarily to the denial campaign.[3] Nonetheless, the relevant literature suggests that the campaign to deny the reality and significance of anthropogenic climate change has been a crucial factor contributing to the current policy stalemate (Jamieson 2014).

HISTORICAL, CULTURAL, AND POLITICO-ECONOMIC ROOTS OF CLIMATE CHANGE DENIAL

It is not surprising that acknowledging the reality and seriousness of climate change has engendered so much opposition, led by the denial campaign. It is widely noted that the development of modern industrial societies has been based on an ever-growing use of energy from fossil fuels—the primary source of GHG emissions—and that the current way of life in such societies is dependent on the availability of fossil fuels (e.g., Foster, Clark, and York 2011; Neubauer 2011). However, the cultural roots of climate change denial run even longer and deeper than efforts to protect modern economies and lifestyles premised on endless and affordable energy.

Numerous analysts suggest that modern, industrial societies embody an anthropocentric view of the natural world stemming from the Judeo-Christian assumption that nature was created for human use. A subsequent "human–nature schism" (historically referred to as "man–nature dualism") was solidified by Enlightenment thinking emphasizing the use of science and technology to master nature and transform the environment into resources for human use, thus creating the promise of unlimited progress. The capitalist-driven Industrial Revolution implemented this instrumental view of nature, generating great wealth and growing prosperity (albeit with widespread poverty) and instigating political developments that elevated individual rights and private property as central values (Barry 2007:34–49). By the mid-twentieth century, wealthy Western nations—epitomized by the United States—came to share (at least loosely) a "dominant social paradigm" that valued individual rights, *laissez-faire* government, private property, and free enterprise. This paradigm also placed great faith in the ability of these politico-economic conditions to combine with science and technology to yield abundant resources, economic growth, and endless progress (Dunlap and Van Liere 1984; Milbrath 1984).

The tremendous success of modern capitalist societies in producing economic growth and increasing prosperity, especially from the late 1940s to the early 1970s, led their citizens, leaders, and most intellectuals to assume that growth and progress were inevitable, resulting in ever-better living standards for ever-more people (Antonio 2009). This mindset, which Douglas (2007:550) refers to as "growthism,"[4] was challenged by energy crises beginning in 1973–1974, economic stagflation, and talk of "limits to growth," but it gained renewed vigor with the staunchly conservative Reagan and Thatcher regimes in the United States and United Kingdom, respectively, and the heightened globalization of capitalism.

The displacement of Keynesian economics, which accepted the necessity of governmental regulation of the economy, by the antiregulatory economics of neoliberalism, which demonized governmental intervention in principle, represented a transformation of conservatism. The goal of limited government was replaced by a staunch antigovernment orientation, representing a fundamental shift in governing philosophy that significantly reduced constraints on capital accumulation and growth (Bockman 2013; Neubauer 2011; Palley 2005). The rise of neoliberalism, along with the global spread of capitalism in an increasingly unregulated world (facilitated by neoliberalism), spawned what Antonio (2009) terms the "global growth imperative." The global entrenchment of capitalism in recent decades has been fueled not only by the unquenchable desire for profit but by an unlimited faith in technological development (seemingly always referred to as "advances") to overcome any potential resource limits and readily solve any (minor) environmental problems. Indeed, as reflected by influential techno-optimists such as Julian Simon (1981) and more recently Bjorn Lomborg (2001), there is a widespread belief than modern industrial societies have become "exempt" from nature's constraints (Catton and Dunlap 1980). As Antonio (2009:30) notes, "Exemptionalist presuppositions anchor the growth imperative and many other facets of capitalist political economy, lifestyle habits, built environments, consumer culture, and development schemes"—that is, our modern way of life.

The problem of climate change has emerged within this historical, cultural, and political context, and it therefore comes as no surprise that powerful forces have denied the reality and significance of this global environmental problem. As leading social critic Naomi Klein (2011:14) argues, "The expansionist, extractive mindset, which has so long governed our relationship to nature, is what the climate crisis calls into question so fundamentally." Unfortunately, this exemptionalist paradigm (Catton and Dunlap 1980), which assumes the possibility and desirability of endless growth and human freedom from ecological constraints, provides a valuable resource to those promoting denial. These widely held and deeply embedded beliefs furnish denialists a rich cultural toolkit (Swidler 1986) that allows them to label efforts to reduce GHG emissions as threatening economic growth and prosperity, the free-market system, individual rights, the American way of life, and even Western civilization—discursive resources that they readily employ (e.g., Jacques, Dunlap, and Freeman 2008; McCright and Dunlap 2000).

THE CONSERVATIVE MOVEMENT, THE REAGAN PRESIDENCY, AND THE ASCENDANCE OF NEOLIBERALISM

The emergence of neoliberalism and its impact on the political culture of the United States and other industrialized nations (especially the United Kingdom) did not happen by chance but was the result of a well-funded and coordinated effort (Lapham 2004). The progressive social movements of the 1960s and implementation of federal government programs seen as creating a "welfare state" provoked a reaction among American conservative elites. Corporations and wealthy donors such as Joseph Coors banded together to fund a "countermovement" to combat these trends, building upon earlier efforts to foster conservatism and neoliberal ideology (Bellant 1990; Mirowski and Plehwe 2009), particularly in the United Kingdom (Cockett 1995). They funded conservative think tanks (CTTs) such as the Heritage Foundation, which function as countermovement organizations, with the goal of developing an apparatus for the incubation and diffusion of conservative ideas. CTTs quickly developed into a powerful political force widely recognized to have shifted American politics significantly rightward (Blumenthal 1986; Stefanic and Delgado 1996).

These CTTs (and their counterparts in other industrial nations) promoted an extreme version of the neoliberal ideas of economic thinkers such as Friedrich Hayek and Ludwig von Mises, one that jettisoned recognition of the necessity of a strong state to facilitate a successful market economy (Mirowski and Plehwe 2009; Neubauer 2011). In the process, they denigrated governmental regulations writ large, arguing that "unhindered markets are best able to generate economic growth and social welfare"— an ideology often termed "market fundamentalism" (Bockman 2013:14; also see Oreskes and Conway 2010). The inauguration of Ronald Reagan as president in 1981 represented the success of this revitalized conservative movement in altering the U.S. political landscape, as Margaret Thatcher's 1979 election as prime minister had done in the United Kingdom (Antonio and Brulle 2011). The institutionalization of neoliberalism not only signified a major cultural shift, but also produced structural changes by reducing governmental regulations on corporations and by facilitating the accumulation of capital (Harvey 2005).

In the 1960s the growing environmental movement called attention to the negative consequences of unrestrained economic growth, especially the unintended side effects of technological developments such as pesticide contamination and air and water pollution. The increasing visibility of such problems led to calls for governmental regulations to limit environmental degradation and promote environmental quality. Environmentalists achieved considerable success in passing environmental laws and setting up agencies such as the Environmental Protection Agency in the 1960s and early 1970s, when Keynesian thinking was still prevalent and the U.S. economy was strong (Neubauer 2011). But industry and its allies viewed these developments with alarm, and opposition to environmentalism quickly developed—primarily in the American West, where battles over access to natural resources raged (Switzer 1997)—and became a component of the wider conservative countermovement against the progressivism of the 1960s (Jacques et al. 2008; Lo 1982).

The Reagan administration rode the crest of antigovernment sentiment by implementing a neoliberal program of reducing governmental regulations, especially environmental ones. In short, conservative activists in the administration largely exercised what scholars have termed the "first dimension of power" (Lukes 1974) to oppose environmental policy and regulations via overt political actions, including appointing staunch antienvironmentalists to lead key agencies like the Environmental Protection Agency and the Department of the Interior (Kraft and Vig 1984). However, use of this blunt form of power attracted scrutiny. Appalled by the strong antienvironmental actions of the administration, several environmental organizations mounted campaigns to mobilize public opposition to the administration's efforts to weaken environmental laws and inhibit their implementation (Dunlap 1987). Their successful efforts at fostering a public backlash helped a Democratic Congress temper, but certainly not stop, Reagan's antienvironmental initiatives (Kraft and Vig 1984). Yet, by demonstrating that despite their growing antigovernment sentiment most Americans still supported measures to protect air and water quality and the environment more broadly (Dunlap 1987), the backlash served as a warning to conservatives (and business interests) hoping to dismantle federal (and state) environmental regulations to "get government off the back of industry."

FROM ANTIENVIRONMENTALISM TO CLIMATE CHANGE DENIAL: EMPLOYING THE SECOND DIMENSION OF POWER TO MANUFACTURE UNCERTAINTY AND CONTROVERSY

The placement of global environmental problems (e.g., climate change, biodiversity loss, ozone depletion) on the international policy agenda in the early 1990s—symbolized by the 1992 Rio "Earth Summit"—presented a much greater and more sustained challenge to conservatives' and industry's neoliberal agenda, especially the spread of privatization and unfettered markets globally, than had the typically localized environmental problems of earlier eras. Further, the dissolution of the Soviet Union in 1991 and the consequent decline in communism (conservatives' number-one enemy for decades) led the conservative movement to substitute a "green threat" for the declining "red threat" (Jacques et al. 2008).[5] Climate change became the *bête noir* of conservatives due to its sweeping regulatory implications (Antonio and Brulle 2011). It is not surprising, then, that the conservative movement and industry began to mobilize in the early 1990s[6] to block attempts at climate change policymaking—largely by denying the reality and seriousness of climate change (Gelbspan 1997, 2004).

While mobilizing against climate policy early on, conservative activists learned from the Reagan administration's experience that it was unwise to attack environmental protection directly (exercising the first dimension of power), given that Americans are generally supportive of environmental protection and see it as a governmental responsibility. Conservative activists shifted to a more subtle form of power characterized by non-decision-making and agenda setting—what Lukes (1974), following Bachrach and Baratz (1970; see also Molotch 1970), refers to as the "second dimension of power." Briefly, conservatives and their industry allies learned to prevent the implementation of environmental policies that might threaten their political and economic interests by opposing the inputs to, and undermining the foundations of, such policy proposals earlier in the policy formation process (Bonds 2010).

The conservative movement recognized that those pushing for environmental policies—often coalitions of environmentalists, scientists, and policymakers—typically build their cases on the basis of scientific evidence of alleged environmental risks and hazards (Yearley 1991). Sociologists refer to this type of science, which examines the impacts of technologies and economic activities on the natural environment and public health, as "impact science" to contrast it with the more predominant "production science" that yields technological innovations and economic growth (Schnaiberg 1980; also see McCright and Dunlap 2010). Conservatives seized upon the strategy of "manufacturing uncertainty" that had been effectively employed for several decades by corporations and entire industries, most notably the tobacco industry, in efforts to protect their products from regulations and lawsuits by questioning the adequacy of evidence suggesting the products were hazardous (e.g., Michaels 2008; Oreskes and Conway 2010). Manufacturing uncertainty about the scientific evidence documenting environmental problems—labeled "junk science"—became the favored strategy employed by conservatives (and their industry allies) in promoting antienvironmentalism, particularly when they focused their attention on climate change in the early 1990s (Dunlap and Jacques 2013; Jacques et al. 2008).

Manufacturing uncertainty about climate change is especially attractive given several features of climate science and the IPCC. First, the inherent interdisciplinarity and complexity of climate science make paradigmatic disagreements over appropriate methods and data collection/analyses endemic to the field. This increases the likelihood of scientific disputes and makes "consensus" more difficult to achieve, heightening the challenges already present in translating science to policymakers and the general public. Second, we are almost entirely dependent upon scientific experts to tell us about the reality, extent, and impacts of climate change. Such a strong dependence upon expertise amplifies the importance of trust in science and scientists—but also makes this trust more vulnerable to challenge.

Third, from its creation by the World Meteorological Organization and the United Nations Environmental Program in 1988, the IPCC has epitomized the marriage of science and policymaking, making climate science an attractive target for opponents of climate policy. Consequently, even though Powell (2011:47) argues that "From the get-go . . . the IPCC was a conservative organization predestined for understatement," and subsequent analyses lend credibility to his claim (Brysse et al. 2013; Freudenburg and Muselli 2010, 2013), denialists commonly refer to the IPCC as "alarmist." Finally, and more generally, additional scientific investigation and evidence tend to add complexity and uncertainty to most science-based policy issues; thus, the goal of settling disputes with more science seems unreachable (Sarewitz 2004; Yearley 2005). This is particularly the case when some "investigators" (such as "contrarian scientists") have the explicit goal of generating uncertainty (e.g., McCright 2007).

The strategy of manufacturing uncertainty, which capitalizes upon these characteristics of climate science and the IPCC, has been employed by conservative activists and the wider range of actors in the denial campaign. In fact, Powell (2011) notes that unlike earlier corporate campaigns to manufacture uncertainty about scientific evidence regarding hazardous products, the climate change denial campaign has waged an unprecedented war on the entire field of climate science. For instance, the various elements of organized climate change denial have questioned the validity of climate models, challenged the use of paleoclimate data to establish historical trends, criticized statistical techniques employed by climatologists, argued that modern records of temperature trends have been "adjusted" to show false warming, and criticized many other aspects of climate science—all in an effort to show that evidence supporting anthropogenic global warming and its consequences is "uncertain" at best and "fraudulent" at worst (Dunlap and

Jacques 2013; McCright and Dunlap 2000; Rahmstorf 2004; Washington and Cook 2011).[7]

The strategy of manufacturing uncertainty has two related dimensions: challenging or undermining the validity and legitimacy of the field of climate science and attacking the integrity and authority of individual, or groups of, climate scientists (e.g., Hess 2014; McCright and Dunlap 2010; Michaels 2008).[8] These tactics—labeled the "swift-boating" of climate science and scientists (Mann 2012)—have broadened over time to entail criticisms of pillars of modern science, including the peer-review process, the objectivity of refereed journals, the fairness of governmental grant making, and the credibility of institutions such the American Association for the Advancement of Science and the U.S. National Academy of Sciences (Powell 2011). Members of the denial campaign employ both arms of this strategy within the government (e.g., in executive branch agencies and in Congress) and within the media (Hogan and Littlemore 2009; Powell 2011), depending on whether they are in positions of power or are operating outside of institutional channels.

Over time, manufacturing uncertainty has evolved into "manufacturing controversy," creating the impression that there is major debate and dissent *within* the scientific community over the reality of anthropogenic climate change (Ceccarelli 2011).[9] To accomplish this, corporations and especially CTTs have supported a small number of contrarian scientists (many with no formal training in climate science) and other self-styled "experts" (often social scientists affiliated with CTTs) to produce non–peer-reviewed reports and books, publish in a handful of marginal journals, hold "scientific" conferences, compile dubious lists of supposed scientists who question climate change, and in general mimic the workings of conventional science. The goal is to produce not just criticisms of, but alternatives to, mainstream science (Dunlap and Jacques 2013; Tollefson 2011)[10], fashioning a "parallel scientific universe" (Mirowski 2008) that serves to generate confusion among the public and policymakers. Essentially, the denial campaign engages climate science in framing contests, continually countering the manner in which the threat of climate change is framed with ever-evolving counterclaims (Brulle 2014; McCright and Dunlap 2000).

These techniques are exemplified by the Heartland Institute forming a "Nongovernmental International Panel on Climate Change" or NIPCC as a direct counter to the IPCC, with the goal of making the basic claims of climate science—the Earth is warming, due in significant part to human activities, resulting in negative consequences—appear not only uncertain but the subject of major scientific controversy (Plehwe 2014). These counterclaims, along with the alleged misconduct of climate scientists, are heavily publicized to create the impression of serious disagreement on climate change among scientists in the eyes of the public, media, and policymakers, thereby undermining any need to take action (Dunlap and Jacques 2013; Tollefson 2011).

As Ceccarelli (2013:762) puts it, a "manufactured scientific controversy" is not "a *real* scientific controversy, one that currently exists within a scientific community, rather one that only *appears* to exist" in the eyes of the public and policymakers. By creating the appearance of controversy within the *public realm*, denialists are able to appeal to values such as freedom of speech, fairness to both sides, and respecting minority viewpoints to add legitimacy to their claims—thereby bypassing the scientific realm in which peer review and accumulating knowledge eventually lead to the rejection of discredited claims (Ceccarelli 2011; Dunlap and Jacques 2013). There is no better example of a manufactured controversy than climate change, as illustrated by the fact that only a small minority of the American public believes that over 90 percent of climate scientists think that human-caused global warming is happening (Leiserowitz et al. 2014) even though the actual percentage likely exceeds 90 percent (Anderegg et al. 2010).

THE CONCEPTUALIZATION AND KEY COMPONENTS OF ORGANIZED CLIMATE CHANGE DENIAL

Journalists (e.g., Gelbspan 1997) and environmental advocacy organizations (e.g., Ozone Action 1996a, 1996b) were the first to describe organized climate change denial in the United States, followed shortly by analytical work by social scientists (e.g., Beder 1999; Lahsen 1999; McCright and Dunlap 2000, 2003). While the former focused primarily on the activities of various actors representing the fossil fuels industry (e.g., corporations, industry associations, and front groups), the latter directed more attention to the activities of CTTs and their affiliated contrarian scientists. In recent years, especially since 2010, growing numbers of social scientists, journalists, and science and environmental advocacy organizations have analyzed the denial campaign—typically its key components and their strategies, but increasingly the coordinated efforts of the growing number of actors involved in denying climate change.

Early analyses portrayed the denial campaign as driven primarily by industry (Beder 1999) or as a new focus of the growing antienvironmental countermovement launched by the conservative movement (McCright and Dunlap 2000, 2003). As more information has emerged about the complex set of actors involved in denying climate change, analysts have begun to refer to a "denial machine"[11] (Dunlap and McCright 2011), "climate denial movement" (Boykoff and Olson 2013), "sceptic movement" (Knight and

Greenberg 2011), "climate denier movement" (Hoffman 2011a), and "climate change countermovement" (Brulle 2014) in efforts to apply conceptual tools (e.g., framing, mobilization, political opportunity structure) from social movements and political sociology scholarship. Going forward, we employ "denial countermovement" to quickly denote the overall campaign to counteract the efforts of scientists, policymakers, and citizen interest groups to stimulate awareness of and action on climate change.

Describing the denial countermovement is challenging, as it has evolved and expanded greatly over the past quarter-century. There have been major changes in its key actors, supporters, and tactics, while the basic strategy of manufacturing uncertainty has expanded into manufacturing controversy. New components and tactics typically supplement rather than replace their predecessors, resulting in an increasingly complex and multifaceted countermovement that is ever changing in response to developments in climate science and policymaking (Dunlap and McCright 2010, 2011). Consequently, analyses by social scientists, journalists, and science and environmental advocacy organizations become dated rather quickly. No work to date has been sufficiently longitudinal to fully capture the evolution of the structure and dynamics of the denial countermovement over the past quarter-century—a priority for future research we address in our conclusion.

Nevertheless, there do exist in-depth, though partial, analyses of the denial countermovement, by Brulle (2014), Dunlap and McCright (2010, 2011), Greenpeace (2013), Hoggan and Littlemore (2009), McCright and Dunlap (2000, 2003), and Powell (2011). In what follows, we draw upon these works and other analyses to identify the core components of the denial countermovement and provide a sense of how they have emerged and evolved over the last quarter-century.[12]

Fossil Fuels Industry and Corporate America

By the early 1990s individual fossil fuels corporations (most notably ExxonMobil and Peabody Coal) and industry associations (e.g., American Petroleum Institute and Western Fuels Association) were leading efforts to deny the significance of climate change—funding contrarian scientists, CTTs, and various front groups promoting denial (Beder 1999; Union of Concerned Scientists 2007). These fossil fuels industry actors were joined by a wide range of other corporations and business associations like the National Association of Manufacturers and U.S. Chamber of Commerce in front groups such as the Global Climate Coalition, as industry mobilized to successfully oppose U.S. ratification of the 1997 Kyoto Protocol (Gelbspan 1997). With the 2000 election of George W. Bush, whose administration not only opposed action on climate change but also suppressed climate science and institutionalized climate change denial (Lynch, Burns, and Stretesky 2010; McCright and Dunlap 2010), industry had less need to attack climate science—although corporations like ExxonMobil and Koch Industries continued to do so (Greenpeace 2007, 2010).

The situation began to change, however, following the publication of Al Gore's *An Inconvenient Truth* (2006), the wide release and Academy Awards for the documentary by the same name, and the awarding of the 2007 Nobel Peace Prize to the IPCC and Gore. Climate change was in the spotlight (Dunlap and Jacques 2013; Elsasser and Dunlap 2013) and was the subject of growing public attention (Brulle, Carmichael, and Jenkins 2012; Skocpol 2012). The 2008 election of Barack Obama and a Democratic Congress stoked corporate apprehension of both national legislation and international action (at the upcoming December 2009 UN Copenhagen climate conference in December 2009) (Pooley 2010). The oil and coal industries, with help from other corporations and industry associations, launched massive oppositional efforts, from Congressional lobbying to generating public opposition via front groups, "astroturf" campaigns, and advertising (Goddell 2010).

Although several large environmental organizations formed a coalition with a few major corporations to support industry-friendly cap-and-trade legislation in 2009–2010, intense opposition from industry clearly contributed to its demise (Pooley 2010; Skocpol 2012). Much of corporate America continues to lobby against action on climate change and has increased its support for candidates opposed to climate policymaking, with notable success in recent elections (Atkin 2014). Overall, despite a good deal of pro-climate marketing, it appears that many major corporations and business associations (especially the U.S. Chamber of Commerce) continue to oppose action to reduce or even limit GHG emissions—sometimes openly via lobbying but often behind the scenes, employing both the first and second dimensions of power (Cray and Montague 2014; Goldman and Carlson 2014; Union of Concerned Scientists 2012). Perhaps because of the difficulty of obtaining relevant information, the corporate world's role in climate change denial has been largely neglected by sociologists, with Perrow (2010) being an exception (also see Chapter 3).

Conservative Think Tanks and Foundations

CTTs are arguably the most visible component of the denial countermovement and have been the focus of considerable investigation by citizen interest groups (Greenpeace 2013), journalists (Mooney 2005), and social scientists (see below). This likely stems from their activities being relatively transparent and because they have played an integral role in denying climate change since the late 1980s. In the late 1990s, when several major corporations retreated from overt sponsorship of denial activities in response both to negative publicity and the growing credibility of climate science, CTTs stepped up their efforts (Layzer 2007). In fact, their staunch commitment to promoting neoliberalism engenders an ideologically based antipathy toward climate change (and its regulatory implications) relatively autonomous from material interests (McCright and Dunlap 2010; Neubauer 2011). Of course, they nonetheless provide "cover" for corporate interests who can fund CTTs to attack climate science while remaining cloaked.

Although CTTs function as countermovement organizations, providing an intelligentsia for promoting and protecting neoliberal interests (Neubauer 2011), they portray themselves as an alternative academia and their spokespersons as unbiased experts (Beder 2001)—making them excellent weapons in the war on climate change and science. CTTs spawn an enormous amount of denial information, and their status provides it and their "climate experts" (both in-house and sponsored) considerable credence in the media and within policy circles (McCright and Dunlap 2000, 2003). CTTs provide the "connective tissue" that helps hold the denial countermovement together and serve as vehicles for broadening its reach (Neubauer 2011; Plehwe 2014).

Numerous CTTs have been involved in denying climate change at least since the early 1990s, ranging from large ones central to the conservative movement (e.g., the Heritage Foundation, Hoover Institution, and Competitive Enterprise Institute) to smaller ones that specialize in climate change (such as the Heartland Institute and the Marshall Institute) (Lahsen 2008; McCright and Dunlap 2000, 2003; Oreskes and Conway 2010). Their crucial roles encompass the production of a vast literature challenging climate science (McCright and Dunlap 2000), including books that often attract considerable attention (Dunlap and Jacques 2013), as well as sponsoring contrarian scientists, holding events for sympathetic members of Congress, providing "experts" for congressional testimony and the media, helping organize front groups for specific campaigns, and harassing climate scientists via freedom-of-information requests (McCright and Dunlap 2003; Powell 2011)—all designed to enhance the appearance of scientific credibility for the denial countermovement while questioning the credibility of mainstream scientists (Oreskes and Conway 2010).

Besides analyzing these broad roles of CTTs, social scientists have provided in-depth analyses of individual think tanks, especially the Marshall Institute (Lahsen 2008; Oreskes and Conway 2010) and the Heartland Institute (Boykoff and Olson 2013; Hoffman 2011a; Plehwe 2014).[13] Studies of the discourse and framing used by the denial community are becoming more common following McCright and Dunlap's (2000) early effort, ranging from examinations of specific organizations (Greenberg, Knight, and Westersund 2011; Knight and Greenberg 2011; McKewon 2012b) to analyses of societal debates over climate change (Hoffman 2011a, 2011b; Malone 2009).

Most recently, Brulle (2014) provides a path-breaking analysis of the overall structure and funding of the denial countermovement, examining 91 countermovement organizations (CTTs and advocacy organizations) and the 140 conservative foundations that have funded them. His network analysis reveals the foundations that provide greatest support for the denial countermovement, the CTTs that receive disproportionate funding from them, and trends over time—the central one being the increasing distribution of untraceable "dark money" through Donors Trust. Consequently, a majority of the funding cannot be traced to other foundations; nevertheless, Brulle's (2014) analysis clearly demonstrates a high degree of cohesion among core elements of the denial countermovement as presumed by other analysts and paves the way for future research.

Contrarian Scientists

A small number of contrarian scientists have played a vital role in promoting climate change denial from the outset, when the physicists who founded the Marshall Institute—Robert Jastrow, William Nierenberg, and Frederick Seitz—issued a report that the George H. W. Bush administration relied on to downplay global warming (Lahsen 2008; Oreskes and Conway 2010). Individuals with scientific credentials (whether relevant to climate science or not) are essential for manufacturing uncertainty and controversy, and corporations, industry associations, and CTTs have made heavy use of such contrarian scientists. While some, like Patrick Michaels, have had strong links to industry, negative publicity over such ties has driven most contrarians to affiliating with CTTs (e.g., Gelbspan 1997; Hoggan and Littlemore 2009). In fact, at present most highly visible contrarian scientists have affiliations with one or more CTTs, ranging from full-time appointments (e.g., Patrick Michaels at the CATO Institute) to serving on boards, advisory committees, and expert panels; speaking at CTTs and their conferences; participating in press and congressional briefings; and especially publishing denial material. These affiliations enable contrarian scientists to reach larger audiences and provide access to conservative policymakers while avoiding the stigma of being spokespersons for the fossil fuels industry (Lahsen 2005; McCright 2007; McCright and Dunlap 2003; Oreskes and Conway 2010).

The academic credentials of contrarian scientists vary greatly, both in terms of the relevance of their degrees to climate science and their scholarly records.[14] Richard Lindzen of MIT is a rare example of a contrarian climate scientist with excellent credentials, while others have strong credentials but

backgrounds marginal to climate science (like the Marshall Institute found-
ers), training relevant to climate science but marginal scholarly records, or
neither relevant training nor scholarly credentials (Anderegg et al. 2010;
McCright 2007; Dunlap and Jacques 2013; Lahsen 2005, 2008; Oreskes and
Conway 2010; Powell 2011). However, sponsorship by CTTs can turn even
the latter into scientific "experts" who receive significant media attention
and thus contribute to the "controversy" about climate change in the public
and policy realms. While their credentials vary widely, the vast number of
contrarian scientists share two common traits. They are overwhelmingly
older, white males, and they are staunch conservatives, sharing a commitment to neoliberal fears of governmental control (Lahsen 1999; McCright
and Dunlap 2011a; Oreskes and Conway 2010). Their shared ideology pre-
disposes them to question the significance of a problem that will inevitably
require governmental action and draws them to CTTs.

Front Groups and Astroturf Campaigns

Corporations and industry associations not only support CTTs to shield
their denial activities, but on their own or in conjunction with CTTs continually set up front groups and shorter-term, more narrowly focused "astroturf"
campaigns to mask their efforts to deny climate change and undermine
climate legislation. Most of what we know about front groups and astroturf
campaigns involved in the denial countermovement comes from journalists (especially Gelbspan 1997, 2004; Hoggan and Littlemore 2009; Pooley
2010; Powell 2011) and citizen interest groups (Cray and Montague 2014;
Greenpeace 2010, 2013), as relatively few social scientists (Lahsen 2005;
Oreskes 2010; Oreskes and Conway 2010) have offered analyses of these
efforts.[15] Our review will therefore be brief (see Dunlap and McCright 2011
for more detail).

The earliest front group was arguably the most significant, as the Global
Climate Coalition—formed in 1989 in response to the establishment of the
IPCC—was sponsored by major fossil fuels companies, automobile manufacturers, and industry associations and immediately challenged climate
science and the need for climate policies. The Kyoto Protocol was a primary target, and the coalition played a critical role in blocking U.S. ratification (McCright and Dunlap 2000). It disbanded in 2002 after several
corporations dropped out and it became clear its goals were shared by the
George W. Bush administration. Next came the Information Council on
the Environment, established in 1991 by coal and utility interests to undermine climate science. It disappeared when its plan to "reposition global
warming as a theory (not fact)" was leaked to the press (Pooley 2010:41),
only to be followed by the Greening Earth Society. Set up by the Western

Fuels Association and relying on a handful of contrarians (including Patrick Michaels), it launched a sophisticated campaign to undermine climate science by arguing that increased levels of atmospheric CO_2 would be beneficial (Oreskes 2010). Reflecting the earlier-noted shift from industry to CTTs as the core of the denial countermovement, the Cooler Heads Coalition emerged in 1997. It is a loose coalition of key CTTs, centered at the Competitive Enterprise Institute, and has been particularly aggressive in attacking climate science and scientists.[16] More recently, the American Legislative Exchange Council, an industry-funded group that provides sympathetic politicians with ready-made legislation, has promoted measures in several states to restrict climate change education (Forecast the Facts 2014).

Two front groups closely associated with the Koch brothers, Americans for Prosperity and Freedom Works, have played central roles in sponsoring short-term astroturf campaigns aimed at blocking climate legislation and Obama's agenda more generally. For example, Americans for Prosperity sponsored "Hot Air Tours" in 2008, and Freedom Works joined with the American Petroleum Institute in promoting 2009 "Energy Citizens" rallies to promote opposition to climate legislation. These types of campaigns, organized with the help of public relations firms, have long been used by industry and CTTs to mimic grassroots campaigns of ordinary citizens but are clearly top-down efforts (Beder 1998).

A final example of front groups and "astroturf" campaigns comes from the Religious Right. In response to growing concern over climate change within the Christian community, reflected by the Evangelical Climate Initiative (Wilkinson 2012), conservative Christians began to embrace climate change denial (McCammack 2007). A front group termed the Interfaith Stewardship Alliance was set up with corporate support (it was renamed the Cornwall Alliance for Stewardship of Creation) and has engaged in a wide range of antienvironmental activities. Headed by Calvin Beisner, who has a long history of involvement with industry, CTTs, and conservative causes, the Cornwall Alliance has issued an "Evangelical Declaration on Global Warming" that was laden with denialist claims and designed to counteract progressive Christians' efforts to generate support for dealing with climate change (People for the American Way 2011). Despite its "Christian face," the Cornwall Alliance is a staunch supporter of neoliberal ideas and causes.

Conservative Politicians

Republican politicians' traditionally strong embrace of neoliberal ideology has made them less likely than Democrats to support environmental protection, and the rightward shift in the Republican Party over the past two decades—recently enhanced by the Tea Party—has created a chasm between them and their Democratic counterparts on environmental issues (McCright, Xiao, and Dunlap 2014), most notably climate change. Republican members of Congress have held hearings to attack climate scientists and challenge evidence for climate change since the 1990s (Mann 2012; McCright and Dunlap 2003). While Republicans historically have slighted scientific expertise in hearings on climate change (Park, Liu, and Vedlitz 2014), in recent years they have been inviting larger numbers of contrarian scientists (and other members of the denial countermovement) to testify, helping create the illusion of scientific controversy (Koebler 2014). Most strikingly, Senator James Inhofe, who called for criminal investigation of leading climate scientists following "Climategate" (Nature 2010), uses the Committee for Environment and Public Works (which he again chairs following the 2014 election) as a megaphone for denialists.

Republicans' ideological predisposition to deny climate change has been enhanced by heavy contributions from the fossil fuels industry (Atkin 2014). In the 113th Congress a majority of Republican House members, and 90 percent of Republican leaders in both the Senate and the House, were on record denying basic tenets of climate science (Germain 2014). Furthermore, via Americans for Prosperity, the Koch brothers have induced hundreds of Republican politicians, including most leaders of the House, to sign their "No Climate Tax" pledge (Mayer 2013). These developments suggest that the Republican-controlled Congress may be even more successful in institutionalizing climate change denial than was the George W. Bush administration—which turned the White House into a key component of the denial countermovement (Lynch et al. 2010; McCright and Dunlap 2010). Ongoing efforts to undermine the Obama administration's climate initiatives, including combating climate science via funding cuts, represent a reliance on the first face of power reminiscent of the early Reagan administration (Hess 2014).

Conservative Media

Another essential component of the denial countermovement is the conservative media, which has been vital in conveying denial to the American public and beyond (Boykoff 2013). For much of the past quarter-century, key outlets for climate change denial have been conservative newspapers (e.g., *Wall Street Journal, New York Post, Washington Times*), magazines (e.g., *The Weekly Standard, National Review*, and *The American Spectator*), talk radio (e.g., Rush Limbaugh), television (e.g., Fox News reporters and popular commentators such as Bill O'Reilly and Sean Hannity), and prominent syndicated columnists and personalities whose influence transcend a single medium (e.g., George Will, Charles Krauthammer, Glenn Beck). In crediting conservative media for helping to move the United States rightward in recent decades, Jamieson and Cappella (2008) view it as an "echo chamber" that endlessly circulates and amplifies ideologically driven claims to its audiences. This seems to be an apt metaphor here, as the barrage of assaults on climate science—and increasingly climate scientists—from the conservative media not only inundates committed conservative audiences but also reaches a large segment of the general public.

In the expansive literature on climate change media coverage, a few robust findings demonstrating the influence of conservative media have emerged. As expected, conservative media outlets promote climate change denial more strongly and more often than do nonconservative outlets (Feldman et al. 2012; McKewon 2012b; McKnight 2010; Painter and Ashe 2012). This pattern holds for news coverage and especially for editorial and opinion pieces (Painter and Ashe 2012). In fact, climate change denial appears quite prominently in editorial and opinion columns of newspapers in general—more so than in news sections (Elsasser and Dunlap 2013; McKewon 2012a, 2012b; Young 2013). Perhaps best known for promoting climate change denial are the Rupert Murdoch–owned News Corporation media outlets in Australia, Britain, and the United States (McKnight 2010; see also Feldman et al. 2012 on the United States and McKewon 2012b on Australia).

Likely due to the success of the conservative echo chamber in influencing the American media environment, U.S. media in general give significantly more coverage to climate change denial claims and claims makers than do media in any other country (Boykoff 2013; Grundmann 2007; Grundmann and Scott 2014; McKnight 2010; Painter and Ashe 2012). Outside of the United States, media outlets in other Anglo countries such as Australia and the United Kingdom feature climate change denial more prominently than do those in other developed and developing nations—where denial claims and claims makers receive only scant attention (Grundmann and Scott 2014; Painter and Ashe 2012; Sharman 2014).

Denial Blogosphere

The last decade has seen the emergence of a crucial new extension of the conservative echo chamber and major addition to the climate change denial countermovement: the denial blogosphere, or what Pooley (2010) terms the "denialosphere." These online blogs, on which self-described climate skeptics and contrarian scientists question the reality and significance of climate change and dispute climate science per se, increasingly supplement—and to some degree supplant—the more traditional media outlets above. While a few denial blogs are hosted by contrarian scientists (e.g., Judith Curry), the most popular North American blogs are run by a retired TV meteorologist (wattsupwiththat.com), a retired mining executive and dedicated critic of the "hockey stick" model of historical climate trends (climateaudit.org), and a self-styled "warrior" in the climate wars (www.climatedepot.com). The skeptical bloggers tend to have large and avid audiences, with new posts often stimulating hundreds of comments that frequently express intense vitriol for climate scientists and climate change activists and policy proponents. The number of English-language denial blogs has grown dramatically since their first appearance in 2005; they now total over 170 (Sharman 2014), so we can expect their impact to increase.[17]

These blogs came to the fore in 2009 when several played key roles in generating and then publicizing the controversies over "Climategate" and a few relatively minor errors in the 2007 IPCC Fourth Assessment Report.[18] Some bloggers have become celebrities in the denial community and are routinely granted access to broader media, especially conservative media. The "auditing" of climate science on the blogosphere has opened the field up to intense scrutiny and made climate scientists the targets of severe harassment (Mann 2012; Powell 2011). As testament to the international influence of the U.S.-based climate change denial countermovement, several recent analyses document that activists and organizations in the U.S. denial campaign have strong ties to much of the denial blogosphere based in the United Kingdom (Gavin and Marshall 2011; Lockwood 2011), France (Niederer 2013), the Netherlands (Niederer 2013), and a range of European and non-European countries (Sharman 2014).

The denial blogosphere stimulates participation in climate change denial across social media more generally (Berkhout 2010). For instance, denial blogosphere posts are regularly circulated on social media websites such as Twitter and Facebook. Further, climate change denial claims seem to be especially conspicuous in online reader comments on newspaper websites (DeKraker et al. 2014; Koteyko et al. 2012). The increasing use of social media helps disseminate denial claims well beyond the actors in the denial movement to broader audiences. Overall, the Web, blogosphere, and social media all facilitate the international spread of climate change denial.

THE INTERNATIONAL DIFFUSION OF DENIAL AND EMERGENCE OF A GLOBAL ADVOCACY NETWORK

We have focused heavily on the structure and strategy of the denial countermovement in the United States, where it originated and remains most firmly rooted and thus has been the subject of a considerable amount of investigation and research. However, climate change denial has diffused internationally, frequently assisted by American CTTs and contrarian scientists and of course more recently facilitated by the global communication technologies just reviewed.

In addition, several international CTT networks have played leading roles in diffusing denial. These include the Atlas Economic Research Foundation (originating in the United Kingdom but based in the United States), the Economic Freedom Network (established by Canada's Fraser Institute) and the U.K.-based International Policy Network. The latter has established the "Civil Society Coalition on Climate Change" consisting of supposed "independent civil society organizations" (often tiny front groups consisting of a few individuals) in over forty nations, all sharing the goal of denying the reality of climate change and the need to act on it (Dunlap and McCright 2011). Much of what we know about the networks that have spread denial comes from nonacademics (Greenpeace 2013; Harkinson 2009), although a recent analysis of links between CTTs and books espousing climate change denial demonstrates the pioneering role of American CTTs and contrarian scientists and the critical role now being played by CTTs in other nations in the publication of these books (Dunlap and Jacques 2013).

Not surprisingly, climate change denial is most firmly rooted in nations with very strong commitments to neoliberalism and a powerful fossil fuels industry, as the United Kingdom, Australia, and Canada follow the United States as strongholds of denialism. This is reflected not only in their production of denial books (Dunlap and Jacques 2013) but also in the rapidly growing number of social science analyses of the denial countermovement within these nations. For instance, Hamilton's early (2007) in-depth analysis of denial in Australia is supplemented by McKewon's (2012a, 2012b) studies, Lack's (2013) detailed portrait of denial actors in the United Kingdom extends earlier work by Douglas (2009) and Gavin and Marshall (2011), and numerous studies are emerging in Canada in recent years (Greenberg, Knight, and Westersund 2011; Knight and Greenberg 2011; Young 2013; Young and Coutinho 2014).

While not as strong as in the Anglo world, climate change denial is clearly spreading to many other nations, particularly across Europe, leading to analyses in France (Zaccai, Gemenne, and Decroly 2012), Sweden (Anselm and Hultman 2013), and the Netherlands (van Soest 2011).[19] We need more studies in other nations, and especially cross-national comparisons, as undoubtedly the sources and nature of denial vary across national contexts.

At the same time, more attention needs to be paid to the international coordination of denial activities, beyond the roles of key actors from the United States, United Kingdom, and Canada in stimulating denial organizations abroad (as noted above). In recognition of the organizational complexity and increasingly global reach of the denial countermovement, Donald (2011) argues for using the concept of "global advocacy network" developed by Farquharson (2003) to analyze the international debate over tobacco restrictions. Building upon and extending two well-established social science perspectives—advocacy coalition framework (Sabatier and Jenkins-Smith 1993) and transnational advocacy networks (Keck and Sikkink 1999)—the global advocacy network concept offers an excellent tool for conceptualizing the full range of denial countermovement actors (organizations and individuals), strategies, and goals. It also clarifies the essence of climate change denial—that it is designed specifically to counteract a competing global advocacy network: the IPCC, civil society organizations, policymakers, and others trying to promote efforts to deal with climate change. Thus, like other countermovements, climate change denial constantly evolves in response to its opponent.

CONCLUSION AND SUGGESTIONS FOR FUTURE RESEARCH

This chapter has provided an overview of the state of knowledge on the emergence, trajectory, and dynamics of organized climate change denial, paying attention to its evolution and international diffusion. It has demonstrated that by constantly challenging the reality and significance of climate change, the denial countermovement represents a powerful obstacle to mobilizing societal action aimed at reducing GHG emissions. Indeed, a successful climate change denial campaign—and the rejection of impact science more generally—may undermine long-term societal resilience in a warming world. By substituting ideology for science to protect the neoliberal order in the global capitalist system, a successful denial countermovement inhibits efforts to mitigate and/or adapt to climate change in a warming world (Dunlap 2014; McCright 2011).[20]

Given the significance of what is at stake, we end by identifying a few key research priorities to guide scholarship in this area in the future. First, we generally lack much theorizing that explains the emergence, structure, and impacts of organized climate change denial. One exception to date is the Anti-Reflexivity Thesis (e.g., McCright and Dunlap 2010; also see Antonio and Brulle 2011; Young and Coutinho 2014), which conceptualizes the conservative movement and industrial sector as a rearguard force defending the industrial capitalist system from widespread scientific, political, and public acknowledgment of the system's unintended and unanticipated consequences, such as climate change. While this perspective is still in its infancy and can benefit from elaboration, it nevertheless appears to offer considerable purchase for understanding how and why key forces of organized climate change denial employ the strategies of manufacturing uncertainty and controversy to oppose the claims of environmentalists and scientists about climate change.

Hess (2014) introduces the term "epistemic rift" to conceptualize how the denial countermovement has created a rupture between science and policymaking. While such efforts that build upon or complement the Anti-Reflexivity Thesis are crucial, our sociological understanding of organized climate change denial also will benefit from alternative conceptualizations and theoretical efforts. For example, advances may be made by drawing insights from the social movements literature to more fully develop the theoretical notions of organized climate change denial as a countermovement and global advocacy network. Further advances may be made by drawing insights from the sociology of science literature to clarify how other climate change denial machine components engage with, and provide resources and venues for, contrarian scientists and attempt to mimic the authority of mainstream climate science while undermining it in the process.

Second, much of what we know about the major components of the denial countermovement in recent years has been provided by journalists and science and environmental advocacy organizations. While these efforts are highly valuable, the time is ripe for social scientists, especially sociologists, to step up our contributions. Four such endeavors seem especially important. Briefly, we need network analyses of the key components of the denial countermovement, extending Brulle's (2014) examination of connections between foundations and CTTs to include, for example, connections between CTTs and contrarian scientists. Also, we need research on patterns of funding for components of the denial countermovement—again, beyond Brulle's (2014) work—especially focusing on funding from corporations, which is more challenging to examine than that from foundations. Further, we need continued analyses of the discourses, claims, and frames employed by key components in the denial countermovement. Lastly, we should investigate denial countermovement components, strategies, and tactics not only within individual nations but also across nations to analyze the evolution of the denial countermovement into a full-fledged global advocacy network.

Third and finally, no work to date has been sufficiently longitudinal to fully capture the evolution of the structure, dynamics, and tactics of the denial countermovement. A few works imply that different "eras" of climate change denial may be discerned (Aykut et al. 2012; Levy and Spicer 2013). Evidence suggests that key components of the denial countermovement have emerged and become more or less central to the denial campaign over time, as we attempted to characterize in earlier sections. Yet, a more complete understanding of organized climate change denial over the last quarter-century, including the international diffusion of denial around the world, demands the collection and rigorous, systematic examination of longitudinal data.

NOTES

1. We use "global warming" and "climate change" interchangeably, despite the broader focus of the latter, as both are used in policy and public debates about climate science—with global warming being the more common term in the 1980s and 1990s. We do so with the clear implication that we are referring to *anthropogenic* or human-caused warming and consequent climatic changes.

2. Denial and skepticism are often treated as a continuum, from outright denial of the reality of anthropogenic climate change and refusal to accept evidence of its existence to skepticism about various aspects of climate change—especially its significance, degree of human causation, potential negative impacts, and the necessity/possibility of ameliorating it. The former tends to be held by key actors in the denial machine, while the latter is more prevalent among broader sectors of the general public. Denialists (some of whom argue that the "denier" label is derogatory) like to label themselves as open-minded skeptics who are willing to revise their beliefs in lieu of additional evidence (Brin 2010; Diethelm and McKee 2009). Yet, Washington and Cook (2011) argue that denial and skepticism, especially when the latter is claimed by denialists trying to don the skeptical hat worn by all good scientists, are in fact opposing perspectives.

3. Indeed, scholars point to additional factors that contribute to this failure. First, as a "super-wicked" problem, climate change is an especially difficult subject for effective policymaking (Lazarus 2009). Second, despite reversing the denialist stance of the George W. Bush administration (McCright and Dunlap 2010), the Obama administration did not prioritize carbon emission reduction policies during the open "policy window" when Democrats briefly controlled the presidency and Congress (Pooley 2010). And third, there are many factors besides lack of leadership by the United States that have blocked effective international action, especially international and intergenerational equity considerations (McCright and Dunlap 2003; Roberts and Park 2007).

4. Douglas (2007:550) describes growthism as follows: "Its central tenet is that the global economy can keep on expanding indefinitely—for all practical purposes, forever. Its second principle is that growth is good, and that more growth is always better than less. Its third principle is that increasing growth should be society's over-riding priority."

5. Growing ideological and partisan polarization over environmental protection, first among political elites (institutionalized by the Gingrich-led Congress) and then the general public, began in the first half of the 1990s (McCright, Xiao, and Dunlap 2014).

6. Actually, the earliest efforts began at the end of the 1980s, following James Hansen's 1988 Senate testimony, but they clearly escalated in the early 1990s.

7. For a continually updated list of specious claims used to attack climate science, see http://www.skepticalscience.com.

8. See McCright and Dunlap (2010) for an extensive discussion of the specific non-decision-making techniques associated with manufacturing uncertainty that conservative activists have employed since the mid-1990s. Also, Hess (2014) focuses on key non-decision-making techniques by Republicans in control of Congress in recent years (e.g., trying to cut government research funding on climate change).

9. Perhaps the strongest evidence of the relatively strong degree of scientific consensus on the basic claims of climate science is that presented by Anderegg, Prall, Harold, and Schneider (2010).

10. Bonds (2010) notes how these efforts involve not just contesting but producing "knowledge," the latter referring to the creation of a body of (mis)information that is used to counter scientific findings.

11. Science journalist Sharon Begley (2007) introduced the term in her *Newsweek* cover story and others have adopted it.

12. For relatively brief but more in-depth (than we provide here) descriptions of most of the key components of the denial countermovement, see Dunlap and McCright (2010; 2011). Figure 10.1 in Dunlap and McCright (2011:147) is an attempt to visually portray the structure of, and interrelationship among, key components within the denial machine.

13. Social critic Naomi Klein (2011) also provides an in-depth look at the Heartland Institute's "International Conference on Climate Change."

14. Various lists of contrarian scientists have been published, but perhaps the most extensive (and updated) list can be found on the Research Database of DeSmog blog (which includes many other individuals in the denial countermovement): www.desmogblog.com.

15. For a rare study of astroturfing *within* the fossil fuels industry, see Mix and Waldo's (2014) insightful analysis of how a natural gas company employed an astroturf campaign to prevent construction of a coal-fired electrical plant that it viewed as a potential competitor.

16. In affiliation with a smaller front group (the American Tradition Institute), the Competitive Enterprise Institute has specialized in harassing scientists by filing Freedom of Information Act requests to gain access to their email and records (see Mann 2012:367).

17. There are nearly as many in other languages as well. A Portuguese blog identifies many of the leading denial websites in some European, Asian, and Latin American countries: http://ecotretas.blogspot.com/p/skeptical-views-in-non-english.html.

18. Having the denialosphere at its service provides the denial campaign a highly effective means of spreading its message, as reflected by its success in turning a tiny and highly unrepresentative sample of thirteen years' worth of personal e-mails hacked from the Climate Research Unit at the University of East Anglia into a major scandal that generated a decline in public belief in climate change and trust in climate scientists (Leiserowitz et al. 2012)—despite the fact that several investigations have concluded the e-mails neither demonstrate unethical behavior nor

undermine climate science (Holliman 2011; Powell 2011:Chapter 12). See Sheppard (2011) and Greenpeace (2011) for insightful analyses of the key players involved in manufacturing Climategate.

19. Dutch economist Labohm (2012), a leading figure in the denial countermovement, provides a upbeat overview of what he calls "climate scepticism" in European nations, noting individuals and/or groups in several additional nations, including Russia, Portugal, Denmark, Italy, Spain, the Czech Republic, Switzerland, and Germany. It is published in a small journal that caters to climate change denialists.

20. See Chapters 6 and 7 of this volume for the importance of adaptation and mitigation efforts.

REFERENCES

Anderegg, William R. L., James W. Prall, Jacob Harold, and Stephen H. Schneider. 2010. "Expert Credibility in Climate Change." *Proceedings of the National Academy of Sciences USA* 107:12107–12109.

Anshelm, Jonas and Martin Hultman. 2013. "A Green Fatwā: Climate Change as a Threat to the Masculinity of Industrial Modernity." *NORMA: International Journal for Masculinity Studies* 9:84–96.

Antonio, Robert. 2009. "Climate Change, the Resource Crunch, and the Global Growth Imperative." *Current Perspectives in Social Theory* 26:3–73.

Antonio, Robert J. and Robert J. Brulle. 2011. "The Unbearable Lightness of Politics: Climate Change Denial and Political Polarization." *Sociological Inquiry* 52:195–202.

Atkin, Emily. 2014. "The Fossil Fuel Industry Spent More Than $721 Million During 2014's Midterm Elections" (http://thinkprogress.org/climate/2014/12/23/3606630/fossil-fuel-spending-midterm-elections/).

Aykut, Stefan Cihan, Jean-Baptiste Comby, and Hélène Guillemot. 2012. "Climate Change Controversies in French Mass Media 1990–2010." *Journalism Studies* 13(2):157–174.

Bachrach, Peter and Morton S. Baratz. 1970. *Power and Poverty: Theory and Practice.* New York: Oxford University Press.

Barry, John. 2007. *Environment and Social Theory*, 2nd ed. London and New York: Routledge.

Beder, Sharon. 1998. "Public Relations' Role in Manufacturing Artificial Grass Roots Coalitions." *Public Relations Quarterly* 43:20–23.

Beder, Sharon. 1999. "Corporate Hijacking of the Greenhouse Debate." *The Ecologist* 29 (March/April):119–122.

Beder, Sharon. 2001. "Neoliberal Think Tanks and Free Market Fundamentalism." *Environmental Politics* 10:128–133.

Begley, Sharon. 2007. "The Truth about Denial." *Newsweek* April 13. CL(7):20–29.

Bellant, Russ. 1990. *The Coors Connection: How Coors Family Philanthropy Undermines Democratic Pluralism.* Cambridge, MA: Political Research Associates.

Berkhout, Frans. 2010. "Reconstructing Boundaries and Reason in the Climate Debate." *Global Environmental Change* 20:565–569.

Blumenthal, Sidney. 1986. *The Rise of the Counter-Establishment*. New York: Times Books.

Bockman, Johanna. 2013. "Neoliberalism." *Contexts* 12:14–15.

Bonds, Eric. 2010. "The Knowledge-Shaping Process: Elite Mobilization and Environmental Policy." *Critical Sociology* 37:429–446.

Boykoff, Maxwell T. 2013. "Public Enemy No. 1? Understanding Media Representations of Outlier Views on Climate Change." *American Behavioral Scientist* 57:796–817.

Boykoff, Maxwell T. and Shawn K. Olson. 2013. "Wise Contrarians: A Keystone Species in Contemporary Climate Science, Politics and Policy." *Celebrity Studies* 4:276–291.

Brin, David. 2010. "Climate Skeptics v. Deniers." *Skeptic* 15(4):13–17.

Brulle, Robert J. 2014. "Institutionalizing Delay: Foundation Founding and the Creation of U.S. Climate Change Counter-movement Organizations." *Climatic Change* 122:681–694.

Brulle, Robert J., Jason Carmichael and J. Craig Jenkins. 2012. "Shifting Public Opinion on Climate Change: An Empirical Assessment of Factors Influencing Concern over Climate Change in the U.S., 2002–2010." *Climatic Change* 114:169–188.

Brysse, Keynyn, Naomi Oreskes, Jessica O'Reilly, and Michael Oppenheimer. 2013. "Climate Change Prediction: Erring on the Side of Least Drama?" *Global Environmental Change* 23:327–337.

Catton, William R. Jr., and Riley E. Dunlap. 1980. "A New Ecological Paradigm for Post-Exuberant Sociology." *American Behavioral Scientist* 24:15–47.

Ceccarelli, Leah. 2011. "Manufactured Scientific Controversy: Science, Rhetoric, and Public Debate." *Rhetoric & Public Affairs* 14:195–228.

Ceccarelli, Leah. 2013. "Controversy over Manufactured Scientific Controversy: A Rejoinder to Fuller." *Rhetoric & Public Affairs* 16:761–766.

Cockett, Richard. 1995. *Thinking the Unthinkable: Think-Tanks and the Economic Counter-Revolution, 1931–83*. London: HarperCollins.

Cray, Charlie and Peter Montague. 2014. *The Kingpins of Carbon and Their War on Democracy*. Greenpeace (http://www.greenpeace.org/usa/Global/usa/planet3/PDFs/Kingpins-of-Carbon.pdf).

DeKraker, Joop, Sacha Juijs, Ron Corvers, and Astrid Offermans. 2014. "Internet Public Opinion on Climate Change: A World Views Analysis of Online Reader Comments." *International Journal of Climate Change Strategies and Management* 6:19–33.

Diethelm, Pascal and Martin McKee. 2009. "Denialism: What Is It and How Should Scientists Respond?" *European Journal of Public Health* 19(1):2–4.

Donald, Rosalind. 2011. "Cool Dudes and Libtards: How Online Incivility, Polarization and Politicisation are Influencing Climate Science and Policy." Unpublished MA Thesis, Center for International Studies and Diplomacy, School of Oriental and African Studies, University of London.

Douglas, Richard. 2007. "Growthism and the Green Backlash." *Political Quarterly* 78:547–555.

Douglas, Richard. 2009. "The Green Backlash: Scepticism or Scientism?" *Social Epistemology* 23:145–163.

Dunlap, Riley E. 1987. "Polls, Pollution, and Politics Revisited: Public Opinion on the Environment in the Reagan Era." *Environment* 29:6–11, 32–37.

Dunlap, Riley E. 2014. "Clarifying Anti-Reflexivity: Conservative Opposition to Impact Science and Scientific Evidence." *Environmental Research Letters* 9; doi:10.1088/1748-9326/9/2/021001.

Dunlap, Riley E. and Peter J. Jacques. 2013. "Climate Change Denial Books and Conservative Think Tanks: Exploring the Connection." *American Behavioral Scientist* 57:699–731.

Dunlap, Riley E. and Aaron M. McCright. 2010. "Climate Change Denial: Sources, Actors and Strategies." Pp. 240–259 in *Routledge Handbook of Climate Change and Society*, edited by C. Lever-Tracy. New York and London: Routledge.

Dunlap, Riley E. and Aaron M. McCright. 2011. "Organized Climate Change Denial." Pp. 144–160 in *Oxford Handbook of Climate Change and Society*, edited by J. Dryzek, R. Norgaard, and D. Schlosberg. Cambridge: Oxford University Press.

Dunlap, Riley E. and Kent Van Liere. 1984. "Commitment to the Dominant Social Paradigm and Concern for Environmental Quality." *Social Science Quarterly* 65:1013–1028.

Elsasser, Shaun W. and Riley E. Dunlap. 2013. "Leading Voices in the Denier Choir: Conservative Columnists' Dismissal of Global Warming and Denigration of Climate Science." *American Behavioral Scientist* 57:754–776.

Farquharson, Karen. 2003. "Influencing Policy Transnationally: Pro- and Anti-Tobacco Global Advocacy Networks." *Australian Journal of Public Administration* 62(4):80–92.

Feldman, Lauren, Edward W. Maibach, Connie Roser-Renouf, and Anthony Leiserowitz. 2012. "Climate on Cable: The Nature and Impact of Warming Coverage on Fox News, CNN, and MSNBC." *International Journal of Press/Politics* 17:3–31.

Forecast the Facts. 2014. "5 Ways ALEC Denies the Facts of Climate Change." Forecast the Facts (https://s3.amazonaws.com/s3.forecastthefacts.org/images/ResponsetoALECPressStatement.pdf).

Foster, John Bellamy, Brett Clark, and Richard York. 2011. *The Ecological Rift: Capitalism's War on Earth*. New York: Monthly Review Press.

Freudenburg, William R. and Violetta Muselli. 2010. "Global Warming Estimates, Media Expectations, and the Asymmetry of Scientific Challenge." *Global Environmental Change* 20:483–491.

Freudenburg, William R. and Violetta Muselli. 2013. "Reexamining Climate Change Debates: Scientific Disagreement or Scientific Certainty Argumentation Methods (SCAMs)?" *American Behavioral Scientist* 57:777–785.

Gavin, Neil T. and Tom Marshall. 2011. "Mediated Climate Change in Britain: Skepticism on the Web and on Television around Copenhagen." *Global Environmental Change* 21:1035–1044.

Gelbspan, Ross. 1997. *The Heat is On*. Reading, MA: Addison-Wesley.

Gelbspan, Ross. 2004. *Boiling Point*. New York: Basic Books.

Germain, Tiffany. 2014. "The Anti-Science Climate Denier Caucus: 113th Congress Edition." ClimateProgress (http://thinkprogress.org/climate/2013/06/26/2202141/anti-science-climate-denier-caucus-113th-congress-edition).

Goddell, Jeff. 2010. "As the World Burns." *Rolling Stone* No. 1096 (January 21):30–34, 62.

Goldman, Gretchen and Christina Carlson. 2014. *Tricks of the Trade—How Companies Anonymously Influence Climate Policy through their Business and Trade Associations*. Center for Science and Democracy at the Union of Concerned Scientists (http://www.ucsusa.org/sites/default/files/legacy/assets/documents/center-for-science-and-democracy/tricks-of-the-trade.pdf).

Gore, Al. 2006. *An Inconvenient Truth*. Emmaus, PA: Rodale Books.

Greenberg, Josh, Graham Knight, and Elizabeth Westersund. 2011. "Spinning Climate Change: Corporate and NGO Public Relations Strategies in Canada and the United States." *International Communication Gazette* 73:65–82.

Greenpeace. 2007. *ExxonMobil's Continued Funding of Global Warming Denial Industry*. Washington, DC: Greenpeace USA (http://www.greenpeace.org/usa/global/usa/binaries/2007/5/exxon-secrets-analysis-of-fun.pdf).

Greenpeace. 2010. *Koch Industries: Secretly Funding the Climate Denial Machine*. Washington, DC: Greenpeace (http://www.greenpeace.org/usa/en/media-center/reports/koch-industries-secretly-fund/).

Greenpeace. 2011. *Koch Industries: Still Fueling Climate Denial—2011 Update*. Washington, DC: Greenpeace USA (http://www.greenpeace.org/usa/en/media-center/reports/Koch-Industries-Still-Fueling-Climate-Denial-2011-Update/).

Greenpeace. 2013. *Dealing in Doubt: The Climate Denial Machine vs Climate Science* (http://www.greenpeace.org/usa/en/campaigns/global-warming-and-energy/polluterwatch/Dealing-in-Doubt---the-Climate-Denial-Machine-vs-Climate-Science/).

Grundmann, Reiner. 2007. "Climate Change and Knowledge Politics." *Environmental Politics* 16:414–432.

Grundmann, Reiner and Mike Scott. 2014. "Disputed Climate Science in the Media: Do Countries Matter?" *Public Understanding of Science* 23:220–235.

Hamilton, Clive. 2007. *Scorcher: The Dirty Politics of Climate Change*. Melbourne, Australia: Black Inc. Agenda.

Harkinson, Josh. 2009. "Climate Change Deniers Without Borders." *Mother Jones* (http://motherjones.com/print/33941).

Harvey, David. 2005. *A Brief History of Neoliberalism*. Oxford: Oxford University Press.

Hess, David J. 2014. "When Green Became Blue: Epistemic Rift and the Corralling of Climate Science." *Political Power and Social Theory* 27:123–153.

Hoffman, Andrew J. 2011a. "Talking Past Each Other? Cultural Framing of Skeptical and Convinced Logics in the Climate Change Debate." *Organization and Environment* 24:3–33.

Hoffman, Andrew J. 2011b. "The Culture and Discourse of Climate Skepticism." *Strategic Organization* 9:1–8.

Hoggan, James and Richard Littlemore. 2009. *Climate Cover-Up: The Crusade to Deny Global Warming*. Vancouver, BC: Greystone Books.

Holliman, Richard. 2011. "Advocacy in the Tail: Exploring the Implications of 'Climategate' for Science Journalist and Public Debate in the Digital Age." *Journalism* 12:832–846.

IPCC. 2014. *Climate Change 2014: Mitigation of Climate Change*. Contribution of Working Group III to the Fifth Assessment Report of the Intergovernmental Panel on Climate Change. Cambridge: Cambridge University Press.

Jacques, Peter J., Riley E. Dunlap, and Mark Freeman. 2008. "The Organization of Denial: Conservative Think Tanks and Environmental Scepticism." *Environmental Politics* 17:349–385.

Jamieson, Dale. 2014. *Reason in a Dark Time*. New York: Oxford.

Jamieson, Kathleen Hall and Joseph N. Cappella. 2008. *Echo Chamber: Rush Limbaugh and the Conservative Media Establishment*. New York: Oxford.

328 Climate Change and Society

Keck, Margaret E. and Kathryn Sikkink. 1999. *Activists Beyond Borders: Advocacy Networks in International Politics*. Ithaca, NY: Cornell University Press.

Klein, Naomi. 2011. "Capitalism vs. The Climate." *The Nation* 293(22):11–20.

Knight, Graham and Josh Greenberg. 2011. "Talk of the Enemy: Adversarial Framing and Climate Change Discourse." *Social Movement Studies* 10:323–340.

Koebler, Jason. 2014. "The House Science Committee Spent Today in a Climate Change Denial Echo Chamber." *Motherboard* (http://motherboard.vice.com/read/the-house-science-committee-spent-all-day-proudly-denying-climate-change).

Koteyko, Nelya, Rusi Jaspal, and Brigitte Nerlich. 2012. "Climate Change and 'Climategate' in Online Reader Comments: A Mixed Methods Study." *Geography Journal* 179:74–86.

Kraft, Michael E. and Norman E. Vig. 1984. "Environmental Policy in the Reagan Presidency." *Political Science Quarterly* 99:415–439.

Labohm, Hans. 2012. "Climate Scepticism in Europe." *Energy & Environment* 23:1311–1317.

Lack, Martin. 2013. *The Denial of Science: Analysing Climate Change Scepticism in the UK*. Bloomington, IN: AuthorHouse.

Lahsen, Myanna. 1999. "The Detection and Attribution of Conspiracies: The Controversy over Chapter 8." Pp. 111–136 in *Paranoia Within Reason: A Casebook on Conspiracy as Explanation*, edited by G. E. Marcus. Chicago: University of Chicago Press.

Lahsen, Myanna. 2005. "Seductive Simulations? Uncertainty Distribution Around Climate Models." *Social Studies of Science* 35:895–922.

Lahsen, Myanna. 2008. "Experiences of Modernity in the Greenhouse: A Cultural Analysis of a Physicist 'Trio' Supporting the Backlash Against Global Warming." *Global Environmental Change* 18:204–219.

Lapham, Lewis H. 2004. "Tentacles of Rage." *Harper's Magazine* 309:31–41.

Layzer, J. 2007. "Deep Freeze." Pp. 93–125 in *Business and Environmental Policy*, edited by M. E. Kraft and S. Kamieniecki. Cambridge, MA: MIT Press.

Lazarus, Richard L. 2009. "Super Wicked Problems and Climate Change." *Cornell Law Review* 94:1153–1234.

Leiserowitz, Anthony, Edward Maibach, Connie Roser-Renouf, Geoff Feinberg, and Seth Rosenthal. 2014. *Climate Change in the American Mind: April, 2014*. New Haven, CT: Yale Project on Climate Change Communication.

Leiserowitz, Anthony A., Edward W. Maibach, Connie Roser-Renouf, Nicholas Smith, and Erica Dawson. 2012. "Climategate, Public Opinion, and the Loss of Trust." *American Behavioral Scientist* 57:818–837.

Levy, David L. and Andre Spicer. 2013. "Contested Imaginaries and the Cultural Political Economy of Climate Change." *Organization* 20:659–678.

Lo, Clarence Y. H. 1982. "Countermovements and Conservative Movements in the Contemporary U.S." *Annual Review of Sociology* 8:107–134.

Lockwood, Alex. 2011. "Seeding Doubt: How Skeptics Have Used New Media to Delay Action on Climate Change." *Geopolitics, History, and International Relations* 2:136–164.

Lomborg, Bjorn. 2001. *The Skeptical Environmentalist: Measuring the Real State of the World*. Cambridge: Cambridge University Press.

Lukes, Steven. 1974. *Power: A Radical View*. London: Macmillan Press.

Lynch, Michael J., Ronald G. Burns, and Paul B. Stretesky. 2010. "Global Warming and State-Corporate Crime: The Politicization of Global Warming under the Bush Administration." *Crime, Law and Social Change* 50:213–239.

Malone, Elizabeth. 2009. *Debating Climate Change*. London: Earthscan.

Mann, Michael E. 2012. *The Hockey Stick and the Climate Wars*. New York: Columbia University Press.

Mayer, Jean. 2013. "Koch Pledge Tied to Congressional Climate Inaction." *The New Yorker* (http://www.newyorker.com/news/news-desk/koch-pledge-tied-to-congressional-climate-inaction).

McCammack, Brian. 2007. "Hot Damned America: Evangelicalism and the Climate Change Policy Debate." *American Quarterly* 59:645–668.

McCright, Aaron M. 2007. "Dealing With Climate Change Contrarians." Pp. 200–212 in *Creating a Climate for Change: Communicating Climate Change and Facilitating Social Change*, edited by S. C. Moser and L. Dilling. New York: Cambridge University Press.

McCright, Aaron M. 2011. "Political Orientation Moderates Americans' Beliefs and Concern about Climate Change." *Climatic Change* 104:243–253.

McCright, Aaron M. and Riley E. Dunlap. 2000. "Challenging Global Warming as a Social Problem: An Analysis of the Conservative Movement's Counter Claims." *Social Problems* 47(4):499–522.

McCright, Aaron M. and Riley E. Dunlap. 2003. "Defeating Kyoto: The Conservative Movement's Impact on U.S. Climate Change Policy." *Social Problems* 50(3):348–373.

McCright, Aaron M. and Riley E. Dunlap. 2010. "Anti-Reflexivity: The American Conservative Movement's Success in Undermining Climate Science and Policy." *Theory, Culture, and Society* 27(2–3):100–133.

McCright, Aaron M. and Riley E. Dunlap. 2011a. "Cool Dudes: The Denial of Climate Change among Conservative White Males." *Global Environmental Change* 21:1163–1172.

McCright, Aaron M. and Riley E. Dunlap. 2011b. "The Politicization of Climate Change and Polarization in the American Public's Views of Global Warming, 2001–2010." *Sociological Quarterly* 52:155–194.

McCright, Aaron M., Chenyang Xiao, and Riley E. Dunlap. 2014. "Political Polarization on Support for Government Spending on Environmental Protection in the USA, 1974–2012." *Social Science Research* 48:251–260.

McKewon, Elaine. 2012a. "Dueling Realities: Conspiracy Theories *vs* Climate Science in Regional Newspaper Coverage of Ian Plimer's Book, *Heaven and Earth*." *Rural Society* 21:99–115.

McKewon, Elaine. 2012b. "Talking Points Ammo: The Use of Neoliberal Think Tank Fantasy Themes to Delegitimise Scientific Knowledge of Climate Change in Australian Newspapers." *Journalism Studies* 13:277–297.

McKnight, David. 2010. "A Change in the Climate? The Journalism of Opinion at News Corporation." *Journalism* 11:693–706.

Michaels, David. 2008. *Doubt Is Their Product*. Oxford: Oxford University Press.

Milbrath, Lester. 1984. *Environmentalists: Vanguard for a New Society*. Albany: SUNY Press.

Mirowski, Philip. 2008. "The Rise of the Dedicated Natural Science Think Tank." Unpublished manuscript, Social Science Research Council. New York: SSRC.

Mirowski, Philip and Dieter Plehwe (Eds.). 2009. *The Road from Mont Pelerin: The Making of the Neoliberal Thought Collective.* Cambridge, MA: Harvard University Press.

Mix, Tamara L. and Kristin G. Waldo. 2014. "Know(ing) Your Power: Risk Society, Astroturf Campaigns, and the Battle over the Red Rock Coal-Fired Plant." *Sociological Quarterly* 56:125–151.

Molotch, Harvey. 1970. "Oil in Santa Barbara and Power in America." *Sociological Inquiry* 40:131–144.

Mooney, Chris. 2005. "Some Like It Hot." *Mother Jones* 30(3):36–49.

Nature. 2010. "Climate of Fear." *Nature* 464(7286):141.

Neubauer, Robert. 2011. "Manufacturing Junk: Think Tanks, Climate Denial, and Neoliberal Hegemony." *Australian Journal of Communication* 38:65–88.

Niederer, Sabine. 2013. "'Global Warming is Not a Crisis!' Studying Climate Change Skepticism on the Web." *European Journal of Media Studies* No. 3 (Spring) (http:www.necsus-ejms.org/global-warming-is-not-a-crisis-studying-climate-change-skepticism-on-the-web).

Oreskes, Naomi. 2010. "My Facts are Better than Your Facts: Spreading Good News about Global Warming." Pp. 135–166 in *How Well Do Facts Travel?* edited by P. Howlett and M. S. Morgan. Cambridge: Cambridge University Press.

Oreskes, Naomi and Erik M. Conway. 2010. *Merchants of Doubt.* New York: Bloomsbury Press.

Ozone Action. 1996a. "Ties That Blind I: Case Studies of Corporate Influence on Climate Change Policy." *Ozone Action Report.* Washington, DC: Ozone Action.

Ozone Action. 1996b. "Ties That Blind II: Parading Opinion as Scientific Truth." *Ozone Action Report.* Washington, DC: Ozone Action.

Painter, James and Teresa Ashe. 2012. "Cross-national Comparison of the Presence of Climate Skepticism in the Print Media in Six Countries, 2007–2010." *Environmental Research Letters* 7; doi: 10.1088/1748-9326/7/4/044005.

Palley, Thomas I. 2005. "From Keynesianism to Neoliberalism: Shifting Paradigms in Economics." Pp. 20–30 in *Neoliberalism: A Critical Reader*, edited by A. Saad-Filho and D. Johnston. London: Pluto Press.

Park, Huung Sam, Xinsheng Liu, and Arnold Vedlitz. 2014. "Analyzing Climate Change Debates in the U.S. Congress: Party Control and Mobilizing Networks." *Risk, Hazards & Crisis in Public Policy* 5:239–258.

People for the American Way. 2011. *The 'Green Dragon' Slayers: How the Religious Right and the Corporate Right Are Joining Forces to Fight Environmental Protection.* Washington, DC: People for the American Way.

Perrow, Charles. 2010. "Organizations and Global Warming." Pp. 59–77 in *Routledge Handbook of Climate Change and Society*, edited by C. Lever-Tracy. New York and London: Routledge.

Plehwe, Dieter. 2014. "Think Tank Networks and the Knowledge-Interest Nexus: The Case of Climate Change." *Critical Policy Studies* 8:101–115.

Pooley, Eric. 2010. *The Climate War: True Believers, Power Brokers, and the Fight to Save the Earth.* New York: Hyperion.

Powell, James Lawrence. 2011. *The Inquisition of Climate Science.* New York: Columbia University Press.

Rahmstorf, Stefan. 2004. "The Climate Skeptics." Pp. 76–83 in *Weather Catastrophes and Climate Change*, edited by Munich Re. Munich, Germany: Munich Re.

Roberts, J. Timmons and Bradley C. Parks. 2007. *A Climate of Injustice: Global Inequality, North-South Politics, and Climate Policy.* Cambridge, MA: MIT Press.

Sabatier, Paul A. and Hank C. Jenkins-Smith. 1993. *Policy Change and Learning: An Advocacy Coalition Approach.* Boulder, CO: Westview Press.

Sarewitz, Daniel. 2004. "How Science Makes Environmental Controversies Worse." *Environmental Science and Policy* 7:385–403.

Schnaiberg, Allan. 1980. *The Environment: From Surplus to Scarcity.* New York: Oxford University Press.

Sharman, Amelia. 2014. "Mapping the Climate Skeptical Blogosphere." *Global Environmental Change* 26:159–170.

Sheppard, Kate. 2011. "The Hackers and the Hockey Stick." *Mother Jones* 36 (May/June):33–39, 65.

Simon, Julian. 1981. *The Ultimate Resource.* Princeton, NJ: Princeton University Press.

Skocpol, Theda. 2012. "Naming the Problem: What It Will Take to Counter Extremism and Engage Americans in the Fight Against Global Warming." Prepared for symposium on *The Politics of America's Fight Against Global Warming.* Department of Government, Harvard University, Cambridge, MA, February 14, 2013.

Stefancic, Jean and Richard Delgado. 1996. *No Mercy: How Conservative Think Tanks and Foundations Changed America's Social Agenda.* Philadelphia: Temple University Press.

Swidler, Ann. 1986. "Culture in Action: Symbols and Strategies." *American Sociological Review* 51:273–286.

Switzer, Jacqueline Vaughn. 1997. *Green Backlash.* Boulder, CO: Lynne Rienner Publishers.

Tollefson, Jeff. 2011. "The Sceptic Meets His Match." *Nature* 475:440–441.

Union of Concerned Scientists. 2007. *Smoke, Mirrors, and Hot Air.* Cambridge, MA: Union of Concerned Scientists (http://www.ucsusa.org/sites/default/files/legacy/assets/documents/global_warming/exxon_report.pdf).

Union of Concerned Scientists. 2012. *A Climate of Corporate Control: How Corporations Have Influenced the U.S. Dialogue on Climate Science and Policy.* Cambridge, MA: Union of Concerned Scientists (http://www.ucsusa.org/sites/default/files/legacy/assets/documents/scientific_integrity/a-climate-of-corporate-control-report.pdf).

U.S. National Research Council. 2010. *Informing an Effective Response to Climate Change.* Washington, DC: The National Academies Press.

Van Soest, Jan Paul. 2011. *Clogs in the Works: The Transition to a Sustainable Energy Economy and Its Deliberate and Unconscious Sabotage.* The Netherlands: De Gemeynt, Klarenbeek. PB2011-007.

Washington, Haydn and John Cook. 2011. *Climate Change Denial: Heads in the Sand.* London: Earthscan.

Wilkinson, Katharine K. 2012. *Between God and Green: How Evangelicals Are Cultivating a Middle Ground on Climate Change.* New York: Oxford University Press.

Yearley, Steven. 1991. *The Green Case.* London: Harper Collins Academic.

Yearley, Steven. 2005. "The Sociology of the Environment and Nature." Pp. 314–326 in *The Sage Handbook of Sociology,* edited by C. Calhoun, C. Rojek, and B. Turner. Thousand Oaks, CA: Sage.

Young, Nathan. 2013. "Working the Fringes: The Role of Letters to the Editor in Advancing Non-standard Media Narratives about Climate Change." *Public Understanding of Science* 22:443–459.

Young, Nathan and Aline Coutinho. 2014. "Government, Anti-Reflexivity, and the Construction of Public Ignorance about Climate Change: Australia and Canada Compared." *Global Environmental Politics* 13:89–108.

Zaccai, Edwin, Fracois Gemenne, and Jean-Michel Decroly. 2012. *Controverses Climatiques: Sciences et Politique.* Paris: Sciences Po Press.

11

The Climate Change Divide in Social Theory

Robert J. Antonio and Brett Clark

INTRODUCTION

Climate scientists hold that climate change has anthropogenic causes, or is driven by socially organized human activities, especially production and consumption practices, which have increased atmospheric greenhouse gas (GHG) emissions. Direct drivers are embedded in much wider social conditions (e.g., economic, technical, industrial, and governance systems, cultural traditions, social values, and ideal and material interests), which vary locally, regionally, and nationally. Climate change's biophysical impacts (e.g., higher sea levels and storm surges, greater storm intensity, increased vulnerability to floods in some regions and droughts in others) are projected to have substantial and potentially catastrophic social consequences (Stocker and Qin 2013). Some localities already suffer severe impacts (e.g., forced migrations in certain island nations, eroded traditional economies among circumpolar First Peoples, and damaged homes and infrastructure in seaside areas in many nations). Climate change's social costs are anticipated to grow substantially given the lagged impacts of current GHG levels (due to "thermal inertia" and other climate system facets) and additional GHG emissions, accelerated by continuing globalization. Efforts to mitigate and adapt to climate change involve myriad social processes. All its facets are socially mediated; even its status as a public "problem" is socially constructed in multiple ways and contested by different groups. Climate change has enormous social complexity as well as naturalistic complexity, demanding plural types of inquiry and knowledge, and social theories to integrate specialized research from diverse sociological subareas, illuminate connections between sociological work and interdisciplinary knowledge, and assess critically how this knowledge complex bears on policy debates.

Possibly reflecting on conditions that inspired the "Objectivity" essay, which the following passage helped conclude, Max Weber ([1904] 2011:112) asserted

> All research in the cultural sciences in an age of specialization, once it is oriented towards a given subject matter through particular settings of problems and has established its methodological principles, will consider the analysis of the data as an end in itself. It will discontinue assessing the value of the individual facts in terms of their relationships to ultimate value-ideas. Indeed, it will lose its awareness of its ultimate rootedness in the value-ideas in general. And it is well that should be so. But there comes a moment when the atmosphere changes. The significance of the unreflectively utilized viewpoints becomes uncertain and the road is lost in the twilight. The light of the great cultural problems moves on. Then science too prepares to change its standpoint and its thinking apparatus and to view the streams of events from the heights of thought . . .

Weber anticipated that sweeping sociocultural ruptures would one day stir specialized social scientists to bring their taken-for-granted presuppositions into view, think critically about their practices' directions, and entertain major switches of tracks, preparing the way for new methods, theories, and policies. He also declared famously that the modern economy's expansionary, materialist cultural ethos, the "spirit of capitalism" wed to "machine-based production," might exert its inexorable determining social force "until the last ton of fossil fuel has burnt to ashes." He penned these words about the "iron cage's" ecological consequences a few months after his 1904 trip to the United States, where he was awed by the intensity of capitalist enterprise and warned, at the St. Louis World's Fair, about its heedless expenditure of nonrenewable resources (Antonio 2009:3–4, 34 [note 2]; Weber [1918] 1958, [1920] 2011:177).[1] Leading climate scientists' projections (e.g., Hansen 2009:ix–x, 223–277) about the scope, diversity, and severity of climate change impacts, social consequences, and urgency of mitigation and adaptation portend the type of rupture that Weber spoke of more than a century ago. Whether or not such a moment already is upon us is at the center of scientific, policy, and social theory debates about climate change.

Natural science climate change theory and research have long been robust; potential ecological and social impacts and related policy issues have been discussed and debated intensely for more than two decades by scientists and in worldwide policy circles, think tanks, nongovernmental organizations, and other social sites and media. Climate change's substantial risks and uncertainties, potentially enormous mitigation and adaptation costs, and possible major sociopolitical adjustments spur sharp value and interest conflicts and divergent visions of policy. Sociologists and other social thinkers have framed broad empirically informed, policy-oriented social theories to weigh in on these debates (e.g., Foster, Clark, and York 2010; Giddens 2011; Hulme 2009; Koch 2012; Stehr and von Storch 2010; Szerszynski 2010; Urry 2011). More narrowly focused empirical models and conceptual frameworks, employed in specialized sociological research programs about climate change and discussed in other chapters, are mostly beyond the scope of this chapter. We focus on social theories that criticize or formulate alternatives to the normative directions of research, related sociopolitical policies, and their epistemic presuppositions.[2]

The social theories of climate change discussed here address varied methodological and substantive issues from divergent normative standpoints. Consequently, the discourse has numerous rifts, major and minor. Some of the more intense splits have been over humanity's embeddedness in the biosphere, relative autonomy from natural limits, and capacity to culturally construct nature. These epistemic divisions have been intertwined with intense substantive disagreements over the certainty and social relations of climate science, state-driven versus market-oriented climate change policies, the role of civil society in climate governance, modernization's and globalization's impact on climate and climate policies, and the degree of path dependency of a carbon-based economy and society. The sharpest splits concern the urgency and scale of mitigation and adaptation efforts, the relative emphasis given to these respective strategies, and ultimately their relevance for maintaining or altering economic growth and the policy regimes that drive it. However, disagreement about the normative priority and prudence of capitalism's growth imperative and empirical veracity of its sustainability arguably constitutes the most fundamental, albeit often tacit, source of division among social theories of climate change. Many thinkers who doubt the perpetuity of infinite unplanned, exponential growth hesitate to confront the topic directly because it leads in politically "unrealistic" or "divisive" directions, especially to the question of the sustainability of capitalism as we have known it. Divisions over whether global growth is the ultimate driver of climate change and its relationship to neoliberalism and capitalism per se and the significance of these matters for catastrophic risk, mitigation, adaptation, and overall sociopolitical reform are the most intellectually and culturally significant facet of social theory debates over warming.

INTERROGATING THE GROWTH IMPERATIVE:
THE PRECURSORY METATHEORETICAL MOVE

Over thirty years ago, William R. Catton Jr. and Riley E. Dunlap (1978, 1980:25–34) made a consequential, enduring social theory intervention affecting substantially environmental sociology (then a new, marginal disciplinary subarea). Warning about the consequences of failing to attend to limits to growth and exceeding global carrying capacity, they stressed the need to strive for a sustainable economy and democratic culture, which they asserted depend also on efforts to forge a more equal society. Catton and Dunlap identified the "human exemptionalist paradigm"—a complex of presuppositions and beliefs that treat culture as if it were independent from the finite "web of nature" and our unique capacity for intelligent, creative mediation of natural environments (via a multitude of symbolic and material instrumentalities) as if it could allow us to transcend the biophysical limits that constrain all other organisms. They contested exemptionalist claims that markets and science can drive exponential growth ad infinitum and simultaneously eliminate ecological problems or at least keep them within manageable limits.

The highly influential, University of Chicago–trained business economist Julian Simon (1981) epitomized the human exemptionalist paradigm, wedding a vision of nature as an infinitely pliable, blank tablet for cultural inscription to emphatic, arrant antiregulatory politics. Simon is one of the most important roots of today's conception of the "neoliberal entrepreneurial agent" that freely shapes a nonresistant biophysical world (Pellizzoni 2011:797). After the long postwar economic expansion ended, Simon argued that human creative capacities, liberated from Keynesian political regulation and motivated by unconstrained market competition, ensure that wealth and human well-being will grow, natural resources will become less scarce, and the environment will be made cleaner. His ideas held sway in the Reagan administration and among ascendant policy elites who opposed environmentalism. Simon championed the then resurgent neoliberal *growth imperative*, which valorizes unplanned, exponential economic expansion, or ever-increasing production to fulfill ever-expanding individual appetites and ever-growing numbers of consumers. The conventional standard for measuring growth is gross domestic product (GDP), or aggregated market value of all final goods and services. Although GDP does not take account of economic inequality, human well-being, or ecological sustainability, its continuous expansion is equated with development per se. Neoliberalism's extension of the deregulated market logic into nearly all areas of life increases dependence on growth ad infinitum. Catton and Dunlap's critique of the human exemptionalist paradigm contradicts the neoliberal growth imperative, rejecting outright its Simonian presuppositions, which they implied would likely rule in the short run but cannot be sustained.

Catton and Dunlap (1978, 1980) called for a new ecological paradigm to take account of society's embeddedness in nature and generate interdisciplinary inquiries into culture–nature relations. They hoped to bring environmental sociology's core focus on this interaction to the attention of sociology and social science per se, and ultimately sensitize policymakers and the public to the biophysical impacts and limits of socially organized human activities. They intended their metatheoretical move to be a precursory step toward envisioning a postgrowth, "steady-state society" capable of preserving the Holocene-like ecology that has sustained complex civilizations and other species with which we share the planet. Anticipating a major ecological rupture and social rupture driven by excessive growth, Catton and Dunlap called for a "Copernican turn" to reconstruct accordingly social science's epistemic presuppositions, bring biophysical problems and their social roots and consequences to the center of sociological inquiry, and transform public policies and ways of life. They implied an approaching moment like Weber forewarned. Although Dunlap and Catton did not have climate change in mind when they formulated the human exemptionalist paradigm and the new ecological paradigm (personal communication from Dunlap, May 29, 2013), increasing globalization, rising GHG emissions and temperatures, and growing climatic and social impacts make their project all the more urgent today. Their aims are still works in progress, but the need for inquiry about climate change's interactive complex of sociocultural and natural factors, exceptionally serious risks, and planetary scale require scuttling exemptionalism and addressing the sustainability of the growth imperative and political-economic regimes that drive it. The theories of climate change discussed here engage and debate various facets of the Catton and Dunlap project.[3]

THE REALIST–CONSTRUCTIVIST SPLIT OVER CLIMATE CHANGE SCIENCE AND CLIMATE POLICY

In 1987, the Brundtland Report called for global sustainable development. A year later, the Intergovernmental Panel on Climate Change (IPCC) was formed, and James Hansen delivered his dramatic testimony before the U.S. Congress, declaring that climate change had begun and is driven primarily by burning fossil fuel. Environmental sociologists began to address climate change during the early 1990s, in the midst of the major political-economic and cultural transformations that ended the Cold War era. Climate change debates in environmental sociology and the social sciences arose in the context of the new finance-led transnational capitalism, featuring fluid national borders; reduced sovereignty; geographically mobile, hybrid people and cultures; and new global risks. After communism, the Right and the Left attacked state redistribution and planning. Blurring differences between labor and social democratic parties and their conservative opponents, the

new "centrism" stressed bottom-up policymaking led by civil society organizations and market forces, facilitated by postinterventionist states and
public–private partnerships. The neoliberal policy regime was fashioned
to unleash markets and accelerate stalled growth by promoting free trade,
weakening labor unions, and cutting welfare state provision, worker protections, and environmental regulation. Although implemented unevenly,
neoliberal policymaking has been hegemonic in transnational, economic
governance organizations and has had global impact. Neoliberals have held
that environmental regulation and especially climate change mitigation
would require draconian state intervention, choking economic growth and
freedom (e.g., Will 2008).

Entwined with globalization, cultural postmodernization stressed new
sensibilities—difference, locality, fragmentation, uncertainty, and perspec-
tivism. Postmodernist theorists favored culture over structure, linguistic
deconstruction over political-economic explanation, parts over totalities,
and dispersed authority over centralized power (Antonio 1998). The new
cultural theories stirred sharp splits in the humanities and social sciences,
giving rise in the 1990s to "science wars" featuring battles over "realism"
and "social construction" and conflicts over science's nature, normative
directions, and social impacts (e.g., Bhaskar 1986; Labinger and Collins
2001; Latour and Woolgar 1979; Sokal and Bricmont 1998; Woolgar 1988).
The new sociopolitical context and wider social theory debate gave rise to
the "realism–constructivism" split in environmental sociology and other
social sciences, especially geography (e.g., Burningham and Cooper 1999;
Buttel and Taylor 1992; Carolan 2005a, 2005b; Demeritt 2001; Dunlap 2010;
Hannigan 1995; Murphy 1994b; 2006; Yearley 1991).

Realists stressed the urgency of climate change and other global environmental problems (Dunlap and Catton 1994). They generally trusted scientists' warnings about potential catastrophic risks, arguing that unregulated,
exponential economic growth, accelerated by globalization, is a core driver
of these problems or *the* main driver of them. They questioned the sustainability of the growth imperative and thereby, at least indirectly, capitalism as
we have known it. They also criticized skeptics' efforts to undermine public
reception of climate science by spreading pseudoscientific claims and con-
spiratorial fears about collusion among scientists, environmentalists, and
bureaucrats. Realists supported international collective action to cope with
climate change and other serious ecological problems, but, in fending off
anticapitalist and socialist charges by skeptics, they were often indirect about
the likely economic and political consequences of their mitigation schemes
for sustaining growth and transforming the hegemonic political-economic
regime. By contrast, constructivists held that realists disregard locality; oversimplify causality; exaggerate risks; cultivate alarmism; justify technocratic,

top-down policy formation; and treat Third World nations unfairly. They charged that realists ignore cultural differences, conflictive interests, and social forces that determine which environmental issues become scientific and public problems and their implications for political responses. Constructivists alleged realist biases in scientific methods, peer reviews, and treatment of climate change skeptics (e.g., Demeritt 2001; Funtowicz and Ravetz 1993; Taylor 1997; Taylor and Buttel 1992). Realists countercharged that constructivists' epistemic relativism exaggerates the role of cultural construction, understates biophysical constraints, and ignores organized antienvironmentalist forces (e.g., Dunlap and Catton 1994; Murphy 1995; Rosa and Dietz 1998).

Even in the heat of the battles in the 1990s, realists and constructivists often qualified their arguments, holding that climate change is a biophysical reality yet also socially defined, shaped, and managed. Constructivist Steven Yearley (2005a:198) argues that disagreement has been largely of emphasis: realists stress mainly inquiry into climate change's biophysical facets and social forces that drive and condition it, while constructivists address primarily "successful moral entrepreneurship by problem claims makers such as environmental pressure groups." Each side, says Yearley, believes that its topic is most significant. By contrast to realists, however, constructivists are "agnostic" about "scientific truth" and, in varying degrees, about the reality and seriousness of climate change (Demeritt 2006:454–455; Dunlap 2010). Thus, sharp differences in substance, tone, and sense of urgency persist between the two camps long after the intense debate over ontological and epistemic issues subsided. They diverge over the immediacy, scope, and intensity of climate change's biophysical risks and how extensively, how rapidly, and by what means they should be addressed. Realists generally stress that the speed of climate change drivers and processes and scope of likely consequences dictate rapid, concerted collective action, while constructivists usually emphasize the social costs of scientific error and de-democratization of hastily arranged, science-driven political policies. Such splits can generate a "productive tension" that leads to fruitful research questions and sheds light on different sides of public policy issues (Soper 1995). Some thinkers argue that this has been the case in sociological work on climate change (e.g., Rosa and Dietz 1998:443–446). And if realists are right and catastrophic climate change risks are imminent, then constructivist studies that make climate change inquiries and collective actions to cope with the process more self-reflexive and self-corrective are all the more necessary. However, as in earlier realist–constructivist debates, divergent normative beliefs about the roles of expertise, science, public participation, state regulation, and economic growth, often left tacit or sublimated in methodological critiques, cut across the theories discussed below. Realists and constructivists usually

disagree—often sharply—over the need, effectiveness, and social consequences of national or transnational regulation and planning, and therefore
about climate change mitigation and adaptation strategies. Ultimately, they
tend to have divergent views about growth and capitalism, but, as we will
argue, some thinkers combine facets of both approaches, forging a middle
ground of sorts.

TOP-DOWN VERSUS BOTTOM-UP CLIMATE
POLICIES: CONSTRUCTIVIST AND REALIST VIEWS

"Science studies" scholars have been central figures in the
constructivism–realism split and in debates over culture, politics, and climate
change. Analyzing social processes that shape scientific problems, methods,
objects, and applications, they pose critical questions about science's autonomy and authority and challenge conventional boundaries between science
and nonscience and experts and publics, issues that constructivists have
brought forward in climate change inquiries and theories (e.g., Collins and
Evans 2002; Demeritt 2006; Jasanoff 2003; Yearley 2009). Bruno Latour's
"actor-network theory" and views of science and politics have been influential in constructivist discussions of climate change (e.g., Latour 1987, 2004,
2012). Another important root of these arguments is Mary Douglas and
Aaron Wildavsky's (1982) "cultural theory," which holds that environmental
risks become public problems via battles over subjective views of contested
ways of life and related social psychological orientations, not on the basis of
scientific evidence. Parallel to premodern purification rituals, they argued,
cultural constructions of ecological risk stress "cleansing" by scapegoating
persons, groups, and organizations and targeting them for resentment,
hostility, and social control. Also influential in constructivist circles, Silvio
O. Funtowicz and Jerome Ravetz (1993) have argued that major ecological
risks, especially climate change, involve uncertain facts, disputed values,
high stakes, and urgent decisions, which cannot be resolved by scientific
knowledge alone. To cope with the mix of technical issues, social values, and
public policies, they called for a "postnormal science" that would democratize expertise by including stakeholders as well as experts in extended peer
review processes and decisions about public interventions.

Ulrich Beck's (1992a, 1992b, 1997) theorizing about the shift from "industrial society" to "risk society" has influenced constructivist theories as well
as neomodernization, ecological modernization, and critical theories (discussed later in the chapter). Beck held that invisible, possibly irreversible,
catastrophic global risks generate pervasive public doubts about scientifi-
cally orchestrated state planning. He contended that consequent skepticism

and related sociocultural changes are ushering in a new era of "reflexive modernization," or a "second modernity," characterized by an ascendant, participatory "subpolitics" (e.g., citizen initiatives, new social movements) operating in civil society beyond electoral politics, state bureaucracies, and official hierarchies. Beck argued that top-down, state-bureaucratic rationalization and scientific planning failed, lost public trust, and were "demonopolized." Ecological risks and environmental movements are central to his theory, which envisions a possible "utopia of ecological democracy" animated by activist citizenry and democratized science. Beck held that scientific issues are increasingly being engaged by alternative "public-oriented scientific experts" and other stakeholder voices, forging extended review processes and constituting a public "upper chamber" that vets scientific findings and their applications in light of how "we [the public] wish to live" (Beck 1992a:118–120, 1992b:158–173). Beck (1992b:230–231) warned, however, that democratization could be overtaken by "ecologically oriented state interventionism," regenerating postwar states' *"scientific authoritarianism and ... excessive Bureaucracy."*[4] Beck did not address climate change per se in his early work, but his point that reflexivity and democracy increase in proportion to the scale of rapidly growing ecological risks set the stage for his later emphasis on enormous climate change risk and its potential to consolidate bottom-up global democratic publics in response.

Constructivist cultural theorists often treat climate change as an archetypical "wicked problem"—exceptionally complex, rife with uncertainties, and notoriously hard or impossible to solve (Hulme 2009:326–340; Rittel and Webber 1973). They are skeptical about the capacity of science-driven, top-down public policy to cope with climate change and other serious ecological problems. Focusing on sociocultural and political facets shaping science and policy, they criticize climate scientists' and other environmental scientists' universalistic models, methods, and theories for ignoring divergent values, interests, and needs; favoring unduly bureaucratic or repressive policies; and producing unfavorable sociopolitical outcomes (Yearley 1996:100–141, 2005b:160–173). Constructivist theorists are highly critical of state-planning regimes, holding that coercive, science-centered coalitions between states and environmentalists should be replaced by bottom-up, participatory science and politics (Taylor 1997; Taylor and Buttel 1992). They charge that treadmill of production, ecological rift, and other approaches, which portray capitalism as the main driver of climate change and other major ecological problems, oversimplify causal matrices; ignore local conditions, voices, and adaptive capacities; harbor Western biases; and ultimately justify nondemocratic top-down policies (White 2006). Constructivists challenge claims about science's certainty, neutrality, and beneficence and contend that ordinary people have "expertise" based on local knowledge of the ecological conditions affecting them. They advocate publics having a wider role in policy deliberations over climate change to improve and democratize decision making and cultivate trust among scientists, public officials, and citizens (e.g., Yearley 1997, 2005b:113–142).

Yearley (2009) acknowledges climate change's biophysical reality but argues that the climate system and complexes of sociocultural conditions that affect it are so complex that future scenarios are highly uncertain. He contends that climate science projections and general circulation models depend on claims and assumptions about how governments, consumers, and other social groups and organizations will act and that climate scientists' beliefs about these matters are shaped by their epistemic communities and intellectual and social networks. Yearley stresses the need for sociological inquiry into climate science's social determinants (e.g., impacting peer review and scientific prestige and success) and into the sociocultural content embedded in its models and methods. He argues that social forces substantially influence the way climate change is constructed, analyzed, and presented but usually are ignored by natural scientists and realist social scientists. Yearley acknowledges organized opposition to climate science but holds that these distorted messages do not reduce the salience of critical sociological inquiries about climate science practices.

Influential constructivist Mike Hulme stresses climate science uncertainty, questions its authority, and berates science-driven policymaking. He sees climate change as a biophysical process with serious risks but criticizes climate scientists' catastrophic projections, universalistic claims, and inattention to locality. Hulme (2009:xxxiii) explains that his studies in the history of science led to a "*Cultural Enlightenment*" and culminated in a break with his former conventional realist standpoint. Hulme focuses on the conflictive cultural constructions of climate change that block collective action to deal with it—divergent media frames about the process and what is at stake, contrary spiritual beliefs about nature and God, discrepant estimates of climate change's economic costs and policy tradeoffs, and contradictory conceptions of development and governance. Drawing heavily on Douglas and Wildavsky's (1982) cultural theory, he contends that "fatalist," "hierarchist," "individualist," and "egalitarian" social psychological orientations and related climate change "myths" generate fundamental disagreements over the process and help make it an irresolvable wicked problem. Hulme (2009:6, 2010b) asserts that complex civilizations persist in radically different climates (e.g., Saudi Arabia and Iceland), that climate fluctuates and cannot be stabilized, and that our climate future is open and should not be reduced to physical predictions. Opposing the United Nations Framework Convention on Climate Change, Kyoto Protocol, and GHG emission reduction targets, Hulme (2009:362) argues that it really is not about stopping climate chaos. Instead, we need to see how we can use the idea of climate change—the matrix of ecological functions, power relationships, cultural discourses and material flows that climate change reveals—to rethink how we can take forward our political, social, economic and personal projects over the decades to come.

He says that climate change must be managed but urges us to learn to live with "recreated climates," embrace their "novelty," and employ their uncertainties and biophysical consequences to motivate rethinking our current and future social arrangements, framing fresh cultural narratives, and innovating new technologies and ways of life that enhance our "social resilience" (Hulme 2008, 2009, 2010a, 2010b). Although acknowledging the reality of climate change, Hulme disagrees profoundly with realist views about the seriousness of its likely impacts, the urgency and prudence of mitigation, and the threat to civilization as we have known it.

Hulme (2006, 2008) excoriates climate scientists' use of the words "chaotic," "irreversible," "rapid," and "catastrophic" to describe climate change. He contends that they should just report findings and avoid alarmist language, which he alleges discourages climate action, politicizes science, and justifies top-down statist strategies, doomed to fail. Some research supports his view that dire messages have contradictory effects (e.g., Feinberg and Willer 2011). Other research, examining changes in public opinion over time, finds that such messages and advocacy increase knowledge and heighten concerns regarding climate change (e.g., Brulle, Carmichael, and Jenkins 2012). Criticizing Hulme, James S. Risbey (2008) holds that climate scientists have a civic duty to be frank and to employ language appropriate to their estimates of climate change's likely rapidity, intensity, and impacts. Projections vary, Risbey says, but most climate scientists agree that the risks are urgent and the stakes are high. Hulme (2009:232–233) sees his differences with Risbey to be characteristic of myriad paralyzing disagreements over climate change and fundamentally divergent orientations portrayed by Douglas and Wildavsky. Stressing uncertainties manifesting climate change's biophysical complexity and contingent, multilayered sociocultural facets, Hulme and many other science studies analysts are skeptical about climate science projections (e.g., carbon dioxide levels, atmospheric temperature increases, sea level rises) (Demeritt 2006). By contrast, climate scientists such as Risbey, and most social scientists, studying climate change drivers, trust the veracity of likelihood estimates and warn that GHG emissions are increasing so fast that slack mitigation could allow runaway warming that would preclude effective human adaptation (Anderson 2012). This fundamental split resists mediation.

Hulme criticizes climate scientists and realist social scientists for political advocacy, but this view can be turned around. Rejecting the Kyoto Accord's "green governmentality" or state-imposed "universalism," Hulme (2009:297, 309–313, 359–364) favors Prins and Rayner's (2007) "non-hierarchist," market-centered approach. Hulme is a coauthor of the first Hartwell Paper (Prins et al. 2010), which treats climate change as a wicked problem, favors adaptation and resilience over mitigation, and embraces the priority of economic growth and bottom-up markets over top-down state regulation and planning. It contends that climate policy limiting carbon dioxide emissions has failed and advocates prioritizing abundant cheap energy supplies, deemphasizing carbon dioxide forcings, and increasing public investment in alternative energy sources until clean energy is cheaper than carbon-based energy.[5] Hulme's views are inspired in part by Latour's critique of Western modernity's hubristic ideas of science, progress, and domesticated nature. Rooted in his science studies research and actor-network theory, Latour's (2004, 2010, 2011, 2013) view of climate change is couched in abundant metaphor and postmodern tones. He advocates a "postenvironmentalist" break with American environmentalism's "gloomy asceticism," demands to "limit ourselves," and fearful view of technology's unintended consequences. He criticizes the preservation of wilderness and attacks the precautionary principle, which he says manifests misplaced faith in state power and scientific certainty (Latour 2012). Latour is a mercurial thinker who has shifted views and has had diverse impacts, but his recent position on climate change policy converges at key points with Hulme and the Hartwell Paper.[6] Constructivist social theorists, like their realist counterparts, belong to epistemic communities, which share normative as well as analytical presuppositions, and thereby take critical standpoints about climate change and climate science that manifest their support for or opposition to particular policy regimes and related visions of growth.

Constructivist sociologists Reiner Grundmann and Nico Stehr (2012:119–194) theorize how scientific knowledge attains power and drives creation of regulatory rules, laws, and policies. They hold that scientists provide interpretive frames, which offer problems to be solved, but argue that scientific evidence is not decisive in practical affairs. They assert that "experts" and "policy entrepreneurs" must identify political "levers for action" and that social prestige and influence are decisive factors determining their success or failure. Their theory poses basic science studies' questions, which could provide resources for realists aiming to make their enterprise more self-critical. Applying the approach to climate science, Grundmann and Stehr criticize realists' uncritical acceptance of climate science, alarmism, and science-driven policy recommendations. They contend that the IPCC and other climate change experts and entrepreneurs overestimated the force of scientific evidence and did not find effective policy levers, and thus the United Nations Framework Convention on Climate Change,

Kyoto Protocol, and Copenhagen Accord failed. They charge that climate scientists lost prestige and integrity after being exposed for exaggerating climate science's certainty and climate change's urgency and for participating in "Climategate" e-mails. Grundmann and Stehr (2012:173) speculate that Americans' doubt about the seriousness of the climate change threat and belief that the media exaggerate the consequent risks manifests public resistance to being bullied by climate science experts' and policy entrepreneurs' alarmism.

Cultural and political differences, manifesting divergent, national contexts, likely have influenced the realist and constructionist split. In particular, the well-funded, politically powerful network of denialist think tanks, foundations, lobbyists, corporate opponents, and nearly unanimous, aggressive Republican Party opposition to climate science has been a distinctive, central feature of U.S. climate politics (e.g., Dunlap and McCright 2011; Lahsen 2005; McCright and Dunlap 2010; 2011; Oreskes and Conway 2010). Moreover, the traditional strength of the more recently, forcefully reasserted market-liberal, "small government," political culture precluded, in the postwar United States, European-style state-centered, social democratic regulatory and planning regimes, criticized by constructivists (Hodgson 1978:67–98; Judt 2010). These context-related differences *may* explain some facets of the realist–constructivist split over top-down and bottom-up policies. However, Grundmann and Stehr discuss the U.S. context but do not address substantially the conservative mobilization against climate change policies. They discuss events (e.g., the hockey stick controversy) and actors (e.g., Steve McIntyre and Ross McKitrick) entwined with the political polarization but do not locate them specifically within the political battles, address counterarguments, or engage denialists' pseudoscience and their powerful institutional agents. Grundman and Stehr are coauthors of the two Hartwell papers (Prins et al. 2010, 2013), which oppose top-down, state-centered approaches, identify market-centered strategies with bottom-up ones, and stress the priority of growth. These policy statements, which are compatible with their views about climate change in their book, may reflect, in part, differences in historical and sociopolitical context and preference for market-liberal value orientations as well as their estimation of failed science-led policy.

Realist Diana Liverman (2011:135) holds that Hulme "underplays the significance of power and political economy and the role of key actors who support and oppose climate policies." In her view, reducing the regulatory state's role in climate policy hardly ensures a bottom-up regime, given the size, power, and cultural influence of the many corporate actors. Realists would identify the same problem in Grundmann and Stehr's book, which purports to theorize power and knowledge but does not engage the political-economic forces staunchly opposed to climate science and to increased regulation and adaptation and mitigation costs likely to follow from acknowledgment of climate change risks. Other realists argue that constructivist science studies have a one-sided affinity for climate skepticism, blur borders between science and pseudoscience, and privilege averting Type 1 errors (declaring benign conditions risky) over avoiding Type 2 errors (declaring risky issues benign), which may be exceptionally costly and dangerous in ecological matters and is a core point of contestation over climate change (e.g., Dunlap and Catton 1994:20–23; Freudenburg, Gramling, and Davidson 2008; Rosa and Dietz 1998). By contrast, constructivists charge that realists take science's claims for granted and ignore social factors that condition them (e.g., Grundmann and Stehr's [2010] critique of Lever-Tracy [2008]). However, theoretical one-sidedness does not preclude conceptual and research fruitfulness, if the theoretical moves are properly qualified and honest; a one-sided specialized focus can illuminate important matters. Also, contrary theories, animated by normative splits, can sometimes clarify and contribute to deliberation of contested public problems. Realists Rosa and Dietz (1998:440, 445–446) are critical of science studies but concede that the core lines of constructivist inquiry about climate science generate fruitful research questions.

Some constructivists aim to close the gap with realists. David Demeritt's (2001:311–312) antidualistic "heterogenous constructionism" stresses the entwinement of "knowing and being, subjects and objects, and nature and society" and science's "constitutive role" in revealing climate change realities. Demeritt (2001:327–329, 2006) acknowledges that climate change is a serious problem and that constructivist studies of climate science are sometimes deployed by skeptics and their corporate sponsors to deny or trivialize climate science evidence. However, he holds that constructivist critiques and science studies methods expose and could help avert physicalist reductionism and make climate science more reflexive about its flaws, limits, and uncertainties; averse to overreach; and able to counter more effectively skeptics' claims "that global warming is merely a social construction." Demeritt seeks to blunt the technocratic tendencies of physicalist climate science, overcome inattentiveness to its social facets, and democratize top-down, science-led climate policy.

In contrast to constructivist arguments holding that realist universalism (e.g., global climate models, state-centered policymaking, conceptions of global environmental problems) ignores sociocultural factors and locality, realists often express constructionist awareness of culture and locality. For example, critical social theorist Robert J. Brulle (2000:15–48, passim) fuses Habermas's "communicative action" theory with a realist approach to climate change and other environmental problems. Making reference to

Beck and Rayner, Brulle (2000:8, 19–32, 53–56) stresses avoidance of physicalist reductionism and engagement of sociocultural factors. By contrast to Rayner and the Hartwell Group, however, Brulle warns of what Habermas has called "colonization" of the lifeworld by the system, in this case genuinely democratic politics and civic responsibility in ecological matters overtaken by unregulated markets and the capitalist growth imperative. Brulle stresses the importance of culture in unifying theory and practice in environmental politics. Feminist theorist Sherilyn MacGregor insists that social analysis must examine both the material and cultural dimensions of climate change. She explains that gender analysis enhances our overall conception of climate change, helps us grasp diverse people's everyday experience of the process, and improves assessment of the responses to it. MacGregor explains that the gender division of labor makes poor women in the Global South among the people most heavily harmed by climate change. She holds that they tend to be critical of approaches that propose technological fixes and favor structural changes to adequately address issues of justice and climate change. She and other feminists argue that climate discussions and policies must address these gendered vulnerabilities and inequalities, which should be part of reclaiming the democratic process for a sustainable world (MacGregor 2009, 2013; Salleh 2011; Whyte 2014). Kyle Whyte (2013) explains that climate science must also integrate "traditional ecological knowledge" from indigenous peoples to address climate justice, responsibility to the earth, and cultural survival. He contends that plural types of cultural knowledge and experience deepen our understanding of the culture–nature relationship, without undermining climate science, and potentially illuminate its divergent local impacts.

Realist sociologist and sharp critic of the growth imperative Wayne A. White (2013) offers a study of biosequestration that meshes scientific analysis of global climate change and its political-economic drivers with a finely attuned appreciation of regional and local differences and for adaptation and mitigation strategies that take account of them. Illuminating social roots of climate change denial in an otherwise politically progressive Norwegian community, Kari Marie Norgaard (2011) addresses social construction, culture, and locality within a realist framework that takes account of political power, political economy, and climate change's catastrophic global risks. Adger et al. (2013:112) state that culture "is embedded in the dominant modes of production, consumption, lifestyles, and social organization that give rise to emission of greenhouse gases" and that climate change impacts are culturally interpreted by scientists, policymakers, and citizens; therefore, culture should be addressed in studies of climate change drivers, mitigation, and adaptation. They provide examples of large-scale climate assessments that have taken account of culture, and discuss the role of social science research on climate issues, address related methods of inquiry, and stress the importance of culture for climate policy. This work bridges the realist–constructivist split by recognizing processes that contribute to the social construction of knowledge, while at the same time accounting for how construction is tempered by the "materialist principle," which highlights how human society is subject to natural laws, cycles, and conditions. Such approaches grapple with earthly questions regarding climate change and culture–nature relations (Foster and Clark 2008; Soper 1995). They share realist visions of climate science, while their emphases on social and political differences and democratic inclusion counter undemocratic top-down policies (e.g., geoengineering).

NEOMODERNIZATION THEORIES AND CLIMATE CHANGE: SUSTAINABLE NEW ECONOMY OR GROWTH METASTASIZED?

The sociocultural rupture generated by industrial capitalism and nation-states gave rise to modern social theory and sociology. The consequent many-phased modernization debate, regenerated by globalization and the "new economy," has influenced theoretical debates over climate change and other ecological problems (Beck 2009; Henwood 2003). The focus among these theorists is on the relationship of climate change to the overall institutional structure of society and culture and less on the role of climate science and related politics of science. This axis of discussion about modernization and climate change overlaps, in part, with the realist–constructivist debate, especially the divide over top-down versus bottom-up strategies and questions regarding sustainable growth. The split in this case, however, is between those who hold that modernizing tendencies, reshaped by a new democratized political-economic regime and related reflexive cultural politics, moderate or overcome tensions among democracy, capitalism, and ecology and others who stress that the capitalist growth imperative (and amplification of deregulation policies by revived market liberalism) accretes the tensions and threatens general ecological and climate change catastrophe and consequent social and political disasters. The following theories vary widely in their intellectual roots, substance, and normative thrust. Broadly, they focus on facets of the political-economic system and on social relations at different scales, which influence the long-term prospects for environmental sustainability.

Drawing on Beck's and Anthony Giddens's reflexive modernity arguments, ecological modernization theorists contend that the rise of an active, intelligent citizenry and participatory civil society, supplanting postwar-era top-down, state bureaucratic governance, generate "environmental progress" and "sustainable development." Ecological modernization theorists con-verge with constructivists about the importance of culture and agree that markets and publics, rather than state planners, should drive environmental policy. However, ecological modernizationalists suggest that environmental sustainability and limits on growth could be harmonized with capitalism. They hold that ecological problems are unavoidable in modernization's early stages, when economic growth, necessary to relieve poverty, comes at the cost of sharp inequalities and ecological degradation. They argue that narrow "economic rationality" once dominated policy decisions, but today advanced reflexive capacities provide social systems with enhanced self-corrective powers that facilitate ecological sustainability. In their view, enlightened publics, social movement leaders, scientists, and government officials forge increasingly ecologically rational environmental regulations, audits, laws, and reforms. Ecological modernizationalists also hold that "environmental concerns" become central to private-sector decision mak-ing, generating environmentally sustainable productive practices, techno-logical innovations, and products (e.g., organic paints) (Mol 1995, 2002; Mol and Jänicke 2010; Mol, Spaargaren, and Sonnenfeld 2010).

Ecological modernization theorists claim that modernizing forces, espe-cially economic growth, which helped create climate change and other ecological problems, increasingly are redirected to serve "environmental reform" and green the economy. Arthur P. J. Mol (1997:141) argues that "environmental improvement can go together with economic develop-ment via a process of delinking economic growth from natural resource inputs and outputs of emissions and wastes." Technological innovation throughout the productive system, ecological modernizationalists claim, will allow industries to respond to environmental problems rather than rely on "end-of-pipe technology" (e.g., smokestack filters), which reduces pollu-tion only after it has been produced (Huber 2010:18). They also contend that reflexive modernization stimulates technological diffusion from the Global North to the South. These institutional, technological, and sociocultural changes, they believe, are dematerializing society. Mol (2002:93) argues that social-environmental reforms could culminate in "an absolute decline in the use of natural resources and discharge of emissions, regardless of economic growth in financial or material terms (product output)." Spurred by ecologi-cal modernization, Joseph Huber (2010:335) contends, technological inno-vations reduce "quantities of resources and sinks used," whether measured by intensity, per unit, or by "absolute volumes." These thinkers believe that reflexive modernization generates comprehensive environmental reforms, resource efficiency, and ecologically benign economic growth (Mol and Jänicke 2010). While ecological modernization theorists do not challenge climate change science, they do parallel other constructivist arguments and

Beck's early work by not treating climate change and other ecological issues as urgent problems requiring comprehensive state initiatives.

Beck's recent work expresses confidence in climate science and IPCC assessments and treats climate change as an urgent problem. Critical of neoliberal hegemony and deeply troubled by the financial crash, Beck moderates his earlier criticism of state intervention, but he remains optimistic about social movements and progressive facets of global civil society. Theorizing the "world risk society" and shift from "nationalism" to "cosmopolitanism," he invokes John Dewey's political theory of democratic publics forming, communicating, and coping with indirect social consequences of corporately organized society and transforming mechanistic interdependence into critical, participatory political communities and democratized states. Beck holds that globalization's impacts and catastrophic threats are generating global "risk communities," potentially able to bind diverse peoples across space and cultivate "cosmopolitan 'collective consciousness,'" or awareness of interdependence and collective responsibility for the species and planet. He believes that mounting perceived risk from climate change is generating critical thought and communication about neoliberalism's extensive role in ecological and social degradation and need for an alternative regime (Beck 2009, 2012, 2013; Beck and Levy 2013). Beck (2010:261) declares that "climate change may yet prove to be the most powerful of forces summoning a civilizational community of fate into existence" and helping spur global grassroots democratic actions.

Giddens's *The Politics of Climate Change* (2011) has received mixed reviews and some scathing criticism.[7] Nevertheless, Giddens makes an important intervention to climate change discussion. Scaling back substantially earlier optimism about reflexive modernity and concurring with Kevin Anderson about the dangers of rapidly increasing GHG emissions and failing governance, Giddens (2011:21–32, 194) argues that urgent climatic risks require aggressive action and that powerful resistance by organized denialist interests undermines public understanding and political action. Although acknowledging that nongovernmental organizations and businesses play vital roles in green politics and that elected officials must work cooperatively with them, Giddens argues emphatically that states must lead in "policy formation and enactment" with regard to climate change mitigation and adaptation. He devotes an entire chapter to a "return to planning" based on the idea that states must *"counter business interests which seek to block climate change initiatives," "keep climate change at the top of the political agenda,"* and forge economic policies that move *"towards a low-carbon economy"* (Giddens 2011:96, 94–128 passim). Giddens retains elements of reflexive modernization theory, but his ideas about propagandistic distortion of civil society and support for state intervention and planning diverge substantially from his earlier views and from ecological modernization and science studies theories of climate change. This is a notable move, because even Left-leaning realists have been loath to mention the word "planning" due to its association with socialism and "failed" postwar, top-down administration. Whatever the weakness of Giddens's overall views on climate change, he challenges us not to dismiss but to discuss seriously planning again, some form of which is implied in all arguments about regional government, nation-state, and interstate mitigation and adaptation.

Other sociologists, much more skeptical about progressive modernization theories, see climate change as modernization gone profoundly awry. Michael Mann (2013:366–399, 432) concludes the final volume of his magisterial *The Sources of Power* with a substantial chapter on climate change, which he warns threatens global social catastrophe ("wars, massive refugee flows, and new extremist ideologies"). He states that "the problem has been created by capitalism ably assisted by both nation-states and individual consuming citizens. These, unfortunately, are the three most fundamental social actors of our time" (Mann 2013:362, 364–365). Mann (2013:365) sees climate change as the most dangerous of multiple, severe growth-driven ecological problems, manifesting an "ideology of modernization in which nature is explicitly subordinated to culture." He argues that climate change mitigation is stymied by capitalism's emphases on short-term profit and fossil fuel–dependent, exponential growth, bolstered by citizens unwilling to reduce consumption and politicians depending on GDP growth for reelection. Hegemonic neoliberalism and the Great Recession, he adds, harden resistance to ecological reform. Mann (2013:368) contends that neoclassical economists dominate policy circles and that consequently climate science emission scenarios and official reports understate risks. Although supporting diverse strategies to cut emissions, he believes any mix must include "government imposing radical restrictions on businesses and consumers" to be effective (Mann 2013:386). Mann does not rule out "salvation" by revolutionary new technologies, global social movements, or other unanticipated historical forces but stresses that there are massive, systemic sociopolitical and sociocultural blockages to mitigation, which must be overcome to avoid eventual ecological and societal disaster.

New forms of marketing, stressing image and identity, play an enormous role in highly mediatized, global consumerism (Mander 2012). Urry (2011:48–87) holds that the postmodern "experience economy" with themed regional "cathedrals of consumption" and travel to global consumer spaces (Dubai, Las Vegas, Macao) typify energy-intensive neoliberalism and provide a cultural engine of climate change. Juliet Schor (2010:41) contends that commodities are now promoted on the basis of abstract qualities rather than their usefulness, giving rise to the "materiality paradox" whereby "when consumers are most hotly in pursuit of nonmaterial meanings, their use of material resources is greatest." She holds that as symbolic values of consumer goods increase and are diffused, environmental demands are multiplied. A society dominated by consumption oriented to abstract values, Schor argues, favors a throwaway consumer culture rife with ecological waste, which globalization has greatly extended and intensified. Other theorists hold that focusing centrally on producing high-quality durable goods would increase production costs and decrease sales (Dawson 2003). They argue that product obsolescence, characteristic of high-consumption cultures, increases energy demands and production of GHGs (Foster et al. 2010). They explain that environmental costs are exacerbated by marketing, packaging, and overall promotion to persuade consumers to buy products that they did not know about or need (Leonard 2010). Schor (2010) proposes that a sustainable, less carbon-intensive society necessitates breaking with neoliberalism and creating an economy and culture that valorizes quality over quantity, durable over disposable goods, and individuality anchored in community, rather than in consumer niches.

CLIMATE CHANGE AND CAPITALISM: PATH DEPENDENCY OF THE CAPITALIST GROWTH IMPERATIVE

The theories discussed in this section suggest that climate and other ecological crises are rooted in capitalism per se. Deriding ecological modernizationalists, John Bellamy Foster (2012) charges that these "new exemptionalists" understate environmental risks and substitute vain reformist hopes and toothless "self-regulation" for compulsory state regulation, needed to ensure reduction of environmentally destructive externalities. Foster and his colleagues see capitalism as inherently "a crisis-ridden, cyclical economic system" that needs to be fundamentally transformed or overcome to deal with climate change and other serious ecological and social problems (Foster et al. 2010:423; 215–442 passim). Both Marx and Weber described modern capitalism as a growth-dependent economic system that must constantly renew itself on an ever-larger scale. Preindustrial societies generally relied on limited supplies of biomass for energy, which restricted growth and development of complex social organization. Weber proposed that capitalism was predicated on the exploitation of fossil fuels (Foster and Holleman 2012; Murphy 1994a). Energy-intensive machinery was introduced during the Industrial Revolution, increasing the fossil fuel required to power production (Smil 1994). This development contributed to massive economic expansion of production, consumption, and markets, which fundamentally transformed the culture–nature relationship. Herman Daly ([1977] 1991:23) argues that capitalist production has "broken the solar-income budget constraint [removing stored energy from the earth] and has thrown [society] out of ecological equilibrium with the rest of the biosphere." Political-economy theorists contend that the scale and intensity of this transgression against ecological limits is driven by globalization's enormous expansion of capitalism and the reach of its growth imperative.

Treadmill of production theory, metabolic theory, regulation theory, and Marxian-rooted globalization theories address how the historical development of modern capitalism has contributed to environmental degradation and an array of ecological problems. These theories focus on the path dependency of the capitalist growth imperative, stressing that there is an "enduring conflict" between capitalism as we have known it and the environment (Foster 2000; Schnaiberg 1980). They argue that capitalist growth requires an ever-increasing expansion of resources (i.e., matter and energy) to meet its insatiable appetite, especially in developed countries with massive consumer markets. The highly stratified global interstate system imposes a division of nature that replicates the flow of economic surplus, while it also creates the structural conditions that accelerate climate change. These theories propose that environmental sustainability requires a productive system that operates within natural limits, protecting the conditions that sustain ecological cycles and ecosystem services. Together these theories offer an extensive analysis of the workings and development of the modern economic system.

Treadmill of production theorists propose that capitalists must constantly seek to increase profits, which are reinvested to enlarge and intensify the scale of production; accumulation takes precedence and drives a cycle of growth that necessitates ever-greater production (Schnaiberg 1980). Treadmill theorists suggest that this growth imperative heavily influences the organization of production and consumption and drives culture–nature relations. Focusing largely on post–World War II economic development, they hold that private capital, the state, and labor depend on economic growth for profits, taxes, and wages. The constant pursuit of profit and expansion has "direct implications for natural resource extraction," pollution generation, and overall environmental conditions (Gould, Pellow, and Schnaiberg 2004:297). Treadmill theorists explain that nature, in the form of matter and energy, fuels industry and animates commodity production; each expansion in the production process to sustain economic operations on a larger, more intensive scale generates higher natural resource demand, often at rates that exceed the ecosystem's regenerative capacity (Burkett 1999; Foster et al. 2010; Gould et al. 2004). Moreover, they contend that energy-intensive materials, such as plastics and chemicals, are incorporated into manufacture, generating widespread waste and pollution that producers externalize

(Foster 1994; Gould, Pellow, and Schnaiberg 2008; Pellow 2007; Schnaiberg and Gould 1994). Departing from the work of Thorstein Veblen and James Kenneth Galbraith, treadmill theorists argue that the rise of monopoly capital contributed to the creation of modern marketing (Foster and Clark 2012). Veblen ([1923] 1964) explained that "salesmanship" was necessary to create customers for the commodities produced, helping to reproduce and expand capitalism. Similarly, Galbraith (1958) proposed that corporations exercise "producer sovereignty," thereby dominating both production and consumption. Given these dynamics, treadmill theorists see capitalist growth as a primary cause of climate change and a major barrier to mitigation.

Drawing upon Marx's critique of political economy as well as insights from natural scientists, metabolic theorists stress socioeconomic systems' ecological embeddedness and examine closely the interchange of matter and energy between human societies and the larger environment (Foster 1999). They see capitalism as a historically specific regime of accumulation that drives the growth imperative; it is a *social* metabolic system that operates in accord with its own logic, reducing labor and nature to serve capital accumulation, shaping material exchanges with the environment, and increasing demands on ecosystems and natural cycles. In their view, capitalism's social metabolism exceeds natural limits, producing "metabolic rifts" in various cycles and processes (e.g., the soil nutrient cycle), which are necessary for ecosystem regeneration (Foster 1999, 2000; Foster et al. 2010; Mészáros 1995). Metabolic theorists contend that the capitalist growth imperative generates ecological rifts through intensification of social metabolism, resulting in a carbon rift that drives climate change; it locks in dependence on burning massive quantities of coal, natural gas, and oil (Clark and York 2005; Foster and Clark 2012). They hold that this process breaks the solar-income budget, releasing enormous quantities of carbon, which previously had been removed from the atmosphere, and that consequent growth-driven, ecological degradation (e.g., deforestation) reduces substantially carbon sinks, further contributing to accumulation of atmospheric carbon dioxide.

Treadmill theorists and metabolic theorists propose that technological innovation plays a crucial role in capitalist development, rationalizing labor processes and generating cost reductions via automated production. Treadmill theorists hold that new technologies often make energy and raw material usage more efficient, but, *contra* ecological modernizationalists and neoclassical environmental economists, they contend that innovation does not dematerialize society or contribute to an absolute decoupling of economic development from energy and resources. Treadmill and metabolic theorists point to the "Jevons paradox"—that more efficient resource usage increases overall consumption of that particular resource so that expanded production outstrips gains made in energy efficiency (Clark and Foster 2001; Jevons [1865] 1906; Jorgenson 2009; Polimeni et al. 2008). They argue that efficient operations produce savings that expand investment in production within the larger economic system, and thereby increase consumption and total energy consumed, raw materials used, and carbon dioxide produced. These theorists note that technological rationalization must be situated within global capitalism's overall social relations and dynamics. The growth imperative, as suggested by treadmill of production theorists, is geared to maximize throughput of energy and matter, and thus conservation does not take place at the macro scale of the economy. As a product of capitalism, the Jevons paradox illustrates that purely technological means cannot solve ecological problems such as climate change (Foster et al. 2010; York 2010a, 2010b).

Sociologist Max Koch (2012) theorizes climate change in relation to the transition from Fordism, characterized by welfare state redistribution and regulation and Keynesian socioeconomic management, to finance-driven neoliberalism, which extended deregulated capitalism nearly worldwide, unleashed global growth, and accelerated climate change and other ecological problems. Koch employs "regulation theory," rooted in a fusion of Marxian ideas with institutional economics. He argues that Kyoto Accord carbon markets failed, worked against poor nations' interests, and benefited mainly finance capital institutions and big investors. He contends that market-based mitigation mechanisms and technological innovations cannot provide alternatives to the current path-dependent, carbon-based economy and society. Stressing neoliberalism's extreme economic inequalities and concentrated economic and political power, Koch calls for a regime change to restructure relations among society, state, and market; impose state regulatory power; and establish an international regulatory regime to institute climate mitigation and sustainable growth. He believes that runaway climate change will undercut accumulation, destabilize society, and ultimately force a fundamental transition, but that could come from the radical right (anticapitalist populism) as well as from other political directions. Koch implies that coping with climate change now would increase the possibility of a democratic transition, but, like Mann, he stresses institutional and citizen consent to business as usual and powerful political-economic interests opposed to costly mitigation.

After decades of neoliberal globalization, John Urry doubts that an environmentally sustainable society can be created. He argues that the U.S.-led "carbon military-industrial complex" is resistant to change and deploys expert public relations policies and "global spin," which feign support for reduced carbon dioxide emissions while they do the opposite. He adds that leading GHG emitter, China, also opposes global accords on emission reductions (Urry 2011:110–112). Urry highlights excesses of global consumerism, celebrated in theme-built environments, which generate public consent for business as usual. He holds that our fate depends on displacing neoliberalism and forging a new policy regime that decarbonizes economy and society (Urry 2011:16–17). Urry urges technological innovation, cultural change, and collective action in support of a "low carbon future," but, deeply pessimistic about the prospects, he contends that global capitalism's path-dependent, carbon-based productive system and societal infrastructure would be extremely costly to decarbonize and would require enormous sociocultural and political change. His pessimism about decarbonization and mitigation converges with that of Hulme, Grundmann and Stehr, and the Hartwell Group, but Urry does not think adaptation and resilience can save us. Except "to *prepare* for various catastrophes," he doubts anything can be done to preclude a rise in temperature of 4 degrees Celsius in this century (Urry 2011:158–160, 166, passim).

Pioneer political-economic theorist Charles H. Anderson (1976:153) argued that "nothing grows faster in the growth society than energy consumption," given the diversification and expansion occurring throughout all sectors of capitalism. Globalization theorists hold that this tendency is amplified by finance-driven, global capitalism, which has accelerated economic expansion, concentrated economic power, and increased global environmental inequalities. They assert that neoliberal globalization undercuts environmental regulation; restructures economic development, sites of production, technology transfers, and the international division of labor; and consequently produces more intense, widespread, and unevenly distributed global ecological degradation (Davis 2002, 2006; Harvey 1996, 2005, 2006; Hornborg 2011; Koch 2012; Parr 2013; Pellow 2007). Focusing on unequal ecological exchange, globalization theorists propose that highly developed, militarily powerful nations externalize environmental costs to less-developed countries, creating a "vertical flow of exports" benefiting rich nations at the expense of poor ones (Frey 2003; Hornborg 2006; Jorgenson 2006; Rice 2007). They contend that a significant proportion of foreign direct investment in the Global South finances carbon-intensive agriculture, forestry operations, and extractive enterprises, which produce goods for export, increase carbon dioxide emissions, and recast national carbon emission patterns (e.g., Grimes and Kentor 2003; Jorgenson 2007; Jorgenson, Dick, and Mahutga 2007). Although the Global North is disproportionately responsible for accelerated accumulation of carbon dioxide, globalization theorists argue that the emissions of nations such as China and India are rapidly increasing, in part as a consequence of the Global North offshoring energy-intensive production, which displaces environmental degradation, slows the North's rate of increase in carbon dioxide emissions, and makes emissions reduction treaties work in their favor (Malm 2012; Roberts and Parks 2007). These theories see the capitalist growth imperative as the main driver of climate change and the primary challenge to international action to mitigate it.

CONCLUSION: THE CLIMATE CHANGE DIVIDE
AND POSTEXEMPTIONALIST VISION

Following Catton and Dunlap, realistic engagement of climate change and other serious ecological problems requires acknowledging that human society is embedded in the biosphere and, like all other creaturely populations, faces natural limits. Such a move does not dismiss constructivism. The two relatively independent realist and constructivist vantage points generate divergent questions and productive tensions. Science studies research is not in inherent conflict with climate science. Critical reflexivity is a vital resource for coming to terms with science's problems and limits, addressing its relation to centralized organizational power, honing its questions and methods of inquiry, and sensitizing us to uncertain and unexpected conditions and events. Moreover, constructivist questions and criticism will be all the more valuable and necessary if climate change and other serious environmental problems grow worse, as realists predict, and as transnational entities, nation-states, and other large organizations, with help from scientists, technocrats, and planners, are forced to embark on major mitigation and adaptation schemes. Nevertheless, recognition of a dynamic ecological reality is foundational for grappling with the culture–nature relationship, ecological crises, and environmental sustainability.

The IPCC (Stocker and Qin 2013) warns of a wide variety of likely or nearly certain serious climate change impacts. Scientists contend that global growth's huge resource spikes and severe ecological problems threaten to undercut Holocene-type conditions, which heretofore have sustained complex civilizations, well-being, and economic growth (Steffen et al. 2011). Scientific teams also warn of "tipping points," which when breached cause irreversible qualitative transformation of "earth subsystems" (e.g., Arctic summer ice loss) (Lenton et al. 2008), and of shifting "earth-system processes," threatening naturalistic interdependence *in toto* (climate change, biodiversity loss, nitrogen cycle, phosphorus cycle, stratospheric ozone depletion, ocean acidification, freshwater usage, atmospheric aerosol pollution, chemical pollution, land use change) (Rockström et al. 2009). The speed and extent of mounting anthropogenic "global forcings" (e.g., energy production and consumption, population growth, resource consumption, habitat destruction and fragmentation), they say, could generate a global "state shift" and mass extinctions. With such risks in mind, Tyndall Centre climate scientists Kevin Anderson and Alice Bows (2012:640) declare: "The elephant in the room sits undisturbed while collective acquiescence and cognitive dissonance trample all who dare to ask difficult questions." They point especially to the enormous "discontinuity" between climate science findings and neoclassical economic emphases on exponential growth and unregulated markets and assurances about future technical fixes and feasibility of adaptation and resilience. Contesting charges of being political, biased, and alarmist for contravening prevailing economic beliefs and interests of corporate and public policy elites, Anderson and Bows contend, climate scientists "repeatedly and severely underplay implications of their analyses." Responding to the chorus of climate science critics and policy entrepreneurs declaring rapid mitigation impossible, Anderson (2012:35) asks: "Is living with a 4°C global temperature rise by 2050 or 2070 less impossible?" The highly probable "4°C and beyond" temperature rise, given current GHG emissions trends, Anderson and Bows (2011) insist, will likely threaten life on the planet and open the way for geoengineering, authoritarianism, and other panic scenarios. That a rise of 0.8 degree Celsius in the global atmospheric temperature has destabilized and likely has begun a long-term collapse of the West Antarctic Ice Sheet should be a sobering warning of the harsh consequences of a climate future even substantially less severe than Anderson and Bows predict (Joughin, Smith, and Medley 2014; Rignot et al. 2014).

Constructivist portrayals of climate change as a wicked problem are apt. Hartwell co-author Roger Pielke Jr. (2010:59–60, 71, passim) claims that sustaining economic growth is *the* precondition of democratically legitimate ecological policy ("iron law of climate policy"), precluding timely decarbonization and effectively removing the growth imperative and capitalism from critical discussion. Realist critics reject assertions about the growth imperative's normative priority and sustainability, but they acknowledge the enormous sociocultural and political blockages to transforming beliefs about growth and the carbon-based political-economic system and society. Most realists concur with constructivist critics that climate change mitigation via coordinated international collective action, thus far, has failed and that attainment of public support for extremely costly mitigation and adaptation is unlikely in the foreseeable future. These matters cry out for serious debate and new political-economic and sociocultural vision.

Luigi Pellizzoni (2011) argues that neoliberalism's view of the biophysical realm as a passive cultural tablet for its world-shaping Simonian, entrepreneurial subject has affected the culture so deeply that it, in part, permeates other theoretical approaches to ecology and climate change and weakens their critical force. Erik Swyngedouw (2010) asserts that it is easier for today's people to envision ecological catastrophe and the end of civilization than to imagine and engage seriously the possibility of an end of capitalism as we have known it. He contends that beliefs about capitalism's inevitability distort climate change discussions and short-circuit the visionary thinking needed to break gridlock about coming to terms with the process and conditions that drive it. Swyengedouw provides a speculative explanation of the limits of the theoretical discourse we have described and of blockages to Catton and Dunlap's hoped-for Copernican turn. The major ecological and social problems arising from the reductive vision of nature, constricted imagination about alternative policy regimes, and consequent political inaction or misdirected public policy will likely disrupt the stasis in the future and may have already begun to do so. Increasing public discussion of major market failures (e.g., financial crises and unemployment), polarized incomes, and corporate power diminish the force of employing simple references to failed postwar planning regimes and to Hayekian claims about the state's inherent repressiveness and inefficiency and the superiority of markets to deny the possibility of effective collective action to constrain growth, regulate or transform capitalism, and cope with climate change and other severe ecological problems.

Yet those who advocate climate change mitigation via state-led decarbonization must deal with fundamental questions of how growth can be stemmed without immiseration and de-democratization. Both sides must come to terms with the limits and possibilities of the enormously expanded global economy and its embeddedness in a finite biosphere. The question of whether global economic growth is the fundamental driver of climate change must be made a more explicit, central topic of debate in theory and policy circles. Theorists should entertain reconstruction of the current regime or alternatives to it in a fresh manner beyond the limits of the postwar capitalism/socialism binary and in relation to the complex problem of sustaining democracy and grappling with its meaning in the new context. It would be presumptuous to argue that theorists can "solve" biophysical, social, or political problems, but they can be a vital source of postexemptionalist vision: frame big pictures that cultivate critical ecological thought and new research agendas; enrich debate about how climate change articulates with other ecological and social problems; imagine alternative sociocultural and political-economic regimes; and, overall, provide cultural resources, "messages in a bottle" to be reappropriated and deployed in the intellectual and sociopolitical struggles likely to accompany the types of ruptures that Weber foresaw, Catton and Dunlap implied, and the IPCC and other scientific and ecological reports prefigure. As we stated in note 2, efforts to formulate "social theory" should support sweeping arguments about mapping large-scale social structures and processes and advancing normative claims about policy reforms or regime changes with empirical-historical evidence. Raymond Murphy (2015) explains that Northern European social democracies, without reducing democracy or economic well-being, have instituted much more successful ecological policies (i.e., greater carbon emission reductions and better climate change and environmental performance indices) and maintained higher levels of economic equality and well-being than neoliberal regimes (e.g., the United States, Canada). Impending climate and ecological crises will require greater departures from "business as usual" and more comprehensive, innovative socioeconomic and ecological strategies than practiced anywhere today, but social democratic regimes offer real-world bases or departure points to theorize possible post-carbon pathways and focus debate over their potentials and limits. Framing visionary postexemptionalist social theory should be informed by warranted knowledge about existent or nascent social institutions, structures, and processes or social experiments that contain immanent possibilities for a more socially and ecologically just, sustainable society, which thereby escapes the pitfalls of dystopian paralysis and utopian whimsy.[8]

NOTES

1. Weber addressed culture–nature relations in his works, holding that sociocultural transformations are often entwined with biophysical changes or are even driven by them. See Foster and Holleman (2012) and Murphy (1994a).

2. *Social theorists* draw on empirical and conceptual knowledge from specialized science to justify their policy positions and contextualize their claims by situating them in cognitive maps of large-scale social processes and structures (sometimes of entire societies or trans-societal orders). They make normative claims public and open to contestation on consequential grounds, providing intelligent means to debate policy options. Properly practiced social theory requires the theorist to be honest in his or her evidential judgment, or consider judiciously competing theoretical claims and information and avert traditionalist positions and other views that rely strictly on appeals to authority. Social theory's veracity rides on the strength of empirical-historical argument about consequences of following existent normative and epistemic directions or charting new ones as well as on systematic normative argument. Scientific and sociological good reasons distinguish social theory's post-traditional normative argumentation from approaches that defend normative positions via incontestable transcendental or absolutist claims. Although not "value free" *in toto*, properly practiced *sociological theory* (usually more specialized and narrowly focused in scope than social theory) is constructed with "objective" intent to minimize normative elements of empirical-historical and analytical inquiries. Empirical analysis and conceptualization should engage all accessible data pertinent to the hypothesized relations or within an inquiry's scope. By contrast, social theorists

marshal the best empirical-historical evidence to defend normative claims, albeit with due consideration to contrary arguments and information. Sociological theory and social theory are polar ideal types; theoretical practices usually have mixed attributes and often fluid, blurred borders, falling on various points along a continuum (Antonio 2005). *Most broad theorizing about climate change by sociologists and other social thinkers qualifies as social theory, or is closer to that pole.*

3. For an update on the impacts of Catton and Dunlap's classic argument, see *Organization & Environment* 21(2008):446–487.

4. Here and below italicized phrases, within quotations, manifest emphases in the original.

5. Gwyn Prins and Steve Rayner's (2007:9–10, 25–31) earlier policy statement, discussed by Hulme (2009:313–315), was a forerunner to the three Hartwell papers (Atkinson et al. 2011; Prins et al. 2010, 2013); it framed the group's core ideas stressing climate change as a wicked problem; the errors of universalist, science-driven, state-orchestrated climate policy; and the superiority of bottom-up, market-centered policies over top-down, command and control ones (i.e., state planning). A former student of Mary Douglas's, a leading figure of the Hartwell Group, and the co-author of two of its policy papers, Rayner wrote the foreword to Hulme's (2009:xxi–xxiv) book. Hulme (2009: xxxiv) identifies as a "democratic socialist," but this political inclination is hard to detect in his book or in the Hartwell Paper.

6. Latour's (2012) elliptical commentary about climate change and critique of American environmentalism appeared in the Breakthrough Institute's journal and as the lead article in its Kindle e-book collection (named after his article), co-edited by Institute heads, Michel Shellenberger and Ted Nordhaus, who are co-authors of two Hartwell papers (Atkinson et al. 2011; Prins et al. 2010). Their e-book and institute advocate a "postenvironmentalist" strategy that rejects aggressive mitigation of climate change and advocates technological fixes and market-centered approaches to cope with this and other ecological problems. Latour has been a Senior Fellow at their policy institute.

7. Critics said the book was superficial, atheoretical, and repeated bankrupt Blair-era centrism (e.g., Castree 2010; Hamilton 2012; Wainwright 2011; for a favorable review, see Kitcher 2010).

8. Subheads have been left out for most articles and books, because our bibliography is very long and we needed to comply with the length requirement for the essay.

REFERENCES

Adger, W. Neil, Jon Barnett, Katrina Brown, Nadine Marshall, and Karen O'Brian. 2013. "Cultural Dimensions of Climate Change Impacts and Adaptation." *Nature Climate Change* 3(February):112–117.

Anderson, Charles H. 1976. *The Sociology of Survival.* Homewood, IL: Dorsey Press.

Anderson, Kevin. 2012. "Climate Change Going Beyond Dangerous." *Development Dialogue* III (September):16–40.

Anderson, Kevin and Alice Bows. 2011. "Beyond 'Dangerous' Climate Change." *Philosophical Transactions of the Royal Society* 369(1934):20–44.

Anderson, Kevin and Alice Bows. 2012. "A New Paradigm for Climate Change." *Nature Climate Change* 2(September): 639–640.

Antonio, Robert J. 1998. "Mapping Postmodern Social Theory." Pp. 22–75 in *What is Social Theory?*, edited by A. Sica. Oxford: Blackwell.

Antonio, Robert J. 2005. "For Social Theory." Pp. 71–129 in *Current Perspectives in Social Theory*, Vol. 23, edited by J. Lehmann. Amsterdam: Elsevier.

Antonio, Robert J. 2009. "Climate Change, the Resource Crunch, and the Global Growth Imperative." Pp. 3–73 in *Current Perspectives in Social Theory* Vol. 26, edited by H. F. Dahms. Bingley, UK: Emerald.

Atkinson, Rob, et al. 2011. "Climate Pragmatism." The Hartwell Group and The Breakthrough Institute. (July 25). Retrieved February 8, 2014 (http://thebreakthrough.org/archive/climate_pragmatism_innovation).

Beck, Ulrich. 1992a. "From Industrial Society to the Risk Society." *Theory, Culture and Society* 9(1):97–123.

Beck, Ulrich. 1992b. *Risk Society*. London: Sage Publications.

Beck, Ulrich. 1997. *The Reinvention of Politics*. Cambridge: Polity.

Beck, Ulrich. 2009. "Critical Theory of World Risk Society." *Constellations* 16(1):3–22.

Beck, Ulrich. 2010. "Climate for Change, or How to Create a Green Modernity." *Theory, Culture & Society* 27(2–3):254–266.

Beck, Ulrich. 2012. "Redefining the Sociological Project: The Cosmopolitan Challenge." *Sociology* 46(1):7–12.

Beck, Ulrich. 2013. "Redefining Power in the Global Age." Pp. 491–500 in *The Meta-Power Paradigm*, edited by T. Burns and P. Hall. Frankfurt am Main, Germany: Peter Lang.

Beck, Ulrich and Daniel Levy. 2013. "Cosmopolitanized Nations." *Theory, Culture & Society* 30(2):3–31.

Bhaskar, Roy. 1986. *Scientific Realism and Human Emancipation*. London: Verso.

Brulle, Robert J. 2000. *Agency, Democracy, and Nature*. Boston: MIT.

Brulle, Robert J., Jason Carmichael, and J. Craig Jenkins. 2012. "Shifting Public Opinion on Climate Change." *Climatic Change* 114(2):169–188.

Burkett, Paul. 1999. *Marx and Nature*. New York: St. Martin's Press.

Burningham, Kate and Geoff Cooper. 1999. "Being Constructive." *Sociology* 33:297–316.

Buttel, Frederick H. and Peter J. Taylor. 1992, "Environmental Sociology and Global Environmental Change." *Society & Natural Resources* 5:211–230.

Carolan, Michael S. 2005a. "Realism without Reductionism." *Human Ecology Review* 20(1):1–20.

Carolan, Michael S. 2005b. "Society, Biology and Ecology." *Organization & Environment* 18:393–421.

Castree, Noel. 2010. "The Paradox of Professor Giddens." *Sociological Review* 58(1):156–162.

Catton, William R., Jr. and Riley E. Dunlap. 1978. "Environmental Sociology: A New Paradigm." *American Sociologist* 13:41–49.

Catton, William R., Jr. and Riley E. Dunlap. 1980. "A New Ecological Paradigm for Post-Exuberant Sociology." *American Behavioral Scientist* 24(1):15–47.

Clark, Brett and John Bellamy Foster. 2001. "William Stanley Jevons and *The Coal Question*." *Organization & Environment* 14(1):93–98.

Clark, Brett and Richard York. 2005. "Carbon Metabolism." *Theory and Society* 34(4):391–428.

Collins, H. M. and Robert Evans. 2002. "The Third Wave of Science Studies." *Social Studies of Science* 32(2):235–296.

Daly, Herman E. [1977] 1991. *Steady-State Economics.* Washington, DC: Island Press.

Davis, Mike. 2002. *Late Victorian Holocausts.* London: Verso.

Davis, Mike. 2006. *Planet of Slums.* London: Verso.

Dawson, Michael. 2003. *The Consumer Trap.* Urbana: University of Illinois Press.

Demeritt, David. 2001. "The Construction of Global Warming and the Politics of Science." *Annals of the Association of American Geographers* 91(2):307–337.

Demeritt, David. 2006. "Science Studies, Climate Change and the Prospects for Constructivist Critique." *Economy and Society* 35(3):453–479.

Douglas, Mary and Aaron Wildavsky. 1982. *Risk and Culture.* Berkeley and Los Angeles: University of California Press.

Dunlap, Riley E. 2010. "The Maturation and Diversification of Environmental Sociology." Pp. 15–32 in *International Handbook of Environmental Sociology*, 2nd ed., edited by M. Redclift and G. Woodgate. Cheltenham, UK: Edward Elgar.

Dunlap, Riley E. and William Catton, Jr. 1994. "Struggling with Human Exemptionalism." *American Sociologist* 25(1):5–30.

Dunlap, Riley, E. and Aaron M. McCright. 2011. "Organized Climate Change Denial." Pp. 144–160 in *Oxford Handbook of Climate Change and Society*, edited by J. Dryzek, R. Norgaard, and D. Schlosberg. Cambridge: Oxford University Press.

Feinberg, Mathhew and Robb Willer. 2011. "Apocalypse Soon? Dire Messages Reduce Belief in Global Warming by Contradicting Just World Beliefs." *Psychological Science* 22(1):34–38.

Foster, John Bellamy. 1994. *The Vulnerable Planet.* New York: Monthly Review Press.

Foster, John Bellamy. 1999. "Marx's Theory of Metabolic Rift." *American Journal of Sociology* 105(2):366–405.

Foster, John Bellamy. 2000. *Marx's Ecology.* New York: Monthly Review Press.

Foster, John Bellamy. 2012. "The Planetary Rift and the New Human Exemptionalism." *Organization & Environment* 25(3):211–237.

Foster, John Bellamy and Brett Clark. 2008. "The Sociology of Ecology." *Organization & Environment* 21(3):311–352.

Foster, John Bellamy and Brett Clark. 2012. "The Planetary Emergency." *Monthly Review* 64(7):1–25.

Foster, John Bellamy, Brett Clark, and Richard York. 2010. *The Ecological Rift.* New York: Monthly Review Press.

Foster, John Bellamy and Hannah Holleman. 2012. "Weber and the Environment." *American Journal of Sociology* 117(6):1625–1673.

Freudenburg, William R., Robert Gramling, and Debra J. Davidson. 2008. "Scientific Certainty Argumentation Methods (SCAMs)." *Sociological Inquiry* 78(1):2–38.

Frey, R. Scott. 2003. "The Transfer of Core-Based Hazardous Production Processes to the Export Processing Zones of the Periphery." *Journal of World-Systems Research* 19(2):317–354.

Funtowicz, Silvio and Jerome Ravetz. 1993. "Science for a Post-Normal Age." *Futures* 25(7):735–755.

Galbraith, John Kenneth. 1958. *The Affluent Society.* Boston: Houghton Mifflin.

Giddens, Anthony. 2011. *The Politics of Climate Change*, 2nd ed. Cambridge: Polity.

Gould, Kenneth A., David N. Pellow, and Allan Schnaiberg. 2004. "Interrogating the Treadmill of Production." *Organization & Environment* 17(3):296–316.

Gould, Kenneth A., David N. Pellow, and Allan Schnaiberg. 2008. *The Treadmill of Production*. Boulder, CO: Paradigm Publishers.

Grimes, Peter and Jeffrey Kentor. 2003. "Exporting the Greenhouse." *Journal of World-Systems Research* 19:261–275.

Grundmann, Reiner and Nico Stehr. 2010. "Climate Change: What Role for Sociology?: A Response to Constance Lever-Tracy." *Current Sociology* 58(6):897–910.

Grundmann, Reiner and Nico Stehr. 2012. *The Power of Scientific Knowledge*. Cambridge: Cambridge University Press.

Hamilton, Clive. 2012. "Theories of Climate Change." *Australian Journal of Political Science* 47(4):721–729.

Hannigan, John A. 1995. *Environmental Sociology: A Social Constructionist Perspective*. London: Routledge.

Hansen, James. 2009. *Storms of My Grandchildren*. New York: Bloomsbury.

Harvey, David. 1996. *Justice, Nature, and the Geography of Difference*. Oxford: Blackwell.

Harvey, David. 2005. *A Brief History of Neoliberalism*. Oxford: Oxford University Press.

Harvey, David. 2006. *Spaces of Global Capitalism*. London: Verso.

Henwood, Doug. 2003. *After the New Economy*. New York: The New Press.

Hodgson, Godfrey. 1978. *America in Our Time*. New York: Vintage.

Hornborg, Alf. 2006. "Footprints in the Cotton Fields." *Ecological Economics* 59:74–81.

Hornborg, Alf. 2011. *Global Ecology and Unequal Exchange*. New York: Routledge.

Huber, Joseph. 2010. "Upstreaming Environmental Action." Pp. 334–55 in *The Ecological Modernisation Reader*, edited by A. Mol, G. Spaargaren, and D. Sonnenfeld. London: Routledge.

Hulme, Mike. 2006. "Chaotic World of Climate Truth." *BBC News* November 4. Retrieved November 27, 201 (http://news.bbc.co.uk/2/hi/science/nature/6115644.stm).

Hulme, Mike. 2008. "The Conquering of Climate Change." *Geographical Journal* 174(1):5–16.

Hulme, Mike. 2009. *Why We Disagree About Climate Change*. Cambridge: Cambridge University Press.

Hulme, Mike. 2010a. "Learning to Live with Recreated Climates." *Nature and Culture* 5(2):117–122.

Hulme, Mike. 2010b. "Cosmopolitan Climates." *Theory, Culture and Society* 27(2–3):267–276.

Jasanoff, Sheila. 2003. "Breaking the Waves of Science Studies." *Social Studies of Science* 33(3):389–400.

Jevons, William Stanley. [1865] 1906. *The Coal Question*. London: Macmillan.

Jorgenson, Andrew K. 2006. "Unequal Ecological Exchange and Environmental Degradation." *Rural Sociology* 71:685–712.

Jorgenson, Andrew K. 2007. "Does Foreign Investment Harm the Air We Breathe and the Water We Drink?" *Organization & Environment* 20(2):137–156.

Jorgenson, Andrew K. 2009. "The Transnational Organization of Production, the Scale of Degradation, and Ecoefficiency." *Human Ecology Review* 16:64–74.

Jorgenson, Andrew K., Christopher Dick, and Matthew Mahutga. 2007. "Foreign Investment Dependence and the Environment." *Social Problems* 54:371–394.

Joughin, Ian, Benjamin E. Smith, and Brook Medley. 2014. "Marine Ice Sheet Collapse Potentially Underway for the Thwaites Glacier Basin, West Antarctica." *Science* 344:735–738.

Judt, Tony. 2010. *Ill Fares the Land*. London: Penguin.

Kitcher, Philip. 2010. "The Climate Change Debates." *Science* 328(June 4):1230–1234.

Koch, Max. 2012. *Capitalism and Climate Change*. New York: Palgrave Macmillan.

Labinger, Jay A. and Harry Collins, eds. 2001. *The One Culture?* Chicago: University of Chicago Press.

Lahsen, Myanna. 2005. "Technocracy, Democracy, and U.S. Climate Politics." *Science, Technology, & Human Values* 30(1):137–169.

Latour, Bruno. 1987. *Science in Action*. Cambridge, MA: Harvard.

Latour, Bruno. 2004. *Politics of Nature*. Cambridge, MA: Harvard.

Latour, Bruno. 2010. "An Attempt at a "Compositionist Manifesto.'" *New Literary History* 41:471–490.

Latour, Bruno. 2011. "Waiting for Gaia." A lecture at the French Institute, London (November). Retrieved June 30, 2014 (http://www.bruno-latour.fr/sites/default/files/124-GAIA-LONDON-SPEAP_0.pdf).

Latour, Bruno. 2012. "Love Your Monsters." *Breakthrough Journal* (Winter). Retrieved June 12, 2013 (http://thebreakthrough.org/index.php/journal/past-issues/issue-2/love-your-monsters/).

Latour, Bruno. 2013. "Which Language Should We Speak with Gaia?" Holberg Prize Lecture, Bergen, June 4. Retrieved June 30, 2013 (http://www.bruno-latour.fr/sites/default/files/128-GAIA-HOLBERG.pdf).

Latour, Bruno and Steve Woolgar. 1979. *Laboratory Life*. Beverly Hills, CA: Sage.

Lenton, Timothy, Hermann Held, Elmar Kriegler, Jim W. Hall, Wolfgang Lucht, Stefan Rahmstorf, and Hans Joachim Schellnhuber. 2008. "Tipping Elements in the Earth's Climate System." *Proceedings of the National Academy of Sciences USA* 105(6):1786–1793.

Leonard, Annie. 2010. *The Story of Stuff*. New York: Free Press.

Lever-Tracy, Constance. 2008. "Global Warming and Sociology." *Current Sociology* 56(3):445–466.

Liverman, Diana. 2011. "Commentary 2: Book Review Symposium: Hulme M (2009) *Why we Disagree about Climate Change*." *Progress in Human Geography* 35(1):134–136.

MacGregor, Sherilyn. 2009. "A Stranger Silence Still: The Need for Feminist Social Research on Climate Change." *Sociological Review* 57:124–140.

MacGregor, Sherilyn. 2013. "Only Resist: Feminist Ecological Citizenship and the Post-politics of Climate Change." *Hypatia* DOI: 10.1111/hypa.12065.

Malm, Andreas. 2012. "China as Chimney of the World." *Organization & Environment* 25(2):146–177.

Mander, Jerry. 2012. "Privatization of Consciousness." *Monthly Review* 64(5):18–41.

Mann, Michael. 2013. *The Sources of Social Power*, Vol. 4. New York: Cambridge University Press.

McCright, Aaron M. and Riley E. Dunlap. 2010. "Anti-Reflexivity." *Theory, Culture and Society* 27(2–3):100–133.

McCright, Aaron M. and Riley E. Dunlap. 2011. "The Politicization of Climate Change and Polarization in the American Public's Views of Global Warming, 2001–2010." *Sociological Quarterly* 52(2):155–194.

Mészáros, István. 1995. *Beyond Capital*. New York: Monthly Review Press.

Mol, Arthur P. J. 1995. *The Refinement of Production*. Utrecht: Van Arkel.

Mol, Arthur P. J. 1997. "Ecological Modernization." Pp. 138–149 in *The International Handbook of Environmental Sociology*, edited by M. Redclift and G. Woodgate. Northampton, MA: Edward Elgar.

Mol, Arthur P. J. 2002. "Ecological Modernization and the Global Economy." *Global Environmental Politics* 2(2):92–115.

Mol, Arthur P. J. and Martin Jänicke. 2010. "The Origins and Theoretical Foundations of Ecological Modernisation Theory." Pp. 17–27 in *The Ecological Modernisation Reader*, edited by A. Mol, G. Spaargaren, and D. Sonnenfeld. London: Routledge.

Mol, Arthur P. J., Gert Spaargaren, and David Sonnenfeld. 2010. "Ecological Modernisation." Pp. 3–14 in *The Ecological Modernisation Reader*, edited by A. Mol, G. Spaargaren, and D. Sonnenfeld. London: Routledge.

Murphy, Raymond. 1994a. *Rationality and Nature*. Boulder, CO: Westview Press.

Murphy, Raymond. 1994b. "The Sociological Construction of Science without Nature." *Sociology* 28(4):957–974.

Murphy, Raymond. 1995. "Sociology as if Nature Did Not Matter." *British Journal of Sociology* 46(4):688–707.

Murphy, Raymond. 2006. "Environmental Realism: From Apologetics to Substance." *Nature and Culture* 1(2):181–204.

Murphy, Raymond. 2015. "The Emerging Hypercarbon Reality, Technological and Post-carbon Utopias, and Social Innovation to Low-carbon Societies." *Current Sociology* 63(3):317–338.

Norgaard, Kari Marie. 2011. *Living in Denial*. Cambridge, MA: MIT.

Oreskes, Naomi and Erik M. Conway. 2010. *Merchants of Doubt*. New York: Bloomsbury.

Parr, Adrian. 2013. *The Wrath of Capital*. New York: Columbia University Press.

Pellizzoni, Luigi. 2011. "Governing Through Disorder: Neoliberal Environmental Governance and Social Theory." *Global Environmental Change* 21:795–803.

Pellow, David. 2007. *Resisting Global Toxins*. Cambridge, MA: MIT.

Pielke, Roger, Jr. 2010. *The Climate Fix*. New York: Basic Books.

Polimeni, John, Kozo Mayumi, Mario Giampietro, and Blake Alcott. 2008. *The Jevons Paradox and the Myth of Resource Efficiency Improvements*. London: Earthscan.

Prins, Gwyn and Steve Rayner. 2007. "The Wrong Trousers: Radically Rethinking Climate Policy." James Martin Institute for Science and Civilization and Mackinder Centre for Study of Long-Wave Effects. Retrieved February 8, 2014 (http://eureka.bodleian.ox.ac.uk/66/1/TheWrongTrousers.pdf).

Prins, Gwyn, et al. 2010. "The Hartwell Paper." Institute for Science Innovation and Society, University of Oxford and LSE MacKinder Programme (May). Retrieved November 3, 2012 (http://eprints.lse.ac.uk/27939/1/HartwellPaper_English_version.pdf).

Prins, Gwythian, et al. 2013. *The Vital Spark (The Third Hartwell Paper)*. London: LSE Publishing.

Rice, James. 2007. "Ecological Unequal Exchange." *Social Forces* 85:1369–1392.

Rignot, Eric, J. Mouginot, M. Morlighem, H. Seroussi, and B. Scheuchl. 2014. "Widespread, Rapid Grounding Line Retreat of Pine Island, Thwaites, Smith, and Kohler Glaciers, West Antarctica, from 1992–2011." *Geophysical Research Letters* 41:3502–3509, doi:10.1002/2014GL060140.

Risbey, James S. 2008. "The New Climate Discourse: Alarmist or Alarming?" *Global Environmental Change* 18:26–37.

Rittel, Horst W. J. and Melvin M. Webber. 1973. "Dilemmas in a General Theory of Planning." *Policy Sciences* 4:155–169.

Roberts, J. Timmons and Bradley Parks. 2007. *A Climate of Injustice*. Cambridge, MA: MIT.

Rockström, Johan, Will Steffen, Kevin Noone, Åsa Persson, F. Stuart Chapin III, Eric F. Lambin, Timothy M. Lenton, Marten Scheffer, Carl Folke, Hans Joachim Schellnhuber, Björn Nykvist, Cynthia A. de Wit, Terry Hughes, Sander van der Leeuw, Henning Rodhe, Sverker Sörlin, Peter K. Snyder, Robert Costanza, Uno Svedin, Malin Falkenmark, Louise Karlberg, Robert W. Worell, Victoria J. Fabry, James Hansen, Brian Walker, Diana Liverman, Katherine Richardson, Paul Crutzen, and Jonathan A. Foley. 2009. "A Safe Operating Space for Humanity." *Nature* 461(24):472–475.

Rosa, Eugene A. and Thomas Dietz. 1998. "Climate Change and Society." *International Sociology* 13(4):421–455.

Salleh, Ariel. 2011. "Climate Strategy." *Journal of Australian Political Economy* 66:124–149.

Schnaiberg, Allan. 1980. *The Environment.* New York: Oxford University Press.

Schnaiberg, Allan and Kenneth A. Gould. 1994. *Environment and Society.* New York: St. Martin's Press.

Schor, Juliet. 2010. *Plenitude.* New York: Penguin.

Simon, Julian L. 1981. *The Ultimate Resource.* Princeton, NJ: Princeton University Press.

Smil, Vaclav. 1994. *Energy in World History.* Boulder, CO: Westview.

Sokal, Alan and Jean Bricmont. 1998. *Fashionable Nonsense.* New York: Picador USA.

Soper, Kate. 1995. *What Is Nature?* Oxford: Blackwell.

Steffen, Will, Åsa Persson, Lisa Deutsch, Jan Zalasiewicz, Mark Williams, Katherine Richardson, Carole Crumley, Paul Crutzen, Carl Folke, Line Gordon, Mario Molina, Veerabhadran Ramanathan, Johan Rockström, Marten Scheffer, Hans Joachim Schellnhuber, and Uno Svedin. 2011. "The Anthropocene: From Global Change to Planetary Stewardship." *Ambio* 40:739–761.

Stehr, Nico and Hans von Storch. 2010. *Climate and Society: Climate as Resource, Climate as Risk.* Singapore: World Scientific Publishing.

Stocker, Thomas S. and Dahe Qin (Eds.). 2013. *Climate Change 2013 The Physical Science Basis: Summery for Policy Makers (WG1).* IPCC Fifth Assessment Report (October 13). Retrieved November 24, 2013 (http://www.climatechange2013.org/images/uploads/WGI_AR5_SPM_brochure.pdf).

Swyngedouw, Erik. 2010. "Apocalypse Forever." *Theory, Culture & Society* 27(2–3):213–232.

Szerszynski, Bronislaw. 2010. "Reading and Writing the Weather." *Theory, Culture & Society* 27(2–3):9–30.

Taylor, Peter J. 1997. "How Do We Know We Have Global Environmental Problems? Undifferentiated Science-Politics and its Potential Reconstruction." Pp. 149–174 in *Changing Life: Genomes-Ecologies-Bodies-Commodities,* edited by P. Taylor, S. Haflon, and P. Edwards. Minneapolis: University of Minnesota Press.

Taylor, Peter J. and Frederick H. Buttel. 1992. "How Do We Know We Have Global Environmental Problems?" *Geoforum* 23(3):405–416.

Urry, John. 2011. *Climate Change & Society.* Malden, MA: Polity.

Veblen, Thorstein. [1923] 1964. *Absentee Ownership and Business Enterprise in Modern Times.* New York: Augustus M. Kelley.

Wainwright, Steven P. 2011. "Is Sociology Warming to Climate Change?" *Sociology* 45(1):173–177.

Weber, Max. [1904] 2011. "Objectivity in the Social Sciences." Pp. 49–112 in *Methodology of the Social Sciences,* edited by Edward A. Shils and Henry A. Finch. New Brunswick, NJ: Transaction Publishers.

Weber, Max. [1918] 1958. "Science as a Vocation." Pp. 129–156 in *From Max Weber*, edited by H. H. Gerth and C. Wright Mills. New York: Oxford University Press.

Weber, Max. [1920] 2011. *The Protestant Ethic and the Spirit of Capitalism*. New York: Oxford University Press.

White, Damian Finbar. 2006. "A Political Sociology of Socionatures." *Environmental Politics* 15(1):59–77.

White, Wayne A. 2013. *Biosequestration and Ecological Diversity*. Boca Raton, FL: CRC Press.

Whyte, Kyle P. 2013. "Justice Forward: Tribes, Climate Adaptation and Responsibility." *Climatic Change* 120:517–530.

Whyte, Kyle P. 2014. "Indigenous Women, Climate Change Impacts and Collective Action." *Hypatia* DOI: 10.1111/hypa.12089.

Will, George. 2008. "An Environmental 'License to Intrude.'" *LJ World.com* (May 23). Retrieved November 20, 2013 (http://www2.ljworld.com/news/2008/may/23/environmental_license_intrude/).

Woolgar, Steve. 1988. *Science: The Very Idea!* London: Tavistock.

Yearley, Steven. 1991. *The Green Case*. London: Harper Collins.

Yearley, Steven. 1996. *Sociology, Environmentalism, Globalization*. London: Sage.

Yearley, Steven. 1997. "The Changing Social Authority of Science." *Science Studies* 11(1):65–75.

Yearley, Steven. 2005a. "The 'End' or the 'Humanization' of Nature? *Organization & Environment* 18(2):198–201.

Yearley, Steven. 2005b. *Making Sense of Science*. Los Angeles: Sage.

Yearley, Steven. 2009. "Sociology and Climate Change after Kyoto." *Current Sociology* 57(3):389–405.

York, Richard. 2010a. "The Paradox at the Heart of Modernity." *International Journal of Sociology* 40(2):6–22.

York, Richard. 2010b. "Three Lessons from Trends in CO_2 Emissions and Energy Use in the United States." *Society and Natural Resources* 23(12):1244–1252.

12

Methodological Approaches for Sociological Research on Climate Change

Sandra T. Marquart-Pyatt, Andrew K. Jorgenson, and Lawrence C. Hamilton

INTRODUCTION

Sociological Research at the Society–Environment Interface

Research that analyzes both social and environmental data is central to understanding the human causes and consequences of climate change. Historically, the first discussions of environmental sociology as a distinct subfield (Catton and Dunlap 1978) emphasized the importance of society–environment interactions, which underscores the need to develop and continually refine research designs and modeling techniques that allow for rigorous analyses of the dynamic relations between human and natural systems (Liu et al. 2007a, 2007b). This chapter begins with a brief overview of methods used for earlier society–environment investigations, providing context for contemporary sociological research on climate change. Next, we provide examples from recent work that illustrate how spatial and temporal dimensions are considered in tandem with the integration of environmental and climate data for sociological research. We then present overviews of two dominant streams of climate-related research in which sociologists are currently engaged: human drivers of greenhouse gas (GHG) emissions and public opinion on climate change. Since Chapters 2 and 10 in this volume provide thorough summaries of their substantive findings, we focus here on the employed methods and research designs. We conclude by discussing additional methods sociologists currently use that are well positioned to contribute to future sociological research on climate change, meeting disciplinary and cross-disciplinary challenges.

Historical Context Within Environmental Sociology

The pioneering work of environmental sociologists Catton and Dunlap (1978) laid the foundation for emphasizing the relevance of the biophysical environment for sociological research. Their core idea for environmental sociology involves studying reciprocal relations between environment and society; that is, how humans affect and are affected by their surrounding natural and built environments. Because societies are embedded in ecological systems, both ecological and social factors affect social structures and individual behavior (Dunlap and Catton 1979). The early arguments of Dunlap and Catton called for a fundamental epistemological shift within the discipline of sociology: humans are no longer viewed as exempt from the environmental constraints affecting all other living species. Such a shift demands that environmental variables be included in sociological analyses of social–environmental problems, such as air and water pollution (Catton and Dunlap 1978).

The late William Freudenburg and Robert Gramling's research on society–environment interactions provide clear examples of the utility in pursuing sociological investigations that answer Dunlap and Catton's paradigm-shifting challenge. Often employing comparative-historical methods, this work shows reciprocity between social and physical facts as part of their contextual grounding or embeddedness (Freudenburg and Gramling 1993, 2010; Gramling and Freudenburg 1996). Gramling and Freudenburg (1996:352) argued for greater integration of environmental variables into sociological scholarship, urging scholars to not merely establish society–environment links but to delve further into the "nature, causes, and extent of those connections." Many scholars have responded to this call, employing techniques ranging from comparative-historical methods to spatial analysis. For the latter, a main premise is that location and distance are important attributes, in addition to standard sociological variables.

Such work embodies the spirit of integrative scholarship central to this chapter and to future research on climate change. In the following, we provide examples of sociologists and social scientists working on society–environment interactions focusing explicitly on climate change. Integrative research on climate change is becoming increasingly prominent within environmental sociology. Methodological issues can be formidable, however, requiring new approaches or creative application of existing techniques. Substantial progress has been made to date, but challenges remain; both are discussed in this chapter. Global changes in the last few decades across social, economic, political, and environmental domains broaden the research agenda for environmental sociology and intensify the need for a better understanding of these issues. Social and environmental data are collected for units of analysis that range from individuals, families, households, or census tracts through cities, states, and nations. Relevant theories and available data at each level shape the possibilities for research.

Integrated modeling approaches in the coupled human and natural/ecological systems tradition emphasize the interrelations and complexities of both human and natural systems rather than treating social and environmental domains as topics for separate, disciplinary analysis. Thus, work on coupled human and natural/ecological systems (Liu et al. 2007a, 2007b) is often interdisciplinary, and environmental sociology has an important role to play in such sustainability investigations. Both quantitative and qualitative analytical techniques are applicable to the examination of biophysical and social factors holistically, as illustrated in this chapter. Given space limitations, in the following sections we emphasize quantitative techniques but note recent advances in qualitative research designs (e.g., Norgaard 2011; Rudel 2009). In the next section we provide examples of recent work that uses spatial and temporal dimensions to effectively integrate social science with environmental and climate data.

INTEGRATION OF SOCIAL AND ENVIRONMENTAL DATA: TEMPORAL AND SPATIAL DYNAMICS

Integrated analyses of social and environmental data are essential for advancing our understanding of climate change mitigation and adaptation. Integrated studies have been encouraged, in principle, from the earliest days of environmental sociology (Catton and Dunlap 1978). Qualitative research can accomplish such integration informally, by linking human and environmental domains through case studies, historical narrative, and text analysis (Freudenburg and Gramling 1994). Quantitative research, though clearly necessary, continues to face challenges given the diverse scales and units of analysis for both social and physical data as well as in identification of analytical methods suitable for combining them. Social scientists' experience in working with complex data structures, together with their broad and evolving statistical toolkit, has led recently to an upsurge in rigorous integrated research traditions (Jacobs and Frickel 2009).

For environmental variables to enter into sociological analysis, they must vary—spatially, temporally, and often both (Freudenburg and Gramling 1993). Consequently, we might merge social and environmental data using place and/or time as integrating dimensions. The next few sections discuss some basic challenges and provide examples for each approach. These examples reveal practical ways to harmonize diverse data structures and effectively conduct analyses across disciplinary boundaries.

Spatial Integration: Many Places, One Time

Using place as the integrating dimension, scholars can construct datasets with social indicators (e.g., survey results, socioeconomic measures, voting data, policies) for specific places (e.g., communities, cities, counties, states, nations) at discrete points or intervals of time. These social indicators could then be merged with physical data (e.g., pollution levels, emissions, vulnerability, resource use, climate) for the same set of places, and the relations analyzed between social and physical variables, controlling for other factors as appropriate. Some examples include explanations for regional variations based on resource characteristics (Fisher 2006), physical vulnerability (Brody et al. 2008, 2009; Zahran et al. 2006), and climate change (Hamilton and Keim 2009). Other examples include health and temperature in urban heat islands (Harlan et al. 2013) and exposure to pollution and inequality from the environmental justice literature (Brulle and Pellow 2006; Downey 2003).

Many complexities arise when seeking to integrate environmental and social dimensions of research. Merging social with environmental data is straightforward if both are defined for the same observational units such as individuals, households, communities, or states. More often, however, the social and environmental data are defined for quite different units. Environmental monitoring and climate records, for example, commonly originate from point locations such as weather stations rather than administratively defined polygons like counties or states. If the social places studied are relatively small, such as individual communities or most U.S. counties, then nearby station data could provide a reasonable summary for the location. Looking forward, however, future climate projections have very high uncertainty as they are downscaled to specific regions, let alone communities. Thus, in the case of climate applications, harmonizing units of analysis becomes a major part of the research effort.

Figures 12.1a and 12.1b give a simple example of the many places/one time (*cross-sectional*) approach. Social data for this example were derived from more than nine thousand interviews in a series of rural-area surveys across nine U.S. states (Hamilton and Keim 2009). Survey questions include an item about local impacts of climate change. Physical variables include thirty-year trends in seasonal temperature recorded by weather stations in or near each rural area. Winter warming in snow country proved to be a significant predictor of perceived climate impacts, as visualized in aggregate form for Figures 12.1a and 12.1b but confirmed through individual-level analysis applying mixed-effects logistic regression (Rabe-Hesketh and Skrondal 2012) using the full survey dataset. The mixed-effects logistic models controlled for individual characteristics including a political party × education interaction, replicating findings from an earlier nonintegrated study

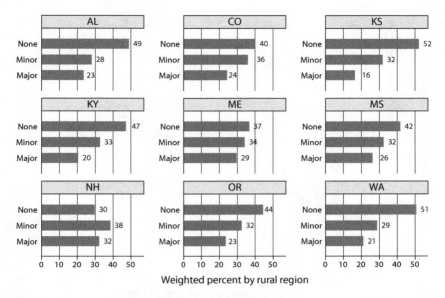

Weighted percent by rural region

Figure 12.1a Perceived Local Effects of Climate Change From Surveys in Nine Rural Regions

From: Hamilton and Keim 2009

(Hamilton 2008). The models also included random intercepts to represent "everything else" (not specified in the model) that might cause systematic variation between regions.

Other researchers have also found associations between objectively measured climate or weather events and survey response about climate change (Goebbert et al. 2012; Howe and Leiserowitz 2013; Shao et al. 2014; Zaval et al. 2014). Three recent studies take integration to the daily level by merging survey responses with local weather data, testing for effects of weather around the day of interview on respondents' views about climate change. All found that short-term (two-day) weather has measurable effects (Egan and Mullin 2012; Hamilton and Lemcke-Stampone 2013; Hamilton and Stampone 2013). Although basically cross-sectional, such survey/daily-weather studies may incorporate elements of temporal as well as spatial variation when the surveys are repeated and integration is done by interview date.

Figure 12.2 shows one such survey/weather analysis, from Hamilton and Stampone (2013). In this study, daily weather data for the state of New Hampshire were merged with about five thousand random-sample survey interviews conducted on ninety-nine days over 2010–2012. Mixed-effects logistic regression found that temperature anomaly (departure from normal

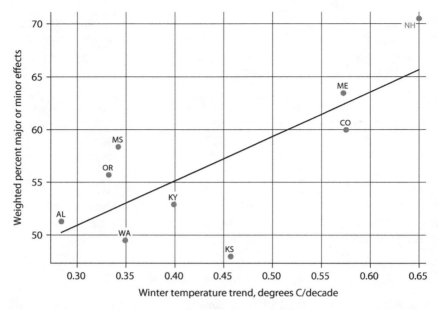

Figure 12.1b Perceived Local Effects of Climate Change Merged with Winter Temperature Trends at Weather Stations Within Each Region

From: Hamilton and Keim 2009

for that date) on the interview day and the previous day significantly affected belief in anthropogenic climate change, particularly among independent voters. Although this finding is not yet replicated, it serves to focus the next frontier of integrated survey/weather research: not *whether* weather affects climate views (which appears to be increasingly well established at least in the case of immediate weather experiences), but *how*—over what time scales, with what functional forms, and for which individuals and which combinations of outcome variables, weather indicators, and physical/ social environments. This matters because unusual weather might rein- force scientific reports linking global change with extreme weather events (e.g., Tang et al. 2013), influencing public opinion (e.g., Hamilton and Lemcke-Stampone 2013).

A different example of spatial integration investigates urban heat islands using neighborhoods in Maricopa County, Arizona, the greater Phoenix metropolitan area. This work considers individual and aggregate factors as mechanisms affecting heat-related deaths, using both remotely sensed data and actual mortality data, with neighborhoods as their unit of spatial orga- nization. Harlan et al. (2013) examine how social variables (socioeconomic status and age) and environmental variables (land cover and microclimates)

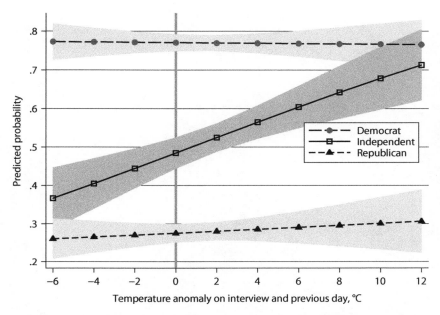

Figure 12.2 Belief in Anthropogenic Climate Change By Political Party and Temperature Anomaly

From: Hamilton and Stampone 2013

affect the incidence of heat-related deaths in more than two thousand census block groups over a nine-year period using a cross-sectional design (considering such deaths as rare events).[1] Using U.S. Census population data and a Landsat image to calculate amount of green vegetation and land surface temperature, they created neighborhood measures, modifying a heat vulnerability index (Reid et al. 2009) to the census block level. By mapping decedents' home addresses, they were able to analyze binary logistic regressions at the census block group level and conduct spatial analyses using a geocoded outcome variable created from public records identifying heat-related deaths. Results indicated the presence of spatial clustering among neighborhoods relating to vulnerability to heat, showing variability between neighborhoods. Figure 12.3 shows these relations. Inner-city neighborhoods had higher vulnerability to heat and more deaths, whereas suburban neighborhoods had fewer deaths and lower vulnerability scores.

This study extended earlier research investigating risk and exposure to extreme heat in microclimates in greater Phoenix (Ruddell et al. 2010). Using data from a regional atmospheric model designed to simulate a four-day heatwave in 2005, researchers compared simulated air temperatures with population characteristics from census block groups. Analyses revealed that

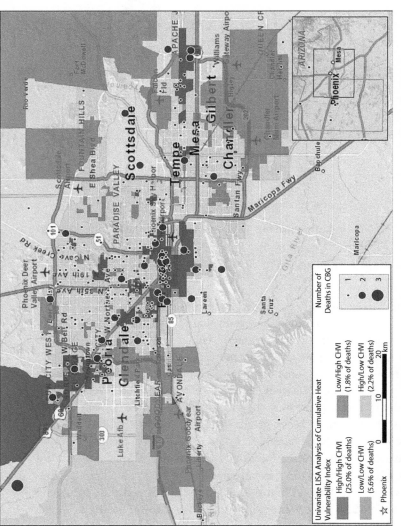

Figure 12.3 Univariate analysis of the Local Indicator of Spatial Association (LISA)-identified clusters of census block groups (CBGs) in Maricopa County, AZ, with similar or dissimilar Heat Vulnerability Index scores ($p \leq 0.05$). High/high areas in the map are clusters of neighboring CBGs with uniformly high vulnerability scores; low/low areas are clusters with low vulnerability scores; low/high areas represent a CBG with a low vulnerability score neighbored by high-vulnerability CBGs; high/low areas represent a CBG with a high vulnerability score neighbored by low-vulnerability CBGs. Entries in the boxes (next to the legend) also show the percentages of 2000–2008 heat-related decedents who were residents in each type of cluster.

From: Harlan et al. 2013. http://ehp.niehs.nih.gov/1104625/

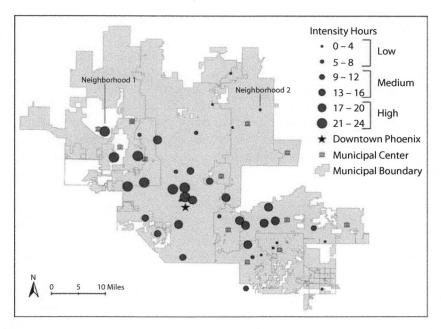

Figure 12.4 Hours of Exposure to Threshold Temperatures at or Above the 97.5 Percentile from July 15 to 19, 2005, by Neighborhood

From: Ruddel et al. 2010

exposure to temperature extremes differs significantly within the urbanized area, with low-income neighborhoods and those with larger minority and elderly populations more likely to be exposed to intense heat for longer time periods. Figure 12.4 shows that some neighborhoods exceeded threshold temperatures for less than four hours whereas others exceeded threshold for twenty-one to twenty-four hours out of the ninety-six-hour heatwave. This study reinforces previous research showing that land use/land cover characteristics, specifically vegetation abundance, are significant drivers of local air temperature, and vegetation has an inverse relation with household income at the block group level (Ruddell et al. 2010).

Temporal Integration: One Place, Many Times

Climate change has a temporal dimension that is necessarily flattened for cross-sectional analyses like Figure 12.1a and b (where change is represented by rates) or Figure 12.2 (temperatures matched to interviews by date). More dynamic studies of climate–society interaction become possible when we follow climate and societal phenomena over time. Historical case studies often have done this informally by incorporating past environmental change into narrative accounts.

Instrumental climatic, oceanographic, and other physical records exist back through the twentieth century for many regions. In some cases the historical time series go back farther, or long-term climate reconstructions have been derived from proxies such as sediment or ice cores. Plotting such physical data can help social scientists to frame qualitative accounts or match environmental timelines with sociohistorical records. Figure 12.5 illustrates informal integration applied to case studies of Icelandic fishing communities (Hamilton 2007; Hamilton et al. 2004; Hamilton, Otterstad, and Ögmundardóttir 2006).

Discovery and commercial development of rich cod and herring fishing grounds propelled Iceland's rapid twentieth-century emergence from poverty into economic and political independence, and affluence. An early, labor-intensive phase of the herring fishery particularly benefited the North Iceland town of Siglufjörður, elevating it for a time to the "herring capital of the world." Later, as overfishing depleted the resource, the herring fishery became more industrial and long distance, at locations better pursued

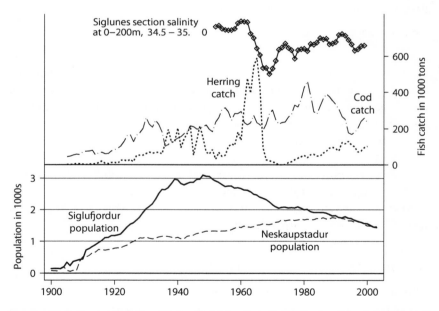

Figure 12.5 Ocean Salinity off North Iceland, Iceland Region Herring and Cod Catches, and the Population of Two Fishing Towns Over the Twentieth Century

From Hamilton 2007; see Hamilton et al. 2004, 2006 for data sources and historical narrative

from East Iceland towns such as Neskaupstaður. Social and environmental factors combined in unique fashion. A "killer spike" of unsustainable overfishing in the 1960s coincided with adverse environmental changes, marked by a pulse of cold low-salinity surface water in the seas north of Iceland that suppressed the herring's food supply. This salinity anomaly resulted from unusual wind conditions that exported a large volume of cold, fresh Arctic ice and water south through Fram Strait and into the northern Atlantic in the late 1960s, with widespread impacts on fisheries. In the case of Iceland herring, the resource collapsed below 1 percent of its former biomass. The human population of Siglufjörður declined through outmigration beginning in the 1950s, with correspondingly broad social change. Neskaupstaður, which unlike Siglufjörður also had access to cod fishing, did not begin declining until cod stocks too became depleted in the 1980s. The "rise and fall of the herring towns" illustrates a common pattern in fisheries-dependent regions: collapse driven by the unfortunate synergy of resource overuse and climate variation. Similar historical patterns of synergistic collapse occurred in small- and large-scale societies and will likely characterize the future impacts of climate change. Climate impacts can be magnified where they coincide with unsustainable resource or land use decisions. Integrated time plots like Figure 12.5 pull diverse physical, biological, and human elements together, framing participant interviews and other qualitative or quantitative elements in the historical account.

A second example of informal integration appears in Figure 12.6. The relics around New England of hundreds of abandoned ski areas, many in locations that now rarely have skiable snow, inspired this study of "Warming Winters and New Hampshire's Lost Ski Areas" (Hamilton et al. 2003). These areas had opened during decades when cold winters were more common, then shut down in stages as the regional winters warmed. The ski areas surviving today are mostly at high-elevation northern locations and require major capital investment in snowmaking, supported by real estate development, seasonal diversification, and rising prices. Smaller areas without such resources and geographical advantages could not keep up. The trend continues as many ski or winter recreation areas worldwide face difficulties from warming winters (Burakowski and Magnusson 2012; Hamilton, Brown, and Keim 2007; Scott and McBoyle 2007).

Figure 12.6 graphs statewide temperatures based on Climate Divisional Data supplied by the National Climate Data Center. Even in this small state, the geographic range from seacoast to mountain summits creates difficulties for defining "statewide" temperatures, and the Climate Divisional approach has potential for bias (Keim et al. 2003). A better method used in subsequent work (e.g., Fig. 12.2) relies on temperature anomalies from the

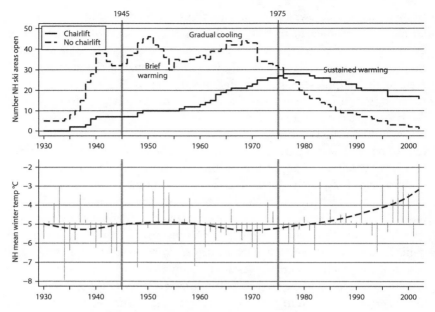

Figure 12.6 Warming Winters (bottom) and New Hampshire's Lost Ski Areas (top), 1930–2002.

Annual December–March temperatures as mean deviations, with locally weighted, smoothed regression curve. Small areas that never possessed a chairlift or gondola are shown separately (dashed line).

From: Hamilton et al. 2003

local average on each date, across the state's U.S. Historical Climatology Network (USHCN) weather stations.

USHCN station data provide well-screened records of yearly, monthly, or daily temperature and precipitation for hundreds of U.S. locations back to 1895. With economically important and climate-sensitive social activities such as fisheries or winter recreation, those detailed records open possibilities for more formally integrated *time series analysis*. Time series analysis has a long history in econometrics but less so among sociologists, who (until recently) seldom dealt with multivariate series containing hundreds of observations. Figure 12.7 gives a sociological example involving weather and climate effects on skier attendance at one New Hampshire ski area. This graph depicts daily variations over a single ski season, drawn from a larger analysis that encompassed daily weather and skier data over nine winters and a thousand skiable days at one area (and also seven winters and eight hundred skiable days at a second area, not shown). The gray "mountains" in Figure 12.7 visualize daily snow depth measured in the town of Lakeport,

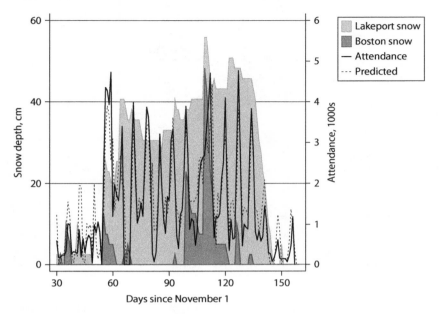

Figure 12.7 Observed and ARMAX Model Predictions of Attendance at Gunstock Ski Area During the 2002–2003 Ski Season, Graphed with Snow Depth in Boston and in the Nearby Town of Lakeport

From: Hamilton et al. 2007

near this particular ski area, and the city of Boston which is 161 kilometers away. Lines show the close fit between observed and predicted number of skier visits.

Predictions in Figure 12.7 derive from an ARMAX (autoregressive moving average with exogenous variables) model that includes as predictors the lagged values of temperature, snowfall, and snow depth in Boston and Lakeport, along with indicators for day of the week and the availability of nighttime skiing. Disturbances or error terms are modeled as a first-order autoregressive and moving average process—ARMA(1,1), in time series notation. Residuals pass statistical tests for white noise. A salient finding from this model is confirmation of a long-suspected "backyard effect": in terms of skier visits, a centimeter of snow on the ground in Boston is worth more than a centimeter of snow in the mountains. Quantifying the snow–skier relationship, this analysis shows a gap between high- and low-snow seasons on the order of 100,000 skier-days for these two areas alone (Fig. 12.8). A trend toward less snowy winters causes problems for the regional economy that cannot be offset by artificial snowmaking.

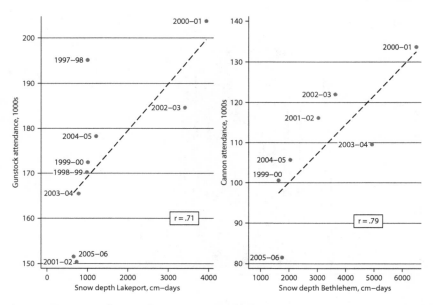

Figure 12.8 Annual Attendance at Two Ski Areas Versus Cumulative Annual Centimeter-Days of Snow Cover at Nearby Towns

From: Hamilton et al. 2007

ARMAX modeling tends to be data-intensive, working better when there are hundreds of observations—as with the daily data in Figure 12.8. Although ARMAX modeling could in principle be applied to annual data where far fewer observations exist, less formal integration as in Figures 12.5 and 12.6 remains more attractive in that case. For sufficiently long records of monthly or daily resolved data, however, these methods provide well-developed and general tools for exploring relations between society and climate. Formal time series modeling requires the analyst to think clearly about causal lags and the dynamic behavior of "everything else" (i.e., disturbances) affecting dependent variables—complications that are submerged but not absent when taking a cross-sectional or qualitative approach.

Spatial/Temporal Integration: Many Places, Many Times

A further step toward complexity and realism involves the analysis of separate, parallel time series across many different places. The term *panel data* could apply to such structures, but in sociological practice that term more often connotes designs with just a few points in time, such as a repeated experiment or survey. A cumbersome but more descriptive term for data with many parallel time series is *cross-sectional/time series*. Mixed-effects modeling, mentioned earlier in connection with 12.1a, 12.1b, and 12.2, provides general tools for this purpose. Although the main analysis in Figure 12.1a and 12.1b is cross-sectional, with climate dynamics simplified to rates of change in each place, those rates of change were estimated through a separate mixed-effects regression of seasonal temperature anomalies on year. Random intercepts and slopes from the mixed-effects model characterize the different linear trends of each place (Hamilton and Keim 2009).

Figure 12.9 depicts results from a more elaborate cross-sectional/time series analysis. The observations for this analysis are forty-two predominantly Native towns and villages in Arctic Alaska, over the years 1990 to 2009. Climate change is already a strong force in the Arctic and is felt nowhere more keenly than in such remote and indigenous communities. This analysis tested methods for the integrated analysis of community-level human and physical data. For readability, only nine of the forty-two study communities appear in this graphic. Annual electricity use in each community forms the dependent variable, of particular importance for these places because most electricity is generated locally by burning diesel fuel. Rising fuel costs, transporting oil to remote roadless areas, and the maintenance of critical infrastructure under Arctic conditions all highlight the significance of this resource. Observed electricity use is marked by squares in each graph. Dashed curves show electricity use predicted from a model based on community population (subject to natural increase along with interannual variations in net migration), electricity cost, seasonal temperature and precipitation, and longer-term trend.

Population dominates annual variations in electricity use, showing both general and community-specific (random) effects. Higher prices reduce electricity use, as expected. In southern latitudes where air conditioning is common, warmer summers tend to have higher electricity use, but in these Arctic communities (where there is little need for air conditioning) the opposite seems true: towns and villages use more electricity in cool, wet summers when people spend more time indoors. Net of population, seasonal climate, and price, there has been a general upward trend. Disturbances for this model are characterized as a first-order autoregressive process, AR(1). Although the original electricity time series are nonstationary, model residuals pass statistical tests for white noise.

Many of these same Alaskan communities face serious erosion threats linked to climate change through such things as permafrost failure, reduced shoreline protection from sea ice, or increased precipitation and river flows. A next step in this integrated social/physical research will apply the mixed-effects methods described here to testing and quantifying the demographic impacts (e.g., on net migration) from environmental changes measured in these places. This mixed-effects approach could be applied much

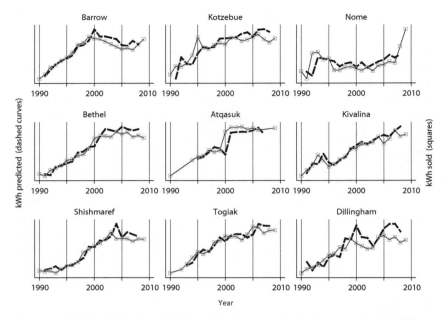

Figure 12.9 Observed Electricity Use in Nine Arctic Alaska Towns and Villages (Squares), Together with Electricity Use Predicted from a Mixed-Effects Model

From: Hamilton, White, et al. 2012

more widely to other kinds of cross-sectional/time series data, such as multiple state- or city-level series.

Mixed-effects modeling (Rabe-Hesketh and Skrondal 2012) and the closely related methods of multilevel modeling (Raudenbush and Bryk 2002; Snijders and Bosker 2012), discussed later, have advanced substantially in recent years. They provide much-needed new tools for integrated analysis of social and physical data, explicitly modeling variations in level and relations across data clusters or places. Like the ARMAX models discussed earlier, mixed models tend to be data-intensive. For sociological applications they may require many hundreds if not thousands of observations. The rising popularity of these methods among environmental sociologists reflects their suitability for key research questions and the growing availability of large, structurally complex datasets. For example, the models behind Figure 12.9 employed 742 place-years of data (42 places, the years 1990–2008 with some gaps). Updating these data through 2013 would bring in 210 additional place-years, with corresponding gains in statistical power and precision. Similarly, the interviews that underlie Figure 12.2 have since been supplemented by thousands of additional interviews asking the same questions.

SOCIOLOGICAL RESEARCH ON THE HUMAN DRIVERS
OF GREENHOUSE GAS EMISSIONS

Developing a better understanding of how human actions lead to environmental change such as global warming is an essential task. Rosa and Dietz (2012) note that sociologists have made significant contributions to such understanding, particularly on the anthropogenic drivers of GHG emissions. The term "anthropogenic drivers" refers to the range of human activities and broader societal characteristics and relations leading to the emissions of GHG into the atmosphere. As a discipline, sociology is well equipped to investigate how human activities, social institutions, and relations between social units at multiple scales contribute to GHG emissions. In this section we discuss some methodological aspects of these works, such as the most common outcome variables, units of analysis, and analytical techniques used in empirical investigations. Chapter 2 in this volume provides a thorough overview of the theoretical advances and empirical findings in sociological research on drivers of emissions. Here we focus on some of the key methodological contributions of this body of work.

How Cross-National Human Drivers Research
Incorporates Place and Time

The nation-state is the unit of analysis in the majority of sociological studies of the drivers of anthropogenic GHG emissions,[2] thus representing the principal way place is incorporated. There are two main reasons for this. First, abundant data are available for a wide range of potential outcome and predictor variables at the nation-state level, suitable for theoretically grounded empirical analyses of GHG emissions that are primary causes of modern climate change. Second, numerous well-established perspectives focus on country-level characteristics and interrelations (Jorgenson and Clark 2012). While most research on drivers of GHG emissions to date has been cross-national, as better data became available researchers have also begun conducting comparative quantitative studies at lower levels such as U.S. states (e.g., Clement and Schultz 2011) and power plants (Grant, Jorgenson, and Longhofer 2013).

The outcome variable for the vast majority of research on drivers of GHG emissions, regardless of the unit of analysis, is carbon dioxide emissions, measured as total emissions, per capita emissions, or emissions per unit of production (per unit of gross domestic production [GDP]). Carbon emissions constitute the key outcome variables for three reasons. First, there is scientific consensus that anthropogenic carbon dioxide emissions are a primary contributor to climate change (U.S. National Research Council 2010).

Second, most economic activities require the burning of fossil fuels, which results in carbon dioxide emissions. Third, there are many more data available for anthropogenic carbon dioxide emissions than for any other type of GHG or environmentally harmful product of human activities (World Bank 2007).

Some cross-national studies consider anthropogenic methane emissions along with carbon dioxide as dependent variables (Burns, Davis, and Kick 1997; Rosa, York, and Dietz 2004; York, Rosa, and Dietz 2003), while a few focus solely on methane emissions (Jorgenson 2006; Jorgenson and Birkholz 2010). Studies of methane emissions are less common due to limited data, and until quite recently the available data were only suited for cross-sectional analysis. The human drivers of methane emissions are important and somewhat different from those for carbon dioxide, warranting increased attention. While atmospheric carbon dioxide is approximately two hundred times more plentiful than atmospheric methane, molecule-for-molecule methane is at least ten times more effective at absorbing and reradiating infrared energy and heat back to the Earth's surface (Christianson 1999). Like carbon dioxide, anthropogenic methane emissions from many nations have increased substantially during recent decades (World Resources Institute 2010).

Nearly all early studies of anthropogenic GHG emissions by sociologists were national-level and cross-sectional by design, illustrating the "many places, one time" analyses described in the prior section. Cross-sectional models are conducted for a single point in time; caution must be taken when drawing conclusions from such static analyses. These early studies typically employed a form of ordinary least squares (OLS) regression and include analyses in several areas grounded in differing frameworks: the environmental Kuznets' curve (EKC) tradition,[3] analyses that formalized the STIRPAT (Stochastic Impacts by Regression on Population, Affluence, and Technology) approach, and studies testing hypothesis derived from political-economic perspectives.

Early cross-sectional work done by sociologists in the EKC tradition considered the curvilinear (quadratic) effect of level of economic development (GDP per capita) along with relevant control variables (Roberts and Grimes 1997). These studies engaged sociological theories such as modernization theory and world systems theory that provided contrasting explanations for potential curvilinear relationships between emissions and economic development. The results of Roberts and Grimes (1997) suggested that by the early 1990s, the relationship between carbon emissions per unit of GDP and levels of economic development had become curvilinear, resembling an inverted-U distribution, consistent with the EKC tradition. Roberts and Grimes suggested that their results were largely due to efficiency improvements in a small number of wealthy countries combined with worse performance in poor and middle-income countries and concluded that the overall observed relationships were largely conditioned by power differences and inequalities between nations.

Early cross-sectional work on drivers of emissions in the STIRPAT tradition, like Dietz and Rosa (1997), built on the IPAT framework, where environmental impacts are proposed to be a function of population, affluence, and technology. STIRPAT work employed elasticity models where the outcome and all explanatory variables are in logarithmic form. While IPAT is helpful for conceptual framings, it is not well suited for hypothesis testing since by design it assumes the effects of each driver (population, affluence, technology) to be proportional. Unlike IPAT, STIRPAT is not an accounting equation; it is a stochastic model appropriate for hypothesis testing.

The baseline STIRPAT equation includes the outcome or dependent variable as the environmental impact along with an intercept term (a), estimated coefficients b and c linked with population and affluence measures respectively, "technology" representing everything else in the equation, and an error term (e). This equation was designed for testing theoretically derived hypotheses from human ecology concerning the effects of population, affluence, and technology. There is flexibility in the model, as it allows for inclusion of factors and explanatory variables besides population and affluence. Within the model, one can also decompose population, affluence, and technology. Findings from early STIRPAT work found population and affluence to be the primary drivers of total carbon emissions (Dietz and Jorgenson 2013). Technology has proven difficult to measure and is often relegated to the error term. The coefficients of an elasticity model like STIRPAT are straightforward to interpret. Specifically, the coefficient for each continuous independent variable is the estimated percentage change in the dependent variable associated with a 1 percent increase in the independent variable, controlling for all other factors.

Early cross-sectional research on drivers at the national level testing propositions from political-economic perspectives employed either OLS regression or path analysis, with the latter allowing the inclusion of intervening variables. An example that employed both was Burns et al. (1997), which tested propositions derived from world systems and structural human ecology perspectives in path models of carbon dioxide emissions and methane emissions. For both outcomes they found that more powerful core nations exhibited the highest levels of emissions, followed by less-powerful semiperipheral and peripheral nations. Further, and consistent with the STIRPAT tradition, total carbon and methane emissions are both positively associated with population size.

Cross-sectional research on national-level drivers of GHG emissions continued up through roughly the 2005–2006 period, largely due to the scarcity of longitudinal emissions data. During this time, environmental sociologists were successful in publishing these works in sociology journals (e.g., Fisher and Freudenburg 2004; Grimes and Kentor 2003; Jorgenson 2006; Roberts, Grimes, and Manale 2003; Schofer and Hironaka 2005; Shandra et al. 2004; York et al. 2003) and in environmental science journals (e.g., Rosa et al. 2004). As described in detail in Chapter 2, these studies focused on various drivers of GHG emissions, including economic development, population size, urbanization, economic globalization characteristics, and the role of environmental governance and environmental international non-governmental organizations. Cross-sectional approaches are still sometimes employed, reflecting data availability limitations for theoretically important explanatory variables (e.g., Ergas and York 2012).

In recent years, more extensive longitudinal data—especially on carbon dioxide emissions—have become available at national and sometimes subnational (e.g., U.S. state) levels that allow for temporal integration, as outlined earlier, as well as spatial integration. Sociologists have in a short period of time published a large number of studies employing panel (or cross-sectional time series) data analysis. Newer methods can account for heterogeneity bias or the confounding effect of unmeasured time-invariant variables that are omitted from the regression models. Fixed-effects (FE) and random-effects (RE) techniques have gained in popularity in various areas of sociology (Halaby 2004), including environmental sociology. Mixed-effects models such as the one shown in Figure 12.8 represent a generalization of these ideas, allowing models that include both fixed and random intercepts and slopes. Both FE and RE methods treat unmeasured time-invariant factors as case-specific intercepts. The FE model treats the case-specific intercepts as fixed effects to be estimated, while the RE model treats case-specific intercepts as a random component of the error term. The general equation for a panel model is

$$y_{it} = \mu_t + \beta x_{it} + \gamma z_i + a_i + \varepsilon_{it}$$

Here, subscript i represents each unit of analysis (i.e., country), subscript t represents the time period, y_{it} is the dependent variable for each country at each time period, μ_t is an intercept that may be different for each time period, and β and γ are vectors for coefficients. Predictor variables that vary over time are represented by the vector x_{it}, time-invariant predictor variables are represented by the vector z_i, a_i represents the combined effect on y of all unobserved variables that are constant over time, and ε_{it} represents purely random variation at each time point. Since a_i is perfectly collinear with

z_i, the FE model will not produce coefficient estimates for time-invariant predictors.

There are two common approaches to estimated case-specific fixed effects in linear panel models. One approach involves the inclusion of dummy variables for N – 1 cases, known as the dummy-variable FE model (Allison 2009). The other, the within-estimator or conditional approach, involves a mean deviation algorithm for the dependent variable and each time-varying independent variable that is easily implemented in the popular statistical analysis software package Stata. It is also known as the conditional method because it conditions out coefficients for the FE dummy variables (Allison 2009). Employing either approach for case-specific FEs is often referred to as a one-way FE model (Baum 2006).

In either random effects or fixed effects models, researchers can also adjust for unexplained time-to-time (or random place-to-place) variations. A simple approach for doing so, especially in studies with relatively few time points, involves a series of T – 1 dummy variables for each value of time. Representing time as a set of dummy variables allows one to control for potential unobserved heterogeneity that is cross-sectionally invariant within periods, or what are known as period-specific intercepts. The inclusion of the period-specific intercepts in a panel model is equivalent to modeling temporal fixed effects, and including both period-specific intercepts (i.e., temporal FEs) and either form of case-specific FEs is analogous to estimating a two-way FE model (Baum 2006), a common strategy in current sociological research on drivers of emissions.

Random effects models offer several advantages for drivers of emissions research. First, they permit analysis of variation between observational units or cases, as well as within cases through time, whereas FE models focus only on within-case variation. This is an important issue in cross-national studies of human drivers of GHG emissions with relatively few time points, such as ten years or less. Most variation in key variables might occur between cases (e.g., between countries), instead of within countries, over time. Theoretically derived hypotheses tested in such studies often focus on potential differences between cases. Further, RE models can specify time-invariant explanatory variables useful for hypotheses testing. For example, the human ecology tradition asserts that ecological conditions like climate influence how societies and their activities contribute to GHG emissions (Dietz and Rosa 1997; York et al. 2003).

Researchers testing this perspective often employ ecological measures, such as climate proxies of latitude and longitude that are by definition time-invariant characteristics for particular locations. Earlier cross-sectional studies in the human ecology tradition indicate that nations in colder climates tend to exhibit higher levels of carbon emissions. Similarly, some macro political-economic perspectives hypothesize that certain world-economic and transnational factors contribute to emissions. Such independent variables may change only slowly with time but vary more substantially across cases, thus motivating the use of RE models. Research, for instance, examines effects of foreign direct investment in the manufacturing sector on carbon emissions in developing nations. Results indicate that foreign direct investment and other world-economic factors, which vary little through time but do vary across cases, increase carbon emissions in developing nations, while controlling for other relevant factors (Jorgenson, Dick, and Mahutga 2007; Jorgenson, Dick, and Shandra 2011).

One-way and two-way fixed effects panel models have, to date, been more popular than random effects models. Given that they both account for factors unique to each case that do not vary through time, and that two-way models also account for factors varying through time but not across cases, both approaches reduce the likelihood of falsely rejecting the null hypothesis. Also, including period-specific intercepts (or temporal FEs) or a linear time control lessens the likelihood of biased model estimates resulting from outcomes and predictors sharing similar time trends (Wooldridge 2005). Some suggest that FE models more closely approximate experimental conditions than other panel model and cross-sectional approaches (Hsiao 2003). Examples of sociological research on drivers of emissions that employ FE models include Jorgenson (2007, 2012, 2014), Jorgenson and Clark (2010, 2012), Jorgenson, Auerbach, and Clark (2014), and York (2008, 2012a, 2012b). Jorgenson and Birkholz (2010) estimate two-way FE models in a panel study of national-level anthropogenic methane emissions, and Clement and Schultz (2011) estimate FE models in a longitudinal study of U.S. state-level carbon emissions. Findings from cross-national analyses conducted in the former show effects for economic development, export-oriented growth, agricultural indicators, and oil and gas production across nations from 1990 to 2005. Results of the latter indicate that economic growth and urbanization both affect rising total energy use from 1960 to 1990.

Panel analysis techniques can also test whether the effects of explanatory variables change through time. This is done by including interactions between time and predictor variables, which allows more rigorous assessments of society–environment relations and their temporal dynamics. For research on drivers of emissions, this is especially useful for hypothesis testing, since key theories in environmental sociology focus on the possibility that the effects of social factors, such as economic development, urbanization, international trade, and population size, might change over time. Figure 12.10, borrowed from Jorgenson and Clark's (2012) panel study of national-level carbon dioxide emissions, shows the result of including interactions between levels of economic development (GDP per capita) and dummy variables for time (yearly observations). In less-developed countries the estimated effect of level of economic development of per capita carbon dioxide emissions slightly increased in value through time, at least from 1960 to 2005, while for developed countries the effect of development on per capita emissions remained stable. The two-way FE models include various control variables such as urban population and international trade.

Figure 12.11 graphs results focusing on fifteen countries in Asia from 1960 to 2005 and includes interactions between total population size and time in a two-way FE model of total carbon emissions. These findings are from Jorgenson and Clark (2013), who estimate the same model separately for nations in the regions of Africa, Asia, Latin America, and the combined regions of Europe, North America, and Oceania. The model includes

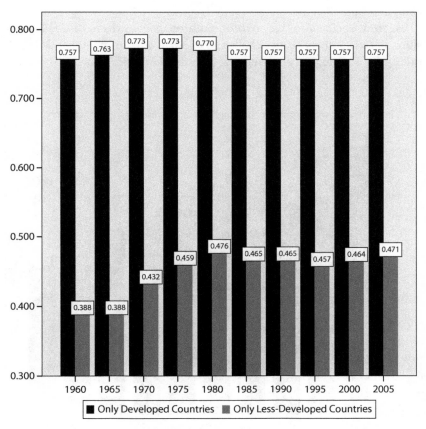

Figure 12.10 Estimated Effects of Per Capita GDP on Per Capita Carbon Dioxide Emissions for Developed and Less-Developed Countries, 1960–2005 (unstandardized coefficients)

relevant controls like levels of economic development and urbanization. Results suggest that the effect of population size on total emissions in the sample of Asian nations varies over time: the estimated coefficient generally decreases from 1960 to 2005. Results in the broader study highlight how the temporal relationship between GHG emissions and population size differs for the world regions, which underscores the importance of temporality and place, and interactions between explanatory variables and time in panel models of anthropogenic carbon emissions have proven useful in other sociological studies on the drivers of national-level anthropogenic emissions (e.g., Jorgenson 2014; Jorgenson et al. 2014; Jorgenson and Clark 2010; Jorgenson and Dietz 2014). With the availability of panel data on emissions and explanatory variables at smaller units of analysis, including U.S. states, cities, and different types of facilities like factories or power plants, future research may routinely consider such interactions between predictor variables and time.

In sum, sociological research on anthropogenic drivers of GHG emissions has responded to the challenge of early environmental sociologists calling for sociological analyses that include biophysical data. This body of

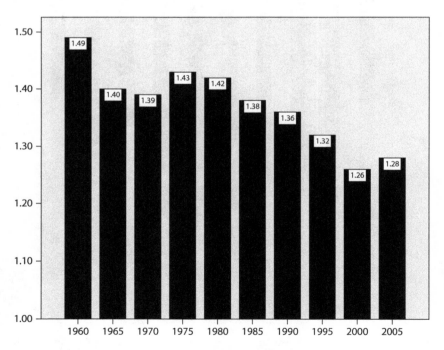

Figure 12.11 Estimated Effects of Population on Carbon Dioxide Emissions for Fifteen Countries in Asia, 1960–2005 (unstandardized coefficients)

sociological work and the methods employed greatly advance humanity's understanding of climate change and the necessity in looking to both the social sciences and the natural sciences for integrated explanations of the drivers of anthropogenic GHG emissions. Scholarship integrating environmental variables with core sociological ideas about the structure and relationships of human societies has built a foundation for future research on the drivers of emissions. With the recent increase in the availability of longitudinal data, sociological work has successfully incorporated time into analyses, allowing researchers to consider how effects of human activities, social institutions, and other societal characteristics on GHG emissions change through time as well as by location. As solid findings accumulate, model estimation techniques become more sophisticated. As longitudinal data for various units of analyses become increasingly available, sociological research on the human drivers of GHG emissions will progress, becoming increasingly detailed, sophisticated, and policy-relevant.

SOCIOLOGICAL RESEARCH ON CLIMATE CHANGE PUBLIC OPINION

Social scientific research analyzing national public opinion surveys on climate change and global warming views has established that belief and concern among the U.S. general public are widespread but also characterized by severe and growing political polarization (e.g., Dunlap and McCright 2008; Marquart-Pyatt et al. 2011; McCright and Dunlap 2011; McCright et al. 2014a).[4] Many studies analyze the social bases of climate change views, focusing on individual-level predictors. The strongest and most consistent predictors are indicators of political orientation, especially ideology and political party. Self-identified liberals and Democrats are more likely to believe that the effects of global warming have already begun and that the seriousness of global warming has been underestimated, compared with their conservative and Republican counterparts (Hamilton 2012; McCright and Dunlap 2011). The Democrat–Republican gap on acceptance of anthropogenic climate change is among the widest now seen on surveys (Hamilton 2014; Marquart-Pyatt et al. 2014), and this gap becomes even wider when Tea Party supporters define one endpoint (Hamilton and Saito 2015).

Interaction effects can be important. For instance, the effects of education or knowledge measures are often moderated by political orientation, such that education and knowledge have opposite effects for liberals and conservatives, or for Democrats and Republicans (Hamilton 2008, 2011, 2012; Hamilton, Cutler, and Schaefer 2012; Hamilton and Keim 2009; Hamilton and Safford 2014; Hamilton and Saito 2015; McCright 2011; McCright and Dunlap 2011). The well-replicated, nonadditive effects of education and politics may account for some differences in anticipated relations among reports where education × politics (or similar) interaction terms are *not* tested in the model. Zero-order relations between education and climate views could then represent an average of positive effects among liberals/ Democrats in the sample, and weak or negative effects among conservatives/Republicans. Main effects of education in models including interactions will depend on specific codes used but are often reported to be positive (Hamilton 2008, 2012; Hamilton et al. 2012, 2013). Less consistent results are reported for gender (McCright 2009, 2010; Zahran et al. 2006) and age (Hamilton 2008, 2011, 2012; McCright 2009). Such inconsistencies might reflect weaker effects, question-specific variations, or as-yet-unidentified interactions. Many studies have established the need to test for knowledge/ politics interactions (e.g., Hamilton 2008, 2011, 2012; Hamilton et al. 2012; McCright 2011; McCright and Dunlap 2011). Future research might uncover other interactions that also prove robust and important.

Spatial and Temporal Integration in Public Opinion Research

Public opinion studies of climate change views vary in a number of dimensions. In addition to spatial and temporal variation, they differ in types of questions asked about climate change, units of analyses, data sources (e.g., polling, survey), geographic coverage, temporal coverage, mode of data collection, sample characteristics (national, regional, community), use of secondary data, and integration of social and natural data. Here we summarize a subset of recent studies, including the spatial and temporal dimensions that are an organizing frame for this chapter, and analytical techniques and data sources.[5]

Much of the social scientific scholarship characterizing public opinion on climate change relies on survey data gathered through mail, telephone, and Web-based surveys. The specific topics of climate change investigated include beliefs, attitudes, knowledge, and concern about climate change as well as various dimensions of climate change policy support, or climate-relevant behavior. Most studies are cross-sectional, a few use panel data, and a handful use repeated cross-sectional surveys to construct quasi-longitudinal/time series data (e.g., Hamilton and Lemcke-Stampone 2013; Hamilton and Stampone 2013). Some research on climate change public opinion uses information about weather and climate, linking it with public opinion survey data.

In many areas, sociological research cannot be very specific about the time structure of causal relations because our data lack fine temporal resolution. Weather and climate measures are an exception, however, being available for many locations at daily resolution going back for years, and at monthly resolution going back for a century or more. On the social side, for telephone interviews or Internet surveys the exact date (even time of day) of completion may be known. Integrating weather/climate with survey data therefore opens up new analytical possibilities that differ from other social research but address questions of practical importance. For example, three studies have reported that daily temperature anomalies over a two-day window affect survey responses more sharply than longer or shorter windows (Egan and Mullin 2012; Hamilton and Lemcke-Stampone 2013; Hamilton and Stampone 2013). Short-lived effects could have lasting impacts, however, to the extent that climate itself continues changing.

Most of these studies employ regression-type analytical methods such as OLS, logit, or probit. Like any statistical methods, such modeling requires assumptions about processes generating the data, where plausibility is a matter of degree. Researchers often choose OLS or its variants to study continuous, measured outcome variables. Categorical variables common to survey research require different techniques, such as variants of logit or probit regression, designed for binary, multicategory or ordered dependent variables. In addition, a few studies use structural equation modeling (SEM) to demonstrate direct, indirect, and total effects between variables hypothesized to affect climate change policy support (Dietz, Dan, and Shwom 2007; Ding et al. 2011; Malka, Krosnick, and Langer 2009) and support for government action (McCright, Dunlap, and Xiao 2013, 2014a, 2014b).

Weather and climate variables often correlate with details of location and timing, raising the risk of spurious conclusions if we attribute to weather or climate some influence that actually reflects geography or timing (e.g., regional variations, current events). Mixed-effects versions of regression-type models provide one way of reducing this risk of spurious conclusions, by including random-effect terms that could represent "everything else" (besides variables in the model) causing systematic place-to-place or time-to-time differences. Several papers (Hamilton and Keim 2009; Hamilton and Stampone 2013; Hamilton et al. 2013) have used mixed-effects modeling this way.

Public Opinion Studies Integrating Natural/Biophysical and Social Data

Studies seeking to integrate natural and social science data are an emerging area of public opinion work, specifically those investigating whether weather or climate affects views of climate change. In this regard, it represents a frontier for examining society–environment interrelations. Of the public opinion studies discussed here, half integrate natural and social science data and use different approaches for harmonizing or matching the data, which are often at different levels of spatial scale or aggregation.

Some of these studies use temperature data such as large-scale indexes from the National Oceanic and Atmospheric Administration (NOAA) or historical weather station data from USHCN and test for relations with beliefs about the reality of climate change, risk perceptions, and support for climate-related policies. Findings have been mixed, with some reporting significant weather or climate effects on beliefs (Egan and Mullin 2012; Hamilton and Keim 2009; Hamilton and Lemcke-Stampone 2013; Hamilton and Stampone 2013; Howe et al. 2013; Scruggs and Benegal 2012; Shao et al. 2014; Zahran et al. 2006; Zaval et al. 2014), while other studies do not reveal effects (Brody, Zahran, Vedlitz, and Grover 2008; Brulle, Carmichael, and Jenkins 2012; Goebbert et al. 2012; Shum 2012). The divergence may largely reflect different temporal resolution and spatial resolution, with weather/climate effects most likely to be observed when survey and climate data are closely matched in place and time, and the climate indicators have high (experientially salient) variance. Moreover, in some places and for some weather/climate phenomena, the influence undoubtedly is stronger than in others, so we can expect sample-to-sample differences in conclusions. Two studies illustrate spatial matching. Hamilton and Keim (2009) analyzed the Community and Environment in Rural America (CERA) survey data from telephone interviews with more than nine thousand residents in rural regions of nine U.S. states. They applied linear mixed-effects modeling to estimate 1970–2007 seasonal temperature trends for each of these nine regions using USHCN data. Trends were defined as the sum of fixed and random slopes for each region. Those trend coefficients were then merged as region-specific variables in the CERA survey dataset, where mixed-effects ordered logit models (suited to the ordinal dependent variables) tested for possible effects of temperature trends on individual beliefs about climate change. These mixed-effects ordered logit models also included individual characteristics (sex, race, age, income, party, religious attendance, and newcomer/old-timer status), education × party interaction terms, and random intercepts to capture "everything else" causing between-region variation in the survey results. Even after controlling for individual and place factors, winter temperature trends showed the hypothesized positive effect on local climate perceptions. Marquart-Pyatt et al. (2014) offer a thorough and rigorous test of the link between climate extremes and climate perceptions in the U.S. public using more than fifty years of NOAA data and eleven years of Gallup data. Summarizing roughly 180 models, the authors conclude that political orientation, namely political ideology and party identification, has the most important effect in shaping public perceptions about the timing and seriousness of climate change. Objective climatic conditions have a negligible effect on perceptions of the seriousness of climate change and no effect on Americans' perceptions of its timing.

Two studies aggregate individual-level opinion data to the survey, making the individual survey the unit of analysis (Brulle et al. 2012; Scruggs and Benegal 2012). Using ten years of polling data, Brulle et al. (2012) find that political and economic factors—specifically elite cues, unemployment, and economic growth—affect public concern about climate change. Their results also show evidence of media effects, which they argue are intertwined with other factors. However, they do not find effects for the biophysical measure of weather extremes from NOAA, included as an aggregate measure of the forty-eight contiguous U.S. states. While their perceived climate change threat index measure created from survey data is indeed robust and indicative of a policy mood (Stimson 1999), it is not intended to capture shifts among specific demographic groups, which is one prominent explanation for recent shifts in U.S. public opinion on climate change. Moreover, they use an aggregate measure of weather instead of local or regional conditions. Scruggs and Benegal (2012) find some support for effects of weather, specifically seasonal and global temperature, in addition to effects for economic conditions, using data covering only two years.

Use of aggregated instead of individual-level data has consequences for the choice of modeling strategy. With aggregation, the number of observations is much smaller than in typical public opinion studies—for example, sample sizes of thirty-six (Brulle et al. 2012) and twenty-seven (Scruggs and Benegal 2012). In aggregating individual-level data to the level of a survey or country, variation at the individual level is lost. In disaggregating data or assigning the same value for a variable measured at the national level to all individual-level data, the individual cases are correlated, which violates the assumption of independence of measures in linear regression. By allowing for random intercepts or other random effects that vary between clusters of the data (e.g., different regions, or a series of polls), mixed-effects modeling and multilevel modeling provide useful tools for integrated analysis (Hamilton and Keim 2009; Hamilton and Stampone 2013; Hamilton et al. 2013; Marquart-Pyatt et al. 2014). A simpler though less flexible approach would be to include regional FEs or dummy variables to explicitly model the intercepts of different clusters. Either mixed-effects or FE approaches can account for mean differences without attempting to explain them (Steenbergen and Jones 2002), which is attractive when we consider there may be countless unmeasured reasons why mean levels of opinion are different in one place than another. Conversely, if we do not make such adjustments, we might spuriously attribute regional differences to some other variables in our model (Raudenbush and Bryk 2002; Snijders and Bosker 2012).

CHALLENGES FOR FUTURE SOCIOLOGICAL WORK
ON CLIMATE CHANGE

A recent overview and synthesis of statistics in sociology provides important context for the consideration of future methodological challenges for
climate change research (Raftery 2001). We are at a critical juncture, tasked
with integrating multiple and continually expanding types of environmental data within the established empirical traditions of sociology. Given this
challenge, we seek to balance research employing analytical techniques and
methodological approaches well entrenched within sociology with broader
forces articulating interdisciplinary approaches to the environment–society
interface. Here, we discuss two analytical techniques—SEM and multilevel modeling—that are well suited for tackling many challenging issues
related to the society–environment interface in general and climate change
in particular.

We can account for complex relations among individual-level predictors
in public opinion studies by applying SEM, an analytical technique well
grounded in empirical sociological research (Raftery 2001). SEM has advantages regarding its ability to incorporate direct and indirect effects, reciprocal relations, feedback loops, and observed and latent variables. However, as
with any technique, when applying SEM, assumptions of the model are critical. We refer readers to specialized SEM texts for more detailed description
(e.g., Bollen 1989; Kaplan 2009; Kline 2011). We provide examples from SEM
for approaching complexities in public opinion research on climate change.
Figures 12.12a and 12.12b show examples of two nonrecursive path models
(Paxton, Hipp, and Marquart-Pyatt 2011). Exogenous variables, denoted with
x, are predictor or independent variables (i.e., not explained by the model).
Endogenous variables, denoted by **y**, are outcomes or dependent variables
determined within the model. Error terms are denoted by ζ (zeta). A path
diagram pictorially displays how variables are related to one another in a
theoretical model. Path diagrams use particular conventions: variables in
rectangles are observed variables, single-headed arrows denote the direction
of influence, and double-headed arrows depict a covariance not explained
in the model. Errors in the equations are typically depicted unenclosed
(though they represent unobserved or latent variables). Figure 12.12a is nonrecursive due to the reciprocal paths between y_1 and y_2 and the correlated
error between ζ_1 and ζ_2. Figure 12.12b is nonrecursive given the feedback
loop among y_1, y_2, and y_3, where y_1 can be traced back to itself through the
change in y_2 and y_3; similarly, y_2 and y_3 can be traced back to themselves in
the diagram.

Regarding modeling relations among variables shown in these path
models, the researcher must appropriately specify the model as recursive or

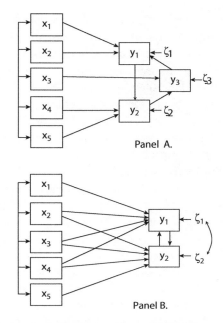

Figure 12.12a and 12.12b Two Nonrecursive Path Models

nonrecursive and then use the appropriate estimator. No published work on climate change views or actions to date approximates the models shown in either Panel A or B. However, with more data being gathered in recent years, increasingly complex models can be examined showing a number of paths through which the social bases of climate change opinion work to affect attitudes, policy views, and even personal actions. For instance, one possible model for Panel A might specify age and education affecting climate change beliefs (y_1) directly and actions (y_3) only indirectly through attitudes (y_2), which then feed back into beliefs. A possible model for Panel B might specify predictors like age, education, and income affecting beliefs about the timing of climate change (y_1) and perceptions of its seriousness (y_2), which are themselves proposed to be mutually reinforcing.

Researchers propose complex relations among values, attitudes, and behaviors including direct and indirect effects (Fishbein and Ajzen 2010). Only a few studies use mediation models or SEM to investigate support for climate change policy (Dietz et al. 2007; Ding et al. 2011; Malka et al. 2009) and support for government action (McCright et al. 2013, 2014a, 2014b). To advance the climate change literature, consider the path model example of climate change actions including both latent and measured variables shown in Figure 12.13 (based on Marquart-Pyatt 2012). As before, variables in boxes are observed and those in ovals are latent variables. This model includes one exogenous latent variable, denoted by ξ (ksi), and three latent endogenous variables, denoted with η (eta). SEM with latent variables has a number of advantages, where latent exogenous (ξ) and endogenous (η) variables are constructed and evaluated using confirmatory factor analysis (CFA) (Bollen 1989; Kaplan 2009; Kline 2011). Specifying a CFA includes measurement error and shifts the investigation to latent levels, which is essential for research on climate change beliefs, attitudes, and actions as these are abstract, multidimensional constructs. Also, the range of fit statistics provided enables a comprehensive assessment of model fit like values assessing the quality of survey items.

In the model in Figure 12.13, demographics, socioeconomic status, knowledge, beliefs (ξ_1), efficacy, attitudes (η_1), and behavioral intentions (η_2) directly affect climate actions (η_3). These relations are shown with solid lines. Many of these variables are also hypothesized to directly influence climate change attitudes and willingness to contribute, effects captured with dotted lines in the figure. Efficacy and attitudes directly affect willingness to contribute. The bolded lines highlight possible mediating effects to explore: how willingness to contribute mediates paths from education and efficacy to climate-related actions. Although some studies use path models (Dietz et al. 2007; Ding et al. 2011; McCright et al. 2013, 2014a, 2014b), a number of properties of these models are underused in prior research, thus presenting an opportunity for future scholarship. SEM with latent variables is vital for highlighting various ways in which individual survey items coalesce to reflect complex, latent constructs like opinions, beliefs, and attitudes about climate change. Currently absent from the pictorial representation in Figure 12.13 is how the biophysical environment or climate itself shapes the expression of these views and how these views may, in turn, affect the environment to model the embeddedness of human views in the climate cycle. This complexity could be included by placing the entire path model within another overlapping concentric circle for the biophysical environment referencing climate specifically. Complex path models like those in SEM with latent variables can illuminate interactions, feedbacks, emergent behavior, and reciprocal relations between human and environmental or climate systems. SEM is also able to include multiple levels of scale or aggregation that can illuminate the society–environment interface. In such instances, spatial and temporal variability can be easily introduced into these models through subscripting to denote groups or clusters and other aspects of model specification. Recent advances in generalized SEM extend the SEM approach to nonlinear models such as the logit or probit methods often needed for survey data analysis.

A second analytical technique rooted in sociology is multilevel or hierarchical linear modeling (mathematically equivalent to the mixed-effects

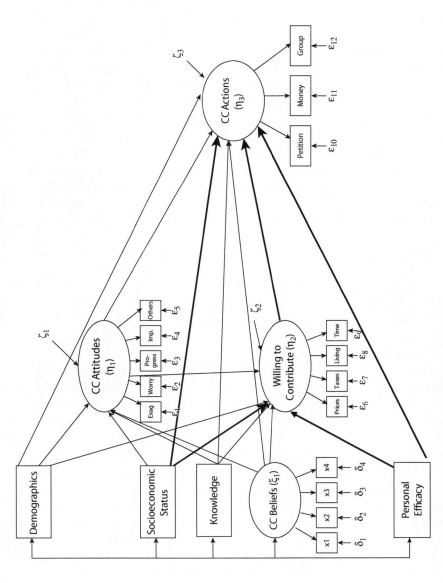

Figure 12.13 Example of Full SEM with Latent Variables Path Model

Based on Marquart-Pyatt 2012

modeling discussed earlier; see Rabe-Hesketh and Skrondal 2012). A multilevel approach is applicable in instances of clustering where there is nesting of observational units at one level within another level of aggregation (Raudenbush and Bryk 2002; Snijders and Bosker 2012). Multilevel models account for micro–macro linkages through a number of model specifications articulating separate within-group and between-group regressions to empirically link parameter estimates with their respective clusters. Across various multilevel specifications, the terminology of within and between groups is essential, as multilevel models allow for estimation of models including fixed and random effects associated with levels of aggregation articulated as groups or clusters. Of the many possible model specifications, here we briefly note two: the typical macro-to-micro proposition and a cross-classified model. Briefly, the former proposes that human systems are nested within natural systems, and human decision making is embedded within a series of processes at increasingly aggregate scales. The latter proposes complex relations among micro-level variables and variables at other levels of aggregation, which are referred to as imperfect hierarchies. These multilevel classifications recognize that there may not be perfect nesting of lower-level units within higher-level units (Snijders and Bosker 2012). For example, an individual may be simultaneously nested within a community, a township or municipality, and a climate region, units that have social and ecological elements that may not perfectly align. Central to these models is the ability to incorporate spatial and/or temporal variability through nested hierarchies linked with groups or clusters for ecological and/or social units.

Finally, we conclude with a brief discussion of the benefits of modeling for understanding the causes and consequences of complex social and environmental issues. As shown by the various techniques described in this chapter, models have a number of advantageous properties for climate change research. Models can illuminate interactions, feedbacks, emergent behavior, and reciprocal relations between human and environmental systems that are not immediately apparent to individual researchers. The tools of path analysis used in SEM make these relations straightforward to depict pictorially. Yet, social scientists might also introduce the role of human behavior, societal conditions, and/or policies into the model as important mechanisms to include as potentially mitigating factors, adding an important observational unit not typically included in a climate model that examines global, climatic processes. This adds complexity to the model as the individual researcher is likely in unfamiliar territory in describing relations among variables spanning both the natural and social sciences. This is true for a social scientist seeking to include environmental data and for natural scientists including social data. Questions arise regarding directionality of proposed relations, including how human societies affect and are affected by natural environments and how to properly account for spatial and temporal effects. An interdisciplinary team of researchers may create a complex model that includes multiple levels of aggregation and reciprocal relations among social and environmental processes across the natural and social sciences. The question of how to visually depict these complexities, including interactions, feedbacks, and reciprocity, is often an iterative process. Working in interdisciplinary teams also reveals how these notions of complexity and uncertainty, central to scholarship on climate change, can be communicated, often in ways not immediately apparent to individual researchers.

CONCLUSION

Sociologists conducting climate change research are often engaged in interdisciplinary and cross-disciplinary research, since such studies involve consideration of social and environmental phenomena jointly. While this poses a unique challenge for sociologists, it is a challenge for which our discipline is well suited given the variety of analytical techniques employed and range of observational units considered in sociological research. Model specification remains crucial for sociological research on climate change. This encompasses not only the theoretical frameworks used in a study but also the use of observational data, the operationalization of key constructs, the selection of units for analysis, choices of predictor and outcome variables, and how spatial and temporal variation are analyzed. Thus, scholars must be well versed in a variety of analytical techniques with sound methodological training as sociologists. Thoughtful and innovative research designs are keys to obtaining interpretable results and laying a foundation on which future scholarship can build. We believe sociological research is well placed to meet the challenges set forth decades ago by Catton and Dunlap, as recent work reveals integrative and innovative research at the society–environment interface. We are optimistic that ongoing and subsequent studies will continue this promising trajectory.

NOTES

1. We gratefully acknowledge contributions from Sharon Harlan at Arizona State University.

2. A related body of research focuses on the human drivers of deforestation. While substantively relevant given the role of forested areas in carbon sequestration,

due to data limitations, nearly all of the research employs cross-sectional OLS regression methods (see Shandra, Shircliff, and London 2011 and references therein; see Jorgenson, Austin, and Dick 2009; Rudel 2005 for notable exceptions).

3. The environmental Kuznets approach, from environmental economics, recognizes that economic development generates environmental harms but argues that further growth can largely correct these problems (Grossman and Krueger 1995). Environmental quality is considered a luxury good, subject to public demand through the workings of an advanced market. During the early stages of economic development, environmental impacts escalate, but as affluence within these societies rises, the value the public places on the environment will increase. If the market is allowed to operate without dramatic interference, continuing economic development will lead to a leveling and eventual decline in the use of natural resources and emission of pollutants. The proposed trend, known as the environmental Kuznets curve, is an inverted U-shaped distribution representing the relation between environmental impacts and economic development.

4. A parallel research stream exists regarding climate change views internationally and comparatively (Brechin 2010; Brechin and Bhandari 2011; Kim and Wolinsky-Nahmias 2014; Kvaloy, Finseraas, and Listhaug 2012; Sandvik 2008; Scruggs and Benegal 2012). Given space considerations, we do not review these in this chapter. We refer readers to Chapter 10.

5. This is not a comprehensive review of the public opinion literature. We selected a subset of studies given our focus on temporal and spatial integration. See Chapter 10 in this volume. The articles include surveys from national polling organizations; studies by university research teams with diverse, multidisciplinary backgrounds; and studies by single investigators in traditional disciplinary domains. They include analyses of the Gallup Poll, which provides the longest time series of climate change indicators, and the General Social Survey gathered by the National Opinion Research Center, which included climate questions in 2006, 2010, and 2012, including a 2012 panel that carried the same climate change belief question analyzed in Figure 12.2.

REFERENCES

Allison, Paul. 2009. *Fixed Effects Regression Models*. Thousand Oaks, CA: Sage.
Baum, Christopher. 2006. *An Introduction to Modern Econometrics Using Stata*. College Station, TX: Stata Press.
Bollen, Kenneth. 1989. *Structural Equations with Latent Variables*. New York: Wiley.
Brechin, Steven R. 2010. "Chapter 10: Public Opinion: A Cross-national View." Pp. 179 in *Routledge Handbook of Climate Change and Society*, edited by C. Lever-Tracy. New York: Routledge.
Brechin, Steven R. and Medani Bhandari. 2011. "Perceptions of Climate Change Worldwide." *Wiley Interdisciplinary Reviews: Climate Change* 2:871–885.
Brody, Samuel, Sammy Zahran, Wesley E. Highfield, Sarah P. Bernhardt, and Arnold Vedlitz. 2009. "Policy Learning for Flood Mitigation: A Longitudinal Assessment of the Community Rating System in Florida." *Risk Analysis* 29(6):912–929.

Brody, Samuel, Sammy Zahran, Arnold Vedlitz, and Himanshu Grover. 2008. "Examining the Relationship between Physical Vulnerability and Public Perceptions of Global Climate Change in the United States." *Environment and Behavior* 41:72–95.

Brulle, Robert J., Jason Carmichael, and J. Craig Jenkins. 2012. "Shifting Public Opinion on Climate Change: An Empirical Assessment of Factors Influencing Concern Over Climate Change in the US, 2002–2010." *Climatic Change*:1–20.

Brulle, Robert J. and David N. Pellow. 2006. "Environmental Justice: Human Health and Environmental Inequalities." *Annual Review of Public Health* 27:103–124.

Burakowski, Elizabeth and Matthew Magnusson. 2012. *Climate Impacts on the Winter Tourism Economy in the United States*. Natural Resources Defense Council. Retrieved April 22, 2014 (http://www.nrdc.org/globalwarming/files/climate-impacts-winter-tourism-report.pdf).

Burns, Thomas, Byron Davis, and Edward Kick. 1997. "Position in the World-System and National Emissions of Greenhouse Gases." *Journal of World-Systems Research* 3:432–466.

Catton, William and Riley Dunlap. 1978. "Environmental Sociology: A New Paradigm." *American Sociologist* 13:41–49.

Christianson, Gale. 1999. *Greenhouse: The 200-Year Story of Global Warming*. New York: Walker and Company.

Clement, Matthew and Jessica Schultz. 2011. "Political Economy, Ecological Modernization, and Energy Use: A Panel Analysis of State-Level Energy Use in the United States, 1960–1990." *Sociological Forum* 26(3):581–600.

Dietz, Thomas, Amy Dan, and Rachael Shwom. 2007. "Support for Climate Change Policy: Some Psychological and Social Structural Influences." *Rural Sociology* 72:185–214.

Dietz, Thomas and Andrew Jorgenson. 2013. *Structural Human Ecology: New Essays in Risk, Energy, and Sustainability*. Pullman, WA: Washington State University Press.

Dietz, Thomas and Eugene Rosa. 1997. "Effects of Population and Affluence on CO_2 Emissions." *Proceedings of the National Academy of Sciences USA* 94:175–179.

Ding, Ding, Edward W. Maibach, Xiaoquan Zhao, Connie Roser-Renouf, and Anthony Leiserowitz. 2011. "Support for Climate Policy and Societal Action are Linked to Perceptions About Scientific Agreement." *Nature Climate Change* 1:462–466.

Downey, Liam. 2003. "Spatial Measurement, Geography, and Urban Racial Inequality." *Social Forces* 81(3):937–952.

Dunlap, Riley and William Catton. 1979. "Environmental Sociology." *Annual Review of Sociology* 5:243–273.

Dunlap, Riley and Aaron McCright. 2008. "A Widening Gap: Republican and Democratic Views on Climate Change." *Environment: Science and Policy for Sustainable Development* 50:26–35.

Egan, Patrick J. and Megan Mullin. 2012. "Turning Personal Experience into Political Attitudes: The Effect of Local Weather on Americans' Perceptions about Global Warming." *Journal of Politics* 74(3):796–809.

Ergas, Christina and Richard York. 2012. "Women's Status and Carbon Dioxide Emissions: A Quantitative Cross-national Analysis." *Social Science Research* 41(4):965–976.

Fishbein, Martin and Icek Ajzen. 2010. *Predicting and Changing Behavior: The Reasoned Action Approach*. New York: Psychology Press.

Fisher, Dana. 2006. "Bringing the Material Back In: Understanding the U.S. Position on Climate Change." *Sociological Forum* 21(3):467–494.

Fisher, Dana and William Freudenburg. 2004. "Post-Industrialization and Environmental Quality: An Empirical Analysis of the Environmental State." *Social Forces* 83:157–188.

Freudenburg William R. and Robert Gramling. 1993. "Socioenvironmental Factors and Development Policy: Understanding Opposition and Support for Offshore Oil." *Sociological Forum* 8:341–364.

Freudenburg, William R. and Robert Gramling. 1994. *Oil in Troubled Waters: Perceptions, Politics, and the Battle over Offshore Drilling.* Albany: State University of New York Press.

Freudenburg, William R. and Robert Gramling. 2010. *Blowout in the Gulf: The BP Oil Spill Disaster and the Future of Energy in America.* Cambridge, MA: MIT Press.

Goebbert, Kevin, Hank C. Jenkins-Smith, Kim Klockow, Matthew C. Nowlin, and Carol L. Silva. 2012. "Weather, Climate, and Worldviews: The Sources and Consequences of Public Perceptions of Changes in Local Weather Patterns." *Weather, Climate, and Society* 4:132–144.

Gramling, Robert and William Freudenburg. 1996. "Environmental Sociology: Toward a Paradigm for the 21st Century." *Sociological Spectrum* 16:347–370.

Grant, Don, Andrew Jorgenson, and Wesley Longhofer. 2013 "Targeting Electricity's Extreme Polluters to Reduce Energy-Related CO_2 Emissions." *Journal of Environmental Studies and Sciences* 3: 376–380.

Grimes, Peter and Jeffrey Kentor. 2003. "Exporting the Greenhouse: Foreign Capital Penetration and CO2 Emissions 1980–1996." *Journal of World-Systems Research* 9:261–275.

Grossman, Gene and Alan Krueger. 1995. "Economic Growth and the Environment." *Quarterly Journal of Economics* 110:353–377.

Halaby, Charles. 2004. "Panel Models in Sociological Research: Theory into Practice." *Annual Review of Sociology* 30:507–544.

Hamilton, Lawrence C. 2007. "Climate, Fishery and Society Interactions: Observations from the North Atlantic." *Deep Sea Research II* 54:2958–2969.

Hamilton, Lawrence C. 2008. "Who Cares about Polar Regions? Results from a Survey of U.S. Public Opinion." *Arctic, Antarctic, and Alpine Research* 40(4):671–678.

Hamilton, Lawrence C. 2011. "Education, Politics, and Opinions about Climate Change: Evidence for Interaction Effects." *Climatic Change* 104(2): 231–242.

Hamilton, Lawrence C. 2012. "Did the Arctic Ice Recover? Demographics of True and False Climate Facts." *Weather, Climate, and Society* 4(4):236–249.

Hamilton, Lawrence C. 2014. *Do You Trust Scientists About the Environment? News Media Sources and Politics Affect New Hampshire Resident Views.* Regional Issue Brief No. 40. Durham, NH: Carsey Institute, University of New Hampshire.

Hamilton, Lawrence C., Cliff Brown, and Barry D. Keim. 2007. "Ski Areas, Weather and Climate: Time Series Models for New England Case Studies." *International Journal of Climatology* 27:2113–2124.

Hamilton, Lawrence C., Matthew J. Cutler, and Andrew Schaefer. 2012. "Public Knowledge and Concern about Polar-Region Warming." *Polar Geography* 35(2):155–168.

Hamilton, Lawrence C., Joel Hartter, Thomas Safford, and Forrest Stevens. 2013. "Rural Environmental Concern: Effects of Position, Partisanship, and Place." *Rural Sociology* 79(2):257–281. DOI: 10.1111/ruso.12023.

Hamilton, Lawrence C., Steingrimur Jónsson, Helga Ögmundardóttir, and Igor M. Belkin. 2004. "Sea Changes Ashore: The Ocean and Iceland's Herring Capital." *Arctic* 57(4):325–335.

Hamilton, Lawrence C. and Barry D. Keim. 2009. "Regional Variation in Perceptions about Climate Change." *International Journal of Climatology* 29(15):2348–2352.

Hamilton, Lawrence C. and Mary Lemcke-Stampone. 2013. "Arctic Warming and Your Weather: Public Belief in the Connection." *International Journal of Climatology* 34:1723–1728. doi: 10.1002/joc.3796.

Hamilton, Lawrence C., Oddmund Otterstad, and Helga Ögmundardóttir. 2006. "Rise and Fall of the Herring Towns: Impacts of Climate and Human Teleconnections." Pp. 100–125 in *Climate Change and the Economics of the World's Fisheries,* edited by R. Hannesson, M. Barange, and S. F. Herrick Jr. Northampton, MA: Edward Elgar.

Hamilton, Lawrence C., David E. Rohall, Benjamin C. Brown, Gregg F. Hayward, and Barry D. Keim. 2003. "Warming Winters and New Hampshire's Lost Ski Areas: An Integrated Case Study." *International Journal of Sociology and Social Policy* 23(10):52–73.

Hamilton, Lawrence C. and Thomas G. Safford. 2014. "Environmental Views From the Coast: Public Concern About Local to Global Marine Issues." *Society and Natural Resources.* doi: 10.1080/08941920.2014.933926.

Hamilton, Lawrence C. and Kei Saito. 2015. "A Four-party View of U.S. Environmental Concern." *Environmental Politics.* doi: 10.1080/09644016.2014.976485.

Hamilton, Lawrence C. and Mary D. Stampone. 2013. "Blowin' in the Wind: Short-term Weather and Beliefs about Climate Change." *Weather, Climate, and Society* 5(2):112–119. doi: 10.1175/WCAS-D-12-00048.1.

Hamilton, Lawrence C., Daniel M. White, Richard B. Lammers, and Greta Myerchin. 2012. "Population, Climate and Electricity Use in the Arctic: Integrated Analysis of Alaska Community Data." *Population and Environment* 33(4):269–283.

Harlan, Sharon L., Juan H. Declet-Barreto, William L. Stefanov, and Diana B. Petitti. 2013. "Neighborhood Effects on Heat Deaths: Social and Environmental Predictors of Vulnerability in Maricopa County, Arizona." *Environmental Health Perspectives* 121(2):197–204.

Howe, Peter D. and Anthony Leiserowitz. 2013. "Who Remembers a Hot Summer or a Cold Winter? The Asymmetric Effect of Beliefs About Global Warming on Perceptions of Local Climate Conditions in the U.S." *Global Environmental Change.* doi: 10.1016/j.gloenvcha.2013.09.014.

Howe, Peter D., Ezra M. Markowitz, Tien Ming Lee, Chia-Ying Ko, and Anthony Leiserowitz. 2013. "Global Perceptions of Local Temperature Change." *Nature Climate Change* 3:352–356.

Hsiao, Cheng. 2003. *Analysis of Panel Data.* New York: Cambridge University Press.

Jacobs, Jerry and Scott Frickel. 2009. "Interdisciplinarity: A Critical Assessment." *Annual Review of Sociology* 35:43–65.

Jorgenson, Andrew. 2006. "Global Warming and the Neglected Greenhouse Gas: A Cross-national Study of Methane Emissions Intensity, 1995." *Social Forces* 84:1777–1796.

Jorgenson, Andrew. 2007. "Does Foreign Investment Harm the Air We Breathe and the Water We Drink? A Cross-national Study of Carbon Dioxide Emissions and Organic Water Pollution in Less-Developed Countries, 1975–2000." *Organization & Environment* 20:137–156.

Jorgenson, Andrew. 2012. "The Sociology of Ecologically Unequal Exchange and Carbon Dioxide Emissions, 1960–2005." *Social Science Research* 41:242–252.

Jorgenson, Andrew. 2014. "Economic Development and the Carbon Intensity of Human Well-Being." *Nature Climate Change* 4:186–189.

Jorgenson, Andrew, Daniel Auerbach, and Brett Clark. 2014. "The (De-) Carbonization of Urbanization, 1960–2010." *Climatic Change* 127:561–575.

Jorgenson, Andrew, Kelly Austin, and Christopher Dick. 2009. "Ecologically Unequal Exchange and the Resource Consumption / Environmental Degradation Paradox: A Panel Study of Less-Developed Countries, 1970–2000." *International Journal of Comparative Sociology* 50:263–284.

Jorgenson, Andrew and Ryan Birkholz. 2010. "Assessing the Causes of Anthropogenic Methane Emissions in Comparative Perspective, 1990–2005." *Ecological Economics* 69:2634–2643.

Jorgenson, Andrew and Brett Clark. 2010. "Assessing the Temporal Stability of the Population/Environment Relationship in Comparative Perspective: A Cross-national Panel Study of Carbon Dioxide Emissions, 1960–2005." *Population & Environment* 32:27–41.

Jorgenson, Andrew and Brett Clark. 2012. "Are the Economy and the Environment Decoupling? A Comparative International Study, 1960–2005." *American Journal of Sociology* 118:1–44.

Jorgenson, Andrew and Brett Clark. 2013. "The Relationship Between National-level Carbon Dioxide Emissions and Population Size: An Assessment of Regional and Temporal Variation, 1960–2005." *PLoS One* 8(2):e57107.

Jorgenson, Andrew, Christopher Dick, and Matthew Mahutga. 2007. "Foreign Investment Dependence and the Environment: An Ecostructural Approach." *Social Problems* 54:371–394.

Jorgenson, Andrew, Christopher Dick, and John Shandra. 2011. "World Economy, World Society, and Environmental Harms in Less-developed Countries." *Sociological Inquiry* 81:53–87.

Jorgenson, Andrew and Thomas Dietz. 2014. "Economic Growth Does Not Reduce the Ecological Intensity of Human Well-Being." *Sustainability Science* DOI:10.1007/s11625-014-0.

Kaplan, David. 2009. *Structural Equation Modeling: Foundations and Extensions*, 2nd ed. Thousand Oaks, CA: Sage.

Keim, Barry D., Adam M. Wilson, Cameron P. Wake, and Thomas G. Huntington. 2003. "Are There Spurious Temperature Trends in the United States Climate Division Database?" *Geophysical Research Letters* 30(7):10.1029/2002GL016295.

Kim, So Young and Yael Wolinsky-Nahmias. 2014. "Cross-National Public Opinion on Climate Change: The Effects of Affluence and Vulnerability." *Global Environmental Politics* 14(1):79–106.

Kline, Rex B. 2011. *Principles and Practice of Structural Equation Modeling*. New York: Guilford Press.

Kvaloy, Berit, Henning Finseraas, and Ola Listhaug. 2012. "The Public's Concerns for Global Warming: A Cross-national Study of 47 Countries." *Journal of Peace Research* 49(1):11–22.

Liu, Jianguo, Thomas Dietz, Stephen R. Carpenter, Marina Alberti, Carl Folke, Emilio Moran, Alice N. Pell, Peter Deadman, Timothy Kratz, Jane Lubchenco, Elinor Ostrom, Zhiyun Ouyang, William Provencher, Charles L. Redman, Stephen

H. Schneider, and William W. Taylor. 2007a. "Complexity of Coupled Human and Natural Systems." *Science* 317:1513–1516.

Liu, Jianguo, Thomas Dietz, Stephen Carpenter, Carl Folke, Marina Alberti, Charles Redman, Stephen H. Schneider, Elinor Ostrom, Alice N. Pell, Jane Lubchenco, William W. Taylor, Zhiyun Ouyang, Peter Deadman, Timothy Kratz, and William Provencher. 2007b. "Coupled Human and Natural Systems." *AMBIO* 36: 639–649.

Malka, Ariel, Jon A. Krosnick, and Gary Langer. 2009. "The Association of Knowledge with Concern about Global Warming: Trusted Information Sources Shape Public Thinking." *Risk Analysis* 29:633–647.

Marquart-Pyatt, Sandra T. 2012. "Explaining Environmental Activism across Countries." *Society and Natural Resources* 25(7):683–699.

Marquart-Pyatt, Sandra T., Aaron M. McCright, T. Dietz, and R. Dunlap. 2014. "Politics Eclipses Climate Extremes for Climate Change Perceptions." *Global Environmental Change* 29:246–257.

Marquart-Pyatt, Sandra T., Rachael L. Shwom, Thomas Dietz, Riley E. Dunlap, Stan A. Kaplowitz, Aaron M. McCright, and Sammy Zahran. 2011. "Understanding Public Opinion on Climate Change: A Call for Research." *Environment: Science and Policy for Sustainable Development* 53:38–42.

McCright, Aaron M. 2009. "The Social Bases of Climate Change Knowledge, Concern, and Policy Support in the U.S. General Public." *Hofstra Law Review* 37(4):1017–1047.

McCright, Aaron M. 2010. "The Effects of Gender on Climate Change Knowledge and Concern in the American Public." *Population and Environment* 32:66–87.

McCright, Aaron M. 2011. "Political Orientation Moderates Americans' Beliefs and Concern about Climate Change." *Climatic Change* 104(2):243–253.

McCright, Aaron M. and Riley E. Dunlap. 2011. "The Politicization of Climate Change and Polarization in the American Public's Views of Global Warming, 2001–2010." *Sociological Quarterly* 52:155–194.

McCright, Aaron M., Riley E. Dunlap, and Chenyang Xiao. 2013. "Perceived Scientific Agreement and Support for Government Action on Climate Change in the USA." *Climatic Change* 119:511–518.

McCright, Aaron M., Riley E. Dunlap, and Chenyang Xiao. 2014a. "Increasing Influence of Party Identification on Perceived Scientific Agreement and Support for Government Action on Climate Change in the United States, 2006–12." *Weather, Climate, and Society* 6:194–201.

McCright, Aaron M., Riley E. Dunlap, and Chenyang Xiao. 2014b. "The Impacts of Temperature Anomalies and Political Orientation on Perceived Winter Warming." *Nature Climate Change* 4:1077–1081.

Norgaard, Kari Marie. 2011. *Living in Denial: Climate Change, Emotions and Everyday Life*. Cambridge, MA: MIT Press.

Paxton, Pamela M., John R. Hipp, and Sandra Marquart-Pyatt. 2011. *Nonrecursive Models: Endogeneity, Reciprocal Relationships, and Feedback Loops*. Quantitative Applications in the Social Science (QASS) Series #07-168. Thousand Oaks, CA: SAGE.

Rabe-Hesketh, Sophia and Anders Skrondal. 2012. *Multilevel and Longitudinal Modeling Using Stata*, 3rd ed. College Station, TX: Stata Press.

Raftery, Adrian E. 2001. "Statistics in Sociology, 1950–2000: A Selective Review." *Sociological Methodology* 31:1–45.

Raudenbush, Stephen W. and Anthony S. Bryk. 2002. *Hierarchical Linear Models*, 2nd ed. Thousand Oaks, CA: SAGE.

Reid, Collen E., Marie S. O'Neill, Carina J. Gronlund, Shannon J. Brines, Daniel G. Brown, Ana V. Diez-Roux, and Joel Schwartz. 2009. "Mapping Community Determinants of Heat Vulnerability." *Environmental Health Perspectives* 117:1730–1736.

Roberts, Timmons and Peter Grimes. 1997. "Carbon Intensity and Economic Development 1962–1991: A Brief Exploration of the Environmental Kuznets Curve." *World Development* 25:181–187.

Roberts, Timmons, Peter Grimes, and Jodie Manale. 2003. "Social Roots of Global Environmental Change: A World-systems Analysis of Carbon Dioxide Emissions." *Journal of World-Systems Research* 9:277–315.

Rosa, Eugene and Thomas M. Dietz. 2012. "Human Drivers of National Greenhouse Gas Emissions." *Nature Climate Change* 2:581–586.

Rosa, Eugene A., Richard York, and Thomas Dietz. 2004. "Tracking the Anthropogenic Drivers of Ecological Impacts." *AMBIO* 33(8):509–512.

Ruddell, Darren M., Sharon L. Harlan, Susanne Grossman-Clarke, and Alexander Buyantuyev. 2010. "Risk and Exposure to Extreme Heat in Microclimates of Phoenix, AZ." Chapter 9 in *Geospatial Techniques in Urban Hazard and Disaster Analysis*, edited by P. S. Showalter and Y. Lu. New York: Springer.

Rudel, Thomas K. 2005. *Tropical Forests: Regional Paths of Destruction and Regeneration in the Late Twentieth Century.* New York: Columbia University Press.

Rudel, Thomas K. 2009. "How Do People Transform Landscapes? A Sociological Perspective on Suburban Sprawl and Tropical Deforestation." *American Journal of Sociology* 115(1):129–154.

Sandvik, Hanno. 2008. "Public Concern Over Global Warming Correlates Negatively with National Wealth." *Climatic Change* 90(3):333–341.

Schofer, Evan and Ann Hironaka. 2005. "The Effects of World Society on Environmental Outcomes." *Social Forces* 84:25–47.

Scott, D. and G. McBoyle. 2007. "Climate Change Adaptation in the Ski Industry." *Mitigation and Adaptation Strategies for Global Change.* doi: 10.1007/s11027-006-9071-4.

Scruggs, Lyle and Salil Benegal. 2012. "Declining Public Concern about Climate Change: Can We Blame the Great Recession?" *Global Environmental Change* 22(2):505–515.

Shandra, John M., Bruce London, Owen P. Whooley, and John B. Williamson. 2004. "International Nongovernmental Organizations and Carbon Dioxide Emissions in the Developing World: A Quantitative, Cross-national Analysis." *Sociological Inquiry* 74(4):520–545.

Shandra, John M., Eric Shircliff, and Bruce London. 2011. "The International Monetary Fund, World Bank, and Structural Adjustment: A Cross-national Analysis of Forest Loss." *Social Science Research* 40:210–225.

Shao, Wanyun, Barry D. Keim, James C. Garland, and Lawrence C. Hamilton. 2014. "Weather, Climate, and the Economy: Explaining Risk Perceptions of Global Warming, 2001–2010." *Weather, Climate, and Society* 6(1):119–134. doi: 10.1175/WCAS-D-13-00029.1.

Shum, Robert Y. 2012. "Effects of Economic Recession and Local Weather on Climate Change Attitudes." *Climate Policy* 12:38–49.

Snijders, Tom and Roel Bosker. 2012. *Multilevel Analysis: An Introduction to Basic and Advanced Multilevel Modeling*, 2nd ed. Thousand Oaks, CA: SAGE.

Steenbergen, Marco R. and Bradford S. Jones. 2002. "Modeling Multilevel Data Structures." *American Journal of Political Science* 46:218–237.

Stimson, James A. 1999. *Public Opinion in America: Moods, Cycles and Swings*, 2nd ed. Boulder, CO: Westview Press.

Tang, Qiuhong, Xuejun Zhang, Xiaohua Yang, and Jennifer A. Francis. 2013. "Cold Winter Extremes in Northern Continents Linked to Arctic Sea Ice Loss." *Environmental Research Letters* 8 (http://www.glisaclimate.org/media/Tang_Francis_Cold_Winter_Sea_Ice_EnvironResLett_2013.pdf).

U.S. National Research Council. 2010. *Advancing the Science of Climate Change*. Washington, DC: National Academies Press.

Wooldridge, Jeffrey M. 2005. *Introductory Econometrics: A Modern Approach*, 3rd ed. Mason, OH: South-Western/Thomson.

World Bank. 2007. *World Development Indicators*. Washington, D.C: World Bank.

World Resources Institute. 2010. *Earth Trends Data CD-ROM*. Washington, DC: World Resources Institute.

York, Richard. 2008. "De-Carbonization in Former Soviet Republics, 1992–2000: The Ecological Consequences of De-Modernization." *Social Problems* 55:370–390.

York, Richard. 2012a. "Asymmetric Effects of Economic Growth and Decline on CO_2 Emissions." *Nature Climate Change* 2:762–764.

York, Richard. 2012b. "Do Alternative Energy Sources Displace Fossil Fuels?" *Nature Climate Change* 2:441–443.

York, Richard, Eugene A. Rosa, and Thomas Dietz. 2003. "A Rift in Modernity? Assessing the Anthropogenic Sources of Global Climate Change with the STIRPAT Model." *International Journal of Sociology and Social Policy* 23(10):31–51.

Zahran, Sammy, Samuel D. Brody, Himanshu Grover, and Arnold Vedlitz. 2006. "Climate Change Vulnerability and Policy Support." *Society and Natural Resources* 19:771–789.

Zaval, Lisa, Elizabeth A. Keenan, Eric J. Johnson, and Elke U. Weber. 2014. "How Warm Days Increase Belief in Global Warming." *Nature Climate Change* 4:143–147.

13

Bringing Sociology into Climate Change Research and Climate Change into Sociology

Concluding Observations

Riley E. Dunlap and Robert J. Brulle

INTRODUCTION

As this collection of essays demonstrates, sociology offers a wide range of important perspectives on key aspects of global climate change. These perspectives include analyses of the primary political-economic and institutional drivers of climate change, the social factors that influence both mitigation of and adaptation to climate change, and the nature of political and cultural struggles over how climate change should be addressed. The sociological research that is summarized in these essays results, for the most part, from independent intellectual inquiry by the sociological community, and thus represents contributions tied to neither research questions developed by the natural science community nor to the priorities of external funders. As a consequence, these contributions move well beyond the limitations of the existing body of social science research to provide new insights into the dynamics of communities, institutions, and nation-states, and how larger social processes systematically produce climate change and shape our responses to it. They open up new theoretical and empirical perspectives on climate change and thus have the potential to provide unique viewpoints that could lead to innovative and more effective strategies for dealing with the challenge of global climate change.

How do we move forward from this point? As we noted in Chapter 1, the dominant global environmental (and climate) change research efforts have typically taken a very narrow view of the potential contributions of sociology and social science more generally. What has resulted is a body of literature that is constrained by a number of factors and unnecessarily limits the role of sociology in climate change research. First, use of the Coupled Human and Natural Systems approach restricts the relevance of social science research to cases where they are part of multidisciplinary research efforts. This ignores the unique contributions that the various social sciences can make via their own disciplinary perspectives, and thus subordinates social science research to fit within natural science approaches. Second, the focus on individualist perspectives, primarily via the disciplines of economics and psychology, marginalizes institutional, societal, and cultural perspectives, and thus limits the range of analyses. At times the assumed role of the social sciences amounts to little more than raising awareness of climate change and promoting behavioral changes to deal with it (Shove 2010). Finally, as noted by the "post-political" critique, the dominant climate change frame used in official reports depoliticizes the discussion of climate change and marginalizes analyses of the socioeconomic processes that generate and per-petuate it (Crist 2007:35). This creates a situation whereby climate change is detached from its social-structural context, resulting in insufficient atten-tion to the political, economic, and cultural contestations surrounding it.

Reflecting concerns over the integration of the social sciences into exist-ing climate change research, two recently created initiatives have aimed at developing new approaches to this issue—those of the U.S. Global Change Research Program (USGCRP) and the International Social Science Council (ISSC), respectively. We begin our assessment of the way forward with a review of these initiatives. We briefly outline these initiatives, comment on their strengths and weaknesses, and then discuss the degree to which the sociological work in this volume accords with the research agendas they provide. We find a reasonable fit with the ISSC research agenda but a sub-stantial divergence with the more applied and individualistic approach of the USGCRP. To help develop a uniquely sociological research agenda for climate change, we provide a series of specific research questions developed by the authors of each of the essays in this volume, and three broad recom-mendations for future sociological engagement with climate change. We then stress the importance of sociological analyses focused on the politically contentious nature of climate change, and end by noting that sociology can play a valuable role by asking questions that are typically ignored in climate change research.

USGCRP

The first effort to integrate the social sciences into climate change and other global environmental research efforts has been the work of the USGCRP.[1] Beginning in 2012, an Ad Hoc Social Sciences Task Force was created to develop a program dedicated to that goal.[2] The Task Force focuses on expanding the social science research capacity of the U.S. government to better integrate the social sciences into efforts to support science-based decision making. A major concern is the development of shared research questions and programs between the social and natural sciences.

Within this program there has been a disjuncture between its stated goals and aims and their subsequent implementation. The overall research plan for the USGCRP (National Science and Technology Council 2012:24) identifies a need to understand the drivers of climate change, focusing on the areas of population dynamics, technology, and economic development. It also emphasizes the need to develop a sound basis for decision making, and to identify how societal responses to global change may be "constrained by institutions, social networks, and political, economic and cultural contexts." In a recent journal article, members of the Social Science Task Force have expanded this perspective to acknowledge that "it is people and their communities, institutions and governments, who are at the centre of the three main aspects of the global change challenge: that is, humans are the drivers of, are affected by, and have the capacity to respond to global change" (Weaver et al. 2014).

However, in practice, the USGCRP suggests a much narrower focus for social science contributions to the climate change agenda. While its goal of promoting social science research is laudable, it seems that the USGCRP envisions a rather limited role for social science (especially sociology)—primarily to supplement and enhance the use of natural science research on climate (and environmental) change (Weaver et al. 2014). The Task Force's PowerPoint notes a "dual role" for social science: (1) "Social science research and expertise [are] an important part of the integrated knowledge base about the causes and consequences of global change" and (2) The social sciences "can also identify principles that help make this knowledge 'work' better for society" (Weaver 2013:6). In reality, however, it appears that the USGCRP places more emphasis on the second role (Weaver et al. 2014).

Overall, the work of the USGCRP can be characterized as emphasizing the roles of individual citizens and decision-makers in dealing with climate change. However, it does not view them as agents promoting the type of fundamental social, economic, and political changes many analysts argue are necessary to make real progress in lowering greenhouse gas (GHG) emissions. In this context, the social sciences are seen as providers of information and techniques to modify individual behaviors rather than a means of transforming social practices and institutions. The USGCRP is no doubt constrained by the political situation in the United States, where anthropogenic climate change is a highly controversial issue—especially within the U.S. Congress, which has budgetary oversight on climate change research (e.g., Germain, Koronowski, and Spross 2013). It is therefore not surprising that the USGCRP envisions social science mainly as a supplement to a natural science agenda and focuses on such "politically safe" issues as providing support for societal decision-making via improved understanding of individuals' risk perceptions and decisions (Melillo, Richmond, and Yohe 2014:Chapter 26).

The work of the USGCRP exemplifies the "post-political" approach outlined in Chapter 1 that views climate change primarily as a technical and managerial problem to be handled by developing and applying the best scientific information in a consensual fashion—albeit with the participation of individual citizens and decision-makers who will become more informed and engaged in part via the work of social scientists.[3] For example, it does not acknowledge the extreme degree of controversy and political polarization surrounding the topic stemming from the power of vested interests such as the fossil fuels industry (Jamieson 2014). The result is that a good deal of sociological work on climate change, including much of that reported in prior chapters of this volume, is unlikely to find a home in the USGCRP without substantial reformulation of its research agenda.

ISSC AND THE FUTURE EARTH PROJECT

A second and much more ambitious effort to raise the profile and contribution of the social sciences to global environmental change research has been led by the ISSC in conjunction with a consortium of major national research funders known as the Belmont Forum (Hackmann, Brown, and Kershaw 2011). The important role of the social sciences in addressing climate change has long been recognized in the international social science community, starting with the International Human Dimensions Program in 1996. Recently, this program was integrated into the newly developed Future Earth: Research for Global Sustainability project (Mooney, Duraiappah, and Larigauderie 2013), and thus efforts to expand and strengthen the role of the social sciences in global environmental change research now center on the contribution of these disciplines to this comprehensive project (O'Riordan 2013). Future Earth was launched in June 2012 at the UN Conference on Sustainable Development (Rio+20) and is overseen by the International Council for Science. It aims to "provide critical knowledge required for societies to face the challenges posed by global environmental change and to identify opportunities for a transition to global sustainability" (Future Earth 2013:10). Thus, this research program seeks to provide science-based solutions to global environmental problems like climate change.

As part of Future Earth, the ISSC spearheaded an effort to develop an agenda focused on the role of social science in this endeavor. Free of the governmental constraints limiting the USGCRP, the ISSC effort is very ambitious and far-reaching. Its aim is to "fundamentally reframe climate and environmental change as a social, rather than physical problem" (Hackmann, Moser, and St. Clair 2014:654). This approach was initially outlined in a report titled *Transformative Cornerstones of Social Science Research for Global Change* (Hackmann and St. Clair 2012:7), which emphasizes that "the social sciences are an essential piece of the [global environmental change] research puzzle, to be fully integrated throughout the research process, starting with the identification of research agendas and the framing of research questions." It further acknowledges that "Existing global change research agendas and the Earth system framework and language in which they have traditionally been embedded, simply do not speak to the concerns and skills of mainstream social scientists" (Hackmann and St. Clair 2012:12).

To remedy the situation, the report lays out six "transformative cornerstones" involving "cross-cutting questions that demonstrate the central importance of social science knowledge for global change research," designed to provide "a fundamental set of lenses for understanding processes of climate change and global environmental change as social processes embedded in specific social systems, past and present" (Hackmann and St. Clair 2012:15). These six cornerstones are as follows:

1. *Historical and contextual complexities*—"understand the political economy of climate and other processes of environmental change, and how these processes relate to a multitude of other social crises" (Hackmann and St. Clair 2012:16)

2. *Consequences*—"Identifying and mapping the full range of actual and unfolding threats and impacts of global change processes on people and communities in diverse locations" (Hackmann and St. Clair 2012:17)

3. *Conditions and visions for change*—"understand what drives individual and collective processes of change, as well as change in social practices" (Hackmann and St. Clair 2012:18)

4. *Interpretation and subjective sense making*—"make sense of the ingrained assumptions and associated blind-spots that underlie choices and priorities, prevent awareness of that which needs to change, and keep systems deadlocked in spirals of inaction" (Hackmann and St. Clair 2012:19)

5. *Responsibilities*—"understand what it takes to foster global and inter-generational solidarity and justice" and bring "these concerns

into the legitimate space of scientific expertise, policy and practice" (Hackmann and St. Clair 2012:19)

6. *Governance and decision making*—"knowledge is needed on how decisions are made in the face of uncertainty, what pathways are available for influencing decision making, what determines the success or failure of political agreements and what drives political will" (Hackmann and St. Clair 2012:20)

Each cornerstone is described via a series of "illustrative questions" that provide more concrete guides to research issues. For example, "How do we account for and track the influence on global change processes of dominant neoliberal thinking and the marketization of all social life?" (Hackmann and St. Clair 2012:17); "To what extent do existing economic, social and political systems, policies and practices promote unjust global relations and inequalities?" (Hackmann and St. Clair 2012:20); and (regarding climate change denial and skepticism) "What power bases and interests are advanced by appeals for inaction and how can such forces be counteracted?" (Hackmann and St. Clair 2012:19). Terms such as "neoliberal thinking," "inequalities," and "power bases" seldom, if ever, appear in calls for social science research on climate change, and thus the ISSC report makes an important contribution that escapes (to some degree at least) the constraints of the prevalent "post-political" approach to climate change.

The ISSC lays out an impressive research agenda, one that speaks to the interests and expertise of many social scientists, especially sociologists, as reflected by the work reported in the current volume.[4] For example, the political-economic approach of the first cornerstone calls for clarifying the drivers of climate change and identifying linkages between climate change and other social crises. These types of issues are treated extensively in Chapters 2 (driving forces), 3 (market organizations), 4 (consumption), 6 (climate justice) and 11 (social theory) and touched on in other chapters, reflecting their importance to sociologists. Likewise, the second cornerstone's focus on the consequences of climate change is the central subject of Chapters 6 (climate justice) and 7 (adaptation), where sociology's historical emphasis on inequality leads to a concern with inequitable consequences. Similarly, the third cornerstone points to the importance of understanding the roles of individual and collective change in response to climate change, issues dealt with extensively in Chapters 4 (consumption), 5 (justice), 8 (civil society), and to some degree 9 (public opinion) and 10 (denial and skepticism). The trend continues with the fourth cornerstone's call to examine the understandings and discourses through which climate change is viewed, as seen in Chapters 8 (civil society), 9 (public opinion), and 10 (denial and skepticism). The fifth cornerstone stresses "responsibilities," and Chapters 5 (climate justice), 6 (adaptation), 7 (mitigation), and 8 (civil society) report highly relevant work, with Chapter 5 being devoted specifically to the justice-related issues emphasized by the ISSC. More broadly, this cornerstone suggests the need for a "normative" approach to climate change that resonates well with growing calls for "public sociology," one of the emphases of the current volume. Finally, the sixth cornerstone calls for more work on "governance and decision making," and Chapters 7 (mitigation), 8 (civil society), 9 (public opinion), and 10 (denial and skepticism) all present pertinent sociological knowledge on this and related topics.

In short, although not designed with the ISSC cornerstones in mind,[5] the current volume provides clear evidence that sociologists are already dealing in varying degrees with the critical research issues and questions the ISSC proposes as crucial for social science research on climate change and global environmental change more generally. Like the ISSC, our volume calls for, and takes steps toward fulfilling, a far more expansive role for social science (and sociology in particular) than merely filling gaps in predetermined research agendas set by natural scientists and government agencies, such as the effort to educate the public and encourage more carbon-friendly lifestyles (Shove 2010).[6] Certainly there is much more to be done, and one of our major goals here is to encourage sociologists to devote more attention to climate change. Indeed, from the outset the American Sociological Association's Task Force on Sociology and Global Climate Change has endorsed the dual goals of demonstrating the value of sociological insights into climate change to the broad scholarly and policy communities interested in the topic and of stimulating more work on climate change by the sociological community.

LIMITATIONS OF THE ISSC AND USGCRP EFFORTS
TO PROMOTE SOCIAL SCIENCE CONTRIBUTIONS

While there is much to be admired in the ISSC's report on transformative cornerstones, it too has limitations. First, both the ISSC and USGCRP treat "social science" as if it were a single field, rather than acknowledging the existence of distinct disciplines spanning psychology, sociology, political science, economics, geography, anthropology, and communications. Each of these disciplines has unique epistemological assumptions and theoretical and methodological approaches that can be applied to studying climate change (e.g., see Barnes et al. [2013] on anthropology and Javeline [2014] on political science). Neither the ISSC nor USGCRP provides summaries (or even acknowledgments) of current research on climate change being conducted *within* specific disciplinary communities (although the ISSC does note the predominance of economics). Instead, the reports offer rather shallow and nonspecific notions of the social sciences, absent any sense of their substantive differences and unique contributions. The failure of the USGCRP and ISSC to engage with these disciplinary communities leads to the development of research agendas that not only are tilted toward the natural sciences, but also fail to tap fully the rich traditions and strengths of the individual social sciences.

Second, despite its far-broader perspective, the ISSC shares with the USGCRP a commitment to the integration of the social sciences with the natural sciences, with the ISSC strongly emphasizing "the need for integrated research" (Hackmann and St. Clair 2012:2). Thus, despite posing the rich research questions noted above, the ISSC still emphasizes research that will contribute to efforts to integrate social and natural science efforts to deal with global environmental problems such as climate change. This approach again ignores the unique contributions that each of the social sciences can make through its own distinct disciplinary approach, and—despite the ISSC's best efforts—still tends to subordinate social science research to priorities set by natural sciences or funding agencies and foundations.

Most crucially, the ISSC and USGCRP efforts focus on what the social sciences can do for global environmental and climate change research, but not vice versa. Ironically, the ISSC report does acknowledge the importance of the latter, noting that while progress has been made in "bringing social science to the heart of global environmental change ... there has not been equal success in bringing global environmental change to the heart of the social sciences" (Hackmann and St. Clair 2012:9–10). This situation is partly due to the fact that such projects are often framed in ways that fail to speak to the concerns and expertise of mainstream social scientists. We doubt that telling sociologists and other social scientists that they should engage primarily in integrative research, working in multidisciplinary teams with natural scientists on government- or foundation-driven research agendas, is the best path for bringing global environmental change to the heart of social science. Numerous sociologists, including several authors in this volume, are already engaged in multidisciplinary projects on climate change involving collaboration with climate and other natural scientists and recognize the benefits of such work, and they frequently publish their work in leading multidisciplinary journals.[7] However, other scholars may not prefer this path, and should be encouraged to examine climate change on their own terms.

In fairness, the modest success thus far in bringing global environmental change to the core of the social sciences also reflects a tendency within the social sciences (especially in the last half of the twentieth century) to ignore or at least downplay the relevance of biophysical conditions in modern industrial societies (Dunlap 1980). As a subsequent ISSC report (2013:38) notes, global environmental change has "failed to capture the attention and imagination of the more traditional, mainstream social sciences ..., disciplines which view the social and human world as their focus." For these disciplines, "social phenomena, relationships, interactions and human behaviours may take place on an environmental stage, but they tend to be understood as being determined by humans alone" (ISSC 2013:38). We agree with this assessment but see considerable progress in overcoming the legacy of "human exemptionalism" (Dunlap 1980) over the past decade or two. First, sociological research on environmental topics is appearing with increased frequency in elite disciplinary journals; second, a growing number of leading sociologists not traditionally identified with environmental sociology are providing informed analyses of climate change (e.g., Antonio 2009; Derber 2010; Mann 2012:Chapter 12; Skocpol 2013). These efforts are reinforced by the fact that sociologists—like other educated and informed citizens—tend to be increasingly aware that climate change and other global environmental problems are disrupting the once seemingly stable stage on which human affairs are enacted, and thus deserve scholarly attention (Urry 2011).

THE U.S. ENVIRONMENTAL SOCIOLOGY COMMUNITY'S RESPONSE TO CLIMATE CHANGE

Efforts to develop a sociological perspective on climate change beyond the work of individual scholars can be traced to a workshop held by the National Science Foundation in 2008 (Nagel, Dietz, and Broadbent 2010). The workshop provided a broad overview of existing sociological research on climate change and offered a number of recommendations on how to move this body of work forward. Included were recommendations on how to (1) catalyze the sociological discipline around climate change research, (2) forge interdisciplinary collaborations, and (3) enhance capacity building in the area of climate change and sociology. It also contained twenty-seven short essays summarizing specific sociological research projects on climate change and identified a number of future research questions. This excellent effort provided a concise overview of sociological research on climate change at that time. However, no further action was taken on the workshop's recommendations. The report was not referenced or used in any of the formal efforts to develop a research agenda for the social sciences and climate change, including the reports from the ISSC or USGCRP.

The current volume extends and amplifies the initial National Science Foundation workshop effort under the auspices of the American Sociological Association. The essays in this volume represent research and theory that diverge in substantial ways from the vision of social science presented in both the ISSC and USGCRP efforts. Rather than being connected to national research agencies and funders, such as the Belmont Forum for the ISSC or the U.S. government in the case of the USGCRP, our volume is independent of any external influences. As noted by Mooney et al. (2013:7), the development of a climate change research agenda "should not be influenced or dictated by vested interests in the forms of governments, private sector, or nongovernmental organizations." Since this report results from an unfunded effort conducted largely by members of the U.S. environmental sociology community and represents their views, it provides an independent perspective on sociological engagement with the issue of climate change. While the authors of each chapter were cognizant of the findings of the Intergovernmental Panel on Climate Change, National Climate Assessment, and U.S. National Research Council reports, these efforts neither defined nor constrained the intellectual focus of the various chapters. The goal of this volume is not to provide a formal agenda for future sociological engagement with climate change (which will surely involve new and unexpected lines of research and theorizing) but to demonstrate the value of current work—with the dual aim of taking stock of current knowledge and stimulating more attention to climate change among sociologists.

Also, we do not assume that sociologists need to contribute to multidisciplinary efforts. While some chapters present work on topics that are inherently multidisciplinary (such as Chapters 2, 4, 6, and 7 on driving forces, consumption, adaptation, and mitigation, respectively), there are a number of sociological perspectives that do not engage the natural sciences directly (e.g., Chapters 8 and 9 on civil society and public opinion). Also, the political nature of the climate change issue is a concern in several chapters (especially 8, 9, 10, and 11 on civil society, public opinion, denial and skepticism, and social theory), introducing elements of political critique and public sociology. Finally, no attempt is made to integrate the various chapters into an overarching perspective. Within sociology, a wide variety of theoretical and empirical approaches are used to address climate change, and any attempt to develop a comprehensive perspective would be both premature and restrictive.

To enhance the value of this book, the authors of each chapter were asked to make recommendations regarding important future research directions in their particular area. While the chapters typically provide extended discussion of recommended research agendas, what follows is a quick summary of these agendas.

Driving Forces

- Extend the analysis of causal chains to include how social forces drive environmental changes and, in turn, how these environmental changes result in further social impacts.
- Conduct evaluations of economic models other than capitalism and how they affect the levels and nature of environmental degradation.
- Investigate the relationship between carbon emissions and human well-being.

Market Organizations

- Develop measures of corporate performance on carbon emissions for use in analysis.
- Examine the political dynamics that allow corporations to continue to profit while emitting carbon levels that are harmful.
- Analyze the institutional arrangements between state policies, capital investment, and labor markets that differentially affect mitigation strategies.

Consumption

- Clarify the social context of relevant consumption decisions and behaviors.
- Examine the relationship between social inequality and carbon emissions using comparative analysis.
- Analyze the malleability of carbon-related habitual behavioral patterns.
- Conduct research into the social development of household practices that influence carbon emissions.

Climate Justice

- Apply theories of inequality to explain differential carbon emissions and the unequal impacts of climate change at varying social levels.
- Include environmental inequality in climate change assessments.
- Analyze the social, economic, and cultural factors that influence the capacity of communities to respond to or recover from disasters.
- Conduct research into the factors that drive the success or failure of climate justice social movements.
- Facilitate the participation of underrepresented voices in climate change politics.

- Contribute to the development of a vision of a sustainable society outside of the taken-for-granted neoliberal worldview.
- Develop and provide guidance to agencies to ensure that climate justice issues are taken into account in the decision-making process.

Adaptation

- Examine the social factors crucial for engaging in adaptation planning.
- Identify the most efficacious approaches for engaging citizens in adaptation planning.
- Develop methods to ensure that all affected populations, including disadvantaged groups, are included and their needs are addressed in adaptation planning.
- Analyze different municipalities' capacities for engaging in adaptation planning and carrying out the necessary actions.

Mitigation

- Conduct research to create an integrated sociological approach to understanding the factors that drive mitigation actions at the individual, community, and societal levels.
- Develop an understanding of the political, economic, and social factors that drive national mitigation policies.
- Conduct cross-national research on mitigation activities to identify social conditions that aid or hinder efforts at climate change mitigation.
- Analyze the social conditions that foster or inhibit mitigation activities that facilitate social change.

Civil Society

- Conduct research to understand the institutional structure of climate change civil society organizations and their interactions with external organizations, including governments, corporations, trade associations, and foundations.
- Analyze the strategies that climate change civil society organizations follow, and evaluate their relative efficacy.
- Examine the resource mobilization practices of civil society organizations and how these practices influence the structure and actions of the organizations.

- Analyze the different framings of climate change by civil society organizations and how this framing influences their sources/levels of external funding and membership participation.
- Study the factors that result in shifts in the political opportunity structure for climate change action by civil society organizations.

Public Opinion

- Conduct research into the relationship between individual opinion regarding climate change and mitigation behavior.
- Research the relationships among acceptance of climate change science, support for climate policies, and participation in political behavior.
- Examine the influence of public opinion regarding climate change on climate policy adoption.
- Analyze the influence of public opinion regarding climate change on corporate behavior.

Climate Denial and Skepticism

- Further examine the ideological and socioeconomic contexts in which climate change denial and skepticism are most likely to flourish.
- Follow the evolution of strategies and tactics employed by the "denial machine," as well as the key actors involved in it.
- Highlight the elements of climate science that make it especially vulnerable to the strategy of "manufacturing uncertainty" employed by the denial machine.
- Conduct more research on the funding of climate change denial actors.
- Analyze the relationship between organized denial and the diffusion of skepticism to the public at large, paying attention to the roles of media and political elites.

Methodology

- Develop integrated models that incorporate measurements of the biological and physical world into social science analyses.
- Specify and test robust integrated models that use a variety of quantitative techniques.
- Build collaborative research efforts that focus on a well-defined research agenda that can build cumulative knowledge over time.

Sociological Theory

- Move beyond the capitalist/socialist binary approach and develop a new perspective capable of adapting to the conditions of planetary biophysical constraints, as well as the sociocultural limits to our political-economic transformation capacity.
- Address how growth can be stemmed without immiseration and de-democratization.
- Examine the nature of global economic growth and its role as a fundamental driver of climate change.
- Expand and enrich our understanding of how climate change engages and overlaps with other ecological and social problems.

The overall picture of sociological research on climate change that emerges from this collection of essays is a multifaceted one. There is no one privileged or "correct" approach, theoretically or methodologically. Rather, there are a wide variety of ways of conducting sociological research on climate change that illuminate different aspects of the phenomenon. First, there is a segment of the environmental sociology community that focuses on human and natural system interactions or, at a minimum, includes natural science data as "variables" in empirical and/or theoretical analyses. Second, there are a number of discipline-specific approaches that do not engage with natural science evidence except as background information regarding the reality of climate change. Finally, there are elements of critical/public sociology that seek to enrich societal attention to climate change via sociological insights.

This diversity of approaches strengthens the current and potential contributions of sociology to understanding climate change. What is needed is much more sociological attention to climate change, including but not limited to the following emphases:

1. Better integration of sociological research with major multidisciplinary research programs on global environmental and climate change. Despite our criticisms of both the ISSC and USGCRP research agendas, we do recognize that efforts to better integrate sociological concerns into such efforts can be valuable to sociology and to these programs. A crucial goal should be the expansion of such research agendas from a singular focus on the integration of social and natural sciences in order to provide intellectual space for discipline-specific research questions and approaches that nonetheless provide insights into climate change useful to scholars in other disciplines, including natural scientists.
2. Generate more attention to climate change within the broader sociological community, stimulating a growing body of analyses employing

a diverse range of theoretical and methodological approaches. Such increased attention needs to be accompanied by efforts to enhance basic knowledge of climate science among sociologists so that their analyses will be well grounded and have credibility outside of the discipline.

3. Expand participation in developing social science research agendas on climate change to ensure that alternative viewpoints—especially those that question the socioeconomic and political status quos, including views provided by disadvantaged communities[8]—are well represented. These agendas should recognize that climate change has become a very controversial issue and should stimulate research on the social, economic, and political sources of conflicts over the topic.

Beyond these emphases, it is essential to underscore that the role of social science, and sociology in particular, in climate change research should not be limited to providing advice for government agencies and input for decision making. While such tasks are important, sociology has a broader responsibility to contribute to society's reflexive capacity for addressing climate change. By openly acknowledging and analyzing the inherently conflict-generating status of climate change, sociology can help expand the locus of decision making from governmental (and corporate) arenas to societal fora that might more effectively engage the broader citizenry. The creation of a research agenda that seeks to help generate informed deliberation on climate change in the public sphere should be a key role for sociology (Habermas 1998) and a major challenge for public sociology.

RECOGNIZING AND ANALYZING THE "POLITICAL" NATURE OF CLIMATE CHANGE

In Chapter 1 we discussed the tendency for climate change to be given a "post-political" frame, one in which an amorphous humanity is viewed as producing excessive CO_2, which in turn becomes a common "enemy" to be defeated. The result is that "what one opposes first and foremost are not necessarily specific, particularly polluting practices, let alone specific social actors who bear responsibility for [them]. Rather, one opposes CO_2 as such, which is the by-product of almost all thinkable practices (even breathing)" (Kenis and Lievens 2014:540). This framing leads to calls for everyone to reduce his or her personal GHG emissions. A further result is that CO_2 becomes a "commodity" that can be priced and thus also dealt with via market-based policies such as carbon emissions trading, making potential solutions to climate change nonthreatening to the existing economic system

(Shaw and Nerlich 2014).[9] Achieving the necessary reductions will require scientific information, technological innovation, and skillful management to guide the battle against excess CO_2 and other GHGs—as well as a societal consensus on the necessity of doing so (Swyngedouw 2011). Epitomizing this view, Giddens (2009:71) suggests that "Responding to climate change must not be seen as a left-right issue. Climate change has to be a question that largely transcends party politics, and about which there is an overall framework of agreement that will endure across changes in government." The ultimate goal is to attain "low carbon" technologies, economies, and societies "where nothing in the world has changed except the amount of carbon emitted by the activities which define late neo-liberal patterns of economic activity" (Shaw and Nerlich 2014:38).

The idea of climate change as a post-political issue is popular in Europe, where it seems to have some validity when applied to how the topic is treated—especially in wealthy, North European nations (and the United Kingdom). These nations have been trendsetters in climate policy, although leading climate scientists see their efforts as clearly inadequate for preventing dangerous levels of climate change (Anderson and Bows 2011) and social scientists increasingly emphasize that they are not close to reaching environmentally sustainable paths (Anshelm and Hultman 2015; Bluhdorn 2013; Kenis and Lievins 2014; Swyngedouw 2011). Strikingly, however, it seems to have no applicability to the contemporary United States, where climate change has become a major "political" issue (McCright and Dunlap 2011). Thus, McCarthy (2013:23) readily agrees that "the perpetuation of capitalism and use of market techniques are never seriously questioned in mainstream discussions," but argues that "the rest [of the post-political frame] could not be further off from the U.S. experience." In fact, in contrast to Western Europe,

> Large sectors of the American public, including very large percentages of professional politicians and the media, accept neither scientific expertise nor consensus . . . Likewise, environmentalism in general, and climate change in particular, are very commonly portrayed as entirely fictional issues of concern, invented by self-interested and unpatriotic scientists and activists either for their own gain, or as an excuse for increased government control over the entire society. (McCarthy 2012:23)

We agree entirely, and point to the substantial body of evidence presented in Chapters 9 and 10 on public opinion and climate change denial and skepticism as justification. The situation can only be expected to worsen following Republican congressional gains in the 2014 U.S. election, given the GOP's growing skepticism about climate change (Germain et al. 2013). Clearly Giddens's hope that climate change will transcend party politics has been dashed by the American experience, and it is increasingly challenged by the growing resistance to climate change policymaking in Australia, Canada, and the United Kingdom under their neo-liberal governments.

The problem is that despite climate change having never come close to being a consensual, post-political issue in the United States—with fossil fuels and many other corporations and now nearly all of the conservative movement challenging its reality and opposed to acting to limit it—key actors concerned about the issue seem to act as if it were. Thus, government agencies sponsoring and funding climate science and response efforts (epitomized by the USGCRP), foundations funding "climate work," most national environmental organizations, many scientific organizations, numerous climate scientists, and countless individual citizens acting to "save the climate" all tend to downplay if not ignore the extreme degree of conflict and polarization surrounding climate change in the United States.[10] This is why in Chapter 1 we described the post-political perspective as a characteristic of current research on climate change (including attempts to integrate natural and social science) and efforts to ameliorate it within the research community, policy circles, and society at large. While European social scientists have done an excellent job of demonstrating the weakness of the post-political approach, a key task for American sociologists and other social scientists is to show its inapplicability to the United States by highlighting and clarifying the sources and nature of conflicts over climate change permeating our society (and, increasingly, nations such as Australia, Canada, and the United Kingdom).

Considerable work along these lines appears in this volume, but much more is needed. A good starting point would be to examine carefully the roles of the fossil fuels industry, and the corporate world more generally, in opposing or at least shaping in corporate-friendly ways action to limit GHG emissions. As noted in Chapter 3, these issues have received insufficient attention, and this is especially so among sociologists. Until now, activist organizations and citizen interest groups have taken the lead in analyzing issues like fossil fuel industry campaign contributions (Atkin 2014) and efforts to influence our political system more generally (Cray and Montague 2014), as well as how the corporate world at large influences debates over climate science and climate policy (Goldman and Carlson 2014; Union of Concerned Scientists 2012). Sociologists are well positioned to provide strong empirical and theoretically informed analyses of corporate lobbying, campaign contributions (via individual companies and their employees, political action committees, and "dark money"), public relations and advertising, and other means of preventing the implementation of governmental efforts to reduce GHG emissions.

We expect to see more research along these lines by environmental sociologists, and fortunately other scholars are also beginning to examine the political conflicts over climate policy and barriers to making progress in reducing GHG emissions. Skocpol (2013), for example, provides a highly informative analysis of the failure of the 2009–2010 "cap-and-trade" legislation in the U.S. Congress (also see Pooley 2010). She faults major environmental organizations in two ways. The first was for supporting a weak measure in order to reach compromise with their corporate partners in the U.S. Climate Action Partnership (who ultimately abandoned them and lobbied against the bill), rather than trying to build grassroots support for a much stronger measure. The second was for naïvely underestimating political opposition from increasingly conservative Republicans due to the emergence of the Tea Party. Taking a more macro-level approach, Mann (2012:Chapter 12) emphasizes the global importance of capitalism and its quest for endless growth, compounded by the eagerness of governments and their leaders to promote such growth to enhance national welfare and personal success, and the desires of individual citizens—both Northern consumers who enjoy their high-carbon lifestyles and Southern consumers eager to emulate them. Mann particularly emphasizes the importance of neoliberal ideology in generating opposition to climate change policies, drawing in part on empirical research showing a significant relationship between nations' free-market orientations and their environmental impacts (Özler and Obach 2009).

At the most elemental level, there is a need to move beyond the tendency, in the United States and elsewhere, to cast climate change as a consensual issue and thereby marginalize efforts to acknowledge its conflictual status. As Kenis and Lievins (2014:542) put it, "Discourses stating that the conflict approach is obsolete and that we all ought to work together are, in a paradoxical fashion, very polemical." The post-political perspective, despite claims to the contrary, "takes sides and engages in conflict . . . Its opponent is not a particular agent but the conflict approach as such (Kenis and Lievins 2014:542)." Unfortunately, one cannot begin to understand the failure of current efforts to reduce GHG emissions without acknowledging the inherently conflictual nature of these efforts, as Jamieson (2014), Mann (2012), and Skocpol (2013) among others have done.

BRINGING CLIMATE CHANGE INTO THE HEART OF SOCIOLOGY

Much more work on the divisive and inherently political nature of climate change in the contemporary world is needed, and sociology has a repertoire of concepts and perspectives well suited to the task. These include the ability to analyze the roles of (1) vested interests, political power, and ideological beliefs in opposing climate change policies, (2) socialization processes, the mass media, and political elites in the creation of individuals' views of climate change, as well as their consumption practices, and (3) social inequalities centered on race, gender, and class differences and how these affect both mitigation efforts and the distribution of adverse impacts of climate change.

One of the most pressing contributions our field can make is to legitimate big questions, especially the ability of the current global economic system to take the steps needed to avoid catastrophic climate change. For example, stressing the incompatibility between continued patterns of economic growth and necessary reductions in GHG emissions, climate scientists Anderson and Bows (2012:64) note that "The elephant in the room [commitment to growth] sits undisturbed while collective acquiescence and cognitive dissonance trample all who dare to ask difficult questions." Similarly, in a recent commentary on scholarly work on global environmental change, Newell (2011:4) notes that "Capitalism forms the context in which most of the world now responds to global environmental change ... And yet curiously students of global environmental change rarely refer to capitalism directly."

Sociology is helping to alter the situation, with a number of analysts directly addressing the nearly taboo subjects of economic growth and capitalism (see, e.g., Antonio 2009; Derber 2010; Foster 2009; Mann 2012) and thereby legitimating open discussion concerning what kind of fundamental socioeconomic changes, and attendant political reforms, may be necessary to achieve a low-carbon and sustainable future. If interrogating capitalism and its growth imperative become legitimate topics,[11] on a par with currently more acceptable (but clearly related) analyses of high-carbon lifestyles, the global environmental change community will be able to entertain a far wider range of research questions than those described at the beginning of this chapter. All aspects of climate change, especially the tremendous challenges it poses, will become legitimate. In our opinion, this is the way to achieving the ISSC's goal of bringing global environmental change into the heart of social science—particularly sociology.

When establishment figures like Gus Speth[12] (2008:9) conclude that "most environmental deterioration is a result of systemic failures of the capitalism that we have today and that long-term solutions must seek transformative change in the key features of this contemporary capitalism," we need not leave analyses of linkages between capitalism and climate change to social critics like Naomi Klein (2014). In fact, in a wide-ranging discussion of the future of capitalism, one that acknowledges climate change as a serious threat, Wallerstein et al. (2013:7) remind us that "Capitalism . . . is only a particular historical configuration of markets and state structures where private economic gain by almost any means is the paramount goal and measure of success. A different and more satisfying organization of markets and human society may yet become possible." They later note that such options become more likely in periods of crisis (as may be generated by climate change) when the status quo is breaking down, adding that "Such times call for a conscious strategy of systemic transformation" (Wallerstein et al. 2013:189). Importantly, they also claim an important role for social science in such a transformation, suggesting that it "should clarify . . . the circumstances and emerging possibilities, especially when the possibilities may be opening and closely rapidly" (Wallerstein et al. 2013:189). Responding to the call by Wallerstein and his eminent colleagues with a reflexive and critical approach to climate change will represent a fundamental contribution by our discipline—one that can enrich current global climate change research agendas and, more importantly, societal deliberations over appropriate responses.

NOTES

1. The USCGRP is a U.S. government entity that coordinates and integrates scientific research relevant to global environmental change across thirteen federal agencies. Along with the National Science and Technology Council, the USGCRP is responsible for producing the U.S. National Climate Assessments (see USGCRP 2014).

2. Thus far the Task Force has produced a "non-citable" draft of a white paper, but its effort is described in a PowerPoint presentation (Weaver 2013). In addition, Weaver et al. (2014) provide insight into the Task Force's view of the role of the social sciences within the USGCRP. While the Task Force's promised "white paper" has yet to be formally issued, the most recent USGCRP budget plan submitted to Congress (USGCRP 2014:8) states that "the task force fulfilled its mandate."

3. Besides the latest National Climate Assessment (Melillo, Richmond, and Yohe 2014) see *The National Global Change Research Plan, 2012–2021* (National Science and Technology Council 2012) for a good sense of the USGCRP's vision of the role of social science.

4. The initial step in implementing the ISSC research agenda was taken in March 2014 when two calls for proposals were issued for research that builds knowledge networks to realize this research agenda (ISSC 2014).

5. The topics and lead authors for the chapters in this volume were selected before we became aware of Hackmann and St. Clair (2012) and the subsequent reports of the ISSC.

6. The contributors to this volume, reflecting a tendency in U.S. environmental sociology more generally (Dunlap 2010), tend to adopt a "realist" stance toward

anthropogenic climate change that broadly speaking assumes that the evidence for increasing climate disruption is compelling—even when examining public opinion, climate change movements and countermovements, and other phenomena involved in the social construction of climate change (Zehr 2014). Other strands of constructivism, including those that adopt a more agnostic stance toward climate science and its findings, also provide important contributions. The work of U.K. sociologist Steven Yearley (2007, 2009), for example, provides insightful analyses of the construction of climate change as problematic, the workings of climate science and controversies over its findings, and development of policy tools to deal with climate change. In contrast to this volume, however, he tends to ignore key topics such as driving forces, impacts, and adaptation. Analyzing the latter requires granting credence to the findings and implications of climate science, which contributors to this volume find reasonable, but which some constructivists are more hesitant to grant (e.g., Grundmann and Stehr 2010). Yet, research suggests that if anything the threat of climate disruption is likely underestimated (Brysse et al. 2013; Freudenburg and Muselli 2010, 2013).

7. We do not have space to list all of the individuals engaged in these activities, and fear we would ignore some. Likewise, we cannot begin to list all relevant publications (most cited in previous chapters) but will note that the contributors to this volume have published in *Science, Proceedings of the National Academy of Sciences USA, Nature Climate Change, Global Environmental Change,* and other leading multidisciplinary journals. In general, environmental sociologists have a higher-than-average (for the discipline) level of involvement in interdisciplinary research, in the United States (Pellow and Brehm 2013) and Europe (Gross and Heinrichs 2010) and probably other regions as well.

8. Featherstone (2013) demonstrates the importance of opening up space for the voices of the disadvantaged.

9. In contrast to market-based approaches like establishing carbon trading markets, a potentially more effective carbon tax has generated little enthusiasm (even in Europe) because it is incompatible with neoliberal ideology (Mann 2012:Chapter 12).

10. For example, the National Climate Assessment (Melillo, Richmond, and Yohe 2014) has a one-word mention of "polarization" as a barrier to adaptation on page 683, ignoring the extensive literature on climate denial (see Chapter 10 of the current volume). Bill McKibben's 350.org and its activities to build a "climate movement" are a partial exception, as are some other (typically small) activist groups (McKibben 2007).

11. As apparent from Chapter 2, a considerable body of empirical work on driving forces does suggest that economic affluence and growth generate growth in GHG emissions, although such work is typically couched in terms of the "treadmill of production" rather than in-depth analyses of the dynamics of capitalism.

12. Speth served as Chairman of the Council on Environmental Quality under President Carter, founded the World Resources Institute, served as Administrator of the United Nations Development Program, and recently was Dean of the Yale School of Forestry and Environmental Studies.

REFERENCES

Anderson, Kevin and Alice Bows. 2011. "Beyond 'Dangerous' Climate Change: Emission Scenarios for a New World." *Philosophical Transactions of the Royal Society A* 369:2044.

Anderson, Kevin and Alice Bows. 2012. "A New Paradigm for Climate Change." *Nature Climate Change* 2:639–640.

Anshelm, Jonas and Martin Hultman. 2015. *Discourses of Global Climate Change.* Cambridge: Polity Press.

Antonio, Robert. 2009. "Climate Change, the Resource Crunch, and the Global Growth Imperative." *Current Perspectives in Social Theory* 26:3–73.

Atkin, Emily. 2014. "The Fossil Fuel Industry Spent More Than $721 Million During 2014's Midterm Elections" (http://thinkprogress.org/climate/2014/12/23/3606630/fossil-fuel-spending-midterm-elections/).

Barnes, Jessica, Michael Dove, Myanna Lahsen, Andrew Mathews, Pamela McElwee, Roderick McIntosh, Frances Moore, Jessica O'Reilly, Ben Orlove, Rajindra Puri, Harvey Weiss, and Karina Yager. 2013. "Contributions of Anthropology to the Study of Climate Change." *Nature Climate Change* 3:541–544.

Blühdorn, Ingolfur. 2013. "The Governance of Unsustainability: Ecology and Democracy after the Post-Democratic Turn." *Ecological Politics* 22:16–36.

Brysse, Keynyn, Namoi Oreskes, Jessica O'Reilly, and Michael Oppenheimer. 2013. "Climate Change Prediction: Erring on the Side of Least Drama?" *Global Environmental Change* 23:327–337.

Cray, Charlie and Peter Montague. 2014. *The Kingpins of Carbon and Their War on Democracy* (http://www.greenpeace.org/usa/Global/usa/planet3/PDFs/Kingpins-of-Carbon.pdf).

Crist, Eileen. 2007. "Beyond the Climate Crisis: A Critique of Climate Change Discourse." *Telos* 141:29–55.

Derber, Charles. 2010. *Greed to Green: Solving Climate Change and Remaking the Economy.* Boulder, CO: Paradigm.

Dunlap, Riley E. 1980. "Paradigmatic Change in Social Science: From Human Exemptionalism to an Ecological Paradigm." *American Behavioral Scientist* 24:5–14.

Dunlap, Riley E. 2010. "The Maturation and Diversification of Environmental Sociology: From Constructivism and Realism to Agnosticism and Pragmatism." Pp. 15–32 in *International Handbook of Environmental Sociology*, 2nd ed., edited by M. Redclift and G. Woodgate. Cheltenham, UK: Edward Elgar.

Featherstone, David. 2013. "The Contested Politics of Climate Change and the Crisis of Neo-liberalism." *ACME: An International E-Journal for Critical Geographies* 12(1):44–64.

Foster, John Bellamy. 2009. *The Ecological Revolution: Making Peace with the Planet.* New York: Monthly Review Press.

Freudenburg, William R. and Violetta Muselli. 2010. "Global Warming Estimates, Media Expectations, and the Asymmetry of Scientific Challenge." *Global Environmental Change* 20:483–491.

Freudenburg, William R. and Violetta Muselli. 2013. "Reexamining Climate Change Debates: Scientific Disagreement or Scientific Certainty Argumentation Methods (SCAMs)? *American Behavioral Scientist* 57:777–785.

Future Earth. 2013. *Future Earth Initial Design: Report of the Transition Team.* Paris: International Council for Science (ICSU).

Germain, Tiffany, Ryan Koronowski and Jeff Spross. 2013. "The Anti-Science Climate Denier Caucus: 113th Congress Edition" (http://thinkprogress. org/climate/2013/06/26/2202141/anti-science-climate-denier-caucus-11 3th-congress-edition/).

Giddens, Anthony. 2009. *The Politics of Climate Change.* Cambridge: Polity Press.

Goldman, Gretchen and Christina Carlson. 2014. *Tricks of the Trade—How Companies Anonymously Influence Climate Policy through their Business and Trade Associations.* Center for Science and Democracy at the Union of Concerned Scientists (http://www.ucsusa.org/sites/default/files/legacy/assets/documents/ center-for-science-and-democracy/tricks-of-the-trade.pdf).

Gross, Matthias and Harald Heinrichs. 2010. *Environmental Sociology: European Perspectives and Interdisciplinary Challenges.* Dordrecht: Springer.

Grundmann, Reiner and Nico Stehr. 2010. "Climate Change: What Role for Sociology?" *Current Sociology* 58:897–910.

Habermas, Jürgen. 1998. *Between Facts and Norms.* Cambridge, MA: MIT Press.

Hackmann, Heidi, Barrett Brown, and Eleanor Kershaw. 2011. *ISSC—Belmont Forum Agenda Setting Workshop: Synthesis Report and Resource Document.* Paris: International Social Science Council.

Hackmann, Heidi, Susanne Moser, and Asuncion St. Clair. 2014. "The Social Heart of Global Environmental Change." *Nature Climate Change* 4:653–655.

Hackmann, Heidi and Asuncion St. Clair. 2012. *Transformative Cornerstones of Social Science Research for Global Change.* Paris: International Social Science Council.

ISSC. 2013. *World Social Science Report 2013: Changing Global Environments.* Paris: OECD Publishing and UNESCO Publishing.

ISSC. 2014. *Call for Seed Funding: Transformations to Sustainability Programs.* Paris: International Social Science Council.

Jamieson, Dale. 2014. *Reason in a Dark Time.* New York: Oxford University Press.

Javeline, Debra. 2014. "The Most Important Topic Political Scientists are Not Studying: Adapting to Climate Change." *Perspectives on Politics* 12:420–434.

Kenis, Anneleen and Matthias Lievens. 2014. "Searching for 'The Political' in Environmental Politics." *Environmental Politics* 23(4):531–548.

Klein, Naomi. 2014. *This Changes Everything: Capitalism Vs. the Climate.* New York: Simon and Schuster.

Mann, Michael. 2012. *The Sources of Power: Volume 4, Globalizations, 1945–2001.* New York: Cambridge University Press.

McCarthy, James. 2013. "We Have Never been 'Post-political.'" *Capitalism Nature Socialism* 24:19–25.

McCright, Aaron M. and Riley E. Dunlap. 2011. "The Politicization of Climate Change and Polarization in the American Public's Views of Global Warming, 2001–2010." *Sociological Quarterly* 52:155–194.

McKibben, Bill. 2007. *Fight Global Warming Now.* New York: St. Martin's Griffin.

Melillo, Jerry M., Terese Richmond, and Gary Yohe, eds. 2014. *Climate Change Impacts in the United States: The Third National Climate Assessment.* U.S. Global Change Research Program, Washington, DC: U.S. Government Printing Office.

Mooney, Harold A., Anantha Duraiappah, and Anne Larigauderie. 2013. "Evolution of Natural and Social Science Interactions in Global Change Research Programs." *Proceeding of the Natural Academy of Sciences USA* 110 (Suppl 1):3665–3672.

Nagel, J., T. Dietz, and J. Broadbent. 2010. *Workshop on Sociological Perspectives on Global Climate Change, May 30–31, 2008.* Washington, DC: National Science Foundation & the American Sociological Association.

National Science and Technology Council. 2012. *The National Global Change Research Plan 2012–2021: A Strategic Plan for the U.S. Global Change Research Program (USGCRP).* Washington, DC: National Science and Technology Council.

Newell, P. 2011. "The Elephant in the Room: Capitalism and Global Environmental Change." *Global Environmental Change* 21(1):4–6.

O'Riordan, Timothy. 2013. "Future Earth and Tipping Points." *Environment* 55(5):31–40.

Özler, S. Ilgü and Brian K. Obach. 2009. "Capitalism, State Economic Policy and Ecological Footprint: An International Comparative Analysis." *Global Environmental Politics* 9:78–109.

Pellow, David N. and Hollie Nyseth Brehm. 2013. "An Environmental Sociology for the Twenty-First Century." *Annual Review of Sociology* 39:229–250.

Pooley, Eric. 2010. *The Climate War.* New York: Hyperion.

Shaw, Christopher and Brigette Nerlich. 2014. "Metaphor as a Mechanism of Global Climate Governance: A Study of International Policies, 1992–2012." *Ecological Economics* 109:34–40.

Shove, Elizabeth. 2010. "Beyond the ABC: Climate Change Policy and Theories of Social Change." *Environment and Planning A* 42:1273–1285.

Skocpol, Theda. 2013. *Naming the Problem: What It Will Take to Counter Extremism and Engage Americans in the Fight against Global Warming.* Prepared for the Symposium on The Politics of America's Fight Against Global Warming, Harvard University (http://www.scholarsstrategynetwork.org/sites/default/files/skocpol_captrade_report_january_2013y.pdf).

Speth, James Gustave. 2008. *The Bridge at the Edge of the World: Capitalism, the Environment, and Crossing from Crisis to Sustainability.* New Haven, CT: Yale University Press.

Swyngedouw, Erik. 2011. "Depoliticized Environments: The End of Nature, Climate Change and the Post-Political Condition." *Royal Institute of Philosophy Supplement* 69:253–273.

Union of Concerned Scientists. 2012. *A Climate of Corporate Control—How Corporations Have Influenced the U.S. Dialogue on Climate Science and Policy.* Cambridge, MA: Union of Concerned Scientists (http://www.ucsusa.org/our-work/center-science-and-democracy/fighting-misinformation/a-climate-of-corporate-control.html#.VF1DVWfLKqk).

Urry, John. 2011. *Climate Change & Society.* Malden, MA: Polity.

USGCRP. 2014. *Our Changing Planet: The U.S. Global Change Research Program for Fiscal Year 2015.* Washington, DC: U.S. Government Printing Office.

Wallerstein, Immanuel, Randall Collins, Michael Mann, Georgi Derluguian, and Craig Calhoun. 2013. *Does Capitalism Have a Future?* New York: Oxford University Press.

Weaver, C. 2013. *The U.S. Global Change Research Program (USGCRP) Social Sciences Task Force PowerPoint Presentation*, July 11, 2013 (http://www.cesu.psu.edu/meetings_of_interest/SCSPS/SCSPS2_Weaver.pdf).

Weaver, C., S. Mooney, D. Allen, N. Beller-Simms, T. Fish, A. Grambsch, W. Hohenstein, K. Jacobs, M. Kenney, M. Lane, L. Lagner, E. Larson, D. McGinnis, R. Moss, L. Nichols, C. Nierenberg, E. Seyller, P. Stern, and R. Winthrop. 2014.

"From Global Change Science to Action with Social Sciences." *Nature Climate Change* 4:656–659.

Yearley, Steven. 2007. "Global Warming." In *Blackwell Encyclopedia of Sociology,* edited by G. Ritzer. Blackwell Publishing, Blackwell Reference Online.

Yearley, Steven. 2009. "Sociology and Climate Change after Kyoto." *Current Sociology* 57:389–405.

Zehr, Stephen. 2014. "The Sociology of Climate Change." *WIREs Climate Change.* doi: 10.1002/wcc.328.

Name Index

Subject Index

Marxian-rooted globalization theories,
 353, 356–357
Mass opinion, 270. *See also* Public
 opinion, on climate change
Materialist principle, 348
Materiality paradox, 351–352
Meat consumption, 61, 98–99, 112
Media
 conservative, in denial countermovement, 316–317
 on consumption, household, 102
Meso-level mitigation, 210–215. *See also*
 under Mitigation
Metabolic rift theory, 39
Metabolic theory, 353, 354–355
Methane emissions, GHG research
 on, 386
Methodological approaches, 369–404
 for anthropogenic drivers of GHG
  emissions research, 385–393
  (*See also* Greenhouse gas emissions (GHG), anthropogenic
  drivers of)
 data integration in, social and
  environmental, 371–384
  fundamentals of, 371
  spatial integration in, 372–377,
   373f–377f
  spatial/temporal integration in,
   375f, 382–384, 384f
  temporal integration in, 373f–375f,
   377–382, 378f, 380f–382f
 future work on, 397–403, 399f, 401f
 modeling in, benefits of, 402–403
 modeling in, multilevel (hierarchical) linear, 400–402
 modeling in, SEM and path,
  397–399, 398f, 401f
 historical context in, 370–371
 public opinion on climate change
  research in, 393–397, 404
  interaction effects in, 393–394
  natural/biophysical and social data
   integration in, 395–397
  predictors in, strongest, 393
  spatial and temporal integration in,
   394–395
  structural equation modeling in, 395
 research on, recommended, 424
 society–environment interface in,
  369–371

Mexico, REDD in, 80
Microfinancing, for climate
 adaptation, 176
Micro-level mitigation, 204–226. *See*
 also under Mitigation
Migration, human, 182–185
 adaptation and, 183
 history of, 182
 social networks in, 185
 studies of, recent, 182, 187–188
 vulnerability and, 184
Migration, industrial, 46
Millennium Development Goals, adaptation in, 187
Millennium Ecosystem Assessment, 33, 49
Mitigation, 199–229
 America's Climate Choices on, 200–201
 definition of, 199
 future research on, 227–228
 interplay of levels in, 226–227
 IPCC report on, 199–200, 228–229
 macro, 215–226
  ecological modernization theory
   and environmental systems in,
   217–218
  global environmental systems
   theory in, 220–221
  international climate agreements
   failure in, 221–224
  Kyoto Protocol and international
   efforts in, 216
  participation in climate policies and mitigation efforts in,
   216–217
  regime theories in, 217
  U.S. national mitigation legislation
   in, 224–226
  world society theory and international environmentalism culture
   in, 218–220
 meso, 210–215
  business cycles in, 214
  cities, states, and regions as miti-
   gators in, 214
  ecological modernization in, 212
  economic opportunities in, 211
  exogenous events in, 214–215
  focusing events in, 210
  interorganization coalitions on status quo in, 213
  new industrial norms in, 215